A PRACTICAL GUIDE TO THE WIRING REGULATIONS

A PRACTICAL GUIDE TO THE WIRING REGULATIONS

17th EDITION IEE WIRING REGULATIONS (BS 7671:2008)

Fourth Edition

Eur Ing Geoffrey Stokes

BSc (Hons), CEng, FIEE, FCIBSE

Eur Ing John Bradley

BSc, CEng, MIEE, FCIBSE

A John Wiley and Sons, Ltd., Publication

Blackwell Publishing was acquired by JohnWiley & Sons in February 2007. Blackwell's publishing programme has been merged with Wiley's global Scientific, Technical, and Medical business to form Wiley-Blackwell.

Registered office
JohnWiley & Sons Ltd, The Atrium, Southern Gate, Chichester,West Sussex, PO19 8SQ, United Kingdom

Editorial offices
9600 Garsington Road, Oxford, OX4 2DQ, United Kingdom
2121 State Avenue, Ames, Iowa 50014-8300, USA

For details of our global editorial offices, for customer services and for information about how to apply for permission to reuse the copyright material in this book please see our website at www.wiley.com/wiley-blackwell.

First edition published 1994

Second edition published 1999

Third edition published 2002

Fourth edition published 2009

Library of Congress Cataloging-in-Publication Data

Stokes, Geoffrey, CEng.
   A practical guide to the wiring regulations/Geoffrey Stokes. – 4th ed.
      p.cm.
   Includes bibliographical references and index.
   ISBN 978-1-4051-7701-6 (cloth)
1. Electric wiring–Insurance requirements–Handbooks, manuals, etc. 2. Electric wiring, Interior–Handbooks, manuals, etc. 3. Electric apparatus and appliances–Installation–Great Britain–Handbooks, manuals, etc.
I. Bradley, John. II. Title.
TK3275.S86 2008
621.319'24021841–dc22

2008047084

A catalogue record for this book is available from the British Library

ISBN 978-1-4051-7701-6

Typeset in 10/12 Times by Laserwords Private Limited, Chennai, India
Printed and bound in Great Britain by TJ International, Padstow, Cornwall

For further information on Blackwell Science, visit our website: www.blackwell-science.com

Contents

About the authors

Geoffrey Stokes BSc (Hons), CEng, FIEE, FCIBSE was the author of the first three editions of this Guide. For many years he was the Principal Engineer of the National Inspection Council for Electrical Installation Contracting. Before that he was Technical Regulations Manager at the Institution of Electrical Engineers, which is now the Institution of Engineering and Technology (IET). Geoffrey has served for many years on many BSI, CENELEC and IEC Technical Committees and now represents the IET on the joint IET/BSI Committee JPEL/64, which is responsible for the technical content of the Wiring Regulations (BS 7671). He is also a Member of the IET's Wiring Regulations Policy Committee.

Geoffrey left the National Inspection Council in 2005 to set up a new electrical safety consultancy (Benchmark Electrical Safety Technology Ltd) and to assist in the development work of a new innovative revolution in automated electrical installation testing technology (Test Marshal).

John Bradley BSc, CEng, MIEE, FCIBSE updated this Guide to the 17th Edition of the Wiring Regulations. He is the Principal Engineer at the Electrical Safety Council (formerly the National Inspection Council for Electrical Installation Contracting), where he has special responsibilities for the Council's technical standards and publications. John has also worked in electrical contracting and as a consulting engineer, putting the requirements of the Wiring Regulations into practice.

He serves on a number of CENELEC and IEC technical committees and on the joint IET/BSI Committee JPEL/64 which is responsible for the technical content of the Wiring Regulations (BS 7671) and European and international standards for electrical installations. John is the Chairman of Panel C of JPEL/64 (protection against electric shock, and isolation and switching).

Preface to the Fourth Edition

This fourth edition of *A Practical Guide to the Wiring Regulations* takes account of the requirements of BS 7671:2008 Requirements for Electrical Installations (*IEE Wiring Regulations Seventeenth Edition*).

BS 7671:2008 was issued on 1 January 2008 and came into effect on 1 July 2008. It replaces BS 7671:2001 (*IEE Wiring Regulations Sixteenth Edition*) as the national standard for electrical installation work, and its requirements are to be complied within all electrical installation work designed after 30 June 2008.

The content of BS 7671:2008 has undergone extensive changes and additions compared with that of BS 7671:2001. The numbering of the regulations has also been revised to follow the pattern and corresponding references of International Electrotechnical Commission (IEC) Standard 60364. Account has been taken in BS 7671:2008 of the technical intent of a significant number of revised and new CENELEC harmonization documents (HDs). Indeed, of the 28 HDs listed in the preface of BS 7671:2008, 17 are revised compared with the versions used in BS 7671:2001, as finally amended, and seven are newly introduced to BS 7671.

The revised HDs have led to changes in, amongst other things, the various protective measures specified in Part 4 of BS 7671 and the requirements for special installations or locations. Not the least of the changes are those affecting the general requirements for protection against electric shock, which have been restructured and are subject to new terminology. Another notable change is that it is now permitted to install general-purpose socket-outlets in locations containing a bath or shower, provided these outlets are at least 3 m horizontally outside the boundary of zone 1 and the circuit supplying them is provided with additional protection by an RCD having specified characteristics (as must be all the circuits of the special location).

The newly introduced HDs have led to the addition of new sections in BS 7671 relating to: marinas and similar locations; exhibitions, shows and stands; solar photovoltaic power supply systems; mobile or transportable units; caravans and motor caravans (previously covered in Section 608 of BS 7671:2001); temporary installations for structures, amusement devices and booths at fairgrounds, amusement parks and circuses; and floor and ceiling heating systems.

While many changes have been associated with CENELEC HDs, a number of modifications made have been initiated in the United Kingdom. These primarily relate to cables concealed in walls and/or partitions in installations that are not intended to be under the supervision of a skilled or instructed person. In many cases, such cables are now required to be provided with additional protection by an RCD having specified characteristics, unless other specified protective provisions are employed.

In their professional experience the authors of this Guide have been asked, and attempted to answer, numerous questions over the years relating to the regulatory requirements and their implementation. While most practitioners will recognise where a proposed solution will not, or does not, meet the requirements, many find it difficult to attribute a precise regulation number to the deficiencies they believe to exist or to decide what action to take in solving the problem. This is not surprising since the subject of electrical installations is vast and complex. Those that believe that all issues are crystal clear (or black and white) and that there is only one possible solution to a design problem are deluding themselves. As there are many ways of killing a cat (besides electrocution) so too are there many design and installation options so long as the basic constraints are met.

An attempt has been made in this Guide to make life a little easier and topics are addressed with the pertinent Regulation numbers listed where appropriate. However, the Guide will be most useful to those who have at least a working knowledge of earlier editions of the National Standard. This Guide is not intended for use by the DIY enthusiast unless, of course, he or she happens to be competent in this field.

Where considered necessary by the authors, some background guidance is given together with worked examples embodied in the text at the appropriate place. It is hoped that this Guide will serve both as a useful aid to designers, installers and verifiers of electrical installations and to others not directly professionals in the industry but who have an interest in the safety aspects of electrical installations perhaps as 'duty holders' as defined by the Electricity at Work Regulations 1989. It is also expected that it will be of use to those students of the industry who are endeavouring to come to grips with all the many facets of electrical safety.

Extensive use has been made of tables which draw together the various relevant Regulations and options. Where appropriate, tables have also been employed as check lists for reference for those who find such listings useful in their day-to-day activities. Similarly, numerous figures have been used to more clearly identify specific points that the authors have thought worthy of mention.

No single book can ever cover all the aspects of this topic and the authors have had to take the view that this Guide should include guidance on the issues that are more frequently encountered, leaving aside some of the more esoteric aspects. Inspection, testing, verification, certification and reporting come in for special attention as the authors believe many electrical contractors would welcome some guidance in this respect.

The views expressed here are the authors' own and should not be regarded as coinciding with those of any authoritative body, though the authors believe they do not differ materially.

Geoffrey Stokes and John Bradley

Acknowledgements

Geoffrey Stokes acknowledges with gratitude the initial encouragement and subsequent support given to him by his dear friend and former colleague Brian D. Jenkins, who also reviewed the final draft of the First Edition of this Guide.

Geoffrey also wishes to record the considerable assistance given to him by his friends in the industry and, in particular, his former NICEIC colleagues both at Head Office and three Inspecting Engineers in the field. He particularly wishes to acknowledge with gratitude the contributions made, by way of constructive criticism and comment on the drafts, by: Bill Holdway, IEng, MIEIE, NICEIC Inspecting Engineer, Staffordshire; Brian D. Jenkins, BSc, CEng, FIEE; Keith Morriss, BSc, CEng, MIEE, Technical Director, AVO International Ltd; Terry Morrow, IEng, MIEIE, NICEIC Inspecting Engineer, Northern Ireland; Nick Piper, NICEIC Inspecting Engineer, West Midlands; Ted Smithson, BSc, CEng, MIEE, Engineering Manager, AVO: Megger Instruments Ltd.

John Bradley acknowledges the encouragement and support given to him by Geoffrey Stokes in the considerable task of updating this Guide to align with the requirements of the Seventeenth Edition of the IEE Wiring Regulations. John also records his appreciation of his employer, The Electrical Safety Council, for its support and permission to reproduce certain copyright material. He also wishes to acknowledge with gratitude the assistance given to him by his colleague Martyn Allen, BEng Hons, CEng, MIET, Senior Engineer at The Electrical Safety Council, who reviewed the final draft of the Fourth Edition of this Guide and provided many helpful comments.

Extracts from British Standards are reproduced with the permission of BSI. Complete copies can be obtained from BSI Shop Support Services (Hardcopy support: +44 (0)20 8996 9001; Softcopy support: +44 (0)20 8996 7555; Fax: +44 (0)20 8996 7001; Email: BSonlineSupport@bsigroup.com).

Grateful acknowledgement is made to the Institution of Engineering and Technology for its permission to reproduce copyright material from the Seventeenth Edition of the IEE Wiring Regulations and related Guidance Notes.

Notation

Acronyms and abbreviations

BCMC	British Cable Makers' Confederation
BS	British Standard
BS EN	Harmonized European Standard
BSI	British Standards Institution
CENELEC	Comité Européen de Normalisation Electrotechnique (European Committee for Electrotechnical Standardization)
CIBSE	Chartered Institution of Building Services Engineers
CONSAC	Concentric Solid Aluminium Core (cable)
ccc	current-carrying capacity
cpc	circuit protective conductor
csa	cross-sectional area
DCL	device for connecting a luminaire
DoE	Department of Environment (now incorporated into Department for Environment, Food and Rural Affairs (Defra))
DOL	direct-on-line
DP	double-pole
ED	electricity distributor
ADS	automatic disconnection of supply
EIILC	Electrical Installation Industry Liaison Committee
EL	emergency lighting
ELV	extra-low voltage
ESC	The Electrical Safety Council
ESQCR	Electricity Safety, Quality and Continuity Regulations 2002, as amended
ESR	Electricity Supply Regulations 1988, as amended
EWR	Electricity at Work Regulations 1989
FA	fire alarm
FELV	functional extra-low voltage (see definition in the Wiring Regulations)
HBC	high breaking capacity
HD	harmonized document
HSE	Health and Safety Executive
HV	high voltage
ICEL	Industry Committee for Emergency Lighting
IEC	International Electrotechnical Commission
IEE	Institution of Electrical Engineers
IET	Institution of Engineering and Technology

L	line
LPG	liquefied petroleum gas
LSC	luminaire supporting coupler
LSHF	low smoke, halogen free
LV	low voltage
M	maintained (emergency lighting)
MCB	miniature circuit-breaker
MCCB	moulded case circuit-breaker
MET	main earthing terminal
MICC	mineral-insulated copper cable
NHBC	National House-Building Council
NICEIC	NICEIC Group Limited
NM	Nonmaintained (emergency lighting)
PELV	protective extra-low voltage (see definition in the Wiring Regulations)
PEN	protective earth and neutral (conductor)
PME	protective multiple earthing
PRCD	portable residual current device
PVC	polyvinyl chloride
RCD	residual current device
RCCB	residual current circuit-breaker
RCBO	combined MCB and RCD
SELV	separated extra-low voltage (see definition in the Wiring Regulations)
SP	single-pole
SP&N	single-pole and neutral
TDM	time-division-multiplexing
TP	triple-pole
TP&N	triple-pole and neutral
TTA	type-tested assembly
XLPE	cross-linked polyethylene or ethylene propylene rubber insulation
4P	four-pole

Variables

α_{20}	resistance/temperature coefficient at $20\,^{\circ}\text{C}$
B	reciprocal of temperature coefficient of resistivity
C_a	rating factor for ambient temperature
C_c	rating factor for the type of protective device or installation condition
C_g	rating factor for grouping of conductors
C_i	rating factor for conductors embedded in thermal insulated materials
$\cos \phi$	power factor
C_t	factor for the operating temperature of the conductor
D_e	cable diameter (mm)
I_a	current causing automatic disconnection within stated time (A)
I_b	design current of circuit (A)
I_d	first-fault fault current of circuit (IT systems) (A)
$I_{\Delta n}$	rated residual current of an RCD (A or mA)
I_F	fault current (both short-circuit and earth fault depending on the context in which it is used) (A)
I_{FLC}	full-load current (A)

I_{inst}	current causing instantaneous (within 100 ms) operation of protective device (A)
I_L	earth-leakage current (mA)
$I_{L(T)}$	total earth-leakage current (mA or A)
I_n	rated current of protective device (A)
I_{pf}	prospective fault current (A)
I_t	tabulated current-carrying capacity (A)
$I_{t(min)}$	minimum tabulated current-carrying capacity required (A)
I^2t	energy let-through (A^2 s)
$I^2t_{(pa)}$	pre-arcing energy let-through (A^2 s)
$I^2t_{(t)}$	total energy let-through (A^2 s)
$I_{(sec)}$	transformer secondary current (A)
I_z	effective current-carrying capacity (A)
I_2	current causing effective operation of protective device on overload (A)
j	imaginary part of a complex variable ($j = \sqrt{-1}$)
k	a constant attributed to a particular conductor
L, l	length (m)
ln	log to the base e
π	pi, geometric constant (3.1416)
P	active power (W or kW)
Φ	diameter (mm)
ϕ	phase angle
Q	reactive power (VA_r or kVA_r)
Q_c	volumetric heat capacity of a conductor ($J/(°C mm^3)$)
Q_{20}	electrical resistivity of conductor material at 20 °C (Ω mm)
ρ	resistivity (Ω m)
R	resistance (generally) (Ω)
R_A	sum of the resistances of the installation earth electrode and protective conductor (Ω)
R_n	circuit neutral conductor resistance (Ω)
R_t	conductor resistance at temperature t (Ω)
R_1	circuit line conductor resistance (Ω)
R_2	circuit protective conductor resistance (Ω)
R_{20}	conductor resistance at temperature of 20 °C (Ω)
S	apparent power (VA or kVA)
S	conductor cross-sectional area (mm^2)
S_p	protective conductor cross-sectional area (mm^2)
θ_i	initial temperature of conductor (°C)
θ_f	final temperature of conductor (°C)
t	temperature (°C)
t	time (s)
T_d	time delay (s)
t_p	maximum permitted operating temperature (°C)
U_n	nominal voltage (V)
U_o	nominal voltage to Earth (V)
U_t	touch voltage (V)
V_d	voltage drop (V or mV)
$V_{d(r)}$	resistive voltage drop (V or mV)

$V_{d(x)}$	reactive voltage drop (V or mV)
V_{FL}	full-load voltage (V)
V_{L-L}	line-to-line voltage (V)
V_{NL}	no-load voltage (V)
Z_{dpc}	impedance of distribution circuit protective conductor (Ω)
Z_e	external line–earth loop impedance (Ω)
Z_L	line impedance (Ω)
Z_{L-N}	line–neutral impedance (Ω)
Z_{pu}	per unit impedance
Z_1	impedance of circuit line conductor (Ω)
Z_2	impedance of circuit protective conductor (Ω)
Z_s	line–earth fault loop impedance (Ω)
Z_s^*	line–earth fault loop impedance (IT systems) (Ω)
$Z_{s(max)}$	maximum permitted line–earth fault loop impedance (Ω)
X	reactance (Ω)
X_L	inductive reactance (Ω)
X_C	capacitive reactance (Ω)

1

Plan and terminology of BS 7671:2008 and supporting publications

1.1 Plan of BS 7671:2008

BS 7671:2008 is based, as was the 2001 version, on the International Electrotechnical Commission's (IEC's) publication 60364, the international rules for electrical installations. The pattern of BS 7671:2008 uses a similar logical plan to that of earlier versions, although there are some minor changes to detail. Table 1.1 illustrates the plan and shows the main routes through the various parts to assimilate all the necessary requirements for the electrical design. The routes shown should not be regarded as exhaustive or the only feasible routes; indeed, reference to other parts and sections, on an iterative basis, will always be necessary.

Although the plan of BS 7671 does follow a logical sequence, it may not follow the generally accepted sequence undertaken by designers. For example, a designer's sequence may take the form of:

- assessment of general characteristics (Part 3);
- assessment of the number and types of circuit (Part 3, Section 314);
- selection of wiring system(s) (Chapter 52);
- assessments of design currents, including taking account of diversity, where appropriate, and any special operating conditions (Sections 311 and 433);
- assessment of environmental conditions, including external influences (Section 522);
- selection of overcurrent protective devices (Section 533);
- selection of devices for fault protection by automatic disconnection of supply (Regulation Groups 531.1–531.6);
- determination of current-carrying capacities (Section 523);
- assessment of voltage drop (Section 525);
- assessment of isolation and switching requirements (Section 537);

A Practical Guide to The Wiring Regulations: 17th Edition IEE Wiring Regulations (BS 7671:2008)
Fourth Edition Geoffrey Stokes and John Bradley
© Geoffrey Stokes and John Bradley. Published by John Wiley & Sons, Ltd

Table 1.1   Plan of BS 7671:2008

PART 1: SCOPE OBJECT AND FUNDAMENTAL REQUIREMENTS FOR SAFETY					
PART 2: DEFINITIONS					
PART 3: ASSESSMENT OF GENERAL CHARACTERISTICS	PART 4: PROTECTION FOR SAFETY	PART 5: SELECTION AND ERECTION OF EQUIPMENT	PART 6: INSPECTION AND TESTING	PART 7: SPECIAL INSTALLATIONS OR LOCATIONS	APPENDICES
Chapter 31: Purposes, supplies and structure	Chapter 41: Protection against electric shock	Chapter 51: Common rules	Chapter 61: Initial verification	Section 700: General	Appendix 1: British Standards to which reference is made in the regulations
Chapter 32: Classification of external influences	Chapter 42: Protection against thermal effects	Chapter 52: Selection and erection of wiring systems	Chapter 62: Periodic inspection and testing	Section 701: Locations containing a bath tub or shower basin	Appendix 2: Statutory regulations and associated memoranda
				Section 702: Swimming pools and other basins	Appendix 3: Time/current characteristics of overcurrent protective devices and residual current devices
				Section 703: Rooms and cabins containing sauna heaters	Appendix 4: Current-carrying capacity and voltage drop for cables and flexible cords

(continued overleaf)

Table 1.1 *(continued)*

Chapter 36: Continuity of Service	*Chapter 56:* Safety services	*Section 712:* Solar voltaic (pc) power supply systems	*Appendix 11:* Effect of harmonic currents on balanced three-phase systems
		Section 717: Mobile or transportable units	*Appendix 12:* Voltage drop in consumers' installations
		Section 721: Electrical installations in caravans and motor caravans	*Appendix 13:* Methods of measuring the insulation resistance/impedance of floors and walls to earth or to the protective conductor system
		Section 740: Temporary electrical installations for structures, amusement devices and booths at fairgrounds, amusement parks and circuses	*Appendix 14:* Measurement of fault loop impedance: consideration of the increase of resistance of conductors with increase of temperature
		Section 753: Floor and ceiling heating systems	*Appendix 15:* Ring final and radial final circuit arrangements, Regulation 433.1

Part	5				
Chapter		3			
Section			4		
Group				2	
Individual identifying number					3	
Further individual identifying number (where present)						1
Individual number for a regulation group[a]	5	3	4 . 2		
Individual number for a regulation[b]	5	3	4 . 2 . 3		
Individual number for a regulation[b]	5	3	4 . 2 . 3 . 1		

[a] May also be used for a regulation.

[b] May also be used for a regulation subgroup.

Figure 1.1   Regulation numbering system

- consideration of the earthing and protective bonding requirements (Chapter 54).

It can be seen that the example of the design process given above does not follow the layout of BS 7671, and even the example shown is an oversimplification of the real process. However, the designer familiar with the earlier versions would have no difficulty in applying BS 7671:2008. In essence, the designing process will inevitably mean numerous references back and forth throughout BS 7671, and only experience will reduce the iterative procedure.

The numbering system in BS 7671 is in line with the IEC and CENELEC (Comité Européen de Normalisation Electrotechnique) numbering system. It uses regulation numbers with mainly one, two or three decimal points (e.g. 411.3, 411.3.1 and 411.3.1.1), although numbers with up to five decimal points are used in Part 7. Additionally, a group of associated regulations is covered by a side heading such as 531.3 – *RCDs in a TN system*. Similarly, a subgroup of associated regulations is covered by a side heading such as 533.1.1 – *Fuses*. Figure 1.1 illustrates how the numbers are formulated and this applies to all parts except Part 7.

In Part 7, sections take the place of chapters, and in the regulation numbers, the number appearing after the section number generally refers to the corresponding chapter, section or regulation in Parts 1–6.

BS 7671 is less a design code of practice than a standard for safe practice and is more of a framework document like the emergent CENELEC Harmonized Documents.

Part 6 addresses inspection and testing in a format with individual chapters allocated to initial verification, periodic inspection and testing, and certification and reporting.

Part 7 – *Special Installations or Locations* – is based on, but not identical to, IEC publication 60364, Part 7.

1.2   Terminology of BS 7671:2008

Part 2 of BS 7671:2008 deals with definitions which are inherently an essential part of any set of rules giving precise meaning to the terms used throughout. The British Standard now contains some 250 definitions, some of which have been added or modified, to a

greater or lesser extent, in the new edition of BS 7671. The reader is strongly encouraged to become familiar with the definitions, which will be beneficial in acquiring a broader understanding of the regulatory requirements.

1.3   Supporting publications

The Institution of Engineering and Technology (IET), formerly the Institution of Electrical Engineers, has published a number of supporting publications which give background information and guidance to the installation designer, installer, verifier and others in the implementation and application of BS 7671. The guidance documents comprise:

- IEE Guidance Note No. 1:   *Selection and erection*;
- IEE Guidance Note No. 2:   *Isolation and switching*;
- IEE Guidance Note No. 3:   *Inspection and testing*;
- IEE Guidance Note No. 4:   *Protection against fire*;
- IEE Guidance Note No. 5:   *Protection against electric shock*;
- IEE Guidance Note No. 6:   *Protection against overcurrent*;
- IEE Guidance Note No. 7:   *Special locations*;
- IEE Guidance Note No. 8:   *Earthing and bonding*;
- IEE On-site Guide: *Guidance for the small installation*.

The Guidance Notes published by the IET are essential reading for all those involved in electrical installation work, as is the Electrical Safety Council's *Technical Manual*. Constant reference to these documents, in addition to BS 7671 and this guide, will assist in achieving safe and economic installation designs.

NICEIC's *Inspection, Testing and Certification* book is also considered to be essential reading for those involved in the verification of new installations, as well as those who carry out periodic inspection and testing of existing installations. NICEIC's *Guide to Periodic Inspection and Testing of Domestic Electrical Installations* is another useful source of guidance, within its particular subject area.

2

Electricity, the law, standards and codes of practice

2.1 General

It is not the intention here to examine in detail all the statutory requirements relating to the hazards of electricity, but merely to draw to the reader's attention that such requirements do exist. The regulations made under the various acts should, therefore, be consulted to enable those involved with electricity to meet their statutory obligations. Table 2.1 summarizes some of the statutory documents relating to electrical installations and serves only as a quick reference. A rigorous commentary and analysis of the implications of these statutory regulations is given in *Electrical Safety and the Law* by Ken Oldham Smith (see Bibliography) and published by Blackwell Science.

References to a British Standard Code of Practice (CP) are intended only to bring to the designer's (and installer's and verifier's) attention that the requirements of the CP need to be met in addition to those embodied in BS 7671.

2.2 Electricity: the hazards

The principal hazards envisaged by the statutory regulations relate both to electrical machines and other current-using equipment and to the fixed installation. BS 7671 (see Regulation 131.1) addresses the requirements for safety for fixed installations, where the hazards are perceived to be:

- shock currents;
- excessive temperatures likely to cause burns, fires and other injurious effects;
- ignition of potentially explosive atmosphere;
- undervoltages, overvoltages and electromagnetic influences likely to cause or result in injury or damage;

A Practical Guide to The Wiring Regulations: 17th Edition IEE Wiring Regulations (BS 7671:2008)
Fourth Edition Geoffrey Stokes and John Bradley
© Geoffrey Stokes and John Bradley. Published by John Wiley & Sons, Ltd

Table 2.1 Acts, statutory regulations and associated legal requirements[a]

Ref	Title	Comment
A	Electricity Safety, Quality and Continuity Regulations 2002 (ESQCR)	Applies to all installations that obtain their supply from the public network. Part VII deals with supply to consumers' installations. Failure to comply with the fundamental requirements for safety (Chapter 13 of BS 7671) may result in the distributor discontinuing supply (See Regulation 26 of ESQCR). Where the supply is protective multiple earthing (PME), the consumer must comply with ESQCR particularly as regards main protective bonding.
B	The Health and Safety at Work etc Act 1974	Applies generally in all places where a work activity is undertaken.
C	The Management of Health and Safety at Work Regulations 1999	Generally applicable.
D	The Provision and Use of Work Equipment Regulations 1998	Generally applicable.
E	Electricity at Work Regulations 1989 (EWR)	Applies to all installations in places where a work activity is undertaken. For guidance see HSE HS(R)25 *Memorandum of Guidance on the Electricity Regulations 1989*. Regulation 15 addresses the requirements for adequate working space and means of access.
F	Electricity at Work Regulations (Northern Ireland) 1991 (EWR)	Applies to all installations in places where a work activity is undertaken. For guidance see Health and Safety Agency's *HAS 55 Memorandum of Guidance on the Electricity at Work Regulations (Northern Ireland) 1991*. Regulation 15 addresses the requirements for adequate working space and means of access.
G	The Personal Protective Equipment at Work Regulations 1992	Generally applicable.
H	The Workplace (Health, Safety and Welfare) Regulations 1992	Generally applicable.

I	The Manual Handling. Operating Regulations 1992	Generally applicable.
J	The Health and Safety (Display Screen Equipment) Regulations 1992	Generally applicable.
K	The Low Voltage Equipment (Safety) Regulations 1994	Equipment must comply with European ENs, IEC standards, national safety provisions, and/or national standards.
L	The Plugs and Sockets etc. (Safety) Regulations 1994	Applies to domestic-type installations. Relates to requirements for safety in plugs, sockets, adaptors and fuses, etc.
M	Cinematograph (Safety) Regulations 1955 (made under the Cinematograph Act 1955 and/or the Cinematograph Act 1952)	Applies to cinemas and the like. Where the general lighting and the safety lighting is by electricity, there must be at least two power sources of supply (i.e. there must be a source to power emergency lighting on failure of general lighting by, for example, mains failure).
N	Caravan Sites and Control of Development Act 1960	Applies to caravan park installations. Model Standards (1977) issued by the Department of Environment.
O	Consumer Protection Act 1987	Generally applicable.
P	Approval of Safety Standards Regulations 1987	Generally applicable.
Q	Consumer Safety Act 1987	Generally applicable.
R	Trading Standards Act 1968	Generally applicable.
S	Restrictive Trade Practices Act 1976	Generally applicable.
T	The Regulatory Reform (Fire Safety) Order 2005	Generally applicable.
U	Agricultural (Stationary Machinery) Regulations 1959, as amended	Applies to agricultural and horticultural installations. Specific requirements relating to starting and stopping rotating and other machines.

(continued overleaf)

Table 2.1 *(continued)*

Ref	Title	Comment
V	Highly Inflammable Liquids and Liquefied Petroleum Gases Regulations 1972	Applies to all installations where such substances are used. See the APEA/IP Guide mentioned in Section 13.4 of this guide for guidance on filling stations.
W	Dangerous substances and Explosive Atmospheres Regulations 2002	Applies to all installations where such substances are used or such atmospheres exist.
X	Petroleum (Consolidation) Act 1928, as amended	Applies to all installations where such substances are used. Under this legislation, local authorities are empowered to grant licences for premises where such substances are stored. Substances other than petroleum are included.
Y	Local Government (Miscellaneous Provisions) Act 1982	Applies to installations in the UK mainland in places of public entertainment. Includes theatres, places for dancing and music, etc. Applies to installations in the UK of HV luminous tube signs. See also BS 559 *Specification for electric signs and high-voltage luminous-discharge-tube installations*, and BS EN 50107 *Signs and luminous-discharge-tube installations operated from a no-load voltage exceeding 1 kV at not exceeding 10 kV, as applicable.*
Z	Local Government (Miscellaneous Provisions) (Northern Ireland) Order 1985	Applies to petrol filling station installations in Northern Ireland.
A1	Civic Government (Scotland) Act 1982	Applies to installations in Northern Ireland in places of public entertainment. Includes public houses, clubs, dance halls and church halls, etc. Applies to installations in Scotland in places of public entertainment. Includes theatres, places for dancing and music, etc.
B1	Building Standards (Scotland) Regulations 2004	Applies to installations in Scotland of HV luminous tube signs. See also BS 559 *Specification for electric signs and high-voltage luminous-discharge-tube installations.* Applies to installations in Scotland. See row C1 of this table.

Code	Title	Description
C1	Technical Handbooks	Technical Handbooks provide guidance on achieving the standards set in the Building (Scotland) Regulations 2004 and are available in two volumes, for Domestic buildings and for Non-domestic buildings, from Scottish Building Standards.
D1	The Building Regulations 2000 as amended.	Applies generally to buildings and indirectly related to electrical installation work except in respect of dwellings. See row E1 of this table.
E1	Approved Documents	Approved Documents provide guidance on meeting the requirements of the Building Regulations 2000 (as amended) for England and Wales. They are available from Communities and Local Government.
F1	Building Regulations (Northern Ireland) 2000	Applies to certain aspects of electrical installations. See row G1 of this table.
G1	Technical booklets	Technical booklets are prepared by the Department of Finance and Personnel and provide for certain methods and standards of building which, if followed, will satisfy the requirements of the Building Regulations (Northern Ireland) 2000.
H1	The Health and Safety (Safety Signs and Signals) Regulations 1996	Applies generally. Intended to remove confusion in the message intended to be conveyed by the sign.
I1	Factories Act 1961	Applies to factories generally.
J1	The Electrical Equipment (Safety) Regulations 1994	Applies generally.
K1	The Construction (Design and Management) Regulations 2007.	Applies generally.

[a]This listing is intended as a guide only to some of the statutory requirements and should not be regarded as exhaustive.

- mechanical movement of electrically actuated equipment, in so far as such injury is intended to be prevented by electrical emergency switching or by electrical switching for mechanical maintenance of nonelectrical parts of such equipment;
- power supply interruptions and/or interruption of safety services;
- arcing, likely to cause blinding effects, excessive pressure and/or toxic gases.

The Health and Safety Executive (HSE), in a note to BS 7671, states that it regards an electrical installation carried out in accordance with the requirements of BS 7671 as likely also to afford compliance with the relevant aspects of the Electricity at Work Regulations 1989.

2.3   The law

2.3.1   Electricity Safety, Quality and Continuity Regulations 2002

The Electricity Safety, Quality and Continuity Regulations 2002, as amended, which replaced the Electricity Supply Regulations 1988, are applicable to both public and private electricity supplies from licensed suppliers. The regulations are intended to prevent danger to persons, livestock or domestic animals. The hazards envisaged are burns, electric shock, injury from mechanical movement, fire and explosion, attendant upon the generation, transformation, transmission, distribution or use of electricity. The requirements apply to all work undertaken from the introduction of the regulations on 31 January 2003. The provisions are not retrospective, except that the new requirements will need to be met when material alterations are carried out to existing works.

The regulations cover, for example, such matters as definitions, voltage ranges and limits, neutral continuity and earthing, underground cables, overhead lines, safety and reliability, maximum voltages, limits on supply voltages, protective measures, inspection, enclosed spaces and consumers' installations.

2.3.2   The Electricity at Work Regulations 1989 (EWR)

The Electricity at Work Regulations 1989, made under the umbrella of the Health and Safety at Work etc. Act 1974, came into force on the mainland on 1 April 1990 and are intended to protect all persons at their work environment except domestic servants in households (alas a dying breed); in rare cases, an exemption certificate may be granted by the HSE. A similar set of regulations, entitled *The Electricity at Work Regulations (Northern Ireland) 1991*, came into force in Northern Ireland on 6 January 1992.

Alleged infringement of certain aspects of these regulations may result in criminal proceedings against those responsible, who would only have Regulation 29 to use in defence; such defence would require the duty holder to establish that they took all reasonable steps and exercised all due diligence to prevent such an offence.

The authoritative guidance to these regulations is given in *Memorandum of Guidance on the Electricity at Work Regulations 1989 (HS[R]25)*, published by the HSE, for the mainland. For Northern Ireland, a similar document is published entitled *Memorandum of Guidance on the Electricity at Work Regulations (Northern Ireland) 1991*. Further guidance in the province may be obtained from the Department of Economic Development.

2.4   Standards and codes of practice

2.4.1   The IEE Wiring Regulations: BS 7671

The requirements of BS 7671 are not statutory requirements but are referred to in the Electricity Safety, Quality and Continuity Regulations 2002 (as amended) as being an acceptable standard which will satisfy those statutory regulations. Additionally, the HSE considers that compliance with BS 7671 will afford compliance with the relevant aspects of the Electricity at Work Regulations 1989. BS 7671 also often forms part of a contract, but when so used it must be cited in its entirety and not selectively quoted. Although the regulations in BS 7671 are recommendations only, the designer and installer should consider their use desirable in order to fulfil all the requirements of the various statutory regulations.

2.4.2   Electric signs and high-voltage luminous-discharge-tube installations: BS 559 and BS EN 50107

The electrical installation of electric signs and high-voltage (HV) luminous-discharge-tube installations is within the scope of BS 7671. Installations, therefore, need to meet those requirements and the recommendations laid down in BS 559 *Specification for design, construction and installation of signs* or in BS EN 50107 *Signs and luminous-discharge-tube installations operating from a no-load rated output voltage exceeding 1 kV but not exceeding 10 kV*, as applicable.

2.4.3   Emergency lighting: BS 5266

Emergency lighting installations are embraced by BS 7671, and such installations will need to comply with both the recommendations of BS 7671 and BS 5266 *Emergency Lighting* and in particular Part 1 *Code of practice for the emergency lighting of premises*.

2.4.4   Electrical equipment for explosive gas atmospheres: BS EN 60079

Electrical installations in explosive gas atmospheres fall within the scope of BS 7671. Such installations, therefore, need to comply with BS EN 60079. In particular, Parts 10, 14 and 17 are applicable to electrical installation work.

2.4.5   Electrical equipment for use in the presence of combustible dust: BS EN 50281 and BS EN 61241

BS EN 50281 and BS EN 61241 apply to the selection, installation and maintenance of equipment protected by enclosures. They also address methods of determining minimum ignition temperatures.

2.4.6   Electrical installations in opencast mines and quarries: BS 6907

This standard addresses, among other things, general recommendations for the protection against direct contact and protection against indirect contact (referred to in BS 7671:2008 as basic protection and fault protection respectively).

2.4.7   Fire detection and alarm systems for buildings: BS 5839

Installations of fire detection and alarm systems for buildings are embraced by BS 7671. Such installations will need to comply both with BS 7671 and the recommendations given in BS 5839 *Fire detection and alarm systems for buildings*, and in particular with Part 1. The standard has the following parts:

- Part 1   *Code of practice for system design, installation, commissioning and maintenance*;
- Part 2   *Specification for manual call points*;
- Part 3   *Specification for automatic release mechanisms for certain fire protection equipment*;
- Part 4   *Specification for control and indicating equipment*;
- Part 5   *Specification for optical beam smoke detectors*;
- Part 6   *Code of practice for the design, installation and maintenance of fire detection and alarm systems in dwellings*;
- Part 8   *Code of practice for the design, installation and servicing of voice alarm systems*;
- Part 9   *Code of practice for the design, installation, commissioning and maintenance of emergency voice communication systems*.

2.4.8   Telecommunications systems: BS 6701

Installations which are subject to the Telecommunications Act 1984 are included within the scope of BS 7671. The electrical installation part of such installations will need to comply with the requirements of BS 7671 and BS 6701 *Telecommunications equipment and telecommunications cabling. Specification for installation, operation and maintenance*.

2.4.9   Electric surface heating: BS 6351

Requirements for the installation of electrical surface heating are included in BS 7671 and BS 6351 *Electric surface heating*. BS 6351 has three parts:

- Part 1   *Specification for electric surface heating devices*;
- Part 2   *Guide to the design of electric surface heating systems*;
- Part 3   *Code of practice for the installation, testing and maintenance of electric surface heating systems*.

2.4.10   Lightning protection: BS EN 62305

BS 7671 does *not* embrace the various aspects of the installation of lightning protection systems which are covered by BS EN 62305 *Protection against lightning*, except to the extent that requirements are given with regard to equipotential bonding of the lightning protection systems to the electrical installation.

2.4.11 Lift installations: BS 5655 and BS EN 81-1

The aspects of lift installations covered by BS 5655 *Lifts and service lifts* and BS EN 81-1 *Safety rules for the construction and installation of lifts* are not addressed by BS 7671.

Electrical installation aspects not covered by the standard should meet the requirements of BS 7671.

2.4.12 Equipment

The legal requirements relating to all equipment, embodied in the Low Voltage Electrical Equipment (Safety) Regulations 1989, demand that the order of precedence of safety standards be

- Harmonized European Standards – Euro Norms (ENs);
- International safety provisions (e.g. IEC Standards);
- National safety provisions:

Additionally, all equipment must be *fit for purpose*, as the law demands, bearing in mind the purpose to which the equipment is to be put, its use and the environmental conditions in which it is to be used.

3

Scope, object and fundamental principles

3.1 General

Part 1 of the BS 7671:2008 is divided into three chapters: Chapter 11 deals with the scope, Chapter 12 addresses the object and effects and Chapter 13 lays down the fundamental principles for safety. As we shall see later, the measures for safety are given in Chapters 41 to 44, though reference to other parts of BS 7671:2008 will, of course, be necessary.

BS 7671 and this Guide commonly use the term 'equipment' as an abbreviation for 'electrical equipment', which is defined in Part 2 of BS 7671:2008 as:

Any item for such purposes as generation, conversion, transmission, distribution or utilisation of electrical energy, such as machines, transformers, apparatus, measuring instruments, protective devices, wiring materials, accessories, appliances and luminaires.

When so used, the term 'equipment' collectively relates to all and, unless further specified, to every item of installation material. Where there is doubt as to the suitability of equipment to the purposes for which it is to be used, the designer must consult the manufacturer in order to establish that the equipment is appropriate for the intended use and compatible with other proposed equipment.

Where compliance with protective measures stipulated in Parts 3–7 is achieved, the broad-based fundamental principles for safety in Chapter 13 are likely to be satisfied. Where alternative measures are used, the designer would need to be able to argue that these provide for no less a degree of safety.

3.2 Scope

3.2.1 General

The first part of Regulation 110.1 gives a fairly comprehensive, but not exhaustive, list of electrical installations to which the regulations apply. The list now includes the additional items of:

A Practical Guide to The Wiring Regulations: 17th Edition IEE Wiring Regulations (BS 7671:2008)
Fourth Edition Geoffrey Stokes and John Bradley
© Geoffrey Stokes and John Bradley. Published by John Wiley & Sons, Ltd

Table 3.1 Modifications to certain types of installation

BS 7671:2001 (as amended)	BS 7671:2008
Construction sites, exhibitions, fairs and other installations in temporary buildings stage and broadcasting applications	Construction sites, exhibitions, shows, fairgrounds and other installations for temporary purposes including professional
Highway power supplies and street furniture, and outdoor lighting	External lighting and similar installations

- marinas;
- mobile and transportable units;
- photovoltaic systems;
- low-voltage (LV) generating sets.

Additionally, certain of the familiar categories have been modified as indicated in Table 3.1.

It is clear that the types of electrical installation listed in the regulation are for illustrative purposes, and that other types of installation may also fall within the scope of BS 7671.

The second section of this regulation gives a list of types of circuit and wiring systems to which BS 7671:2008 applies. The list is virtually the same as in BS 7671:2001, namely:

- Circuits supplied at nominal voltages up to and including 1000 V a.c. or 1500 V d.c. For a.c., the preferred frequencies which are taken into account are 50, 60 and 400 Hz. The use of other frequencies for special purposes is not excluded.
- Circuits other than the internal wiring of apparatus, operating at voltages exceeding 1000 V and derived from an installation having a voltage not exceeding 1000 V a.c. (e.g. discharge lighting, electrostatic precipitators).
- Wiring systems and cables not specifically covered by an appliance standard.
- All consumer installations external to buildings.
- Fixed wiring for communication and information technology, signalling, control and the like (excluding internal wiring of equipment).
- The addition to, or alteration of, existing installations and also those parts of existing installations that are affected by the addition or alteration.

This regulation lists instances where BS 7671 may need to be supplemented by the requirements or recommendations of other British Standards, or by the requirements of the person ordering the work. In this section, in addition to the previously included reference to BS 559, with regard to electric signs and HV luminous discharge-tube installations, a reference has been added to BS EN 50107, relating to such installations. Also, the previous reference to BS EN 50014 for electrical apparatus in potentially explosives gas atmospheres has been updated to BS EN 60079. With regard to electrical apparatus in the presence of combustible dust, a reference has been added to BS EN 61241, in addition to the previously included reference to BS EN 50281. Finally, a reference has been added to BS 7909, relating to temporary distribution systems for entertainment-related purposes.

Table 3.2 Related standards

Standard no.	Subject matter of standard
BS 559 BS EN 50107	Electric signs and HV luminous-tube installations
BS 5266	Emergency lighting
BS EN 60079	Electrical apparatus for explosive gas atmospheres
BS EN 50281 BS EN 61241	Electrical apparatus for use in the presence of combustible dust
BS 5839	Fire detection and alarm systems in buildings
BS 6701	Telecommunications systems
BS 6351	Electric surface heating systems
BS 6907	Electrical installations for opencast mines and quarries
BS 7909	Design and installation of temporary distribution systems delivering a.c. supplies for lighting, technical services and other entertainment-related purposes

Table 3.2 summarizes the standards to which reference is made in the regulation, the requirements of which supplement the requirements of BS 7671 for particular installations. As will be seen from Table 3.2, reference is made to BS 6907 to include electrical installations for opencast mines and quarries. However, Regulation 110.2 confirms that those aspects of mines and quarries specifically covered by statutory regulations are still excluded from the scope of BS 7671.

Very significantly, as with the earlier edition of BS 7671, there is no distinction made, in terms of safety, between permanent and temporary installations. It follows, therefore, that both permanent and temporary installations need to comply fully with the requirements of BS 7671.

3.2.2   Exclusions from the scope

Regulation 110.2 details those installations which have been excluded from the scope. The list of exclusions, which is given below, is essentially the same as in BS 7671:2001, except that the reference to BS 6651 for lightning protection has been updated to BS EN 62305, a reference has been added to BS EN 81-1 for lift installations, and an item for electrical equipment of machines has been added to the list:

- systems for distribution of electricity to the public;
- railway traction equipment, rolling stock and signalling equipment;
- equipment of motor vehicles, except those to which the requirements of the regulations concerning caravans are applicable;
- equipment on board ships covered by BS 8450;
- equipment on mobile and fixed offshore installations;
- equipment of aircraft;
- those aspects of mines and quarries specifically covered by statutory regulations;

- radio interference suppression equipment, except so far as it affects safety of the electrical installation;
- lightning protection of buildings and structures covered by BS EN 62305;
- those aspects of lift installations covered by BS 5655 and BS EN 81-1;
- electrical equipment of machines covered by BS EN 60204.

3.2.3   Equipment

BS 7671 does not apply to equipment except to the extent of the selection and erection of that equipment. Constructional requirements of prefabricated equipment are covered separately by appropriate specifications, usually British Standards, CENELEC Standards and/or IEC Standards. Section 113 covers this point and further demands that the equipment complies with appropriate standards.

3.2.4   Relationship with statutory authorities

As stated in Section 114, the rules in BS 7671 are not statutory requirements. However, they may be cited in a court of law in order to claim compliance with statutory requirements or indeed to claim noncompliance. Reassuringly, Section 114 also states that where the supply has been given in accordance with the *Electricity Safety, Quality and Continuity Regulations 2002*, as amended, it is deemed that the connection with Earth of the neutral of the supply is permanent and, by implication, reliable.

3.2.5   Installations in premises subjected to licensing

Section 115 states that for installations which are subject to licensing conditions or where another authority exercises statutory control, not only have that authority's requirements to be ascertained, but they must also be adhered to both in the design of the installation and its execution. In the case of licensing, the authority's requirements should also be sought with regard to periodic inspection and testing of installations.

3.3   Object and effects

3.3.1   General

Regulation 120.2 states that Chapter 13 of BS 7671 addresses the fundamental principles but does not give the detailed technical requirements. Therefore, it follows that Parts 3–7 give greater detail of the methods of achieving the fundamental principles laid down in Chapter 13.

As indicated in Regulation 120.3, the various parts of BS 7671 giving technical requirements intended to afford compliance with the fundamental principles are

- Part 3: *Assessment of general characteristics*;
- Part 4: *Protection for safety*;

- Part 5: *Selection and erection*;
- Part 6: *Inspection and testing*;
- Part 7: *Special installations or locations*.

The regulation goes on to say that any intended departure from BS 7671 should receive special consideration by the installation designer, and must be recorded on the Electrical Installation Certificate. It is important to note that this 'dispensation' of sanctioning a departure is not afforded to anyone other than the designer, who must accept responsibility for such noncompliance.

3.3.2   New materials and inventions

From time to time, new ideas for equipment are brought to the industry which, naturally, have not been tried and tested over time in service conditions. It is not the intention of BS 7671 to suppress such innovations and their introduction into electrical installations, and this fact is acknowledged in BS 7671. If a new material or invention is used within an installation then the designer is required, by Regulation 120.4, to ascertain that the degree of safety is no less than that which would have prevailed had the new material or new invention not been used. This would, in all probability, involve a third-party certification body, which would be requested to make an assessment of the product and certify the degree of safety provided by it in use. When used, the new product would nevertheless constitute a deviation from BS 7671 and, consequently, would need to be recorded on the Electrical Installation Certificate. It goes without saying that any departure from BS 7671 needs very careful consideration by the designer.

3.4   Fundamental principles

3.4.1   General

Regulation 131.1 makes clear that BS 7671 is intended to protect persons, property and livestock against dangers and damage which may occur when the installation is used as intended. BS 7671 assumes that, in providing protective measures for safety, the installation will be used in the manner intended and with reasonable care, which may include preventative maintenance as well as proper routine attention. The regulation also states that the requirements to provide for the safety of livestock are applicable in locations intended for them.

Chapter 13 of BS 7671:2008 has been reorganized to some extent, with the sections having been renumbered and some new regulations having been introduced. The chapter still sets out the fundamental principles for safety on which all the detailed technical requirements are based.

Table 3.3 shows the renumbering of the sections of Chapter 13.

Regulation 131.1 states in new terms the hazards of electricity addressed by BS 7671. These are summarized in Table 3.4.

Table 3.3 Renumbering of the sections of Chapter 13

BS 7671:2001 (as amended)		BS 7671:2008	
Section no.	Issue addressed	Section no.	Issue addressed
130	Protection for safety	131	Protection for safety
131	Design	132	Design
132	Selection of electrical equipment	133	Selection of electrical equipment
133	Erection, verification, and periodic inspection and testing of electrical installations	134	Erection and initial verification of electrical installations
		135	Periodic inspection and testing

Table 3.4 Changes in terminology of the hazards

BS 7671:2001 (Regulation 130-01-01)	BS 7671:2008 (Regulation 131.1)
Shock currents	Shock currents
Excessive temperatures likely to cause burns, fires and other injurious effects	Excessive temperatures likely to cause burns, fires and other injurious effects
	Ignition of potentially explosive atmosphere
	Undervoltages, overvoltages and electromagnetic influences likely to cause or result in injury or damage
Mechanical movement of electrically actuated equipment, in so far as such injury is intended to be prevented by emergency switching or be electrical switching off for mechanical maintenance of nonelectrical parts of such equipment	Mechanical movement of electrically actuated equipment, in so far as such injury is intended to be prevented by emergency switching or be electrical switching off for mechanical maintenance of nonelectrical parts of such equipment
Explosion	
	Power supply interruptions and/or interruption of safety services
	Arcing or burning, likely to cause blinding effects, excessive pressure, and/or toxic gases

3.4.2 Electric shock: basic protection

BS 7671:2008 uses the term 'basic protection' to refer to protection of persons and livestock against electric shock as a result of making contact with live parts, whereas BS 7671:2001 used the term 'protection against direct contact' for this.

Regulation 131.2.1 refers to two methods by which basic protection can be accomplished:

- preventing a current from passing through the body of any person or any livestock;
- limiting the current which can pass though a body to a nonhazardous value.

3.4.3 Electric shock: fault protection

The term 'fault protection' is used in BS 7671:2008 to refer to protection of persons and livestock against electric shock as a result of making contact with exposed-conductive-parts that have become live under fault conditions, whereas BS 7671:2001 used the term 'protection against indirect contact' for this.

Regulation 131.2.2 refers to the following three methods by which fault protection may be accomplished, the first two of which are essentially the same as given above relating to basic protection:

- preventing a current resulting from a fault from passing through the body of any person or any livestock;
- limiting the current which can pass though a body to a nonhazardous value;
- limiting the duration of a current resulting from a fault which can pass through the body to a nonhazardous time period.

Regulation 131.2.2 goes on to confirm that, in connection with fault protection, the application of the method of protective equipotential bonding is one of the important principles for safety.

3.4.4 Protection against thermal effects

Regulation 131.3.1 requires that the electrical installation shall be so arranged that the risk of ignition or flammable materials due to high temperature or electric arc is minimized and that, during normal operating conditions of the electrical equipment, there shall be minimal risk of burns to persons or livestock.

With regard to protection against harmful thermal effects of heat or thermal radiation emitted by adjacent electrical equipment, Regulation 131.3.2 requires protection to be provided to persons, fixed equipment and materials, particularly where there are likely to be any of the following consequences:

- combustion, ignition or degradation of materials;
- risk of burns;
- impairment of the safe function of installed equipment.

3.4.5 Protection against overcurrent

Regulation 131.4 requires that persons or livestock are protected against injury, and that property is protected against damage, due to excessive temperatures or electromechanical stresses caused by any overcurrents likely to arise in live conductors. A note to the regulation explains that protection may be achieved by limiting the overcurrent to a safe value and/or duration.

3.4.6 Protection against fault current

Regulation 131.5 requires conductors and any other parts likely to carry fault current to be capable of carrying that current without attaining excessive temperature.

A further requirement, which did not appear in the corresponding Regulation 130-05-01 of BS 7671:2001, is that electrical equipment, including conductors, shall be provided with mechanical protection against electromechanical stress of fault currents as necessary to prevent injury or damage to persons, livestock or property.

A note to the regulation explains that, for live conductors, compliance with Regulation 131.4 assures their protection against overcurrents caused by faults.

3.4.7 Protection against voltage disturbances and measures against electromagnetic influences

The scope of Regulation Group 131.6 of BS 7671:2008 is wider than that of the corresponding Regulation Group 130-06 of BS 7671:2001, which related only to protection against overvoltage.

As was required jointly by Regulations 130-06-01 and 130-06-02 of BS 7671:2001, Regulations 131.6.1 and 131.6.2 of BS 7671:2008 require that persons and livestock are protected against injury, and that property is protected against any harmful effects, resulting from:

- a fault between live parts of circuits supplied at different voltages;
- overvoltages likely to arise from atmospheric events or from switching.

A note to Regulation 131.6.2 points out that protection against direct lightning strikes is dealt with in BS EN 62305.

The requirements of Regulations 131.6.3 and 131.6.4 are new to Part 1 of BS 7671. The first of these regulations requires persons to be protected against injury, and property to be protected against damage, as a consequence of undervoltage and any subsequent voltage recovery.

The second of these regulations requires an installation to have an adequate level of immunity against electromagnetic disturbances so that it will function correctly in the specified environment. The regulation also requires the installation design to take account of the anticipated electromagnetic emissions generated by the installation or by the installed equipment.

3.4.8 Protection against supply interruption

The requirements Regulation 131.7 of BS 7671:2008 are new to Part 1. The regulation requires that, where danger or damage is expected to arise due to an interruption of supply, suitable provisions are made in the installation or installed equipment.

Table 3.5 Summary of Section 132 – Design

Regulation Group	Title	Dealing with
132.1	General	Installation to protect persons, livestock and property and to function properly for the intended use
132.2	Characteristics of available supply or supplies	Characteristic of supply(ies) to be determined
132.3	Nature of demand	Number and type of circuits to be established
132.4	Electrical supply systems for safety services or standby electrical supply systems	Characteristics of supply for safety services, and circuits to be supplied, to be determined
132.5	Environmental conditions	Equipment to be suitable for environmental conditions *including risk of fire or explosion*
132.6	Cross-sectional area of conductors	Conductors to be of adequate cross-sectional area, *including for harmonics and thermal insulation*
132.7	Type of wiring and method of installation	Wiring system and method of installation to be suitable for location etc.
132.8	Protective equipment	Protective devices to be suitable for application
132.9	Emergency control	Emergency interrupting devices to be easily recognizable and operable
132.10	Disconnecting devices	Disconnecting devices to be provided for maintenance, testing, fault detection and repair
132.11	Prevention of mutual detrimental influence	Harmful influences between electrical installation and other installations to be avoided, including electromagnetic interference
132.12	Accessibility of electrical equipment	Adequate space to be provided for installation, replacement, maintenance, testing, fault detection and repair
132.13	Documentation for the electrical installation	Appropriate documentation to be provided for the installation
132.14	Protective devices and switches	Single-pole devices in line conductor only and requirements when switches are inserted in earthed neutral conductors
132.15	Isolation and switching	Removal of voltage from equipment including electric motors to prevent danger

3.4.9   Additions and alterations to an installation

Regulation 131.8 requires that for an alteration or addition to an existing installation the rating and condition of the existing equipment which is utilized for the altered or addition installation must be confirmed as being satisfactory, including the supply. Additionally, where the alteration or addition uses automatic disconnection of supply (ADS) for fault protection, the earthing and bonding must also be of sufficient cross-sectional area.

3.4.10   Design

Section 132 of BS 7671:2008, dealing with design, is summarized in Table 3.5.

3.4.11   Selection of electrical equipment

Section 133 of BS 7671:2008, containing 10 regulations, requires equipment to be suitably selected, as summarized in Table 3.6.

3.4.12   Erection, initial verification of electrical installations, and periodic inspection and testing

The requirements for erection and initial verification of electrical installations and those for periodic inspection and testing are contained in Sections 134 and 135 respectively and are summarized in Table 3.7.

As in previous editions of BS 7671, the requirement contained in Regulation 134.1.1 for good workmanship and proper materials to be used is, as one would expect, retained; these essential elements are the cornerstones of sound installation erection work. By

Table 3.6   Summary of Section 133 – Selection of electrical equipment

Regulation Group	Title	Dealing with
133.1.1 133.1.2 133.1.3	General	Equipment to comply with standards unless agreed otherwise
133.2	Characteristics	Equipment to have suitable characteristics to suit the values and conditions on which the design of the installation is based
133.2.1 133.2.2 133.2.3 133.2.4	Voltage Current Frequency Power	These four regulations require equipment to be selected to suit voltage, current, frequency and power
133.3	Conditions of installation	Equipment to be selected to be suitable for the environmental conditions, or otherwise protected
133.4	Prevention of harmful effects	Electrical equipment not to cause harmful effects on other equipment or the supply

Table 3.7 Summary of Section 134 – Erection and initial verification of electrical installations, and Section 135 – Periodic inspection and testing

Regulation	Title	Dealing with
134.1.1		Good workmanship and proper materials
134.1.2		Characteristics of equipment not to be impaired during erection of installation
134.1.3		Conductors to be identified
134.1.4		Electrical joints and connections to be adequate
134.1.5		Design temperatures not to be exceeded
134.1.6		Equipment causing high temperatures or arcs to be suitably located or guarded
134.1.7		Suitable warning signs and/or notices to be provided where necessary for safety purposes
134.2.1	Initial verification	Appropriate inspection and testing to be performed during erection and upon completion of an installation or an addition or alteration, and appropriate certification to be issued
135.1	Periodic inspection and testing	Recommendation for subsequent periodic inspection and testing required

implication, the regulation requires that established methods, equipment and materials are to be used, but it does not rule out the use of new materials provided they offer no less a degree of safety. The regulation also contains the requirement for electrical equipment to be installed in accordance with the instructions provided by the manufacturer. Regulation 134.1.2 goes on to require that equipment must not be impaired by the installation process.

4

Assessment of general characteristics

4.1 General

Part 3 of BS 7671 requires the electrical installation designer to make an assessment of the entire electrical system. An electrical installation forms part of the complete system, the other constituent parts being the LV supply source and the interlinking LV distribution lines.

Regulation 301.1 requires this assessment to include the characteristics of the installation, such as the purpose for which the installation is to be used, its structure generally, the nature of the supply to the installation, the external influences to which equipment will be exposed, the compatibility and maintainability of that equipment in service, any safety services (such as emergency lighting and fire alarm systems) and the need for continuity of service. The need for this assessment is obvious, in that the particular characteristics of the supply and installation will influence, if not dictate, the methods used to protect against the hazards referred to in Chapter 13 (electric shock, excessive temperatures, and so on). They will also have a bearing on the choice of equipment, including wiring systems.

Additionally, before any design of the installation is contemplated, an assessment of the supply characteristics is required and due account will need to be taken of the implications of those characteristics on the installation design.

4.2 Loading, maximum demand and diversity

4.2.1 General

Whilst 'maximum demand' and 'diversity' are not defined in BS 7671, it is necessary to be clear that such terms are understood by all concerned in the design process. For example, 'maximum demand' may mean different things to the supply engineer and the installation design engineer. In this text, the following meanings have been assigned to the terms used:

A Practical Guide to The Wiring Regulations: 17th Edition IEE Wiring Regulations (BS 7671:2008)
Fourth Edition Geoffrey Stokes and John Bradley
© Geoffrey Stokes and John Bradley. Published by John Wiley & Sons, Ltd

- *maximum demand* – the maximum anticipated installation load plus any allowance for future loading (i.e. the potential load);
- *connected load* – the total of all electrical loads;
- *diversity* – the ratio of maximum demand to the connected load.

Diversity, therefore, can be represented by

$$\text{diversity} = \frac{\text{maximum demand}}{\text{connected load}} \tag{4.1}$$

From Equation (4.1) it can be readily seen that diversity is never going to be more than unity, or 100% in percentage terms (the more usual way of expression). Whilst the connected load is fairly straightforward to assess, the prediction of the maximum demand calls for considerable skill and judgement by the electrical designer. To apply a diversity of one (or 100%) would, of course, meet all the safety requirements, but an installation designed on this basis might prove to be exceedingly costly and much 'over-engineered'. To provide an economic design which meets the safety requirements needs design experience, skill and not a little judgement in establishing a figure for the actual load (i.e. maximum demand).

In load assessments, certain assumptions have to be made with regard to actual current drawn by equipment. Fixed loads are easy, but this is not so for socket-outlet circuits, which may have a theoretical load many times that which is likely to be drawn in service. For example, a final ring circuit may have twelve 13 A socket-outlets protected by a 32 A fuse or circuit-breaker, each socket-outlet capable of supplying, at least for short periods, a load of 2.99 kW at 230 V (13 A × 230 V), giving a total circuit current of 156 A (12 × 13 A). Clearly, such a scenario is not likely to occur in the real world, and the current likely to be drawn in this case must be based on all the known information related to usage of the circuit and the loads of portable and fixed equipment connected to it (the maximum permitted sustained load to be drawn by this circuit would, of course, be 32 A).

As mentioned earlier, assumptions have to be made in the assessment of maximum demands and those regarding the load current of current-using equipment and at points of utilization (e.g. socket-outlets) are given in Table 4.1, which is based on Table H1 of IEE Guidance Note 1 *Selection and erection*.

When selecting a distribution board or consumer unit, it is important to do so taking into account its rated current and rated diversity factor. It should also not be forgotten that the rated currents of the outgoing ways do not imply that the ways can take these currents continuously or for substantially long periods. Where continuous or long duration loadings exceed the 'long term' ratings of the outgoing ways, an excessive temperature rise is attained within the equipment, which may cause premature operation of overload devices. In all cases the equipment manufacturer's instructions with regard to loading should be complied with.

4.2.2 Lighting: loading and diversity

To assess the maximum demand for loading of lighting circuits, it is simply necessary to add the sum of all the lamp wattages of luminaires connected to it. The current is obtained by dividing the rated wattage of the lamp(s) by the nominal voltage (i.e. W/U_n).

Table 4.1 Load demand of current-using equipment and points of utilization[a]

Ref	Equipment or point of utilization	Assumed maximum demand	Comments/examples
A	63 A socket-outlet	63 A	In the absence of precise loading, allow for maximum rated load
B	32 A socket-outlet	32 A	In the absence of precise loading, allow for maximum rated load
C	16 A socket-outlet	16 A	In the absence of precise loading, allow for maximum rated load
D	13 A socket-outlet	13 A	Conventional circuits excepted
E	5 A socket-outlet	5 A	Conventional circuits excepted
F	2 A socket-outlet	0.5 A	The value of load demand given should be regarded as a minimum
G	Known luminaries	Actual load	Account to be taken of power factor of luminaire
H	Other lighting outlets (e.g. pendants)	Not less than 100 W (0.43 A at 230 V)	Minimum of 100 W per lamp
J	Electric clocks	Zero	To be ignored
K	Shaver sockets	Zero	To be ignored
L	Bell transformers	Zero	To be ignored
M	Load of not more than 5 VA (22 mA at 230 V)	Zero	To be ignored
N	Cooker (domestic type)	10 A plus 30% of remainder of load	E.g. 20 kW (86.95 A) cooker on 230 V Loading = $10A + (0.3 \times 86.95) = 39A$
O	Cooker (domestic type) with integral socket-outlet on control unit	10 A plus 30% of remainder of load plus 5 A	E.g. 24 kW (104.35 A) cooker on 230 V Loading = $10A + (0.3 \times 104.35) + 5 = 46.3A$
P	Stationary equipment	Rated current	Consult the manufacturers and/or the Standard and/or rating plate

[a]Where loadings have a power factor other than unity, allowance must be made for the current drawn at stated power factor.

Where luminaires operate at a power factor less than unity (e.g. power-factor-uncorrected fluorescent luminaires), and in the absence of more precise information, a multiplying factor of 1.8 (see Section 10.4.2 of this Guide) must be applied to the calculated current derived from the wattage rating. Where the actual wattage of a luminaire is not known, a minimum of 100 W per lamp must be allowed (see item H of Table 4.1 of this Guide).

Diversity for lighting circuits will depend on the type of premises, and in Table H2 of IEE Guidance Note 1 *Selection and erection* these are given as

- domestic premises – households, including flats 66%;
- commercial premises – shops, offices, stores and other business premises 90%;
- commercial premises – guest houses, small hotels, boarding houses 75%.

By way of example, three shop lighting circuits with connected load (and maximum demand) of 15 A, 7 A and 12 A would represent (after applying an allowance for diversity of 0.9) a load of 30.6 A – i.e. $0.9(15 + 7 + 12)$ – on the supplying distribution board.

4.2.3   Heating: loading and diversity

Assessment of heating loads is straightforward and the current is obtained by dividing the rated wattage W of the loads by the nominal voltage of the circuit (i.e. W/U_n). The power factor for such load is normally at unity unless the heater convection is fan assisted, in which case account should be taken of the motor power factor. This is generally likely to be approaching unity for the whole appliance. For thermal storage heating circuits (e.g. off-peak heaters) and for floor-warming installations, no allowance for diversity is permitted.

Diversity for heating circuits will depend on the type of premises, and in Table H2 of IEE Guidance Note 1 *Selection and erection* these are given as

- domestic premises – households, including flats: 100% of first 10 A plus 50% of remainder;
- commercial premises – shops, offices, stores and other business premises: 100% of largest heater plus 75% of remaining heaters;
- commercial premises – guest houses, small hotels, boarding houses: 100% of largest heater plus 80% of next largest heater plus 60% of remaining heaters.

By way of example, five boarding house heaters, three rated at 3 kW and two rated at 2 kW, are fed from a common distribution board. With allowance for diversity, these heaters would represent a load of 9.6 kW, i.e.$(1 \times 3) + (0.8 \times 3) + (0.6 \times 3) + (0.6 \times 2) + (0.6 \times 2)$.

4.2.4   Cookers: loading and diversity

Diversity for cooker circuits will again depend on the type of premises, and in Table H2 of IEE Guidance Note 1 *Selection and erection* these are given as

- domestic premises – households, including flats: 100% of first 10 A plus 30% of remainder plus 5 A if socket-outlet is integral to the cooker control unit;

- commercial premises – shops, offices, stores and other business premises: 100% of largest cooker plus 80% of second largest cooker plus 60% of remaining cookers;
- commercial premises – guest houses, small hotels, boarding houses; 100% of largest cooker plus 80% of second largest cooker plus 60% of remaining cookers.

By way of example, three small hotel cookers, two rated at 12 kW and one rated at 10 kW, are fed from a common distribution board. With allowance for diversity, these cookers would represent a load of 27.6 kW, i.e. $12 + (0.8 \times 12) + (0.6 \times 10)$.

4.2.5   Water heaters: loading and diversity

Diversity for instantaneous water heater circuits will depend on the type of premises, and in Table H2 of IEE Guidance Note 1 *Selection and erection* these are given as

- domestic premises – households including flats: 100% of first and second largest water heater plus 25% of remainder;
- commercial premises – shops, offices, stores and other business premises; 100% of first and second largest water heater plus 25% of remainder;
- commercial premises – guest houses, small hotels, boarding houses: 100% of first and second largest water heater plus 25% of remainder.

There is no allowable diversity for water heaters which are thermostatically controlled (e.g. immersion heaters and storage water heaters).

By way of example, three office instantaneous water heaters, two rated at 7 kW and one rated at 3 kW, are fed from a common distribution board. With allowance for diversity, these water heaters would represent a load of 14.75 kW, i.e. $7 + 7 + (0.25 \times 3)$.

4.2.6   Motors: loading and diversity

Diversity for motor circuits will depend on the type of premises, and in Table H2 of IEE Guidance Note 1 *Selection and erection* these are given as

- commercial premises – shops, offices, stores and other business premises: 100% of first and second largest motor plus 80% of second largest motor plus 60% of remainder;
- commercial premises – guest houses, small hotels, boarding houses: 100% of first and second largest motor plus 50% of remainder.

By way of example, three motors in an office boiler house, two rated at 11 A and one rated at 5 A, (and all operating at 0.8 power factor), are fed from a common distribution board. With allowance for diversity, these motors would represent a load of 32.5 A, i.e. $[11 + 11 + (0.8 \times 5)]/0.8$. This value represents the steady-state loading, and allowance should also be made for the high starting currents associated with the particular motors. The motor manufacturer must be consulted where there is doubt about the starting currents (typically six times the steady-state current). Consideration should also be given to simultaneous starting of motors and the effects on distribution circuits. Sequential starting may need to be adopted to prevent unacceptable cumulative starting currents. Lift motors are the subject of special consideration, and reference should be made to the requirements of BS 5655 or BS EN 81-1, as applicable.

4.2.7 Stationary equipment: loading and diversity

The loading of stationary equipment will be that given by the manufacturers, and any diversity allowance which may apply will depend on the operational requirements. In the absence of precise information, the allowance given for heating may be used with caution, if appropriate.

4.2.8 Conventional circuits: loading and diversity

Conventional circuit arrangements are detailed in Appendix E of the IEE Guidance Note *Selection and erection*. Diversity for conventional circuits will depend on the type of premises, and in Table H2 of IEE Guidance Note 1 *Selection and erection* these are given as

- domestic premises – households, including flats: 100% of largest circuit plus 40% of remainder;
- commercial premises – shops, offices, stores and other business premises: 100% of largest circuit plus 50% of remainder;
- commercial premises – guest houses, small hotels, boarding houses: 100% of largest circuit plus 50% of remainder.

By way of example, three conventional circuits in a shop, two rated at 32 A (ring final circuits) and two rated at 20 A (radial circuits), are fed from a common distribution board. With allowance for diversity, these circuits would represent a load of 68 A, i.e. $32 + 0.5(32 + 20 + 20)$.

4.2.9 Socket-outlet circuits other than conventional circuits: loading and diversity

Diversity for socket-outlet circuits other than conventional circuits will depend on the type of premises, and in Table H2 of IEE Guidance Note 1 *Selection and erection* these are given as

- domestic premises – households, including flats: 100% of largest point plus 40% of remainder;
- commercial premises – shops, offices, stores and other business premises: 100% of largest point plus 75% of remainder;
- commercial premises – guest houses, small hotels, boarding houses: 100% of largest point plus 75% of all other points in main rooms plus 40% of remainder.

There will be many occasions where the designer will find the installation will not fall neatly into any one of the above three categories and will need to use professional judgement and experience to assess diversity.

4.3   Arrangement of live conductors and type of earthing

4.3.1   Arrangement of live conductors

Regulation 312.2.1 requires that an assessment be made of the number and types of live conductor. Examples given in the regulation are single-phase, two-wire a.c. and three-phase, four-wire a.c. Other arrangements which could be cited are two-wire d.c. and three-phase, three-wire a.c. systems, but it is accepted that those quoted in the above regulation are the ones more likely to be encountered in the normal course of events. In the case of public supplies, the proposed arrangements can easily be established by consultation with the electricity distributor (ED) concerned. In the event of a private distributor, consultation again will be sufficient to establish the relevant parameters. In the case of on-site generation, the designer will only need to consult the generator manufacturer or the generator nameplate to determine the arrangements of live conductors of the source.

4.3.2   Type of earthing

It is important that the requirements of Regulation 312.3.1 are met, in that an assessment of the type of earthing has to be made before the design of the installation commences. The type of earthing will have implications with regard to the value of the external earth loop impedance and to the magnitude of prospective earth fault current.

In the case of public supplies, an ED providing a new connection at LV is normally required by Regulations 24(4) and 24(5) of the ESQCR to make available the combined protective earth and neutral (PEN) conductor, or, if appropriate, the protective conductor, of the supply for connection to the earthing conductor of the consumer's installation. (The expression 'new connection' means the first electric line, or the replacement of an existing electric line, to one or more consumer installations.)

However, in certain circumstances the distributor may take the view that such connection to the consumer's earthing conductor could result in danger and, therefore, not make available the PEN conductor or protective conductor of the supply for this purpose. Examples of situations where caution would be warranted are installations where it may prove difficult to attach and maintain all the necessary main protective bonding connections (e.g. farms or building sites), installations at certain wet environments (e.g. swimming pools) and certain installations outside the equipotential zone of buildings (e.g. certain types of street furniture). In addition, in the case of an electrical installation in a caravan or boat, Regulation 9 of the ESQCR forbids an ED from making available the PEN conductor of the supply for connection to the earthing conductor of the installation.

Other than in the case of a new connection at LV, an ED has no obligation, legal or otherwise, to provide facilities for earthing of an electrical installation, this being the sole responsibility of the consumer.

There are five basic types of earthing arrangement embodied in the systems identified by, TN-C, TN-S, TN-C-S, TT and IT, and these are shown in Figures 4.1–4.5 and detailed in Part 2 of BS 7671. In the figures, the line configuration in each case represents a three-phase source with the star/neutral point earthed where appropriate. For single-phase sources, one pole of the secondary circuit would normally be earthed as indicated in the figures for the neutral point. The meanings attributed to each letter are given in Table 4.2.

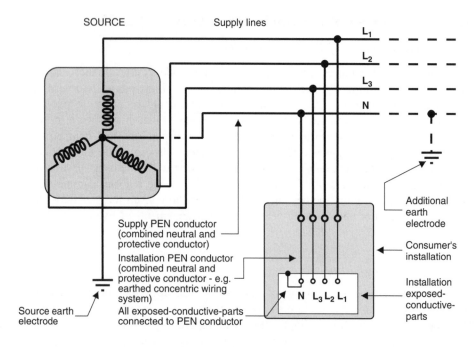

Figure 4.1   TN-C system (three-phase installation)

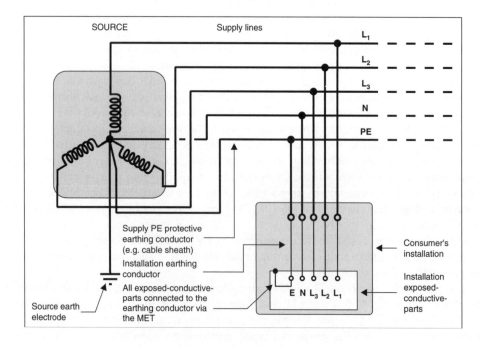

Figure 4.2   TN-S system (three-phase installation)

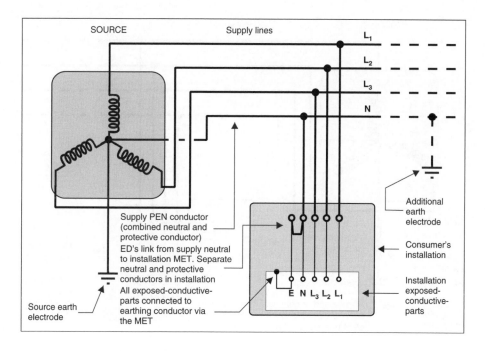

Figure 4.3   TN-C-S system (three-phase installation)

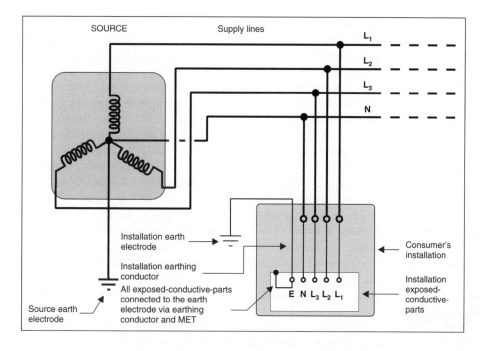

Figure 4.4   TT system (three-phase installation)

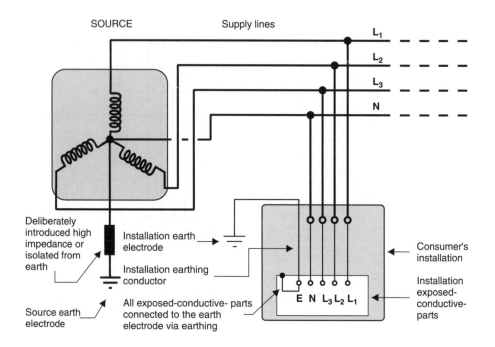

Figure 4.5   IT system (three-phase installation)

In the TN-C system, which is shown in Figure 4.1, the source is earthed through a low-impedance earth electrode effectively 'tying' one pole to Earth in a single-phase system and the star point to Earth in a three-phase system. The functions of neutral conductor and protective conductor are combined in a single conductor, known as a PEN conductor, both in the distribution supply cables and in the consumer's installation. In other words, a TN-C system combines these functions throughout the entire system, and an example of this would be an earthed concentric system.

It is important to appreciate that a TN-C system must not be used where the electricity supply is given in accordance with the ESQCR, as is normally the case. This is because Regulation 9(4) of the ESQCR states that a consumer shall not combine the neutral and protective functions in a single conductor in the consumer's installation.

In the TN-S system, which is shown in Figure 4.2, the source is earthed through a low-impedance earth electrode effectively 'tying' one pole to Earth in a single-phase system and the star point of a three-phase system. The functions of neutral conductor and protective conductor are separate both in the distribution supply cables and in the consumer's installation. In other words, a TN-S system employs separate conductors for neutral and protective functions throughout the entire system, and an example of this would be a supply cable sheath earth connecting directly with the star/neutral point of the source.

In the TN-C-S system, which is shown in Figure 4.3, the source is earthed through a low-impedance earth electrode effectively 'tying' one pole to Earth in a single-phase system and the star point of a three-phase system. The functions of neutral conductor and protective conductor are combined in the distribution supply cables up to the consumer's

Table 4.2 Designation letters and their meanings as they relate to the types of system and earthing arrangements[a]

First letter	Second letter	Subsequent letters
Source earthing arrangements	Arrangement of connection of exposed-conductive-parts of the installation with earth	Arrangement of protective and neutral conductors
		'C'[b] Single conductor provides for both neutral and protective conductor functions
'T' Direct connection of source with Earth at one or more points (e.g. one pole of a single-phase source and the star point of a three-phase source)	'N' exposed-conductive-parts of the installation connected *directly* by protective conductor with the source earth (e.g. source earthed neutral in a.c. systems)	'S' Separate conductors for neutral and protective conductor functions.
		'C-S' Neutral conductor and protective conductor combined in the supply and separate in the installation
'I'[c] All source live parts isolated from Earth or connected by a high impedance to earth	'T' exposed-conductive-parts of the installation connected, via an independent installation earth electrode, by protective conductor with the source earth	

[a]All systems 'reading left to right'. A combination of different systems may be used on the same premise, e.g. TN-C-S and TT.
[b]Earthed concentric wiring is an example of a TN-C system, but specific authorization is required for its use from the appropriate authority.
[c]The ESQCR, as amended, does not permit 'IT' systems for use on public supply networks, although they may be used, where appropriate, on private supplies.

terminals. In other words, a TN-C-S system employs separate conductors for neutral and protective functions throughout the installation but combines the functions in the distribution supply cables. An example of this would be a PME supply on which the neutral is effectively earthed to the star point of the source and via suitable earth electrodes at a number of points along the length of the supply cable.

In the TT system, which is shown in Figure 4.4, the source is earthed through a low-impedance earth electrode system effectively 'tying' one pole to Earth in a single-phase system and the star point of a three-phase system (on a public supply this impedance does not generally exceed 20 Ω). The function of the neutral conductor in the distribution supply cables up to the consumer's terminals is quite separate from the protective function. The path for earth fault current to flow to the earthed point of the source is provided by the general mass of the Earth together with earth electrodes (both source and installation) and separate protective conductors. These TT systems are often encountered in rural locations where the EDs are unable to offer an earthing facility. Additionally, TT systems are sometimes employed where the installation designer declines the ED's offer of an earthing terminal on the grounds of its unsuitability for use on a particular installation or part of an installation; for example, petrol filling station electrical installations. Indeed, many EDs will not offer a PME facility for filling stations.

In the IT system, which is shown in Figure 4.5, the source is isolated from Earth or earthed through a high impedance. This system is seldom used in the UK and is indeed not permitted on public supplies unless special permission is granted for its use by the appropriate Secretary of State.

4.4 Nature of supply

4.4.1 General

Regulation 313.1 lists the particular characteristics of the supply to be assessed: the nominal voltage(s), the nature of current and frequency, the prospective short-circuit current at the origin of the installation, the earth fault loop impedance Z_e of that part of the system external to the installation, the suitability for the requirements of the installation, including maximum demand and, finally, the type and rating of the overcurrent device(s) acting at the origin of the installation. The characteristics must be ascertained for an external supply or determined for a private source. The requirements to do so apply equally to the main supplies and to safety services and standby supplies.

4.4.2 Voltage

Assessment of the nominal voltages and voltage tolerances is straightforward, in that if the source is a private one (for example, a generator), details of the characteristics, including no-load and rated-load voltages, will be available from the generator manufacturer and would usually be given on the generator nameplate together with other relevant design details. If the source is derived from the public supply, then the EDs are required to maintain the voltage of supplies to not exceeding the range +10%/−6% of the declared nominal voltage to meet the statutory requirements under the Electricity Safety, Quality and Continuity Regulations 2002, as amended. For a single-phase supply of nominal voltage of 230 V this would mean a lower limit of 216 V and an upper limit of 253 V. A three-phase supply at a nominal voltage of 400 V at these tolerances would be in the range 376–440 V. In practice, it is unusual for the supply voltage to traverse this range, even over short periods, and in most cases the 'base' voltage is within this range and small variations occur due to the effects of loading of the installations and/or other loads connected to the same source. However, the installation design would need to take account of the whole range, since it can reasonably be anticipated that changes will take place in the supply network which should always be treated as a dynamic (always changing) constituent part of the system.

Part 2 of BS 7671 defines two voltage bands as follows:

- Band I covers:
  - installations where protection against electric shock is provided under certain conditions by the value of voltage;
  - installations where the voltage is limited for operational reasons, such as telecommunications, signalling, bell, control and alarm installations.

Normally, ELV will fall within voltage band I.

- Band II covers:
  - voltages for supplies to household, and most commercial and industrial installations. LV will normally fall within voltage band II. Band II voltages do not exceed 1000 V a.c. rms or 1500 V d.c. between conductors or 600 V a.c. rms or 900 V d.c. between conductors and earth (i.e. LV).

4.4.3   The nature of current and frequency

The most common forms of current are d.c. and a.c. In most power systems, a.c. is one of the sinusoidal or distorted sinusoidal waveforms. Frequency, applying only to a.c., needs to be assessed because of its effect on impedance; the inductive reactance of a constituent part of a system increases and decreases in direct proportion to the frequency; conversely, capacitive reactance varies in inverse proportion to the frequency. EDs are required, again under the Electricity Safety, Quality and Continuity Regulations 2002, as amended, to maintain the frequency of public supplies to 50 Hz ±1%.

4.4.4   Prospective short-circuit current

An assessment of the prospective short-circuit current at the origin of the installation is required by Regulation 313.1 in order to establish that devices, such as overcurrent protective devices, at the origin and elsewhere have short-circuit capacities (breaking capacity for fuses and making/breaking capacities for protective devices other than fuses – for example, circuit-breakers) of not less than the prospective short-circuit currents.

Although the regulation refers only to short-circuit currents at the origin, the designer should not ignore earth fault current, which may in certain circumstances be greater than short-circuit current. Depending on the source and transmission supply cables, earth fault currents can be of the order of 5% greater than short-circuit currents due to the zero-sequence impedances being lower or nonexistent in the earth fault path.

The regulation gives four options for methods of assessment of this and other parameters: *calculation, measurement, enquiry* and *inspection*. The last option is not applicable to the assessment of prospective short-circuit current, it being totally impracticable to be carried out by any amount of inspection.

If the assessment is to be carried out by calculation, then all the constituent impedances of the system need to be known, together with the anticipated fault level of the HV feeder to the power transformer. The fault level of the HV feeder may be obtained from the ED in the case of public supply and in the absence of this information may be taken as

250 MVA for an 11 kV system. This may actually be slightly higher than the true value but would, in any event, tend to give a somewhat exaggerated value of fault current and the approximation would tend to be on the safe side, for prospective fault current considerations.

The method of assessment by measurement is straightforward but, of course, such measurements do need the supply to the installation to be *in situ* and made live. For a new installation this method is often impracticable because the supply is not available sufficiently early to allow decisions to be made, amongst other things, regarding protective device specification. The measurements themselves are straightforward provided good quality instruments are used together with a safe system of work whenever such tests are undertaken. The test instrument will need to be equipped with test probes (suitably insulated, fused, shrouded and with spring-loaded retractable contacts). The measurement taken between line and neutral at the origin will give the line–neutral (L–N) short-circuit current and that between line and the main earthing facility will give the earth fault current. Similarly, line–line (L–L) and three-phase prospective fault currents can be obtained in a comparable way. It must be remembered that the values so obtained relate to the particular time at which the tests were undertaken and that allowance should be made for any increase likely to occur due to modifications of the supply network, which should always be regarded as dynamic – this is particularly important when the supply is derived from the public network.

The third, and final, viable option given is that of enquiry; that is, enquiry of the ED. As required by the Electricity Safety, Quality and Continuity Regulations 2002, as amended, an ED is obliged to provide this information, which is usually done by quoting a maximum value of prospective short-circuit current. Often, the actual value will be less than that quoted, but the EDs maintain this policy for two main reasons: the actual value at each individual supply point would be difficult to give and there is always the possibility, if not the likelihood, that changes will be made to the supply network by, for example, transfer of an installation load to a different, possibly larger or nearer, transformer with perhaps a lower impedance and consequent higher prospective fault level.

For supplies obtained from the public distribution network, information on this subject is given by the Energy Networks Association in their Engineering Recommendation P25/1 entitled *Short-circuit characteristics of public electricity supplier's distribution networks and co-ordination of overcurrent protective devices on 230 V single-phase supplies up to 100 A*. Another publication by the association, P26, gives similar information relating to three-phase supplies. Chapter 7 of this Guide gives further guidance on this subject.

4.4.5   External earth fault loop impedance

Regulation 313.1 also requires an assessment of the *earth fault loop impedance* Z_e *of that part of the system external to the installation*. This impedance is the phasor (vector) sum of all the constituent parts of the earth loop up to the origin of the installation and would include the transformer secondary winding, the line conductor of the supply cable and the earth fault path back to the earthed point of the transformer. In a more rigorous analysis, the contribution made by the HV supply would need to be included by referring its equivalent impedance to the secondary LV loop impedance.

Again there are basically three methods by which this assessment can be made, namely *calculation, measurement* or *enquiry*, all of which can be obtained by similar methods described in the assessment of prospective fault current. Enquiry is generally the only option at the design stage, but confirmation, by testing, is always required as soon as practicable.

It should be noted that where, from enquiry, the maximum prospective fault current and maximum external earth fault loop impedance values are given, these values will be mutually exclusive; they are the worst-case values, which do not occur simultaneously. For example, an ED may quote a Z_e of 0.35 Ω and a prospective fault current of 16 kA which, from a simple application of Ohm's law, will show that the two conditions do not happen at the same time. It follows, therefore, that the design will need to take account of the range of these parameters. For example, for TN-C and TN-C-S systems, the implications are that Z_e will be in the range 14.4 mΩ (230 V/16 000 A) to 0.350 Ω and the corresponding prospective fault level range will be 657 A to 16 kA. The designer will need to take account of these ranges when considering circuit length (limiting Z_s values), short-circuit capacities of protective devices, and discrimination (can the last named be maintained over the range?).

4.4.6 Suitability of supply

Regulation 313.1, item (v), demands an assessment to be made of the suitability of the supply for the requirements of the installation, including maximum demand. This may appear an obvious requirement and one which could be 'taken as read'. However, in practice, all too often this matter is not considered sufficiently early in the installation design and problems arise later at the commissioning stage which are difficult to solve. Where, for example, step loads (suddenly applied loads) have not been properly considered in terms of their effects on other loads, problems can, and do, arise which can result in difficult and sometimes costly remedies. Particular attention should be given to the effects of large loads on, for example, lighting loads, which may, if proper consideration is not given, present problems with luminaires 'dipping', 'flickering' or giving other unwanted and distractingly unsatisfactory visual effects. For example, it is desirable to connect fluorescent luminaire circuits over three phases wherever practicable to avoid the strobe/flicker effects that can sometimes occur.

The installation load profile always needs careful analysis to confirm that the supply characteristics can match the installation loads at all times. To obtain a meaningful profile it is necessary to know the full details of the intended utilization of the installation and, in particular, the times and duration of the various loads, including the times when intermittent loads are switched 'in' and 'out'. Computers can, and do, provide a useful and time-saving function in this analysis, but such profiles need the judgement of a professional engineer who can draw on his experience of similar projects and their performance over a period. This load profile, when completed, should show the maximum demand of the installation load diagrammatically.

In the case where the supplier's tariff includes an availability charge, an overestimation of the maximum demand may prove costly in terms of penalty costs for 'unused' maximum demand; it is very important, therefore, that an accurate assessment is made. However, commercial considerations are not within the ambit of BS 7671, which addresses, in this context, only the safety aspects of the suitability of supply.

4.4.7   Type and rating of overcurrent device at the origin

Regulation 313.1, item (vi), requires the type and rating of the overcurrent protective device at the origin of the installation to be determined. This can only be done by inspection or enquiry, and it is necessary to establish the rated current, short-circuit capacity and its type in terms of British Standard or other performance specification. The purpose of this assessment is to ascertain that the rated current is not less than the connected load, allowing for any permissible diversity, that the short-circuit capacity is adequate for the prospective fault current and that connected wiring systems, including meter tails, are adequately protected against overload and fault currents.

Once the type and rating have been established, it is important to check the characteristics of the device in the relevant British Standard or other standard performance specification so that it can be considered in terms of the protection of wiring systems connected to it.

In the case of public LV supplies, most EDs now use BS 1361 type II (either 60, 80 or 100 A) high rupturing capacity fuses (33 kA short-circuit capacity) on new supplies. However, it should never be assumed that a particular type or rating of device has been used; the details should always be ascertained by inspection or enquiry.

4.5   Supplies for safety services and standby purposes

As called for in Regulation 313.2, supplies for safety services and standby supplies must be determined and confirmed that they are reliable and of adequate capacity. In the case of supplies from an ED, the designer must consult that ED regarding those matters and the switching arrangements, where appropriate, particularly where such services represent a substantial load.

Such safety services would include, for example, fire alarm and detection systems, emergency lighting (including emergency exit signs), fire pumps and sprinkler systems. Wiring systems for these services and their routes should be chosen carefully and the provisions and requirements of Chapter 56 of BS 7671 must be met. The supplies are classified as automatic or nonautomatic, as appropriate. Chapter 14 of this Guide gives further guidance on supplies for safety services.

4.6   Installation circuit arrangements

The four regulations under Section 314 of BS 7671 set out the requirements for the circuit arrangements. Essentially, circuits must be provided in sufficient number so that danger is prevented, and inconvenience minimized, both under normal and fault conditions and arranged so that operation, inspection, maintenance and testing may be carried out in a safe manner. Compliance with other chapters of BS 7671 should affirm that these basic requirements are met, but it is imperative that a sufficient number of circuits are provided so that equipment unaffected by a fault condition remains safe and functioning when another circuit develops a fault. The designer would need to consider the effects, in terms of consequential events, of the operation of a single protective device.

Regulation 314.4 calls for each circuit to be electrically separate and connected individually and detached from all other circuits where they terminate in a distribution board. Implicitly, the connection of each circuit line(s), neutral and circuit protective conductors (cpcs) must follow a logical sequence at their termination in the distribution board.

4.7   External influences

Regulation 301.1 and, more particularly, Chapter 32 and Appendix 5 of BS 7671 deal with external influences, which need to be assessed by the designer and then taken into account in the subsequent installation design and its construction. Section 10.2 of this Guide provides some guidance in this respect.

4.8   Compatibility

Regulation 331.1 demands an assessment be made of the characteristics of all equipment likely to have harmful effects on other equipment or on the supply. Where equipment is liable to have adverse effects on the supply, this regulation calls for the ED to be consulted; such equipment may, for example, include large step loads (suddenly applied loads), electronic switch-mode inverters or any load capable of generating low-order harmonics.

Regulation 331.1 (formerly 331-01-01 of BS 7671:2001) was modified by BS 7671:2008 to the extent that examples of the characteristics which may adversely affect other services or the supply are now given as:

- transient overvoltages;
- undervoltage;
- unbalanced loads;
- rapidly fluctuating loads;
- starting currents;
- harmonic currents;
- leakage current;
- excessive protective conductor current;
- d.c. feedback;
- high-frequency oscillations;
- earth leakage currents;
- necessity for additional connections to Earth;
- power factor.

In effect, there has been no real change in the regulatory requirements for the designer, who would have had to consider all the listed items to meet the previous wording of the regulation.

4.9 Maintainability

Whilst it is important that an installation is designed, constructed and verified to afford compliance with BS 7671 and to meet the operational and functional needs of the user, it is equally imperative that it can be readily and easily maintained throughout the life of the installation (Regulation 341.1). This aspect should not be ignored or overlooked at the design stage; it is of the utmost importance that this issue is addressed by the designer at a very early stage of the design.

Legislation, such as the Electricity at Work Regulations 1989, imposes responsibilities on employers to maintain their installations in a safe condition at all times; this applies to all places of work, including shops, schools, factories, hospitals, etc. It is incumbent on the installation designer, therefore, to be cognizant of all the maintenance requirements and to make due provision for them in the design. Materials used should be of sufficiently high standard to be serviceable throughout the lifetime of the installation. The designer should place little emphasis on the installation user's ability to exercise sound precautions in the use of the installation, and the materials and equipment selected should be suitable for their environment and be of good quality. Equipment manufactured to an appropriate British or CENELEC standard should be used wherever possible.

Unless an installation is under the *strict* supervision of a competent authority, the operation of essential safety measures, such as isolation and switching, should be made absolutely clear by provision of *legible* and *durable* identification labels which leave no room for misunderstanding.

Where equipment is likely to have a limited life expectancy which is less than that of the remainder of the installation equipment, then provision should be made for its periodic replacement prior to its becoming unsafe in use. If, for example, a piece of equipment is intended for a particularly hostile environment, then the designer has the option of selecting that material with the capability of withstanding that environment for the whole lifetime of the installation or, alternatively, the equipment may be selected to have a deliberately short lifespan and arrangements made for its periodic replacement. It is very often a matter of simple economics as to which method is chosen, provided, of course, that safety is not compromised at any time.

BS 7671 demands that, in addition to the initial verification of the installation, periodic inspection and testing are required over the lifetime of the installation. It is of paramount importance, therefore, that access to equipment and wiring system terminations is maintained at all times. Equipment should be so constructed and installed that inspection and testing activities may be carried out safely. Where an item of equipment is less accessible, its degree of reliability must correspondingly be greater in terms of maintaining safety unless its failure cannot cause danger.

Earlier, the installation circuit arrangements were discussed, and, in the context of maintainability, the designer should take account of circuit arrangements also in terms of the implications of maintaining such equipment served by these circuits.

Every installation carried out in a place where persons work requires an Operating and Maintenance Manual to enable the user and person responsible for maintenance to discharge their responsibilities under the Health and Safety at Work etc Act 1974, paragraph 6. This manual should include, amongst other things, instructions for proper operation and list all the items that need maintenance, together with comprehensive details of how the maintenance should be undertaken (see Section 10.11 of this Guide).

5

Protection against electric shock

5.1 General

Before discussing the requirements for protection against electric shock it is important to have a clear perception of the two different risks encountered concerning electric shock.

First, there is the risk of *electric shock under normal conditions*. This is when contact (by a person or livestock) is made directly with a live part and which is likely (almost certain) to cause current to flow through a body to the injury, perhaps fatal, of that person or animal. This hazard was formerly known as direct contact in BS 7671:2001.

Electric shock under fault conditions is when contact is made with an exposed-conductive-parts (e.g. a metal enclosure of a Class I item of equipment) which is not live under normal conditions but which has become live under earth fault conditions. This hazard was formerly known as indirect contact in BS 7671:2001. It arises when the protective provisions against electric shock under normal conditions have ceased to be effective, such as where there is failure of basic insulation.

It is important to recognize that an overcurrent protective device or a residual current device (RCD) will not provide automatic disconnection when human contact is made between two live parts only (e.g. line and neutral), though an RCD may automatically disconnect if contact is also made simultaneously with earthy parts. It should also be noted that a human (or livestock) making contact between live parts, between live parts and exposed-conductive-parts, between live parts and extraneous-conductive-parts or between exposed-conductive-parts and extraneous-conductive-parts, will be subjected to a voltage (a touch voltage or the full line voltage, depending on the particular parts contacted).

Contact by a person with an exposed-conductive-parts whilst a fault occurs on that part, or other electrically connected parts, would surely cause current to flow through their body; but if adequate protective provisions have been applied to the circuit concerned, then injury should be prevented, though it is likely that that person would 'perceive a shock'.

The IEC's rules for electrical installations and BS 7671 take account of the research work done internationally on the effects of electric current through a human body when determining the limits on the magnitude and duration of earth fault currents. IEC TC 64 Working Group 9 has made an in-depth study of disconnection times and related matters

A Practical Guide to The Wiring Regulations: 17th Edition IEE Wiring Regulations (BS 7671:2008)
Fourth Edition Geoffrey Stokes and John Bradley
© Geoffrey Stokes and John Bradley. Published by John Wiley & Sons, Ltd

and their report in this respect has been published by the British Standards Institution as *Effects of current passing through the human body* (DD IEC/TS 60479-1 *General aspects* and PD 6519-2 *Special aspects relating to human beings*). This document addresses the magnitude and duration of electric current through the body and the consequential results in terms of perception and severity of the physiological effects.

The fundamental requirements for protection against electric shock are given in Regulation Group 131.2 in Chapter 13 of BS 7671, *Fundamental principles*. The protective measures to be adopted are given in Chapter 41, *Protection against electric shock*. In addition, Regulation 410.3.2 of that chapter makes it clear that, where there is an increased risk of electric shock due to the conditions of external influence or in certain special locations, additional protection is specified by BS 7671. If the location or installation is dealt with in Part 7 of BS 7671, then the additional requirements will need to be met, by *supplementing* or *modifying* the Chapter 41 requirements as necessary. Section 531 of Chapter 53 sets out requirements for protective devices which may be employed for fault protection by Automatic Disconnection of Supply (ADS). Also, when considering provisions for fault protection, Chapter 54, *Earthing arrangements and protective conductors*, must not be overlooked and indeed is of particular importance when considering the provision for fault protection embodied in the protective measure termed ADS.

The sequence shown, left to right, in Table 5.1, hopefully clarifies the necessary considerations to be made when articulating the requirements for protection against electric shock. The sequence shown should not be regarded as exhaustive; in many instances, reference to other parts, chapters, sections and individual regulations will be necessary.

Chapter 41 outlines the appropriate measures for protection against electric shock and is divided into nine discrete sections as follows:

- Section 410 *Introduction*;
- Section 411 *Protective measure: automatic disconnection of supply*;
- Section 412 *Protection measure: double or reinforced insulation*;
- Section 413 *Protective measure: electrical separation*;
- Section 414 *Protective measure: extra-low voltage provided by SELV or PELV*;
- Section 415 *Additional protection*;
- Section 416 *Provisions for basic protection*;
- Section 417 *Obstacles and placing out of reach*;

Table 5.1 Protection against electric shock: sequence (left to right)

Fundamental requirements	Protective measures	Devices for fault protection	Earthing and protective conductors
Part 1	Part 4	Part 5	Part 5
Chapter 13	Chapter 41	Chapter 53	Chapter 54
131.2.1	(all)	Section 531	(all)
131.2.2		(all)	

Part 7 Special installations or locations (particular requirements)

- Section 418 *Protective measures for application only when the installation is controlled under the supervision of skilled or instructed persons*.

The nine sections will now be dealt with in this order, except that provisions for basic protection (Section 416) will be dealt with immediately after the introduction (Section 410). This is because provisions for basic protection form part of nearly all the protective measures covered in Chapter 41. Therefore, it is advantageous to gain an understanding of these provisions before dealing with the protective measures in detail.

Part 7 *Special installations or locations* is dealt with separately in Chapter 16 of this Guide and it is not the intention here to invoke Part 7 of BS 7671 except to refer to it in passing where appropriate.

The introduction to Chapter 41 explains that the chapter deals with protection against electric shock applied to electrical installations and that it is based on IEC 61140, the basic safety standard for the protection of persons and livestock. This standard is intended to give fundamental principles and requirements that are common to electrical installations and electrical equipment or are necessary for their coordination. The fundamental rule for protection against electric shock in IEC 61140 is that hazardous live parts must not be accessible and that accessible conductive parts must not be hazardous live, either in use without a fault or in single-fault conditions.

Chapter 41 specifies essential requirements for protection against electric shock, including basic protection and fault protection, and deals with the application and coordination of those requirements in relation to external influences.

Regulation 410.3.2 stipulates that a protective measure against electric shock shall consist of:

1. an appropriate combination of a provision for basic protection and an independent provision for fault protection; or
2. an enhanced protective provision which provides both basic protection and fault protection (such as reinforced insulation).

However, as pointed out in a note to Regulation 410.3.2, protective measures which do not follow this concept are permitted for special applications. This is a reference to the protective measures of obstacles, placing out of reach, non-conducting location, earth-free local equipotential bonding and electrical separation for the supply to more than one item of current-using equipment, which are explained respectively in Sections 5.8.1, 5.8.2, 5.9.2 and 5.9.3 of this Guide.

Regulation 410.3.2 points out that, under certain conditions of external influence and in certain locations, additional protection (in accordance with Section 415) is specified as part of a protective measure. Additional protection involves the use of either an RCD having specified characteristics that provide high sensitivity, or by supplementary equipotential bonding. This is explained in more detail in Section 5.7 of this Guide.

In Regulation 410.3.3, the essential requirement is stated that in each part of an installation one or more protective measures against electric shock shall be applied, taking account of the conditions of external influence. The protective measures in Chapter 41 fall into two broad classes: those that are generally permitted, which are listed in Table 5.2, and those that are permitted only for special applications, which are listed in Table 5.3.

Table 5.2 Protective measures that are generally permitted

Section	Protective measure
411	ADS
412	Double or reinforced insulation.
413	Electrical separation for the supply to one item of current-using equipment.
414	Extra-low voltage (SELV or PELV)

Table 5.3 Protective measures permitted only for special applications

Section	Protective measure	Examples of application	Restrictions
417	Obstacles	Factory sub-station switchboards	Installations accessible only to skilled or instructed persons,[a,b] or persons under the supervision of skilled or instructed persons. Not for use in locations of increased shock risk, including those in Part 7 of BS 7671
417	Placing out of reach	Overhead distribution conductors Overhead electric crane conductors	
418	Non-conducting location	Electrical and electronic repair workshops	Installations under the supervision of skilled or instructed persons[a,b] so that unauthorized changes cannot be made to the installation. Not for use in locations of increased shock risk, including those in Part 7 of BS 7671
418	Earth-free local equipotential bonding		
418	Electrical separation for the supply to more than one item of current using equipment		

[a]A skilled person is a person with technical knowledge or sufficient experience to enable him/her to avoid dangers which electricity may raise.
[b]An instructed person is a person adequately trained or supervised by skilled persons to enable him/her to avoid dangers which electricity may create.

It is important that, as required by Regulation 410.3.3, the particular protective measures applied are considered when selecting electrical equipment for an installation. For example, if double or reinforced insulation is used as the sole protective measure for a circuit (which is permitted only where there will be effective supervision during normal use – Regulation 412.1.3 refers), Class I electrical equipment must not be used in the circuit. This is because Class I equipment has exposed-conductive-parts that are required to be earthed but a cpc will not have been provided in the circuit.

As stated in Regulation 410.3.7, if certain conditions of a protective measure cannot be met, then supplementary provisions must be applied so that the protective provisions together achieve the same degree of safety. An example of the application of this regulation is where functional extra-low voltage (FELV) is used where a separated extra-low voltage (SELV) or protective extra-low voltage (PELV) system is not necessary. Because a FELV system does not meet the requirements for a SELV or PELV system, certain supplementary provisions must be made, such as insulation corresponding to the voltage of the primary circuit of the source.

Regulation 410.3.8 requires that different protective measures applied to the same installation or part of an installation or within equipment shall have no influence on each other such that failure of one protective measure could impair the other protective measure or measures. For example, exposed-conductive-parts of a SELV system (such as a metal enclosure of an item of equipment) must not be connected to a protective conductor, such as that of a circuit relying on the protective measure of ADS, as this would impair the safety of the SELV system.

There are a number of instances where fault protection may not be required (Regulation 410.3.9 refers). These relate to where the risk of electric shock is judged to be negligible due to the very low probability of contact by a person or by livestock. Table 5.4 details the extent of this dispensation.

5.2   Provisions for basic protection

5.2.1   General

The purpose of basic protection is to prevent contact by persons or livestock directly with live parts. Section 416 of BS 7671 recognizes two provisions for basic protection: basic insulation of live parts, and barriers or enclosures. Generally, at least one of these provisions needs to be used in every installation, except in the following cases:

- where an enhanced protective measure is used, such as double or reinforced insulation or, in certain circumstances, SELV; or
- where the protective measure of obstacles or the protective measure of placing out of reach is used.

Figure 5.1 shows examples of basic insulation of live parts and of barriers and enclosures.

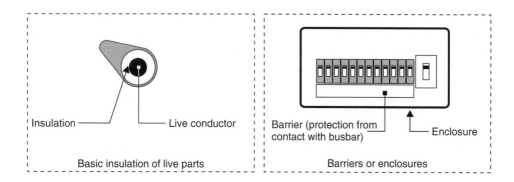

Figure 5.1 Examples of the two provisions for basic protection

5.2.2 Basic insulation of live parts

Basic insulation of live parts is probably the most common way of complying with the requirements for basic protection. It prevents physical and electrical contact with the live parts so insulated. To state the obvious, conductors are usually allocated this protective provision. Conductor insulation must have a suitable voltage rating appropriate to the circuit operating voltage and must be suitable for its intended use in service.

Regulation 416.1 calls for live parts to be covered by insulation that can only be removed by destruction. In other words, a sleeve or tape of insulating material is inadequate if only because it may be removed without destruction. Conductors should be insulated and, where necessary, further protected. This further protection should be regarded as including protection against external influences such as mechanical damage (e.g. from impact, vibration) and ingress of solids and liquids. In this context, except for protective conductors, it is necessary to enclose all insulated-only (non-sheathed) conductors in conduit, trunking or ducting, as required by Regulation 521.10.1.

It is important to ensure that the insulation (and sheathing) of conductors and cables is not damaged during erection of the installation by, for example, rough handling and bad workmanship by electrical operatives and other tradesmen. It will often be necessary to provide additional temporary protection against mechanical damage to cables during the course of construction. When using a cable for the first time, it will be necessary to obtain the manufacturer's advice regarding, for example, installation methods and temperature tolerances both for installation and in service. Some cable sheaths and insulations are liable to fracturing when handled at low temperatures (see also Chapter 10 of this Guide).

Some equipment may contain live parts having functional insulation only (such as varnishes, paint and lacquers as in motors). These coatings should not be regarded as providing the necessary insulation for basic protection against electric shock. The live parts must therefore be within enclosures or behind barriers in order to provide the necessary protection.

5.2.3 Barriers or enclosures

Regulation Group 416.2 deals with this protective provision, which is commonly used (in conjunction with basic insulation of live part) in the manufacture of electrical installation equipment. Regulation 416.2.1 requires that live parts are inside enclosures or behind barriers that provide a degree of protection to at least IPXXB or IP2X (see BS EN 60529 and Table 10.10 of this Guide) ingress protection: preventing ingress of 12 mm diameter objects such as fingers (see the standard test finger in BS 3042). However, there is an exception to this rule, in that larger apertures are permitted if such are necessary to facilitate either replacement of parts or to allow proper functioning of the equipment. This exemption is permitted only if suitable precautions are taken to prevent unintentional contact by persons or livestock with live parts. It must also be ensured that persons, such as operators, are fully aware that such live parts can be touched through the opening and should not be touched. Understandably, because of the practical difficulties involved, this exception to the rule is rarely invoked. Regulation 416.2.2 lays down an additional condition relating to the top surface of an item of equipment that is also readily accessible, in that IPXXD or IP4X degree of protection is required for that surface: preventing ingress by objects of 1 mm diameter, such as small wires, tools, insects and general debris.

Barriers and enclosures must also exhibit the stability and durability to maintain the necessary IP degree of protection for the lifespan of the installation, taking account of the environmental conditions and external influences. It should go without saying (but Regulation 416.2.3 does just that) that barriers and enclosures need to be firmly secured in their intended places in order to fulfil their intended function.

Regulation 416.2.4 lays down the alternative conditions relating to the removal of a barrier or opening of an enclosure. The removal or opening must be achievable only by the use of a key or a tool (e.g. screwdriver). If this arrangement is not available, or otherwise impracticable, then arrangements need to be made within the equipment so that live parts are isolated from the supply *before* it is possible to open the enclosure or remove the barrier. Restoration of supply must be possible only after the refitting of those parts. Alternatively, an intermediate barrier (to IPXXB or IP2X) is to be fitted which again is to be removable only by use of a key or tool. Equipment excepted from this general rule are ceiling roses (BS 67), pull-cord switches (BS 3676), bayonet cap lampholders (BS EN 61184) and Edison screw (ES) lanpholders (BS EN 60238) (i.e. access to live parts is permitted to these accessories without the use of a tool).

The possibility that an item of equipment such as a capacitor that may retain a dangerous voltage after it is switched off is installed behind a barrier is addressed in Regulation 416.2.5. Where this is the case, the regulation requires a warning label to be provided. The label should be fixed in a position where it will be seen by anyone intending to remove the barrier, before they do so. The regulation points out that small capacitors, such as those used for arc extinction and for delaying the response of relays, are not considered dangerous. Also, a note to the regulation points out that unintentional contact (with an electrically charged part) is not considered dangerous if the voltage resulting from the static charge falls below 120 V in less than 5 s after disconnection from the power supply.

5.3 Protective measure: automatic disconnection of supply

5.3.1 General

ADS, covered in Section 411 of BS 7671, is one of the protective measures that are generally permitted by BS 7671, as indicated earlier, in Table 5.2.

In this protective measure, the provision for basic protection is either basic insulation of live parts or barriers and enclosures. The provision for fault protection is protective earthing, protective equipotential bonding and automatic disconnection in the case of a fault. As a further possibility, Class II equipment may be used to provide both basic protection and fault protection where ADS is applied (Regulation 411.1 refers).

Additional protection by means of an RCD having the characteristics stated in Regulation 415.1.1 is also used as part the protective measure of ADS where specified (Regulation 411.1 refers). Section 5.7 of this Guide gives examples of where this is the case. A note to Regulation 411.1 points out that a residual current monitor (RCM), although it is not a protective device, may be used to monitor the residual current in an installation, by giving an audible, or audible and visual, signal when the preselected value of residual current is reached.

5.3.2   Basic protection in ADS

Wiring systems and all other equipment of an installation relying on protection by ADS for protection against electric shock must comply with either Regulation Group 416.1 for protection by basic insulation or Regulation Group 416.2 for protection by barriers or enclosures.

5.3.3   General requirements for fault protection in ADS

As already stated, the provision for fault protection where ADS is applied is protective earthing, protective equipotential bonding and automatic disconnection in the case of a fault. The fundamental principle of this protective provision is that the characteristics of the device(s) providing protection must be coordinated with the protective earthing and equipotential bonding arrangements for the installation and the earth fault loop impedance of the circuit so protected. This is necessary so that the touch voltages occurring between simultaneously accessible exposed-conductive-parts and extraneous-conductive-parts anywhere in the installation during earth fault conditions are limited to such magnitude and duration as not to cause danger. Earth fault current is detected by the circuit protective device, resulting in automatic disconnection of the supply to the line conductor of the circuit or equipment within the maximum time required by the relevant regulations (Regulation 411.3.2.1 refers). Chapter 11 of this Guide deals with protective devices commonly used to provide fault protection.

Figure 5.2 illustrates the touch voltages occurring between exposed-conductive-parts, and between exposed-conductive-parts and extraneous-conductive-parts, during earth fault conditions. Four items of current-using equipment with exposed-conductive-parts (connected to the main earthing terminal (MET) of the installation via cpcs) are shown together with an extraneous-conductive-part (connected to the MET via a main protective bonding conductor). For clarity, only earthing, bonding and cpcs are shown.

An earth fault (of negligible impedance) has occurred on the exposed metalwork of item B. Earth fault current $I_{F(B)}$, therefore, flows from the source, through the line conductor, through the cpc to the MET and on via the earthing conductor to the supply source earthed neutral point. The magnitude of $I_{F(B)}$ will depend on the nominal voltage to Earth U_o and

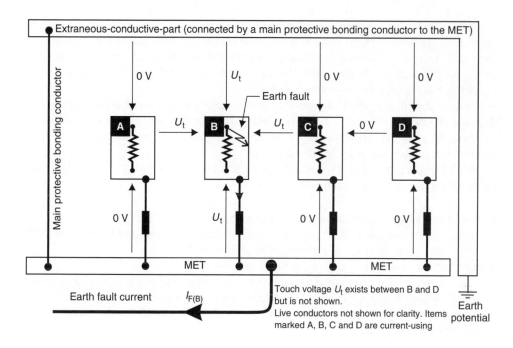

Figure 5.2   Voltage distribution between conductive parts under earth fault conditions

the total earth fault loop impedance $Z_{s(B)}$ at equipment item B, as given by

$$I_{F(B)} = \frac{U_o}{Z_{s(B)}} \tag{5.1}$$

The voltage U_t between the earthed metalwork of equipment B and the MET will be the product of the fault current $I_{F(B)}$ and the cpc impedance $Z_{2(B)}$ of the circuit feeding equipment B given by Equation (5.2) or, if preferred, that given by Equation (5.3):

$$U_t = I_{F(B)}Z_{2(B)} \tag{5.2}$$

$$U_t = \frac{U_o Z_{2(B)}}{Z_{s(B)}} \tag{5.3}$$

By way of example, take the cpc impedance of circuit B to be 0.38 Ω and the total earth fault loop impedance of that circuit to be 0.9 Ω. If the circuit nominal voltage to Earth $U_o = 230$ V, then, by substituting these values in to Equation (5.3), the value of 97.1 V for U_t is obtained:

$$U_t = \frac{230 \times 0.38}{0.9} = 97.1 \text{ V} \tag{5.4}$$

From this example it can be seen that substantial touch voltages can exist during fault conditions and these will be present not only between equipment B and the MET, but also between equipment B and other exposed-conductive-parts and extraneous-conductive-parts, which will be at approximately the same potential as the MET (assuming insignificant fault current flow in the main protective bonding conductor). Voltages, therefore, will also be present between B and A, B and C, B and D, and B and the extraneous-conductive-parts.

It will be seen that, in this example, the touch voltage was at a considerably higher level than is generally considered to be a safe value (50 V for dry conditions). It is necessary, therefore, to establish proper protective device coordination so that the duration of these touch voltages does not give rise to danger.

For all installations using ADS as their protective measure, Regulation 411.3.1.1, relating to protective earthing, requires that:

- exposed-conductive-parts are connected to a cpc; and
- all simultaneously accessible exposed-conductive-parts are connected to the same earthing system, individually, collectively or in groups; and
- the protective conductors comply with Chapter 54; and
- a cpc is run to and terminated at each point in wiring and at each accessory except a lampholder having no exposed-conductive-parts and suspended from such a point (a point in wiring is a termination of the fixed wiring intended for the connection of current-using equipment).

As mentioned earlier and confirmed by Regulation 411.3.1.2, main protective bonding is required where the measure of ADS is to be used. This bonding will need to meet the requirements of Section 544, *Protective bonding conductors*, which is discussed in Chapter 12 of this Guide. Regulation 411.3.1.2 gives a list of examples of extraneous-conductive-parts that need to be bonded to the MET of the installation. An extraneous-conductive-part is defined in Part 2 of BS 7671 as a conductive part liable to introduce a potential, generally earth potential and not forming part of the electrical installation. The examples given in the regulation are:

- water installation pipes;
- gas installation pipes;
- other installation pipework and ducting;
- central heating and air conditioning systems;
- exposed metallic structural parts of the building.

Regulation 411.3.1.2 also requires that any lightning protection system is main bonded in accordance with BS EN 62305, *Protection against lightning*. The regulation also points out it is also necessary to main bond any metallic sheaths of telecommunications cables, but requires the cable owner's or operator's consent to be obtained (preferably in writing) before such bonding is carried out.

The purpose of providing main bonding is so that, within each building, the potential differences, or touch voltages, under earth fault conditions are such that they will not cause electric shock, as defined (a dangerous physiological effect). Earth faults developing in the installation will produce touch voltages between the relevant exposed-conductive-parts of the faulty circuit and other conductive parts (exposed and extraneous). In effect, main bonding provides the circumstances where the maximum touch voltage created by an earth fault within the building is the voltage appearing between the MET and the point at which the earth fault occurs (that is, the product of the earth fault current and the cpc impedance, in most cases). Without main bonding (that is, without the required connection of the extraneous-conductive-parts to the MET), the maximum touch voltage created between the relevant exposed-conductive-parts of the faulty circuit and an extraneous-conductive-part would be higher than that just mentioned. It would be equal to the voltage appearing between Earth (i.e. 'true' earth) and the point at which the earth fault occurs. This voltage is equal to the product of the earth fault current and the impedance of not only the cpc, but of the entire part of the earth fault loop between the point of fault and the earthed point of the distribution transformer or other source of energy, in most cases.

Main bonding also provides protection in the event of an earth fault developing on the supply cable (or distribution circuit) which would result in a fault voltage appearing on the MET, with respect to Earth. By main bonding, all conductive parts are substantially maintained at this potential and there is no dangerous touch voltage between conductive parts and risk of electric shock is prevented, at least within the building.

Further benefit comes from main bonding if the installation uses a PME facility provided by the ED as its means of earthing. In the event of the supply neutral conductor becoming disconnected from Earth, the installation's main bonding conductors will provide a path for load current to return to the source through any extraneous-conductive-parts that are in contact with Earth. The bonding also maintains the exposed-conductive-parts and extraneous-conductive-parts within the building at substantially the same potential under these conditions.

In the context of main bonding to water, gas or other installation pipework (as called for in Regulation 411.3.1.2), it should be noted that Regulation 544.1.2 requires this even if there is an insulating section or insert at the point of entry of the related service into the premises. In this case, the regulation requires the main bonding connection to be made to the consumer's hard metal pipework and before any branch pipework. Different considerations may apply where a gas, water or other pipework installation in a building consists substantially of plastic pipes. In this case, little is gained by main bonding any sections of metallic pipework that form part of the system if these are isolated from Earth or are only connected to it through a metre or more of plastic installation pipe containing tap water (assuming 22 mm diameter plastic pipe and a supply of nominal voltage $U_o = 230$ V). In the UK, a metre of 22 mm diameter plastic filled with cold tap water can be expected to have a resistance in the range of 15 to 65 kΩ or more.

Regulation 411.3.1.2 also calls for each separate building to be subjected to the main bonding requirements; therefore, these separate buildings will each need their own earthing terminal. In many cases, these terminals will need linking by suitable protective conductors to the MET, which is generally in the building where the electricity is supplied (this is discussed further in Chapter 12 of this Guide; see Figure 12.8).

For all installations using ADS as their protective measure, Regulation 411.3.2.1 gives requirements for automatic disconnection in the case of a fault. With two exceptions (given in Regulations 411.3.2.1 and 411.3.2.6, mentioned later in this section of this Guide), the regulation requires that a protective device shall automatically disconnect the supply to the line conductor of a circuit or equipment if a fault of negligible impedance occurs between the line conductor and an exposed-conductive-parts or a protective conductor of the circuit or equipment. Automatic disconnection is required to take place within the disconnection time required by Regulation 411.3.2.2, 411.3.2.3 or 411.3.2.4, as applicable.

Regulation 411.3.2.2 stipulates the maximum disconnection times for final circuits rated at not more than 32 A. The requirements are given in Table 5.5, which is based on Table 41.1 of BS 7671.

The disconnection times in Table 5.5 do not apply to the following types of circuit, for which the maximum permitted disconnection time is 5 s if the circuit forms part of a TN system or 1 s if it forms part of a TT system (Regulations 411.3.2.3 and 411.3.2.4 refer):

- a distribution circuit (otherwise called a sub-main); and
- a circuit not covered by Table 5.4, such as a final circuit rated at more than 32 A.

There are two exceptions to the requirement of Regulation 411.3.2.1 for automatic disconnection to be provided. One is for a system where the supply source will quickly reduce its voltage to 50 V a.c. or 120 V d.c., or less in the event of an earth fault. The other is where automatic disconnection cannot be achieved but where supplementary equipotential bonding is provided. Table 5.6 details both of the exceptions.

For all a.c. installations using ADS as their protective measure, Regulation 411.3.3 requires additional protection by means of an RCD having the characteristics specified in Regulation Group 415.1 for:

- socket-outlets rated at 20 A or less for general use by ordinary persons; and
- mobile equipment (equipment that is moved while in operation or can be easily moved from one place to another while connected to the supply) with a current rating of 32 A or less for use outdoors.

Table 5.4 Dispensation of provisions for fault protection

Conductive part	Qualification (if any)
Metal supports of overhead line insulators attached to a building	Must not be within 'arm's reach' (see Figure 6.4).
Steel-reinforced poles of overhead lines	Steel reinforcement is not accessible
Exposed-conductive-parts, such as: • bolts • rivets • nameplates • cable clips • screws and other fixings	Which because of their reduced dimensions (maximum 50 mm × 50 mm approximately), or their disposition, cannot be gripped or come into significant contact with a part of the human body, and provided that connection with a protective conductor could only be made with difficulty or would be unreliable
Metal enclosures protecting equipment in accordance with Section 412 (e.g. where the protective measure of double or rein-forced insulation is used)	
Unearthed street furniture supplied from an overhead line	Inaccessible in normal use

Table 5.5 Maximum disconnection times for final circuits rated at not more than 32 A (based on Table 41.1 of BS 7671).[a,b]

System	Time (s)							
	$50 < U_o \leq 120$ V		$120 < U_o \leq 230$ V		$230 < U_o \leq 400$ V		$U_o > 400$ V	
	a.c.	d.c.	a.c.	d.c.	a.c.	d.c.	a.c.	d.c.
TN	0.8	– [c]	0.4	5	0.2	0.4	0.1	0.1
TT	0.3	– [c]	0.2	0.4	0.07	0.2	0.04	0.1

[a]Where, in a TT system, disconnection is achieved by an overcurrent protective device and protective equipotential bonding is connected to all the extraneous-conductive-parts within the installation in accordance with Regulation 411.3.1.2, the maximum disconnection times applicable to a TN system may be used.
[b]U_o is the nominal a.c. rms or d.c. voltage between a line conductor and Earth.
[c]Disconnection is not required for protection against electric shock but may be required for other reasons, such as protection against thermal effects.

Table 5.6 Exceptions to the requirement of Regulation 411.3.2.1 such that automatic disconnection need not be provided

Regulation	Application	Qualification[a]
411.3.2.5	A system of nominal voltage greater than 50 V a.c. or 120 V d.c. in which the source will reduce its output voltage to below this value quickly (see opposite) in the event of a fault between a line conductor and an exposed-conductive-part or a protective conductor	The output voltage of the source is reduced to 50 V a.c. or 120 V d.c.(as applicable) or less within: • the time given in Table 5.5 for a final circuit rated at not more than 32 A; or • 5 s (TN system) or 1 s (TT system) for a distribution circuit or circuit not covered by Table 5.5, such as a final circuit rated at more than 32 A
411.3.2.6	Where automatic disconnection within the specified maximum disconnection time cannot be achieved by an overcurrent protective device or an RCD	Supplementary bonding is provided in accordance with Section 415.1

[a]Where the qualification is met, automatic disconnection need not be provided for protection against electric shock. However, automatic disconnection may be required for other reasons, such as protection against thermal effects.

The RCD giving the additional protection must have a rated residual operating current $I_{\Delta n}$ not exceeding 30 mA and an operating time not exceeding 40 ms at a residual current of $5I_{\Delta n}$, as required by Regulation 415.1.1. An ordinary person, in the context of item (i), is defined in Part 2 of BS 7671 as a person who is neither a skilled person nor an instructed person. The definitions of 'skilled person' and 'instructed person' are given in the notes to Table 5.3 of the Guide.

An exception to the requirement of Regulation 411.3.3 for additional protection to be provided is permitted for socket-outlets under the supervision of skilled or instructed persons (such as in some commercial or industrial locations). A further exception is allowed for a specifically labelled or otherwise suitably identified socket-outlet for the connection of a particular item of equipment. This second exception can be useful for a socket-outlet supplying an item of equipment such as a freezer, where considerable inconvenience or cost could result if the RCD giving the additional protection tripped as a result of, say, a defective item of equipment being plugged into another socket-outlet on the same circuit and the ensuing loss of supply going unnoticed for a long period. Note, however, that the exception must not be applied if the circuit cable is concealed in a wall or partition and is required by Regulation 522.6.7 or 522.6.8 to be provided with additional protection by an RCD (see section 5.7 of this Guide).

5.3.4   Fault protection in ADS: for TN systems

In a TN system, the intended return path for earth fault current from the installation to the source of supply (usually a distribution transformer) is through an earthed metallic return path, rather than through the general mass of Earth. The metallic path includes, for example, the metal sheath or armour of a multicore supply cable or a conductor of an overhead supply line. As is indicated Regulation 411.4.1, where the installation receives its electrical energy from a public (or possibly private) distribution network, as is usually the case, provision of this metallic path external to the installation is the responsibility of the distributor.

Regulation 411.4.2 calls for the neutral point or the midpoint of the power supply system to be connected to Earth. Again, this is the responsibility of the distributor if the installation is supplied from a public or private distribution network. The regulation adds that where a neutral point or midpoint is not available or not accessible, a line conductor shall be connected to Earth. However, this is exceptional and does not apply in a UK pubic distribution network, as the neutral (or star) point of the distribution transformer is invariably available and is connected to Earth by the distributor.

Regulation 411.4.2 calls for each exposed-conductive-parts of the installation to be connected by a protective conductor to the MET and then on to the earthed point (neutral) of the supply source (through the metallic return path). Requirements for the protective conductors in the installation are given in Chapter 54 of BS 7671 and dealt with in Chapter 12 of this Guide.

A note to Regulation 411.4.2 points out that it is permissible for the protective conductor of a TN system to be additionally connected with Earth (through an earth electrode), such as at the point of entry to the building. Such a connection with Earth, although it could help reduce the risk of electric shock if an earth fault occurred in an installation at a time when there was a discontinuity in the earthed metallic return path external to the building, is not usually provided in the UK, and is not a requirement of BS 7671. However, advice on the maximum resistance to earth that the earth electrode would have to have, in order for this measure to be effective, is given in IEE Guidance Note 5.

The vast majority of installations forming part of a TN system in the UK are either TN-S or TN-C-S. That is to say, separate protective and neutral conductors are used in the installation. TN-C systems are virtually unheard of because, although Regulation 411.4.3

permits a combined protective and neutral (PEN) conductor to be used in a fixed installation, this is not permitted by law for an installation with a supply given in accordance
with the Electricity Safety, Quality and Continuity Regulations 2002. Regulation 8(4) of
the ESQCR states that 'A consumer shall not combine the neutral and protective functions
in a single conductor in his consumer's installation'. A similar provision was included in
the Electricity Supply Regulations 1988, which were replaced by the ESQCR.

Only two types of protective device, listed in Regulation 411.4.4, are suitable for fault
protection in TN systems:

- overcurrent protective devices (e.g. fuse, miniature circuit-breaker (MCB) or moulded
  case circuit-breaker (MCCB));
- RCDs.

It should be noted that BS 3871 is now an obsolete standard, and MCB is no longer the
correct term, being replaced by the generic term 'circuit-breaker'. However, as 'old habits
die hard' and data for these devices will still be needed when reviewing some existing
installations, both reference to the standard and MCB is retained.

Regulation 411.4.4 demands that an RCD shall not be used in a TN-C system. It also
stipulates that, where an RCD is used in a TN-C-S system, a PEN conductor shall not
be used on the load side of the device and that the load protective conductor connection
must be made on the supply side of the device. Without these provisions of the regulation,
the RCD would not 'see' any earth fault current and would not provide the necessary
detection and consequential automatic disconnection.

The requirement of Regulation 411.3.2 for automatic disconnection in the case of a
fault is satisfied for TN systems, according to Regulation 411.4.5, if, and only if, the
condition in Equation (5.5) is satisfied:

$$Z_s I_a \leq U_o \tag{5.5}$$

where Z_s (Ω) is the impedance of the earth fault loop, comprising:

- the source; and
- the line conductor up to the point of fault; and
- the protective conductor between the point of fault and the source;

I_a (A) is the current causing automatic operation of the disconnecting device within the
time specified in Table 41.1 of BS 7671 (see Table 5.5 of this Guide), or within 5 s where
Regulation 411.3.2.3 applies (a distribution circuit or a circuit not covered by Table 5.5);
and U_o (V) is the nominal a.c. rms or d.c. voltage between a line conductor and Earth.

Table 41.1 of BS 7671 does not relate to reduced low voltage circuits, which are
addressed separately by Regulation Group 411.8 and dealt with later in Section 5.3.8 of
this Guide.

Regulations 411.4.6, 411.4.7 and 411.4.8 set out the references to tables detailing maximum earth fault loop impedances for particular overcurrent protective devices for both
0.4 and 5 s disconnection times. These are summarized in the Table 5.7.

Table 5.7 Tables for maximum Z_s for 0.4 s and 5 s disconnection times for overcurrent protective devices for $U_o = 230$ V

Protective devices	0.4 s disconnection	5 s disconnection
MCB type B (BS EN 60898)[b]	Table 41.3(a)[a]	Table 41.3(a)[a,b]
MCB type C (BS EN 60898)	Table 41.3(b)[a]	Table 41.3(b)[a,b]
MCB type D (BS EN 60898)	Table 41.3(c)[a]	Table 41.3(c)[a,b]
HRC fuses (BS 88-2.2 and BS 88-6 'gG')	Table 41.2(a)	Table 41.4(a)
HRC fuses (BS 88-2.2 and BS 88-6 'gM')	Refer to British Standard	Refer to British Standard
HRC fuses (BS 1361)	Table 41.2(b)	Table 41.4(b)
Cartridge fuses (BS 1362)	Table 41.2(d)	Table 41.4(d)
Semi-enclosed fuses (BS 3036)	Table 41.2(c)	Table 41.4(c)
Other devices	Refer to British Standard or manufacturer	Refer to British Standard or manufacturer

[a]The maximum values of Z_s also apply to the overcurrent characteristics of residual current circuit-breakers (RCCBs) with overcurrent protection (RCBOs) to BS EN 61009.
[b]The device manufacturer may be able to quote lower maximum values of Z_s.

The tabulated maximum values of earth fault loop impedance Z_s given in Tables 41.2, 41.3 and 41.4 of BS 7671, which are referred to in Table 5.7, are derived from the time–current characteristics for the various overcurrent protective devices given in Appendix 3 of BS 7671. For example, for a BS 88-2.2 or BS 88-6 'gG' fuse of nominal rating $I_n = 32$ A, the operating current I_a for 0.4 s disconnection is given as 220 A, for which, if related to a nominal voltage 230 V, the maximum Z_s is 1.045 Ω ($Z = V/I = 230/220$). This derived value, rounded down to 1.04 Ω, will be found in Table 41.2(a) of BS 7671.

For convenience, Table 5.8 lists the maximum earth fault loop impedance values Z_s for common protective devices used for fault protection, at a nominal voltage of 230 V. It should be noted that the Z_s values given relate to the design values, not the measured values, which will need modification due to temperature considerations (see Appendix 14 of BS 7671 and Chapter 17 of this Guide).

It should be noted that the BS 1362 cartridge fuse fitted in a 13 A plug is there for the purpose of protecting the flexible cord against short-circuit. Where used in a 13 A fused connection unit it may be regarded as an acceptable means of fault protection where circuit conditions allow.

Whilst the use of type 4 MCBs to BS 3871 was not precluded as devices for fault protection, their characteristics are such as to place severe restrictions on their use for such purposes owing to the high magnitude of fault current needed to effect operation and the consequential very low Z_s limits. For other protective devices (e.g. MCCBs), reference to the relevant standard and/or manufacturer will be necessary to obtain their operating characteristics.

Table 5.8 Maximum values of earth fault loop impedance (Ω) for common circuit protective devices, for fault protection, at a nominal voltage of 230 V.[a]

Ref	Rated current (A)	Earth fault loop impedance (Ω)													
		Fuses								MCBs to BS 3871 or BS EN 60898[b]					
		BS 88 'gG' Parts 2 and 6		BS 1361		BS 3036		BS 1362		Type 1	Type 2	Type B	Types 3 and C	Type D	Type 4
		0.4 s	5 s	0.4 s	5 s	0.4 s	5 s	0.4 s	5 s	0.1 and 5 s					
A	3	–	–	–	–	–	–	16.40	23.20	–	–	15.33	–	–	–
B	5	8.52	13.50	10.45	16.40	9.58	17.70	–	–	11.50	6.57	9.20	4.60	2.30	0.92
C	6	5.11	7.42	–	–	–	–	–	–	9.58	5.47	7.67	3.83	1.91	0.76
D	10	–	–	–	–	–	–	–	–	5.75	3.28	4.60	2.30	1.15	0.46
E	13	–	–	–	–	–	–	2.42	3.83	–	–	–	–	–	–
F	15	2.70	4.18	3.28	5.00	2.55	5.35	–	–	3.83	2.19	3.06	1.53	0.76	0.30
G	16	1.77	2.91	–	–	–	–	–	–	3.59	2.05	2.87	1.44	0.71	0.28
H	20	1.44	2.30	1.70	2.80	1.77	3.83	–	–	2.87	1.64	2.30	1.15	0.57	0.23
I	25	–	–	–	–	–	–	–	–	2.30	1.31	1.84	0.92	0.46	0.18
J	30	1.04	1.84	1.15	1.84	1.09	2.64	–	–	1.91	1.09	–	–	–	–
K	32	–	1.35	–	–	–	–	–	–	1.79	1.02	1.44	0.72	0.35	0.14
L	40	–	–	–	–	–	–	–	–	1.43	0.82	1.15	0.57	0.28	0.11
M	45	–	1.04	–	0.96	–	1.59	–	–	1.27	0.73	–	–	–	0.10
N	50	–	–	–	–	–	–	–	–	1.15	0.65	0.92	0.46	0.23	0.09
O	60	–	0.82	–	0.70	–	1.12	–	–	–	–	–	–	–	–
P	63	–	0.57	–	–	–	–	–	–	0.91	0.52	0.73	0.36	0.18	–
Q	80	–	0.42	–	0.50	–	–	–	–	0.71	0.41	0.57	0.29	0.14	–
R	100	–	0.33	–	0.36	–	0.53	–	–	0.57	0.32	0.46	0.23	0.11	–
S	125	–	–	–	–	–	–	–	–	–	–	0.37	0.18	0.09	–

(continued overleaf)

Table 5.8 (continued)

Ref	Rated current (A)	Earth fault loop impedance (Ω)													
		Fuses								MCBs to BS 3871 or BS EN 60 898[b]					
		BS 88 'gG' Parts 2 and 6		BS 1361		BS 3036		BS 1362		Type 1	Type 2	Type B	Types 3 and C	Type D	Type 4
		0.4 s	5 s	0.4 s	5 s	0.4 s	5 s	0.4 s	5 s	0.1 and 5 s					
T	160	–	0.25	–	–	–	–	–	–	–	–	–	–	–	–
U	200	–	0.19	–	–	–	–	–	–	–	–	–	–	–	–
V	250	–	0.14	–	–	–	–	–	–	–	–	–	–	–	–
W	315	–	0.10	–	–	–	–	–	–	–	–	–	–	–	–
X	400	–	0.082	–	–	–	–	–	–	–	–	–	–	–	–
Y	500	–	0.060	–	–	–	–	–	–	–	–	–	–	–	–
Z	630	–	0.046	–	–	–	–	–	–	–	–	–	–	–	–
A1	800	–	0.032	–	–	–	–	–	–	–	–	–	–	–	–

[a]Dashes denote either that the device is not commonly available or that by virtue of its characteristics it is not appropriate for fault protection.
[b]The impedance values are based on the 'worst-case' limits allowed by the standard and in certain cases where the manufacturer can claim closer limits than the standard permits the values may be accordingly modified.

It is important to note that the maximum values of Z_s in Tables 41.2, 41.3 and 41.4 of BS 7671 (like those in Table 5.8 of this Guide) relate only to a nominal voltage U_o of 230 V. If other nominal voltages are to be used, then the tabulated Z_s values need modifying accordingly by the application of a factor equal to $U_o/230$. For example, if $U_o = 220$ V, then a factor of 0.956 (i.e. 220/230) would need to be applied to the impedance values given for 230 V. It will be seen that the overriding consideration is the magnitude of the fault current which will flow, not the impedance.

It is also worth noting that the values of Z_s given in Table 41.3 of BS 7671 in respect of MCBs for 0.4 and 5 disconnections are in fact the same. The reason for this will be apparent when the time–current characteristics are viewed and it will be seen that for both disconnection times the device operates instantaneously (defined as within 100 ms) when the operating current I_a is flowing.

Interestingly, Regulation 411.3.2.6 (mentioned earlier, in Table 5.6 of this Guide) addresses the long-recognized problem encountered when prescribed disconnection times cannot be met. This problem is often found on distribution circuits (or sub-mains) where the earth fault loop impedance Z_s is too high to make automatic disconnection within the normal time limit of 5 s impossible when using an overcurrent protective device. For example, take a small distribution circuit having fault protection (and overcurrent protection) by a BS 88 'gG' 200 A fuse. To disconnect within 5 s, the fault current would need to be not less than 1200 A and, consequently, Z_s must not exceed 0.19 Ω. If the earth fault loop impedance was (calculated or measured), say, 0.25 Ω, then disconnection will take longer than 5 s and, therefore, further measures need to be applied. For the larger rated current circuit the situation would be worse.

In such cases, where automatic disconnection times cannot be met by an overcurrent device, then either supplementary bonding must be provided in accordance with Regulation Group 415.2, as referred to in Regulation 411.3.2.6, or an RCD must be used to provide fault protection.

The supplementary bonding requirement is a straightforward option, whilst protection by RCD needs careful consideration. Not least amongst this consideration are the disruptive aspects associated with its operation. Inconvenience may be caused where, for example, an earth fault develops on a final circuit resulting in the operation of the RCD protecting the distribution circuit unless further measures are taken. Discrimination between an RCD and a downstream overcurrent protective device would be difficult, if not impossible, to achieve. The designer would also need to consider RCD protection of final circuits emanating from the distribution board. Discrimination between RCDs in series cannot be achieved on a fault current magnitude basis and can only be designed utilizing the 'time' parameter. A timed delay on the upstream RCD can provide the necessary discrimination with a downstream RCD with no timed-delay operation (or shorter time delay); and if this method is adopted, much of the inconvenience mentioned would be alleviated.

Where an RCD is used for fault protection in a circuit in a TN system, the condition $Z_s I_a \leq U_o$ of Regulation 411.4.5 has to be met, just as it does when an overcurrent protective device is used to provide fault protection. To meet this condition where an RCD is used, the earth fault loop impedance Z_s in the circuit must not exceed the nominal voltage of the supply U_o divided by the current I_a causing automatic operation of the RCD within the specified maximum disconnection time. For example, suppose a disconnection

Table 5.9 Maximum values of earth loop impedances Z_s (Ω) for RCDs, for fault protection in a TN system, operating on 230 V.[a]

Ref	Rated residual operating current $I_{\Delta n}$ (mA)	Earth loop impedance Z_s (Ω)			
		BS 4293 and BS 7288		BS EN 61008-1 and BS EN 61009-1	
		0.4 s	5 s	0.4 s	5 s
RCD without time-delayed operation					
A	10	23 000	23 000	23 000	23 000
B	30	7 667	7 667	7 667	7 667
C	100	2 300	2 300	2 300	2 300
D	300	767	767	767	767
E	500	460	460	460	460
RCD time delayed					
F	100	– [b]	– [b]	1 150	2 300
G	300	– [b]	– [b]	383	767
H	500	– [b]	– [b]	230	460

[a]Measured values of Z_s should be much lower than those in this table, as the earth loop path in a TN system is entirely metallic. The Z_s values in this table apply to TN systems only.
[b]Maximum permitted Z_s depends on the rated (or declared) time delay T_d of the RCD. To determine maximum Z_s, consult RCD time–current characteristics and obtain I_a, the current in amperes causing operation within the required disconnection time, taking account of the time delay. Z_s must not exceed U_o/I_a.

time not exceeding 0.4 s has to be achieved using an RCD to BS EN 61008-1 or BS EN 61009-1 of the nondelay type with a rated residual operating current $I_{\Delta n}$ of 100 mA in a circuit of nominal voltage 230 V. It can be seen from Table 3A in Appendix 3 of BS 7671 that a current of 100 mA is required to cause the RCD to operate within 0.4 s (operation will actually occur within 300 ms at 100 mA). Accordingly, the maximum earth fault loop impedance Z_s that can be tolerated in the circuit is 2300 Ω (i.e. 230 V/0.1 A).

For convenience, Table 5.9 lists the maximum earth fault loop impedance values Z_s for RCDs used for fault protection in TN systems. It is important to note that the Z_s values achieved in the installation should be much lower than those given in the table, as the earth loop path in a TN system is wholly metallic. It should also be noted that the maximum values of Z_s in the table relate only to a nominal voltage U_o of 230 V, and should be multiplied by a factor equal to $U_o/230$ if other nominal voltages are to be used.

It must not be forgotten that where disconnection *cannot* be achieved within the pre-scribed time limits by using overcurrent devices (and either supplementary bonding is provided or an RCD is employed for fault protection), consideration must also be given to protection against overcurrent. For example, for a distribution circuit to disconnect in 12 s may be perfectly acceptable from the viewpoint of fault protection (provided supple-mentary bonding has been undertaken), but the requirements for overcurrent protection also need to be met for this disconnection time; it may be that an earth fault current of 12 s duration will increase the temperature of the circuit conductors above the prescribed limiting temperature.

5.3.5   Fault protection in ADS: for TT systems

In a TT system there is no direct connection between the installation MET and the earthed supply source (earthed neutral or star point) other than through the general mass of the Earth; therefore, particular requirements are necessary.

Regulation 411.5.1 requires that every exposed-conductive-parts protected by a single protective device must be connected to the MET and hence to a common earth electrode. The regulation also recognizes that, where a number of protective devices are employed in series, these may each have their own electrode.

Fault protection in a TT system must be provided by one of two types of protective device as laid down in Regulation 411.5.2:

- a residual current device; or
- an overcurrent protective device.

The regulation states that RCD protection is the preferred option. The reason for this will become apparent when considering the requirements of Regulations 411.5.3 and 411.5.4, for the use of RCDs and overcurrent protective devices respectively.

Regulation 411.5.3, for where an RCD is used for fault protection, requires the following two conditions to be fulfilled:

- the disconnection time must satisfy Regulation 411.3.2.2 (that is, disconnection must be within the time specified in Table 41.1 of BS 7671 for a final circuit rated at not more than 32 A – see Table 5.5 of this Guide), or occur within 1 s where Regulation 411.3.2.4 applies (a distribution circuit or a circuit not covered by Table 5.5, such as a final circuit rated at more than 32 A); and
- $R_A I_{\Delta n} \leq 50$ V

where R_A (Ω) is the sum of the resistances of the earth electrode and the protective conductor connecting it to the exposed-conductive-parts and $I_{\Delta n}$ (A) is the rated residual operating current of the RCD.

For convenience, Table 5.10 lists the maximum earth fault loop impedance values Z_s for RCDs used for fault protection in TT systems. It should also be noted that the maximum values of Z_s in the table relate only to a nominal voltage $U_o = 230$ V and should be multiplied by a factor equal to $U_o/230$ if other nominal voltages are to be used.

Regulation 411.5.4, relating to where an overcurrent protective device is used for fault protection, requires the following condition to be fulfilled:

$$Z_s I_a \leq U_o$$

where Z_s (Ω) is the impedance of the earth fault loop, comprising:

- the source; and
- the line conductor up to the point of fault; and
- the protective conductor connecting the exposed-conductive-parts to the MET of the installation; and

Table 5.10 Maximum values of earth-loop impedances Z_s (Ω) for RCDs, for fault protection in a TT system, operating on 230 V[a]

Ref	Rated residual operating current $I_{\Delta n}$ (mA)	Earth loop impedance Z_s (Ω)			
		BS 4293 and BS 7288		BS EN 61008-1 and BS EN 61009-1	
		0.2 s	1 s	0.2 s	1 s
RCD without time-delayed operation					
A	10	5000[b]	5000[b]	5000[b]	5000[b]
B	30	1667[b]	1667[b]	1667[b]	1667[b]
C	100	500[b]	500[b]	500[b]	500[b]
D	300	167[b]	167[b]	167[b]	167[b]
E	500	100	100	100	100
RCD time delayed					
F	100	– [c]	– [c]	500[b]	500[b]
G	300	– [c]	– [c]	167	167
H	500	– [c]	– [c]	100	100

[a]The values of Z_s in this table meet the conditions $Z_s I_a \leq U_o$ and $R_A I_{\Delta n} \leq$ 50 V to comply with Regulation 411.5.3.
[b]The resistance of the installation earth electrode to earth should be as low as practicable. A value exceeding 200 Ω may not be stable. Refer to Regulation 542.2.2 regarding soil drying and freezing.
[c]Maximum permitted Z_s depends on the rated (or declared) time delay T_d of the RCD. To determine maximum Z_s, consult RCD time–current characteristics and obtain I_a, the current in amperes causing operation within the required disconnection time, taking account of the time delay. Z_s must satisfy both the conditions in table footnote a.

- the earthing conductor of the installation; and
- the earth electrode of the installation; and
- the earth electrode of the source;

I_a (A) is the current causing automatic operation of the disconnecting device within the time specified in Table 41.1 of BS 7671 (see Table 5.5 of this Guide), or within 1 s where Regulation 411.3.2.4 applies (a distribution circuit or a circuit not covered by Table 5.5); and U_o (V) is the nominal a.c. rms or d.c. voltage between a line conductor and Earth.

The maximum value of earth fault loop impedance Z_s that will allow the condition $Z_s I_a \leq U_o$ to be fulfilled for a particular type and rating of overcurrent protective device can be derived from the time–current characteristic for the relevant device given in Appendix 3 of BS 7671. For example, for a BS 88-6 'gG' (Parts 2 and 6) fuse of nominal rating $I_n = 6$ A, the operating current I_a for 0.4 s disconnection is given as 27 A, for which, if related to a nominal voltage 230 V, the maximum value of Z_s is 8.52 Ω ($Z = V/I = 230/27$). Similarly, for a 32 A type C MCB to BS EN 60898, the maximum value of Z_s for 0.4 s disconnection at a nominal voltage of 230 V is 0.72 Ω ($Z = V/I = 230$ V/320 A).

The relatively low values of Z_s that are needed to allow overcurrent protective devices to be used for fault protection cannot usually be achieved in an installation forming part of a TT system, as the combined resistance of the installation earth electrode and the source earth electrode is generally too high for this. Hence, RCDs almost invariably have to be used for fault protection.

5.3.6   Fault protection in ADS: for IT systems

In an IT system, the source is either not connected to Earth or is connected to Earth through a high impedance. Regulation 411.6.1 reinforces this requirement in calling for live parts to be insulated from Earth or connected to Earth through a sufficiently high impedance.

Where live parts are to be connected to Earth through a high impedance, the connection may be made either at the neutral point or midpoint of the system or at an artificial neutral. If there is no neutral point or midpoint, then the connection may be made to one of the line conductors. The value of the high impedance must be such that, if a single fault should occur, the fault current would be of such low magnitude so as not to cause danger from electric shock. A connection to earth is sometimes necessary (via a high impedance) to reduce overvoltage and to dampen voltage oscillations; but when used, it is required to be such that any danger from electric shock from a single fault is obviated.

If the above conditions are met, then automatically disconnecting a single fault occurring to an exposed-conductive-parts or to Earth is not essential. However, as Regulation 411.6.1 points out, precautions still need to be taken to safeguard against the risk of harmful effects to a person in contact with simultaneously accessible exposed-conductive-parts in the event of two faults existing at the same time. This is achieved by employing overcurrent protective devices or RCDs. Regulation 411.6.1 strongly recommends that an IT system with a distributed neutral conductor (e.g. a three-phase four-wire system or a single-phase system) is not used. Distributing the neutral of an IT system increases the probability of the source being inadvertently connected directly with Earth, via a fault between the neutral conductor and Earth, resulting in an increased risk of electric shock.

Regulation 411.6.2 calls for all exposed-conductive-parts of the installation to be earthed individually, in groups or collectively and that condition in Equation (5.6) be satisfied:

$$R_A I_d \leq 50 \text{ V in an a.c. system or 120 V in a d.c. system} \tag{5.6}$$

where R_A is the sum of resistances of the installation earth electrode and the protective conductor(s) connecting it to the exposed-conductive-parts and I_d is the first fault current (fault of negligible impedance between a line conductor and an exposed-conductive-parts, taking account of leakage currents, if any).

Regulation 411.6.3 lists the monitoring devices and protective devices that may be used in an IT system, which are:

- insulation monitoring devices;
- RCM devices;
- insulation fault location systems;

- overcurrent protective devices;
- RCDs.

If an IT system is used for reasons of continuity of supply, then Regulation 411.6.3.1 requires that an insulation monitoring device be employed to detect a first fault condition between a live part and an exposed-conductive-parts or Earth. Also, except where a protective device is installed to interrupt the supply in the event of a first earth fault, either an RCM or an insulation fault location system is required by Regulation 411.6.3.2 to indicate the occurrence of such a fault. In the case of both these regulations, the insulation monitoring device, RCM or insulation fault location system, as applicable, is required to initiate an audible and/or visual signal that continues as long as the fault persists, although it is permitted for an audible alarm to be cancelled if there is also a visual alarm.

The treatment of a second fault condition depends on whether the exposed-conductive-parts are collectively earthed (by a single protective conductor) or earthed separately or in groups. For situations where a single protective conductor is to be employed, the requirements relating to TN systems are applied (Regulation 411.6.4), except that one of the two conditions of Equations (5.7) and (5.8) needs to be satisfied. The condition in Equation (5.7) applies where the neutral conductor of an a.c. system or the midpoint conductor of a d.c. system is *not* distributed. The condition in Equation (5.8) applies where the neutral is distributed (as in single-phase or three-phase, four-wire systems).

$$Z_s \leq \frac{U}{2I_a} \tag{5.7}$$

or

$$Z_s^1 \leq \frac{U_o}{2I_a} \tag{5.8}$$

where U (V) is the nominal a.c. rms or d.c. voltage between line conductors, U_o (V) is the nominal a.c. rms or d.c. voltage between a line conductor and the neutral conductor, Z_s (Ω) is the impedance of the line–earth fault loop (line conductor plus protective conductor of the circuit(s), Z_s^1 (Ω) is the impedance of the neutral–earth loop (neutral conductor plus protective conductor of the circuit(s)) and I_a (A) is the fault current sufficient to cause automatic operation of the protective device within the time specified for TN in Table 41.1 of BS 7671 for a final circuit rated at not more than 32 A (see Table 5.5 of this Guide) or occurring within 5 s where Regulation 411.3.2.3 applies (a distribution circuit or a circuit not covered by Table 5.5, such as a final circuit rated at more than 32 A).

For situations where exposed-conductive-parts are earthed in groups or individually, the requirements relating to TT systems need to be applied, in particular Regulation 411.3.2.2 and the condition in Equation (5.9) needs to be satisfied:

$$R_A I_a \leq 50 \text{ V} \tag{5.9}$$

where R_A (Ω) is the sum of earth electrode and protective conductor resistances and I_a (A) is the current causing automatic operation of the protective device within the time specified for TT in Table 41.1 of BS 7671 for a final circuit rated at not more than 32 A (see Table 5.5 of this Guide) or occurring within 1 s where Regulation 411.3.2.4 applies (a distribution circuit or a circuit not covered by Table 5.5, such as a final circuit rated at more than 32 A).

5.3.7   Functional extra-low voltage

As indicated in Regulation 411.7.1, FELV may be used where, for functional reasons, extra-low voltage (not exceeding 50 V a.c. or 120 V d.c.) is used and all the requirements of Section 414 for SELV or PELV are not fulfilled, and SELV or PELV are not necessary. With an FELV system, provisions need to be made for basic protection and fault protection, in accordance with Regulation Group 411.7.

Basic protection must employ at least one of two options, as required by Regulation 411.7.2:

- basic insulation according to Regulation Group 416.1 (see Section 5.2.2 of this Guide), corresponding with the nominal voltage of the primary circuit of the source; and/or
- barriers and/or enclosures in accordance with Regulation Group 416.2 (see Section 5.2.3 of this Guide).

It is important to note that Regulation 411.7.2 calls for equipment fed from a functional extra-low voltage source to be insulated for the primary voltage. Where equipment does not inherently have this insulation quality, then the accessible insulation of the equipment must be reinforced such that the insulation will meet the test requirements for the voltage of the primary circuit of the source (that is, the test requirements for LV circuits given in Part 6 of BS 7671 (Regulation 612.4.4 refers)).

Regulation 411.7.4 calls for the source of the FELV system to be either a transformer with at least simple separation between windings (that is, a double-wound transformer) or a source complying with Regulation Group 414.3 for the source of an SELV or PELV system. This is a distinct change from the requirements of BS 7671: 2001 for the source of an FELV system, as the use of a source that did not provide at least simple separation, such as an autotransformer, potentiometer or a semiconductor device, would have been permitted by BS 7671:2001.

Regulation 411.7.4 of BS 7671:2008 states that if the extra-low voltage system is supplied from a higher voltage system by equipment that does not provide at least simple separation, then the output circuit is not part of a FELV system. Where this is the case, the extra-low voltage system is deemed to be an extension of the input circuit and is required to be protected by the protective measure applied to that circuit. For example, where the input circuit is protected by ADS, the output circuit is required to meet the same disconnection time requirements as the input circuit. This is a precaution against the risk of electric shock in the event of an earth fault occurring in the output circuit when there has also been a failure in the autotransformer (or other source of extra-LV) that resulted in the input circuit line conductor potential appearing in the output circuit.

In any event, where the source is protected by ADS, all exposed-conductive-parts of the secondary circuit are to be connected to the protective conductor of the primary circuit as mentioned in Regulation 411.7.3. However, except where the source does not provide at least simple separation, automatic disconnection in the event of a fault is not required for electric shock protection purposes in the output circuit, although it may be required in order to provide thermal protection to the circuit conductors.

Finally, Regulation 411.7.5 demands that every socket-outlet, luminaire supporting coupler (LSC), device for connecting a luminaire (DCL) and cable coupler used on a functional extra-low voltage system be dimensionally incompatible with those of any other system installed in the premises.

5.3.8   Reduced low-voltage systems

As indicated in Regulation 411.8.1.1, reduced low voltage systems are permitted where, for functional reasons, the use of extra-low voltage is impracticable and there is no requirement to use SELV or PELV.

The nominal line-to-line voltage of such systems must not exceed 110 V rms a.c. in order to meet the requirements of Regulation 411.8.1.2. This voltage limitation translates to 55 V line to earth on a single-phase centre-tapped system and to 63.5 V to earthed star point of a three-phase system, as indicated in Figure 5.3. Although these voltages to earth exceed LV, they are much lower than the normal 230 V/400 V LV systems and, consequently, pose far less risk of electric shock, both under earth fault conditions, where the voltage on exposed-conductive-parts will not normally exceed 55 V, and under fault-free conditions (direct contact by a person with live parts), where the voltage between live parts is not normally going to exceed 110 V.

As illustrated in Figure 5.3, the neutral or star point of a three-phase and the midpoint of a single-phase transformer or generator must be connected to Earth to afford compliance

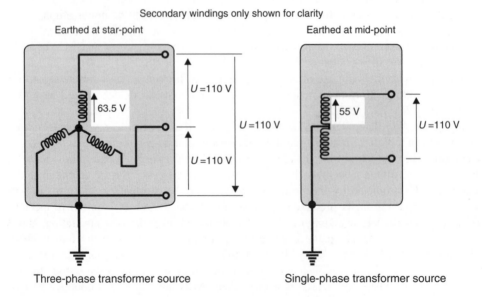

Figure 5.3   Reduced low voltage sources

with the requirement of Regulation 411.8.1.2 that the neutral or midpoint, respectively, shall be earthed.

The source of such a system is required by Regulation 411.8.4.1 to be selected from one of three acceptable options:

- double-wound isolating transformer (to BS EN 61558-1 and BS EN 61558-2-23);
- motor generator (winding isolation equivalent to that provided by an isolating transformer);
- any source independent of other supplies (such as a combustion-engine-driven generator).

It should be noted that, when selecting a generator for this purpose, care should be exercised in selecting one which has its output winding earthed at the midpoint (centre tapped). One which has only one pole earthed, providing an output voltage to Earth twice that of the former, would not meet the requirements for a reduced low voltage source.

No concession is given by Regulation 411.8.2 with regard to basic protection; therefore, basic insulation (Regulation 416.1) and/or barriers and enclosures (Regulation Group 416.2) are necessary as if for an LV system.

Automatic disconnection, to provide for fault protection, must be undertaken, according to Regulation 411.8.3, in one of two ways (or, of course, both):

- by overcurrent device;
- by an RCD.

In both cases, the exposed-conductive-parts of the system must be connected to Earth via a protective conductor. Where an overcurrent device is employed, the line–earth loop impedance at every point must be such as to provide for automatic disconnection within 5 s. This disconnection time limit applies equally to socket-outlets and other points of utilization. It should be noted that both poles must be provided with fault protection (and overcurrent protection), and this is perhaps best achieved by the use of double-pole (or, in the case of three-phase, by triple-pole) overcurrent protective circuit-breakers.

Where a circuit-breaker is employed for automatic disconnection, the line–earth loop impedance Z_s is limited to a value, in ohms, not greater than U_o/I_a or, if the device is listed in Table 41.6 of BS 7671, to the tabulated values corresponding to the device type and nominal rating. Similarly, data for fuses, if listed, may also be obtained from this table, but otherwise may be obtained by applying the formula $Z_s \leq U_o/I_a$, where I_a is the operating current of the device for 5 s disconnection and U_o is the nominal voltage to earth ($U_o = 55$ V for single-phase and 63.5 V for three-phase). For devices not listed, I_a will need to be determined by reference to the relevant BS or CENELEC standards or to the manufacturer. In the case of RCD protection, there is an additional requirement stipulated in Regulation 411.8.3 in that the product of the RCD rated residual operating current and the line–earth loop impedance must not exceed 50 (i.e. $Z_s I_{\Delta n} \leq 50$ V).

Table 41.6 of BS 7671 tabulates values of maximum line–earth loop impedances for MCB types B, C and D, and for 'gG' general-purpose fuses to BS 88-2.2 and BS 88-6, up to and including a rated current of 125 A. The crucial factor in automatic disconnection using overcurrent devices is the current magnitude, and it is important to note that, because

Table 5.11 Limiting impedance Z_s (Ω) for common overcurrent protective devices used for 5 s disconnection for reduced low voltage at 55 V to earthed centre tap and 63.5 V to earthed star point neutral

I_n (A)	Line voltage to Earth U_o (V)	Limiting impedance Z_s (Ω)[a]				
		BS 88 HRC fuse[b]	MCB[c] Type 1	Type B	Type 2	Types 3 and C
5	55.0	–	2.76	2.20	1.58	1.10
5	63.5	–	3.18	2.54	1.82	1.28
6	55.0	3.20	2.30	1.83	1.32	0.92
6	63.5	3.70	2.65	2.12	1.52	1.07
10	55.0	1.77	1.38	1.10	0.79	0.55
10	63.5	2.05	1.59	1.27	0.91	0.64
15	55.0	–	0.92	0.73	0.53	0.37
15	63.5	–	1.06	0.85	0.61	0.43
16	55.0	1.00	0.86	0.69	0.49	0.34
16	63.5	1.15	0.99	0.79	0.57	0.40
20	55.0	0.69	0.69	0.55	0.40	0.28
20	63.5	0.80	0.80	0.64	0.46	0.32
25	55.0	0.55	0.55	0.44	0.32	0.22
25	63.5	0.63	0.64	0.51	0.36	0.26
30	55.0	–	0.46	0.37	0.26	0.18
30	63.5	–	0.53	0.42	0.30	0.21
32	55.0	0.44	0.43	0.34	0.25	0.17
32	63.5	0.51	0.50	0.40	0.28	0.20
40	55.0	0.32	0.35	0.28	0.20	0.14
40	63.5	0.37	0.40	0.32	0.23	0.16
50	55.0	0.25	0.28	0.22	0.16	0.11
50	63.5	0.29	0.32	0.25	0.18	0.13
63	55.0	0.20	0.22	0.17	0.13	0.09
63	63.5	0.23	0.25	0.20	0.14	0.10
80	55.0	0.14	0.17	0.14	0.10	0.07
80	63.5	0.16	0.20	0.16	0.11	0.08
100	55.0	0.10	0.14	0.11	0.08	0.05
100	63.5	0.12	0.16	0.13	0.09	0.06

[a]Dashed entries denote that the rating is either unavailable or inappropriate.
[b]Data for (gG) fuses to BS 88-2.2 and BS 88.6.
[c]Circuit-breakers to BS EN 60898 or BS 3871.

of the lower nominal voltages associated with the systems, there is a consequent lowering of impedance limits. Table 5.11 lists the limiting line–earth loop impedances for common devices operating on 55 V and 63.5 V line to Earth.

As with functional extra-LV systems, there is a requirement (Regulation 411.8.5) for all plugs, socket-outlets and cable couplers of these reduced low voltage systems to be dimensionally incompatible with any such accessories for use on another voltage or frequency in the same installation. Construction-site installations are discussed in Section 16.5 of this Guide.

5.4 Protective measure: double or reinforced insulation

The classification of equipment relating to protection against electric shock is addressed in the IEC publication 60536 (BS 2754). It relates only to protection against electric shock considerations. The international document refers to four such classifications (Classes 0, I, II and III). For completeness, the four IEC Classifications are set out in Table 5.12. Class 0 is not permitted in the United Kingdom (see the Electrical Equipment (Safety) Regulations 1994).

In the protective measure of double or reinforced insulation, either:

- basic protection is provided by basic insulation, and fault protection is provided by supplementary insulation, or
- basic protection and fault protection are provided by reinforced insulation between live parts accessible parts.

The protective measure is intended to prevent the appearance of a dangerous voltage on the accessible metallic parts of electrical equipment due to a failure in the basic insulation.

Regulation 412.1.2 states that this measure is applicable in all situations unless limitations are given in Part 7, relating to special installations or locations. It is more usual in practice for the protective measure of double or reinforced insulation to be applied to individual items of equipment rather than to the whole installation. Where the entire installation is so protected, it is required to be under effective supervision. Unthinking replacement of equipment may at best negate the protective measure or at worst render the installation potentially dangerous.

There are a number of alternatives from which to select equipment, as set out in Regulation Group 412.2.1 and summarized in Table 5.13.

Regulation 412.2.3.1 calls for the installation of Class II equipment having double or reinforced insulation and LV switch and controlgear having total insulation to be selected and erected in such a manner so as not to negate the in-built safety features of that equipment. Where such equipment embodies metalwork, this must not only be protected by basic insulation, but must also be applied with double or reinforced insulation. In such cases, this metalwork would *not* be regarded as an exposed-conductive-parts.

Table 5.12 IEC 60536 classifications (electric shock)

	Class 0	Class I	Class II	Class III
Main equipment characteristics	Means for protective earthing not provided	Provided with means of protective earthing	Additional insulation but no means of protective earthing provided	Designed for supply by SELV.
Safety precautions necessary	Earth-free environment	Connection to protective conductor and Earth	None	Connection to SELV systems only

Table 5.13   Class II or equivalent equipment

Equipment	Comments
Double or reinforced insulated equipment (Class II); type-tested and marked with relevant standard	See definition in Part 2 of BS 7671 and IEC publication 60536
LV switch and controlgear having total insulation; type-tested and marked with relevant standard.	See BS EN 60439-1 *Specification for low-voltage switchgear and controlgear assemblies. Type-tested and partially type-tested assemblies*.
Supplementary insulation applied to equipment as a process in the erection of the installation	Must provide at least an equal degree of safety to above two alternatives; must also comply with Regulations 412.2.2.1 to 412.2.2.3
Reinforced insulation applied to uninsulated live parts in a process of erection of installation	Must provide at least an equal degree of safety to above first two alternatives; must also comply with Regulations 412.2.2.2 to 412.2.2.3. Only to be used where constructional features prevent application of double insulation

Any enclosure for this type of equipment is required, by Regulation 412.2.2.5, not to affect the operation of the equipment in any way. Such operation would involve such considerations of safety, correct functioning and heat dissipation (and temperature rise).

All conductive parts, if only separated from live parts by basic insulation, are required (Regulation 412.2.2.1) to be contained in an insulating enclosure giving at least IPXXB or IP2X degree of protection. It is important that such enclosures are capable of resisting mechanical, electrical and thermal stresses, both for normal operation and under fault conditions.

Great care must be exercised in the installation of Class II equipment having double or reinforced insulation and LV switch and controlgear having total insulation, particularly with regard to piercing or penetrations. Other than for circuit conductors, no penetrations should occur, except possibly by non-conductive parts, in order to comply with Regulation 412.2.2.2. This regulation also calls for the enclosure not to contain insulated screws which could, perhaps inadvertently, be replaced by noninsulating (metallic) screws, thus negating the protection otherwise afforded. If any piercing or penetrations are required for operating handles, screws and the like, then this must be done in such a manner so as not to impair the protection.

As called for in Regulation 412.2.2.3, for enclosures with lids or doors which are openable without the use of a key or tool, all conductive parts that would be accessible on opening the lid or door would need to be located behind an insulating barrier that prevents contact with live parts (to IPXXB or IP2X) and which itself may be opened only by a key or a tool.

Regulation 412.2.3.2 requires a cpc for all circuits supplying Class II equipment except where, as stated in Regulation 412.1.3, this is the only protective measure employed for a whole installation or circuit. Where double or reinforced insulation is the only protective measure, the installation must also be under effective supervision in normal use so that no change is made that would impair the effectiveness of the protective measure. The intention here must be to provide a cpc to each accessory (except lampholders with no exposed-conductive-parts suspended from a point in wiring where a cpc is terminated) so that, in the event that this protective measure is ever abandoned by, for example, the replacement of Class II equipment with Class I equipment (requiring earthing), such a provision of a cpc would facilitate this change without the risk of the omission of earthing. This being so, where cpcs are necessary (i.e. where the whole installation does not employ only this protective measure and is not under effective supervision), other measures, such as ADS, will be required to protect the cpcs from earth faults.

Careful attention is needed for the installation of Class II equipment (Regulation 412.2.3.1) so as not to impair the protection afforded by compliance with the equipment specification. For example, Class II equipment must be installed so that basic insulation is not the only protection between live parts and exposed metalwork of the equipment. Also, metalwork (necessarily unearthed) of Class II equipment must not be, or become, in contact with any part of the installation which is in contact with a protective conductor. This would include, for example, electrically bonded structural metalwork, other exposed-conductive-parts of the installation and other extraneous-conductive-parts in all forms. Regulation 412.1.3 states the obvious by demanding that this protective measure is not to be applied to *any* final circuit feeding a socket-outlets, LSCs, DCLs, or where the user may change items of equipment without authorization.

As stated in Regulation 412.2.2.4, where protective and other conductors run through Class II equipment on their way to other equipment (and this practice should be avoided, if possible), any joints and connections will need to be insulated and marked appropriate to their function. Naturally, where cpcs are installed to Class II equipment, they will be required to be properly terminated.

Regulation 412.2.4.1 indicates that cables of rated voltage not less than the nominal voltage of the system and at least 300/500 V having a nonmetallic sheath or which are enclosed in nonmetallic conduit, trunking or ducting are considered to meet the requirements of Regulation Group 412.2 for basic protection and fault protection.

5.5   Protective measure: electrical separation

The requirements for the use of this protective measure are split into two distinct sets of regulations:

- Regulation Group 413.1, which gives the general requirements for the measure, and also gives specific requirements for where the measure is used to supply only one item of current-using equipment;
- Regulation Group 418.3, which gives additional requirements for where the measure is used to supply more than one item of current-using equipment.

The requirements of Regulation Group 418.3 are dealt with in Section 5.9.3 of this Guide. The requirements of Regulation Group 413.1 are dealt with here.

As explained in Regulation 413.1.1, electrical separation is a protective measure in which basic protection is provided by basic insulation of live parts or by barriers and enclosures in accordance with Section 416, and fault protection is provided by simple separation of the separated circuit from other circuits and from Earth. Not surprisingly, basic protection may also be provided by double or reinforced insulation in accordance with Section 412, as stated in Regulation 413.2.

Where used to supply a single item of current-using equipment, this measure is intended to prevent the possibility of electric shock occurring through contact with exposed-conductive-parts in the event of basic insulation failure.

Regulation 413.3.2 makes demands on the source of supply when this protective measure is used and places an upper voltage limit of 500 V for the electrically separated circuit. The source must provide at least simple separation (the type of separation provided by a double-wound transformer) and there must be *no* connection between the output winding and the body (or enclosure) of the source or to the protective conductor of the primary (or input) circuit.

Regulation 413.3.3 calls for there to be *no* electrical connection, either deliberate or fortuitous, between the live parts of the separated circuit and any point in any other circuit or to Earth or to a protective conductor. In order to avoid the potential risk of an earth fault, special care is needed with regard to the insulation of live parts from Earth. This is particularly important when considering flexible cables and the like. For this reason, all flexible cables and cords, liable to damage from mechanical impact or general wear and tear, must be visible throughout their length so that any damage occurring in service may be spotted and promptly remedied (Regulation 413.3.4 refers). It is always preferable to use separate wiring systems for separated circuits. However, where this is not feasible or economic, multicore cables may be used provided they do not have a metallic sheath. Insulated conductors in an insulated conduit, nonmetallic ducting or nonmetallic trunking are also acceptable wiring systems, but they must be voltage rated for the highest nominal voltage likely to occur and, additionally each separated circuit is required to be protected against overcurrent (Regulation 413.3.5 refers).

All live parts of each separate circuit must be electrically separate from all other circuits. The degree of separation is required by Regulation 413.3.3 to be such that basic insulation is achieved between circuits in accordance with Regulation Group 416.1. This requirement for separation applies equally to relays, contactors and any other equipment where separated circuits may otherwise come into contact with other circuits.

As called for in Regulation 413.3.6, where a separated circuit supplies one item of equipment, care must be taken to determine that no exposed-conductive-parts of that circuit is electrically connected, either deliberately or fortuitously, to a protective conductor or to exposed-conductive-parts of other circuits or to Earth.

5.6 Protective measure: extra-low voltage provided by SELV or PELV

SELV and PELV are protective measures that are generally applicable, as is confirmed by Regulation 414.1.2. A note to that regulation points out that, in some installations or locations of increased shock risk, a reduction in the nominal voltage below the normally permitted maximum nominal voltage of 50 V a.c. or 120 V ripple-free d.c. will be required. Both SELV and PELV systems are entirely separated from higher voltages by virtue of the isolating characteristics of their sources and the method of installation of the downstream equipment. SELV systems are also separated from Earth by the same means.

A safety source suitable for SELV or PELV may be, and commonly is, a safety isolating transformer complying with BS EN 61558-2-6, but other sources may be equally acceptable. An electric-motor-driven generator can be used, provided the motor and generator windings are separated by an equivalent degree to that of a BS EN 61558-2-6 transformer. Batteries, too, are acceptable, as is an engine-driven generator or any other source independent of a higher voltage. If an electronic device is to be used, then special in-built features are required, such as a facility to limit the output voltage. A system supplied from a higher voltage system is not suitable (unless there is the necessary degree of separation) and, therefore, cannot be deemed to comply with the provisions necessary to provide protection against electric shock. Other unacceptable sources include potentiometers (or voltage dividers), transformers not complying with BS EN 61558-2-6 and auto-transformers, semiconductor devices or any other device that does not provide an equal degree of separation to that provided by a BS EN 61558-2-6 transformer. It will be seen that the purpose of this electrical separation is to prevent higher voltages of the 'primary' of a source being transmitted to the 'secondary' during both normal operation and under fault conditions, thus losing what protection there may have otherwise been. Where a mobile source supplied at LV is used (such as a safety isolating transformer or a motor generator), Regulation 414.3 calls for it to be selected and erected so that protection is provided in accordance with the requirements of Section 412 of BS 7671 for protection by double or reinforced insulation. In practice, the most commonly used source will be a transformer complying with BS EN 61558-2-6. Regulation 414.3 details the acceptable SELV and PELV sources, which are summarized in Table 5.14. The table also summarizes the unacceptable sources.

Table 5.14 Summary of acceptable and unacceptable SELV and PELV sources

Acceptable SELV sources	Unacceptable SELV sources	Regulation
Safety isolating transformer to BS EN 61558-2-6	Any transformer without the necessary electrical separation	414.3
Motor/generator with equivalent isolation as BS EN 61558-2-6 transformer	Autotransformer	414.3
Battery or other electrochemical source	Potentiometer (voltage divider)	414.3
Engine driven generator	Semiconductor device	414.3
Electronic device, if proper measures are taken to limit output voltage	Electronic device, if no proper measures are taken to limit output voltage	414.3
Mobile SELV source protected by Class II or equivalent insulation	Mobile SELV source not protected by Class II or equivalent insulation	414.3
	Any source supplied from a higher voltage without the necessary electrical separation	

Regulation 414.4.1 requires SELV and PELV circuits to have:

- basic insulation between live parts and other SELV or PELV circuits;
- protective separation from live parts of circuits other than SELV or PELV by double or reinforced insulation or by basic insulation and protective screening for the higher voltage present.

The same regulation requires that SELV circuits shall have basic insulation between live parts and Earth, but for PELV circuits the regulation permits a live part and/or exposed-conductive-parts to be earthed. In other words, a SELV or PELV circuit must not have any electrical connection with any other circuit. In addition, a SELV system must not have any electrical connection to Earth. In practice, unless only a SELV system was used throughout an installation (which is very unlikely), the requirement to have no connection to Earth would preclude the use of metallic conduit and metallic trunking or bare metallic cable sheaths, since these, if not deliberately connected to Earth, are likely to make fortuitous contact with earthy parts of other systems or equipotentially bonded parts such as, for example, the structural steel of the building.

Regulation 414.4.2 calls for SELV and PELV circuit conductors to have protective separation from live parts of other circuits (which must have at least basic insulation) by one of the following five arrangements:

- SELV and PELV circuit conductors to be enclosed in a nonmetallic sheath or insulating enclosure in addition to their basic insulation;
- SELV and PELV circuit conductors separated from conductors of voltages higher than Band I by an earthed metallic sheath or earthed metallic screen;
- circuit conductors at voltages higher than Band I may be contained in a multiconductor cable or other grouping of conductors if the SELV or PELV conductors are insulated to the highest voltage present;
- the wiring systems of other circuits meet the requirements of Regulation 412.2.4.1 (that is, they are rated for a nominal voltage not less than 300/500 V and have a nonmetallic sheath or nonmetallic enclosure);
- physical separation.

The arrangements for electrically separating the SELV and PELV live parts from those of any other system also extend to include relays, contactors, auxiliary switches and the like.

The requirements of Regulation 414.4.4 reinforce or duplicate those already given in other regulations and confirm that no exposed-conductive-parts of an SELV system is to be connected with Earth or with any exposed-conductive-parts or protective conductor of any other circuit.

Very importantly, the note to Regulation 414.4.4 states that, where contact is made between exposed-conductive-parts of an SELV circuit and exposed-conductive-parts of other circuits, the requirements of the protective measure of SELV have not been met and protection against electric shock no longer depends solely on protection by SELV, but also on the protective provisions to which the latter exposed-conductive-parts are subject. Contact could, for example, be made between exposed-conductive-parts by virtue of independent contacts with Earth which have been made deliberately and intentionally or may have been made quite fortuitously. In other words, when this contact is made, the system should be treated as a PELV system.

Regulation 414.4.5 makes clear that a SELV system or a PELV system with a voltage exceeding 25 V a.c. (or 60 V ripple-free d.c.), or in which equipment is immersed in water, cannot of itself provide the necessary basic protection and stipulates further measures, such as barriers or enclosures to IPXXB or IP2X or better or by the covering of live parts with basic insulation. The same regulation states that basic protection is generally unnecessary in normally dry conditions for SELV circuits where the nominal voltage does not exceed 25 V a.c. or 60 V d.c. and for PELV circuits not exceeding these nominal voltages where exposed-conductive-parts and/or the live parts are connected by a protective conductor to the main earthing terminal of the installation. In any case, as stated in Regulation 414.4.5, basic protection is not required where the nominal voltage of the SELV or PELV system does not exceed 12 V a.c. or 30 V d.c.

Regulation 414.4.3 imposes constraints on installation equipment for use in SELV or PELV systems. The regulation requires that socket-outlets and LSCs in SELV and PELV systems require the use of a plug that is dimensionally incompatible with those of other systems in the same premises.

5.7 Additional protection

Additional protection in accordance with Section 415 of BS 7671 involves the use of an RCD having the characteristics specified in that section (and explained here) and/or supplementary equipotential bonding. Such protection is specified by BS 7671 as part of the protective measures detailed in Chapter 41, for certain applications and under certain conditions of external influence.

For example, where the protective measure of ADS is used, additional protection by means of an RCD is specified for, amongst other things, socket-outlets rated at 20 A or less for use by ordinary persons and intended for general use (Regulation 411.3.3 refers). Additional protection by an RCD is also specified in certain cases for cables concealed in walls or partitions (Regulation 522.6.7 and 522.6.8 refer). Additional protection by RCDs and by supplementary equipotential bonding is specified in certain of the special installations and locations covered in Part 7 of BS 7671.

Where an RCD is used to provide additional protection, it must have a rated residual operating current $I_{\Delta n}$ not exceeding 30 mA and an operating time not exceeding 40 ms at a residual current of $5I_{\Delta n}$, as required by Regulation 415.1.1. The regulation states that the use of such RCDs is recognized in a.c. systems as additional protection in the event of failure of the provision of basic protection and/or the provision of fault protection, or carelessness by users. However, the use of such RCDs is not recognized as a sole means of protection, and does not avert the need to apply one of the protective measures that are generally permitted, specified in Sections 411 to 414 of BS 7671.

There are but two regulations in Chapter 41 relating to additional protection by supplementary equipotential bonding: Regulations 415.2.1 and 415.2.2. The former states that supplementary equipotential bonding (where required) shall include all simultaneously accessible exposed-conductive-parts of fixed equipment and extraneous-conductive-parts, including, where practicable, the main metallic reinforcement of constructional reinforced concrete. The bonding must also connect to the protective conductors of equipment, including those of socket-outlets.

Regulation 415.2.2 provides the formulae $R \leq 50/I_a$, in a.c. systems and $R \leq 120/I_a$, in d.c. systems for the value of resistance R of the supplementary bonding conductor between simultaneously accessible exposed-conductive-parts and extraneous-conductive-parts. Where, as is normal, a number of protective devices have an influence on this bonding, then the worst-case value must be used (i.e. the higher or highest I_a value). The current I_a is the operating current of the device for 5 s disconnection and, in the case of an RCD, $I_{\Delta n}$.

Compliance with Regulation Group 544.2 for supplementary bonding conductors would also be required (dealt with in Chapter 12).

5.8   Obstacles and placing out of reach

Examples of the protective measures of obstacles and placing out of reach are shown in Figure 5.4. The use of these two measures is permitted only in installations where access is restricted to skilled or instructed persons, or to persons under the supervision of skilled or instructed persons (Regulation 410.3.5 refers), provided that the installation is also controlled or supervised by skilled persons (Regulation 417.1 refers). Not surprisingly, these are not measures suitable for applying to special installations or locations, including those addressed in Part 7 of BS 7671.

Both of these protective measures provide basic protection only, and are for application in installations with or without fault protection.

5.8.1   Obstacles

The requirements of Regulations 417.2.1 and 417.2.2 are intended to prevent unintentional contact with live parts by providing a physical obstacle to those parts. It is not the intention to prevent intentional contact by deliberate circumvention of the obstacle.

The former regulation requires that obstacles shall prevent:

- unintentional approach to live parts;
- unintentional contact with live parts during the operation of live equipment in normal service.

The latter regulation requires the obstacle to be secure so as prevent unintentional removal.

5.8.2   Protection by placing out of reach

Regulation Group 417.3 addresses the situation of installing conductors between buildings and structures or, of course, within a building. For an overhead line for distribution between buildings, the regulation group calls for the standards laid down in the Electricity Safety, Quality and Continuity Regulations 2002 (as amended) to be adopted. This requirement applies equally to bare and insulated overhead lines.

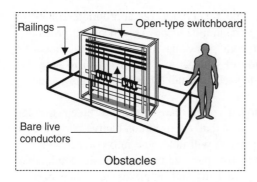
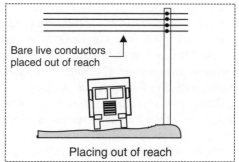

Figure 5.4 Examples of the protective measures of obstacles and placing out of reach

Regulation 417.3.1 stipulates that no live part is to be within 2.5 m or 'arm's reach' of any exposed-conductive-parts, any extraneous-conductive-part or any bare live part of any other circuit (see Figure 6.4 for 'arm's reach'). Regulation 417.3.2 requires 'arm's reach' to be measured from an obstacle formed by a handrail, mesh, screen or the like that restricts access to live parts from a normally occupied position in the horizontal plane, if it affords less than IPXXB or IP2X degree of protection. Where bulky or long conductive objects are handled, there is a recognition in Regulation 417.3.3 that the distances given in the preceding regulations may need to be increased. This would apply, for example, where long metallic ladders or mobile scaffolding are likely to be used in the normal course of events.

As with obstacles, this measure should be treated as a 'last resort' and used only where other measures are wholly impracticable in the particular circumstances. If used within, or in the vicinity of, buildings, particular care is needed in siting conductors. Consideration of the normal activities (including working practices) of the site and maintenance procedures is essential. Where this measure is used, the exposed (bare) conductors would be kept to an absolute minimum required for the operational reasons, positioned at a minimum of 'arm's reach' (see Figure 6.4). A sufficient number of warning notices should be placed in conspicuous positions to warn personnel of the dangers from touching live conductors. Access to such areas should, where necessary, be restricted by, for example, locks and/or permits. Persons authorized to work in such areas would have received sufficient training and have acquired the necessary protective gear. If practical, RCD protection of such conductors should also be considered by the designer in these circumstances.

5.9 Protective measures for application only where the installation is controlled or under the supervision of skilled or instructed persons

5.9.1 Non-conducting location

The very first paragraph of the regulation group relating to this measure, 418.1, makes clear that this protective measure is not recognized for general application. Like the other protective measures in Section 418 of BS 7671, it must only be considered for special situations which are controlled by, or under the supervision of, appropriate skilled or instructed persons so that unauthorized changes cannot be made. This means of protection is intended to prevent simultaneous contact with parts that may be at a difference in potential because of failure of basic insulation of live parts. It is important to note that this measure is not one which may be applied to special installations or locations, including those detailed in Part 7 of BS 7671.

When this protective measure is employed, great care must be taken through design, as mentioned in Regulation 418.1.2, so that exposed-conductive-parts are so arranged that under ordinary circumstances persons will not come into contact with two exposed-conductive-parts or an exposed-conductive-parts and an extraneous-conductive-part. Regulation 418.1.4 reinforces these requirements and amplifies how they may be met. Table 5.15 summarizes the requirements where the location has an insulating floor and insulating walls.

Protective conductors are not permitted in the non-conducting location (Regulation 418.1.3 refers).

Before declaring that floors and walls are or are not insulating, measurements of the resistance of these building components are required to satisfy the requirements of Regulation 418.1.5. Such measurements need to be carried out in accordance with Regulations 612.5.1 and 612.5.2. Measurements are required to be taken at not less than three points

Table 5.15 Protection by non-conducting location. Separating distances (Regulation 418.1.4)

Parts	Minimum separation distance (m)	Alternative arrangements
Exposed-conductive-part and other separated exposed-conductive-part (*Parts within the zone of arm's reach*)	2.50	Interposition of adequate barriers (insulating and isolated) extending the separation distances
Exposed-conductive-part and other separated exposed-conductive-part. (*Parts out of the zone of arm's reach*)	1.25	Interposition of adequate barriers (insulating and isolated) extending the separation distances
Exposed-conductive-part and extraneous conductive part (*Parts within the zone of arm's reach*)	2.50	Interposition of adequate barriers (insulating and isolated) extending the separation distances *or* insulation of extraneous-conductive-parts
Exposed-conductive-part and extraneous conductive part (*Parts out of the zone of arm's reach*)	1.25	Interposition of adequate barriers (insulating and isolated) extending the separation distances *or* insulation of extraneous-conductive-parts

on each and every surface, one of which must be taken at a point approximately 1.0 m from each and every accessible extraneous-conductive-part, the others being taken at greater distances. Where the nominal voltage to Earth U_o is not greater than 500 V, the resistance between points should not be less than 50 kΩ. Where the nominal voltage to Earth U_o is greater than 500 V, the resistance between points should not be less than 100 kΩ. If these values are not achieved, then the floors and walls must be considered to be extraneous-conductive-parts and treated accordingly. Applied insulation or insulating arrangements for extraneous-conductive-parts (to meet Regulation 418.1.4(iii)) must be tested in accordance with Regulation 612.5.2 (see Chapter 17 of this Guide).

Where this protective measure is used, Regulations 418.1.6 and 418.1.7 call for permanent protective arrangements to be made if mobile equipment is used and for precautions to be taken so that extraneous-conductive-parts cannot transmit a potential outside the location.

It has to be said that this method of protection has very limited application and there is always the danger that circumstances may change, rendering the measure ineffective. It is difficult to prevent other conductive parts (e.g. metallic water pipes) being introduced into the area at some later date. Although BS 7671 does not specifically call for a notice, it is always a good idea to have suitable warning signs advising the unwary that this particular measure is in existence in the location. These notices would be located at points of access to the location and would indicate that the location was 'a non-conducting location' and provide the name of the person who is responsible for the supervision.

5.9.2   Protection by earth-free local equipotential bonding

Again, this measure is not generally applicable and is restricted to special circumstances, as stated at the beginning of Regulation Group 418.2. It is intended to prevent the appearance of dangerous voltages between simultaneously accessible parts even under conditions of failure of basic insulation. Wherever it is used, a warning notice, complying with Regulation 514.13.2, must be fixed in a prominent position adjacent to every access point to the location (see Figure 9.10). This measure must not be considered for application in special installations or locations detailed in Part 7 of BS 7671.

The requirements for equipotential bonding are similar to those associated with an ADS-protected installation, *except that* this bonding *must not* be connected to, or be in contact with, Earth (Regulations 418.2.2 and 418.2.3).

Regulation 418.2.4 recognizes that a risk may exist between the earth-free location and another location in which other protective measures are employed. It calls for special precautions to be taken to obviate the risk. This risk may, for example, manifest itself where an appropriately bonded conductive floor in the earth-free location butts up to a floor which is itself connected, either deliberately or fortuitously, to Earth. A person walking over both floors near an access point may experience a difference of potential which may result in electric shock when contact is made with both floor surfaces simultaneously.

Precautions may include, for example, provision of an insulated floor section between the two locations of sufficient width to prevent simultaneous contact between the two surfaces.

5.9.3   Electrical separation for the supply to more than one item of current-using equipment

As stated in Section 5.5 of this Guide, the requirements for the use of this protective measure are split into two distinct sets of regulations:

- Regulation Group 413.1 (dealt with in Section 5.5 of this Guide), which gives the general requirements for the measure, and also gives specific requirements for where the measure is used to supply only one item of current-using equipment.
- Regulation Group 418.3 (dealt with here), which gives additional requirements for where the measure is used to supply more than one item of current-using equipment.

It is important that the guidance given below for the use of electrical separation to supply more than one item of current-using equipment is read in conjunction with that in Section 5.5 of this Guide, relating to the general requirements of BS 7671 for the use of electrical separation, which are given in Regulation Group 413.1.

Where electrical separation is used for the supply of more than one item of current-using equipment from a common source, the intent is the same as where it is used to supply a single item of current-using equipment – namely, to prevent the possibility of electric shock occurring through contact with exposed-conductive-parts in the event of basic insulation failure. However, Regulation 410.3.6 requires that the installation is under the supervision of a skilled or instructed person so that unauthorized changes cannot be made to the installation. Furthermore, Regulation 418.3 requires a warning notice to be fixed in a prominent position adjacent to every point of access to a location where this measure has been used. This warning notice is to be in compliance with Regulation 514.13.2 and is the same as that required for earth-free local equipotential bonding.

Where a source is used to supply a number of items of equipment, Regulation 418.3.3 calls for precautions to be taken to protect the protected circuit from all types of damage and from insulation failure. In addition, the following requirements must be met:

- all exposed-conductive-parts of the separated circuit connected together by an insulated non-earthed protective conductor, which must *not* be connected to the protective conductor or exposed-conductive-parts of any other circuit or to any extraneous-conductive-part (Regulation 418.3.4 refers); and
- all socket-outlets are to be provided with a protective conductor contact which, in turn, is connected to the protective conductor referred to above, and all flexible cables to equipment, other than equipment with double or reinforced insulation, are to include a protective conductor contact to be used as an equipotential bonding conductor (Regulations 418.3.5 and 418.3.6 refer); and
- if two faults, from conductors of different polarity, can occur to exposed-conductive-parts, there is the need to verify that the associated overcurrent device will disconnect

the supply within the disconnection time required by Table 41.1 of BS 7671 for a TN system (see Table 5.5 of this Guide). Regulations 418.3.7 and 612.4.3 refer.

Finally, where the source is used to supply a number of items of equipment, Regulation 418.3.8 recommends that the product of the nominal voltage of the circuit in volts and the length of the wiring in metres should not exceed 100 000 V m, and the length of the wiring should not exceed 500 m.

6

Protection against thermal effects

6.1 General

Chapter 42 of BS 7671 addresses the precautions required in relation to protection against the hazards of fire, thermal effects and burns. The thermal effects resulting from overload and fault current are not embraced in Chapter 42 but are dealt with in the form of requirements for protection against overcurrent in Chapter 43. Provided the measures stipulated there are met in full, further consideration in respect of fire and burns relating to overcurrent is not necessary.

Fires from electrical causes are of great concern, as indicated by statistics published by the Department for Communities and Local Government. Whilst the reasons for such fires are complex and sometimes unclear, some increase can be expected because of the sheer volume of electrical apparatus and appliances now prolific in almost every building. As far as the fixed installation is concerned, the statistics are encouraging. However, many fires could be avoided by the provision of an adequate number of points to serve the ever increasing number of appliances with some provision for anticipated additional equipment.

Many fires are, of course, attributable to appliances and not to the fixed installation, with which Chapter 42 is solely concerned. Appliances used on inadequately rated circuits or connected via badly connected plugs are an obvious cause for concern. Fixed-installation equipment selected without due consideration of the ambient temperature is another. A case in point relates to a standard BS 1363 plug and socket-outlet connection of the immersion heater in the airing cupboard. The concern here is twofold: first, such a socket-outlet, though rated at 13 A, is not suitable to sustain this maximum rate continuously or even for long periods (see BS 1363); and when so loaded, the temperature rise is likely to produce expansion and creepage (in the plug and socket-outlet), finally resulting in loose connections and a real risk of igniting material in close proximity. Second, the airing cupboard temperature is normally much higher, and equipment fixed therein would generally need to be de-rated if limiting temperatures are not going to be exceeded.

Regulation 420.1 sets the scene for Chapter 42. It makes clear that the subsequent requirements relate to protection of people, livestock and property against:

- the harmful effects of heat or thermal radiation developed by electrical equipment;

A Practical Guide to The Wiring Regulations: 17th Edition IEE Wiring Regulations (BS 7671:2008)
Fourth Edition Geoffrey Stokes and John Bradley
© Geoffrey Stokes and John Bradley. Published by John Wiley & Sons, Ltd

- the ignition, combustion or degradation of materials;
- flames and smoke where a fire hazard could be propagated from an electrical installation to other nearby fire compartments;
- safety services being cut off by the failure of electrical equipment.

As previously mentioned, Chapter 42 deals only with the fixed installation, but current-using equipment, including portable appliances, should of course conform to the specifications given in the appropriate British or CENELEC standards. Most British Standards for such equipment include a requirement for a heat test to be carried out under defined conditions. All components are then checked for temperature rise and the limits placed on this are 60 °C for equipment intended for continuous operation and 65 °C for other equipment, with the reference ambient temperature taken to be 20 °C. Other specific tests are also undertaken to confirm that the equipment will perform without risk, if installed and used correctly.

Where fixed equipment has been specified by someone other than the installation designer, the onus to install the equipment so that provisions of Chapter 42 are met still rests with the installer. Of particular importance is the proximity of building fabric to the equipment and the thermal effects that may result.

Bayonet lampholders of types B15 and B22 must have the type T2-rated lampholder (see Table 11.12 of this Guide). Fluorescent and discharge tube luminaires must be mounted to take account of the requirements of BS EN 60598-1 *Luminaires. General requirements and tests*. In particular, it must be noted whether or not the luminaire exhibits the symbol given in Figure 6.1. Where it does not, the standard requires that a suitable warning notice be provided with the luminaire (by the manufacturer) and the designer and the installer should note that such fittings are unsuitable for mounting directly on flammable materials. Most popular fluorescent luminaires are of this type and, therefore, not suitable for direct mounting unless the material to which they are to be mounted is not flammable. Emergency luminaires must be afforded the same consideration, but units bearing the ICEL (Industry

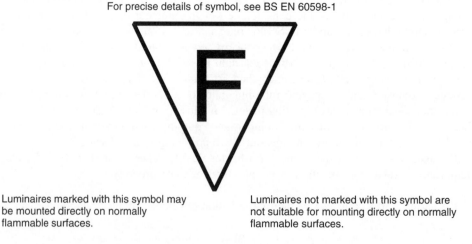

Figure 6.1 Symbol for fluorescent and discharge lamp luminaires

Committee on Emergency Lighting) certification do meet the relevant requirements of BS EN 60598-1 and, therefore, are suitable for direct mounting on flammable materials.

Notwithstanding the requirements of Chapter 42, other parts of BS 7671 contain prerequisite conditions that need to be met with respect to fire, harmful thermal effects and burns, including:

- Regulation Group 527.2
  - sealing of wiring system penetrations (fire barriers)
- Chapter 43 Protection against overcurrent
- Sections 510 and 511
  - requirements relating to equipment standards and compliance with other parts of BS 7671.

In terms of the fire hazard, there are two distinct risks, namely the risk of ignition from equipment, including wiring systems, and the propagation of fire through a wiring system. Compliance with all the regulatory requirements relating to the installation design should obviate the former, and providing that wiring systems are sealed where they pass through walls and floors, the spread of fire should be significantly retarded.

6.2   Fire caused by electrical equipment

6.2.1   Surface temperature

It is important to establish that fixed electrical equipment does not have harmful effects on adjacent materials. Regulation 421.1 makes clear that heat generated by equipment must not be such as to cause a danger in terms of a fire hazard or have detrimental effects on other materials which may be in the proximity of the equipment. This requirement applies to all equipment, including luminaires, under normal and fault conditions and not just equipment intended to be a heat source. The regulation also calls for the equipment manufacturer's installation instructions to be observed. They are best placed to advise on such issues as temperature rise of their equipment and will have the necessary data from tests carried out to meet the specific safety tests laid down in the relevant British or CENELEC standards to which the equipment is designed.

As stated in Regulation 421.2, where equipment may attain a surface temperature sufficient to cause a fire hazard to adjacent materials, then measures must be taken to obviate those risks. This must be done in one or more of three given ways:

- The equipment is to be appropriately mounted on a support that has low thermal conductance or within an enclosure that will withstand, with minimal risk of fire or harmful thermal effect, such temperatures as may be generated.
- The equipment is to be screened by materials of low thermal conductance that can withstand, with minimal risk of fire or harmful thermal effect, the heat emitted by the electrical equipment.
- The equipment is to be mounted so as to allow safe dissipation of heat and at a sufficient distance from adjacent materials on which such temperatures could have a deleterious effect. Any support used must have a low thermal conductance.

In addition to the above-mentioned risks and requirements, protection against burns from such equipment must not be overlooked (see Section 6.4 of this Guide).

6.2.2 Arcs, sparks and high-temperature particles

There are four acceptable methods of protection from arcs, sparks and high-temperature particles from equipment in normal service, as given in Regulation 421.3:

- containment of the equipment in an enclosure of arc-resistant material;
- screening by arc-resistant material so as to prevent harmful effects on materials likely to be damaged;
- equipment to be mounted at a sufficient distance from vulnerable material to provide for safe extinction of high-temperature particles before any damage occurs; or
- equipment to be suitable to provide protection from arcs, sparks and high-temperature particles by virtue of being in compliance with its product standard.

Arc-resistant materials must be nonignitable, of low thermal conductivity and be adequately proportioned so as to provide mechanical stability.

Generally, equipment manufactured to meet the requirements relating to safety in the appropriate standard will be equipped with all the necessary arc-containment or screening. Provided assembly and installation are in accordance with the manufacturer's recommendations, no further consideration is required in this respect. However, there is one common example of failure to meet the requirements in this respect, in that one occasionally encounters evidence of the replacement of a high breaking capacity (HBC) fuse with a semi-enclosed (rewirable) fuse without the necessary replacement of the carrier.

6.2.3 Position of equipment embodying heat sources

Regulation 421.4 gives specific requirements directly relating to the positioning of equipment. It calls for equipment which causes a focusing or a concentration of heat to be sited at sufficient distance from any fixed object or element, such that a dangerous temperature (in the object or building element) is not reached in normal conditions. For example, this regulation would have a direct bearing on the location of heaters and light sources, particularly those with tungsten filaments.

6.2.4 Flammable liquids

Regulation 421.5 states that, where electrical equipment in a single location contains flammable liquids in significant quantity, adequate precautions must be taken to prevent spread of liquid, flame and products of combustion. Equipment containing flammable liquids includes switchgear and transformers containing oil as the insulating and cooling medium.

Note 1 to the regulation gives examples of precautions necessary to meet the requirements of the regulation, including a retention pit to collect any leakage of liquid and ensure extinction in the event of a fire, and the installation of the equipment in a fire-resistant chamber ventilated solely to the external atmosphere with means of preventing burning liquid spreading to other parts of the building.

Notes 2 and 3 clarify that the generally accepted lower limit of a significant quantity of flammable liquid is 25 litres, and that for quantities less than this it is sufficient to take precautions to prevent the escape of the liquid.

Note 4 clarifies that the products of combustion of a liquid are considered to be smoke and gas.

Figure 6.2 illustrates an example of where precautions against the hazard of burning liquid may need to be taken where the transformer shown contains more than 25 litres of oil. The transformer HV and LV cables are shown together with cable transits at the entrance to the cable duct. Figure 6.3 portrays a cable transit seal in more detail, showing the principal components. Such seals are available commercially in all shapes and sizes for cables and pipes and can be constructed to virtually any combination of cable and pipe sizes.

6.2.5   Construction of enclosures

It is important when selecting and installing enclosures for equipment that the material from which the enclosure is made has a resistance to fire and heat that is suitable, and which would meet the requirements of an appropriate product standard. Where no appropriate product standard exists or where an enclosure is constructed during installation, such an enclosure must be capable of withstanding the highest temperature likely to be produced by the enclosed equipment in normal service, as called for by Regulation 421.6.

6.2.6   Live conductors

Irrespective of nominal voltage, Regulation 421.7 calls for all terminations of live conductors and joints between such conductors to be contained within an enclosure. By definition,

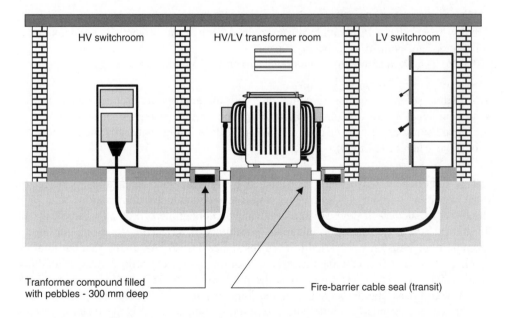

Figure 6.2   Substation layout showing cable transit barriers

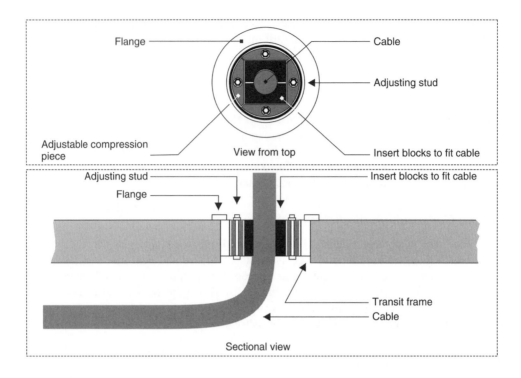

Figure 6.3   Example of cable transit sealing arrangement

a live conductor includes all d.c. conductors and the line and neutral conductors of a.c. circuits, except that a PEN conductor (combined neutral and protective conductor) is, by convention, not defined as a live conductor.

The acceptable methods of containment, given in Regulation 526.5, are:

- a suitable accessory enclosure complying with the appropriate product standard;
- an equipment enclosure complying with the appropriate product standards; or
- an enclosure partially formed or completed with building material which is noncombustible when tested to BS 476-4.

This requirement will usually be satisfied by normal good installation practice, but it is important to note that extra-low voltage live conductor terminations and joints should be treated as if low voltage in terms of containment. For an example of the fire hazard posed by exposed terminations and joints of, say, a 24 V system, consider an unsound, high-resistance, exposed joint of resistance $1.25\,\Omega$. Assuming that the total circuit impedance is $2\,\Omega$ (including the resistance of the faulty connection), the current drawn would be 12 A. The power loss across the joint would be 180 W ($I^2R = 12^2 \times 1.25$), which would cause the temperature of the joint to rise considerably and present a very real fire hazard. Similarly, joints with intermittent contact could ignite material in close proximity by arcing. Hence the requirement to contain all such connections.

6.3   Precautions where particular risks of danger of fire exist

6.3.1   General

Section 422 of BS 7671 gives precautions against thermal effects for installations in locations where particular risks of fire exist. The precautions are additional to the requirements of Section 421 for protection against fire caused by electrical equipment in general, which are covered in Section 6.2 of this Guide. They are based on the principle of selecting and erecting electrical equipment such that its temperature both in normal conditions and in the event of a fault is unlikely to cause a fire, taking due account of external influences. This can be achieved by the construction of the equipment or by additional measures taken during erection. Special measures are not necessary if the surface temperature of equipment is unlikely to cause combustion to nearby substances.

Regulation 422.1.1 requires electrical equipment in the location to be restricted to that necessary for use in the location. However, an exception is made for a wiring system that only passes through the location, to which the requirements of Regulation 422.3.5 must be applied (Section 6.3.3 of this Guide refers).

6.3.2   Conditions for evacuation in an emergency

Regulation Group 422.2 contains requirements intended to protect the means of escape from a location where particular risks of danger of fire exist against thermal effects caused by electrical equipment.

The requirements apply where the conditions of external influence are:

- BD2 – low-density occupation, difficult conditions of evacuation (such as in a high-rise building);
- BD3 – high-density occupation, easy evacuation conditions (such as may be the case in a location open to the public); or
- BD4 – high-density occupation, difficult conditions of evacuation.

A wiring system in premises where any of the above conditions of external influence applies is required by Regulation 422.2.1 not to encroach on an escape route unless the wiring system meets all of the following requirements:

- The wiring has a sheath, such as that of a multicore cable, or an enclosure provided by a cable management system, such as conduit or trunking. (An example of a wiring system not meeting this requirement is bare conductors, such as may be used in an SELV system.)
- The wiring system must not be within arm's reach unless it is provided with protection against mechanical damage likely to occur during an evacuation. (See Figure 6.4 of this Guide for 'arm's reach'.)
- The wiring system must be as short as practicable.

Wiring systems installed where conditions BD2, BD3 or BD4 apply are required to be non-flame propagating. This requirement is met through compliance with the applicable

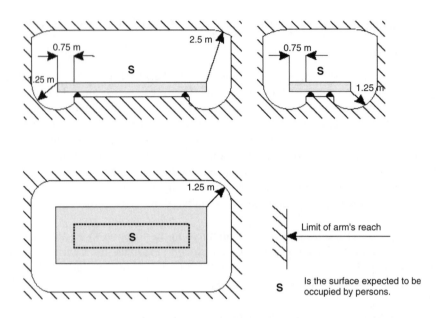

Figure 6.4 Zone of accessibility: arm's reach

Table 6.1 Product standards through which the non-flame-propagating requirement of
Regulation 422.2.1 may be met

Wiring system (or enclosure forming part of wiring system)	Product standard	
Cables	Relevant parts of BS EN 50266 and the requirements of BS EN 61034-2	Common test methods for cables under fire conditions. Measurement of smoke density of cables burning under defined conditions. Test procedure and requirements
Conduit systems	BS EN 61386-1	Conduit systems for cable management. General requirements
Cable trunking systems and cable ducting systems	BS EN 50085	Cable trunking systems and cable ducting systems for electrical installations. Cable trunking systems and cable ducting systems intended for mounting on walls and ceilings

fire test requirement of the relevant product standards listed in Table 6.1 (Regulation
422.2.1 refers).

Wiring in escape routes must have a limited rate of smoke production. The density of
smoke information for a particular type or class of insulated conductor or cable should
preferably be as given in the applicable cable product standard. In the absence of such

information, the adoption of at least a 60% cable light transmission value is recommended for cables tested against BS EN 61034-2 (Regulation 422.2.1 refers).

In addition, wiring systems supplying safety circuits are required to have a fire resistance rating of 2 h or such other duration as may be required for building elements by building regulations.

Except for items installed to facilitate evacuation (such as fire alarm call points), switchgear or controlgear in conditions BD2, BD3, or BD4 is required by Regulation 422.2.2 to be accessible to authorized persons only, and if placed in an escape route must be in a cabinet or enclosure that is not combustible or readily combustible.

Except for individual capacitors in equipment, such as in luminaires, equipment containing flammable liquid is not permitted, by Regulation 422.2.3, in escape routes where conditions BD3 or BD4 apply.

6.3.3   *Locations with risks of fire due to the nature of processed or stored materials*

Installations in locations with risks of fire due to the nature of processed or stored materials are the subject of Regulation Group 422.3, which are additional to those of Section 421 in locations where conditions BE2 (fire risks) exist. The requirements of Regulation Group 422.3 do not, however, apply to installations in locations with explosion risks, which are subject to the particular requirements of BS EN 60079-14. Regulation Group 422.3 requires enclosures to be prevented from attaining too high a temperature where dust and fibres could collect on them. Equipment in the location must be of a type that is suitable for the location.

Luminaires must have enclosures providing a degree of protection of at least IP5X. They must also have a limited surface temperature in accordance with BS EN 60598-2-24 and be of a type that prevents lamp components from falling from the luminaire.

Except where the luminaire product standard specifies otherwise, small spotlights and projectors must generally be installed at not less than the following minimum distances from combustible materials depending on lamp rating (see Table 6.2).

Measures must be taken to prevent the enclosures of equipment such as heaters and resistors from attaining surface temperatures higher than 90 °C under normal conditions, or 115 °C under fault conditions. In addition, measures must be taken to prevent the enclosures of any other type of electrical equipment exceeding these temperatures where materials such as dust and fibres sufficient to cause a fire hazard could accumulate on them.

Table 6.2 Luminaire minimum distances from combustible materials

Luminaire rating	Minimum distance from combustible material (mm)
≤100 W	500
>100 W to ≤300 W	800
>300 W to ≤500 W	1000

Unless suitable for the location and having a degree of protection of at least IP4X, or IP5X in the presence of dust, switchgear or controlgear must be installed outside the location.

Conduit systems, trunking systems and cable tray systems must meet specified fire propagation requirements, as must cables, except where completely embedded in non-combustible material, such as plaster or concrete, or otherwise protected from fire.

Although wiring systems not intended for electricity supply in the location are permitted to pass through the location, they must have no electrical joint within the location unless the joint is placed in a suitable enclosure that does not adversely affect the flame propagation characteristics of the wiring system.

Motors that are automatically or remotely controlled or are not continuously supervised must be protected against excessive temperatures by a protective device with a manual reset. Motors with star–delta starting must be protected against excessive temperature both when operating in star and when operating in delta.

For TN and TT systems, wiring systems must be protected against insulation faults to earth by RCDs having a rated residual operating current $I_{\Delta n}$ not exceeding 300 mA, except for mineral insulated cables and busbar trunking systems and powertrack systems. However, a residual operating current not exceeding 30 mA is required where a resistive fault may cause a fire, such as for overhead heating with film heating elements.

For IT systems, Regulation 422.3.9 calls for insulation monitoring with audible and visual signals and adequate supervision so that manual disconnection may occur as soon as possible. In the event of a second fault, automatic disconnection of the overcurrent protective device must occur within the appropriate time specified in Chapter 41; see Section 5.3.6 of this Guide for information on disconnection time requirements.

All circuits, whether supplying equipment in the location or merely traversing the location, must be protected against overload and fault current by protective devices situated outside the location, on the supply side of the location.

A PEN conductor must not to be used except for a wiring system passing through the location, and every circuit must be capable of being isolated from all live supply conductors by a linked switch or circuit-breaker, except as permitted by Regulation 537.1.2.

Irrespective of the system type, exposed bare and live conductors must not be used and flexible cables must be a heavy duty type rated at not less than 450/750 V or be protected against mechanical damage.

Heating elements in heating and ventilating systems serving the location must not cause a fire hazard and all heating appliances must be fixed, and barriers must be provided to prevent ignition of any combustible materials in close proximity.

Heat storage appliances must be of a type that prevents ignition of combustible dust or fibres by the heat storing core.

6.3.4 Combustible constructional materials

Installations in locations with combustible construction material are subject to the requirements of Regulation Group 422.4. The requirements apply where the location is mainly constructed of combustible materials, such as wood.

The minimum distances between small spotlights and projectors in a location with combustible construction materials are the same as those referred to in Table 6.2 of this

Guide, relating to locations with risks of fire due to the nature of processed or stored materials.

Electrical equipment such as installation boxes and distribution boards installed in a combustible wall must comply with the relevant standards for enclosure temperature rise. Alternatively, the equipment must be enclosed in a suitable thickness of nonflammable material.

Cables and cords must comply with BS EN 60332-1-2 (relating to vertical flame test methods for cables under fire condition). The tests in this part of the standard are more stringent than those required for cables in locations not covered by this new section, where compliance with BS EN 60332-1-1 is generally sufficient.

Conduit and trunking systems must comply with BS EN 61386-1 and BS EN 50085-1 respectively, including compliance with the fire resistance tests specified in the standards.

6.3.5   Fire-propagating structures

Installations in fire-propagating structures are subject to the requirements of Regulation Group 422.5. The requirements apply where a building has a shape and dimensions that facilitate the spread of fire.

Detailed guidance on how the shape and dimensions of a building can influence the spread of fire is beyond the scope of this Guide, and specialist advice may be necessary. Persons or organizations that may need to be approached for such advice include the architect, the fire prevention officer, regulatory bodies and licensing authorities.

The spread of fire, particularly in high-rise buildings, may be facilitated by building features such as ventilation ducts and chimney-like structures (for example, atria, stair wells, lift shafts, refuse chutes and smoke vents). Any forced ventilation system could also be relevant to the spread of fire.

The only requirement of Regulation Group 422.5 is that precautions shall be taken to ensure that electrical installations cannot propagate fire. A note in the introductory part of the regulation group points out that fire detectors may be provided that ensure the implementation of measures for preventing propagation of fire, such as the closing of fire dampers in ducts.

6.3.6   Selection and erection of installations in locations of national, commercial, industrial or public significance

Selection and erection of installations in locations of national, commercial, industrial or public significance is the subject of Regulation 422.6.

The regulation calls for the following measures to be considered for installations in locations that include buildings or rooms with assets of significant value:

- installation of mineral insulated cable according to BS EN 60702;
- installation of cables with improved fire-resisting characteristics in the case of a fire hazard;
- installation of cables in noncombustible walls, ceilings or floors;
- installation of cables in areas with constructional partitions having a fire resistance for a period of 30 min, or 90 min in the case of locations containing staircases and needed for emergency escape.

Table 6.3 Limiting temperature for accessible parts of equipment under normal load conditions

Contact with equipment part	Limiting temperature (°C)	
	Metallic	Non-metallic
Hand-held means of operation	55	65
Part intended to be touched but not hand-held	70	80
Part which need not be touched for normal operation	80	90

Examples that Regulation 422.6 gives of locations where the above measures should be considered are national monuments, museums and other public buildings, railway stations and airports, laboratories, computer centres, and certain industrial and storage facilities.

6.4 Burns

There is only one regulation which addresses the subject of burns, namely Regulation 423.1. The requirement here is that fixed electrical equipment within arm's reach shall not exceed the maximum temperatures given in Table 42.1.

This table differentiates between different types of equipment and between metallic and nonmetallic materials (compare the thermal conductivity of aluminium of the order of 236 $W\,m^{-1}\,K^{-1}$ with that of some plastics at about 0.20 $W\,m^{-1}\,K^{-1}$). Consideration of equipment complying with a British Standard which specifies a maximum temperature for that equipment is not the subject of these requirements. The concept of 'arm's reach', previously restricted to basic protection against electric shock, is now used for the first time in the consideration of protection against burns.

Table 6.3 summarizes the maximum temperatures given in BS 7671. It will be seen that the different temperatures for metallic and nonmetallic materials reflect their thermal conductivity. Where equipment under normal load conditions is likely to reach temperatures exceeding those given, even for short periods, precautions must be taken to guard against accidental contact. This may be achieved, for example, by the use of protective wire-guards or by positioning equipment out of arm's reach (see Figure 6.4).

One problem area exists in the siting of a lighting point in small store and airing cupboards when the luminaire is a pendant or a batten-holder with tungsten lamp. These are often positioned where they can be an ignition hazard (igniting papers or blankets) and a burns risk because of the possibility of human contact on entering the room. Where siting out of 'arm's reach' is impracticable, consideration should be given to the use of a totally enclosed luminaire or a low-energy fluorescent lamped luminaire, thereby obviating both risks.

7

Protection against overcurrent, undervoltage and overvoltage

7.1 General

As ever, it is always useful to start with the definition of overcurrent and definitions relating the various forms of overcurrent, which are summarized in Table 7.1. There are two regulations relating specifically to overcurrent in Chapter 13, which set out the fundamental requirements for safety. Regulation 131.4 addresses the hazards caused by excessive temperatures and mechanical forces associated with overcurrents likely to arise in live conductors. Regulation 131.5 sets out the requirements relating to protection against fault currents in conductors other than live conductors (such as protective conductors) and in other types of electrical equipment. Fault currents can be, and often are, of the order of several thousand amperes and are capable of causing degradation of insulating materials from thermal effects, electromagnetic effects and electromechanical stresses to themselves and adjacent equipment. It is necessary, therefore, to limit the duration of fault current so that the 'destructive' energy released under fault conditions is restrained to acceptable levels.

The basic requirements for protective measures relating to overcurrent are given in Chapter 43, which provides for the protection of live conductors from the effects of overcurrent. Section 530 deals with the common rules for switchgear and Section 533 more specifically includes application conditions for overcurrent protective devices. The designer should not overlook the point that the protective measures detailed in BS 7671 are related to protection of the wiring systems, and that the protection of other equipment may need further consideration.

It is important to note from Table 7.1 that an overload current is *not* a fault current and is, in fact, an overcurrent which occurs *only* in a healthy, or an electrically sound, circuit. Also worthy of note is that both short-circuit current and earth fault current are embraced by the term 'fault current', which when used may mean either or both depending upon the context in which it is used.

Regulation 430.3 requires a protective device to be provided to break any overcurrent (overload current or fault current) in the circuit conductors before this current could

A Practical Guide to The Wiring Regulations: 17th Edition IEE Wiring Regulations (BS 7671:2008)
Fourth Edition Geoffrey Stokes and John Bradley
© Geoffrey Stokes and John Bradley. Published by John Wiley & Sons, Ltd

Table 7.1 Summary of overcurrent definitions

Parameter	Definition
Overcurrent	A current exceeding the rated current (for conductors, the rated current is the current-carrying capacity) embracing all the following terms in this table
Short-circuit current	An overcurrent resulting from a fault of negligible impedance between live conductors having a difference of potential under normal operating conditions
Earth fault current	An overcurrent resulting from a fault of negligible impedance between a line conductor and an exposed-conductive-part or a protective conductor
Fault current	A current resulting from a fault (short-circuit or earth fault)
Overload current	An overcurrent occurring in a circuit which is *electrically* sound

cause danger due to thermal or mechanical effects that may damage insulation, connections, joints terminations or the surroundings of the conductors. The same regulation requires the protection against overload current and the protection against fault current to be coordinated in accordance with Section 435. Generally speaking, every live conductor (line and neutral conductors of an a.c. system and all poles of a d.c. system) must be protected against overcurrent by one or more devices which will automatically interrupt the supply in the event of a fault condition or an overload. In other words, overcurrent protection is specifically intended to protect persons, livestock and property from the hazards associated with a circuit drawing current in excess of its rated current which, normally, is the rated current or preselected setting of the overcurrent protective device.

BS 7671 does not impose specific time limits on the duration of a fault other than to demand that there are no consequential risks. However, where overcurrent protective devices are also used to provide protection against electric shock (fault protection), the maximum duration would not normally exceed 5 s and, for most purposes, this would be considered the maximum time allowed for clearing a fault current. Data for k values given in Table 43.1 of BS 7671 are based on a maximum of 5 s duration (see also Section 7.5.3 of this Guide for the derivation of k).

7.2 Nature of protective devices

The first paragraph of Section 432 calls for the protective devices employed to be suitable for the task that they are being asked to perform (i.e. protection against fault current and/or overload). With certain exceptions, devices providing protection for both fault current and overload need to be (Regulation 432.1) capable of breaking and, in the case of electromechanical devices, of making any fault current likely to occur. This includes the maximum prospective fault current at the point at which they are inserted into a circuit. Where a device is used for protection against overload only, it is permissible (Regulation 432.2) for it to have a fault breaking (and making) capacity of less than the prospective fault current. Where a device is employed for protection against fault current only, it must clearly have a fault current capacity as if it provided protection against both risks (fault current and overload).

Regulation 432.1 recognizes that a single device may provide protection against overload and fault current although separate devices may be employed to serve the two functions. Fault current magnitudes are usually of a much higher order than overload currents (e.g. 500 times higher). Because the heat (I^2Rt) caused by overcurrent conditions is proportional to I^2 and the duration of the condition, the disconnection of fault currents is required in much shorter times than would be the case for an overload condition. An overcurrent protective device (e.g. a fuse or circuit-breaker) when used for all forms of overcurrent will take much longer to operate on overload than that of fault current, as will be seen from its time/current characteristic. In general, there will be abnormal conditions which generate a current which in magnitude is greater than overload but substantially less than fault current. However, selecting a device, or devices, capable of protecting overload and fault current will, in most cases, be all that is required of the designer. Overcurrent protective devices and their performance are discussed in Section 11.4 of this Guide. Figure 7.1 gives a pictorial representation of the various currents and other parameters.

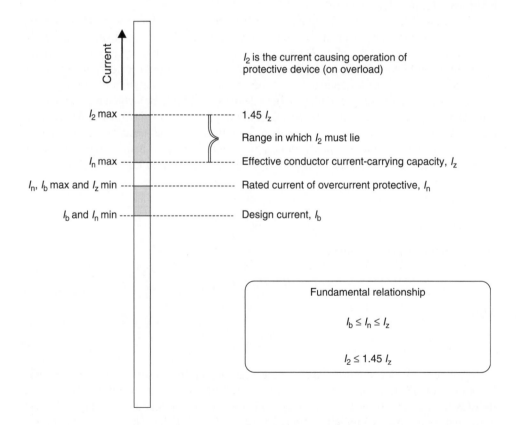

Figure 7.1   Current-carrying capacity and nominal and overload operating currents

7.3   Protection against overload

7.3.1   General

Overload is a current exceeding rated current, which may be provoked by excessive load-ing, either unintentionally or deliberately, of the circuit by the installation user or by some malfunction of load (e.g. motor-driven machinery subjected to mechanical loading in excess of the design load). The purpose of providing overload protection, like any overcurrent protection, is to prevent damage to the cable, conductor insulation, joints and connections and the surrounding area by excessive temperature rise, as indicated in Regulation 430.3. Regulation 433.1 calls for every circuit design to be such that small overloads (those that would not be sufficient to operate the device) do not occur. This is an important aspect in the design, and it is essential that load assessments are suffi-ciently accurate to confirm that currents are not drawn in excess of the rated current. For example, a circuit drawing a current of 10–15% greater than rated current will continue to function for an indefinite period without the operation of the circuit's overcurrent pro-tective device. Whilst overload protective requirements are essentially for the purpose of protecting conductors, the circuit's accessories, switchgear, etc. would normally also be protected provided they had been selected on the basis of complying with the appropriate British or Harmonized Standard.

Regulation 433.1.1 sets out the requirements for the coordination between the protective device and the conductors to be protected against overload. The design load current of the circuit is represented by I_b, while I_z represents the lowest current-carrying capacity of any of the live conductors of the circuit in question. The rated current, or current setting, of the protective device is designated I_n and the current causing operation of the device on overload is I_2. Current setting applies only to devices (e.g. MCCBs) which have adjustable settings for current, which might be less than the rated current of the device. Where used, the arrangement for setting the current of such a device must be inaccessible to unskilled persons and sealed to prevent inadvertent adjustment. The coordination requirements for these parameters can best be summarized by the formulae given in Equations (7.1)–(7.4):

$$I_b \leq I_n \tag{7.1}$$

$$I_n \leq I_z \tag{7.2}$$

or by combining the two equations:

$$I_b \leq I_n \leq I_z \tag{7.3}$$

Also:

$$I_2 \leq 1.45 I_z \tag{7.4}$$

Regulation 433.1.2 confirms that where current-limiting fuses to BS 88-2.2 'gG', BS 88-6 and BS 1361 and circuit-breakers to BS EN 60898 and BS EN 60947-2 and RCBOs to BS EN 61009-1 are used, compliance with the Equation (7.2) ($I_n \leq I_z$) also meets the requirements implicit in Equation (7.4) ($I_2 \leq 1.45 I_z$). This is because the rel-evant British Standards have addressed these aspects in the design constraints of such devices. Type 'gG' fuses must not be confused with type 'gM', used for motor protection, or type 'a' (as in type 'aM'), which have only partial range short-circuit capacity, the latter two types not being suitable for overload protection.

Where a BS 3036 rewirable fuse (or semi-enclosed fuse as it is more accurately termed) is used, compliance is not assured in a similar way to the use of fuses and MCBs as described previously. Regulation 433.1.3 makes further demands in this case, in that it modifies Equation (7.2) to $I_n \leq 0.725I_z$. This is necessary because the rewirable fuse has a fusing-factor of 2, where the fusing factor is the ratio of the minimum fusing current to cause operation ('blowing') in 4 h to the rated current of the fuse. So, for example, for a 30 A circuit protected by a 30 A rewirable fuse and drawing an *overload* current of 60 A, disconnection on overload may take up to 4 h. Clearly, this phenomenon needs to be taken into account and this is done by the application of the 0.725 factor as a multiplier to the current-carrying capacity of the circuit cable. One effect of using a rewirable fuse to BS 3036, therefore, is to increase the required cable cross-sectional area for a given load, other parameters being equal. The factor of 0.725 is derived by application of the two formulae in Equations (7.5) and (7.6):

$$I_2 = 2I_n \tag{7.5}$$

$$I_2 \leq 1.45I_z \tag{7.6}$$

By combining the two foregoing equations:

$$2I_n \leq 1.45I_z \tag{7.7}$$

so that

$$I_n \leq \frac{1.45}{2}I_z \quad \text{or} \quad I_n \leq 0.725I_z \tag{7.8}$$

Overload protective devices are normally positioned at a point where a reduction in the current-carrying capacity of a cable or conductor occurs. As the note to Regulation 433.2.1 points out, there are a number of reasons why a reduction in current-carrying capacity may occur other than the obvious one – a reduction in conductor cross-sectional area. I_z will always be the lower or lowest of the current-carrying capacities of conductors downstream of the overload protective device (unless other protective devices are employed downstream). Factors affecting current-carrying capacity would include:

- change in cross-sectional area, and/or
- change in ambient temperature (on cable route), and/or
- change in type of conductor, and/or
- change in environmental conditions en route (e.g. thermal insulation), and/or
- change in grouping (of conductors or cables).

Whilst overload devices are normally located at the point of reduction of current-carrying capacity, this is not always practical or even desirable, as, for example, in the case of the electric motor where the overload device is often positioned in an enclosure (starter) adjacent to the motor. Regulation 433.2.2 allows such a practice but demands in such cases that there be no branch circuits or outlets for the connection of current-using equipment between the point at which there is a reduction in current-carrying capacity and the device (see Section 7.3.2 of this Guide). However, this relaxation cannot be applied where there is a high risk of fire or explosion, or where special considerations are appropriate.

As envisaged by Regulations 433.3.1 and 433.3.3, there are situations where overload devices need not be provided or may be omitted, an example of the latter being in the secondary winding of a current transformer, where extremely high voltages can occur on secondary-circuit interruption. Table 7.2 details the situations where overload protection need not be provided or may be omitted. Regulation 433.3.2.2 permits the omission of overload protection in one line conductor in an IT system with neutral, provided an RCD is located in each circuit.

Table 7.2 Omission of overload protection[a]

Situations where overload protection may be omitted or need not be provided	Restrictions or qualifications
Secondary circuits of current transformers	May be omitted
At a location of reduction in current-carrying capacity of conductors	Where supply characteristics are such that overload is unlikely to occur
	Where load characteristics are such that overload is unlikely to occur
	Where overload protection is provided by an upstream device
At the origin of installation	Where ED's overload device is used and affords adequate protection (with ED's express agreement)
At any position	Where the current-carrying capacity of the live conductors is greater than the current the source can deliver
Exciter circuits of rotating machines	May be omitted
Supply circuits of lifting magnets	May be omitted
Circuits supplying a fire extinguishing device	May be omitted
Circuits supplying a safety service	May be omitted
Circuits supplying medical equipment for life support in medical locations with an IT system	May be omitted

[a]Regulation 433.3.3 requires consideration to be given to the provision of overload alarm for circuits not provided with protection against overload.

Regulation 433.3.2.1 indicates that the relaxation permitted by Regulations 433.2.2 and 433.3 for the alternative positions and omission of overload protective devices must not be applied to an IT system unless either

- the system is permanently supervised and an insulation monitoring device is used which, in the event of a fault, causes disconnection of the circuit or the operation of a signal, or
- the conductors are protected by either an RCD that will operate on a second fault or by use of double or reinforced insulation.

It should be noted that overload protection may only be omitted where protection is provided by some other means or where a greater danger would be created by the operation of the overload device. In the assessments of other dangers and risks, the designer would need to consider all the scenarios that may occur in practice and weigh the implications of omitting the device very carefully. Section 436 confirms that no further overload protection is necessary where the current-carrying capacity of live conductors is greater than the current that the sources can deliver.

7.3.2 Protection against overload: motors

Requirements for overload protection for circuits supplying motors are generally the same as for other circuits, in that the overload operating current I_2 must not exceed 1.45 I_z. The detailed examination of the many different types of motor and their characteristics is outside the scope of this Guide. However, there is one particular case which is worthy of mention, regarding overload protection, and this is the star–delta starting technique of the three-phase motor.

A commonly accepted 'rule of thumb' has involved the final connection cables, to the motor from the starter, to be selected so as to be half the cross-sectional area of the cables supplying the starter. This method may be too generous in its assumptions and the final connection cables may not be adequately protected against overload. Figure 7.2 shows in (a) the motor windings connected in star configuration for starting mode and (b) the reconnected windings in the delta configuration for the running mode. The starting current when connected in star is much reduced owing mainly to the voltage across each winding being reduced to $1/\sqrt{3}$ (0.577) of the supply line-to-line voltage.

Generally speaking, the starting period in star configuration is relatively short and, although the starting currents can, for example, be as high as 10 times the normal running current, the conductors are unlikely to suffer any damage from excessive temperature rise in the short starting time. Consequently, the overload device would not normally be called upon to operate under starting conditions.

It can be seen from Figure 7.2 that the overload protection is generally in the conductors connecting to the windings, not in the supply lines. In running mode, the current in the supply is $\sqrt{3}$ times the current in each winding; or, to put it another way, the current in each winding is approximately 0.58 times the supply current $(1/\sqrt{3} = 0.577)$. It should also be recognized that because the six cables connecting the motor to the starter are always loaded when the motor is running and are usually run grouped together in the

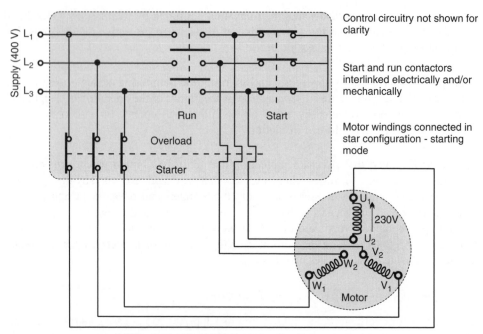

(a) 'Starting' connections in star configuration shown above

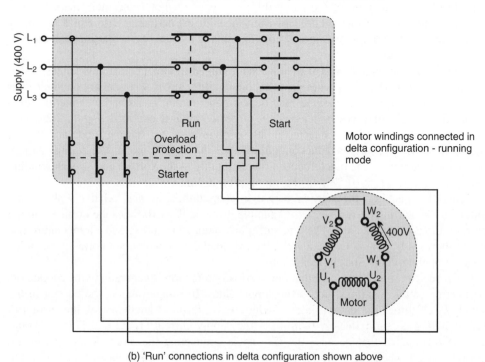

(b) 'Run' connections in delta configuration shown above

Figure 7.2   Star–delta starting of three-phase motor

same conduit or other wiring system, an allowance must be made for such grouping. In effect, these six cables represent two 'circuits', and the appropriate rating factor C_g must be applied. Should the final connection cables be of a different type and/or be subjected to different installation conditions, then other appropriate rating factors must also be applied.

It can be seen that, where the motor connections are grouped in conduit (installation method 4), a rating factor $C_g = 0.80$ from Table 4C1 of Appendix 4 of BS 7671 is called for; this, together with the 0.58 factor, would require these conductors to have an effective current-carrying capacity of approximately 72% (0.58/0.80) of that of the conductors supplying the starter.

In many cases it is just not desirable or even feasible to select a motor starter with sufficiently high fault capacity and, generally, it is not necessary to do so when used in conjunction with a fuse or circuit-breaker capable of providing the function of breaking fault current. However, a motor starter should be capable of closing on to a fault condition (e.g. stalled-rotor condition) without itself sustaining damage and, of course, be capable of interrupting the maximum overload current. It is important, therefore, that the starter does not attempt to operate under fault conditions before the associated overcurrent protective device has had a chance to interrupt the faulted circuit.

The motor 'back-up' overcurrent protective device must be capable of interrupting the fault current but must not operate under normal starting conditions. HBC fuses to BS 88 'gM' have been designed specifically for this purpose, although it is possible to use fuses to BS 88 'gG', of a higher rating than would be suggested by the steady-state load current (i.e. $I_n \geq I_b$), for this purpose provided that the device thermally protects the downstream circuit conductors under fault conditions.

Fuses to BS 88 'gM' have a dual rating, the first of which relates to the maximum continuous current and the second based on the characteristics of the load and, in particular, the starting current draw. The ratings are expressed as a code as, for instance, '100M160'. Take, for example, a direct on-line starter and motor with a full-load current of, say, 99 A; this would require a 160 A fuse to BS 88 'gG' but would also be adequately protected by a 100M160 fuse to BS 88 'gM'.

As regards coordination of protection, BS EN 60947-4-1 identifies three levels of permitted damage to motor starters as a result of fault conditions. Type 2 coordination is required if the starter is expected, after examination and any necessary remedial work, to provide overload protection subsequent to experiencing a fault condition. This type does allow for light contact burning and the possible risk of welding of contacts, but it does not permit any permanent damage or permanent alteration of the starter's overload characteristic.

Motor overload devices are normally arranged to operate between currents exceeding full-load current and the overload limit of the motor as specified by the motor manufacturer. However, the overload device mechanism is usually time delayed (e.g. damped by oil-filled dashpots or a thermal device) to allow normal starting currents to flow without interruption. Fault currents are allowed by the overload device to persist so as to be detected by the upstream overcurrent protective device.

The motor starter overload device can, and usually does, provide overload protection for circuit conductors as well as the motor, but a check is always necessary to confirm that the highest setting of overload operating current I_2 is not more than $1.45 I_z$.

Motors having a duty cycle involving frequent starting and stopping require special consideration to take account of the accumulative heating effects of the recurrent starting currents.

7.3.3   Ring final circuits

The normal requirements relating to protection against overload are relaxed in the case of ring final circuits. Regulation 433.1.5 provides a *deemed to comply* status for protection against overload where certain conditions are a prerequisite:

- the circuit must be protected against overcurrent by a 30 or 32 A overcurrent protective device complying with BS 88-2.2, BS 88-6, BS 1361, BS 3036, BS EN 60898, BS EN 60947-2 or BS EN 61009-1;
- the circuit accessories must be to BS 1363;
- the circuit live conductors must be copper, minimum cross-sectional area of 2.5 mm^2, unless mineral-insulated two-core cable (to BS EN 60702-1) is used, where they may be 1.5 mm^2;
- the current-carrying capacity I_z of the cable must must be not less than 20 A;
- the load current in any part of the circuit must be unlikely to exceed for long periods the current-carrying capacity I_z of the cable.

7.4   Protection against fault current

As previously mentioned, fault current is an overcurrent caused by either a short-circuit (between live conductors) or an earth fault (between a live conductor and an exposed-conductive-part or protective conductor). Faults occur as a result of insulation failure or by the bridging of conductive parts by some conducting material, and there are a number of possible fault conditions which need consideration:

- short-circuit faults between line and neutral;
- short-circuit faults between line and line;
- short-circuit faults between all three lines;
- earth faults between line and exposed-conductive-parts or protective conductor.

Fault currents left unchecked would result in overheating and damage to insulation, and possibly the risk of igniting material adjacent to the conductors either by raising the temperature or by arcing. Therefore, the protective measures are intended to prevent deterioration of insulation, damage to property, including electrical equipment, and the prevention of fire which would be a hazard to both persons and property. Because anticipated fault currents are high, it is necessary to have short disconnection times to prevent the rise in conductor temperature becoming excessive. Acceptable temperature limits for conductors are given in Table 43.1 of BS 7671 for live conductors and Tables 54.2, 54.3, 54.4, 54.5 and 54.6 for protective conductors.

Regulation 430.3 calls for a protective device to break any overcurrent (which includes fault current) in the protected circuit before it causes danger from thermal or mechanical effects. Where devices protect only against fault current (and not overload), the rated current of the device I_n may be greater than the current-carrying capacity of the cable I_z. This would not apply, of course, to where the device performs both functions of protection from overload *and* fault current.

As with overload devices, fault current protective devices generally must be positioned at the point where a reduction in current-carrying capacity occurs (i.e. where the conductor energy-withstand is reduced) as called for in Regulation 434.2. As before, the factors affecting current-carrying capacity would include:

- change in cross-sectional area, and/or
- change in ambient temperature (on cable route), and/or
- change in type of conductor, and/or
- change in environmental condition en route (e.g. thermal insulation), and/or
- change in grouping (of conductors or cables).

There is an exception to this rule of positioning the protective device, which, according to Regulation 434.2.1, may be inserted in the circuit other than at the point where the reduction of current-carrying capacity occurs (see Figure 7.3) providing *all* of the following conditions are met:

- the distance between the device and the point at which the reduction of current-carrying capacity occurs does not exceed 3 m;
- the 'unprotected' circuit cables between points are installed such that the risk of a fault current occurring is minimized;
- 'unprotected' circuit cables are installed in such a manner that the risk of fire is minimized.

A further exception, as indicated in Regulation 434.2.2, is that the fault current protective device may be placed on the supply side of the point where a change occurs (in cross-sectional area, method of installation, type of cable or conductor, or environmental conditions) provided the device affords the necessary protection envisaged by Regulation

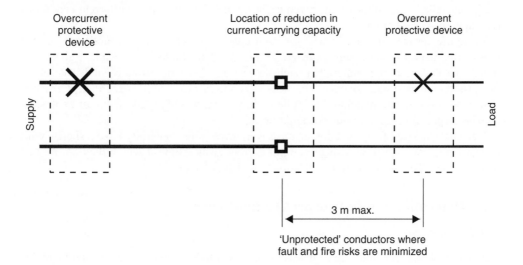

Figure 7.3   Placing of overcurrent device at other than the point of reduction of current-carrying capacity

Table 7.3 Omission of fault current protection

Situations where fault current protection need not be provided	Restrictions or qualifications
A conductor connecting a generator with its control panel	On condition that fault current protection is provided in the panel and interconnected cable is installed so that the risk of fault current occurring and the risk of fire are minimized
A conductor connecting a transformer, rectifier or battery with its control panel	On condition that fault current protection is provided in the panel and interconnected cable is installed so that the risk of fault current occurring and the risk of fire are minimized
In a measuring circuit	Where disconnection would cause greater danger (e.g. secondary circuits of CTs)
In any circuit supplying equipment	Where unexpected interruption causes greater danger than that of a fault current
At the origin of an installation	Provided the ED provides such device(s) *and where agreement is obtained* for its use to protect (part of) the installation

434.5.2 (meeting the adiabatic equation so as to prevent unacceptable conductor temperature rise) to the wiring on the load side of the point.

It is important to note that the exceptions given in Regulations 434.2.1 and 434.2.2 regarding the position of the protective device must not be applied to an installation in a location presenting a fire risk or risk of explosion or where the requirements for special installations and locations, given in Part 7 of BS 7671, specify different conditions (Regulation 434.2.1 refers).

There are certain other cases where a fault current protective device need not be provided (Regulation 434.3), and these are summarized in Table 7.3. Again, it should be noted that the regulation states that such devices *need not* be provided and it is left to the installation designer to make the necessary engineering judgement in this respect. However, this relaxation cannot be applied where there is a high risk of fire or explosion, or where special considerations are appropriate.

Devices employed for protection against overcurrent must generally have a short-circuit capacity not less than the prospective fault current at the point at which they are located in the circuit. This is discussed in Section 7.6 of this Guide.

7.5   Determination of prospective fault current

7.5.1   General

Regulation 434.1 demands that an assessment be made at every relevant point in the entire installation (i.e. at every point where a fault current protective device is positioned). This may be done by either calculation or measurement, depending on the circumstances. Neither of these methods should be viewed as highly accurate, and a margin of error must be assumed and allowances made accordingly. Where measurement is used, this will need to be undertaken using a suitable test instrument (preferably giving direct prospective fault current readings) that is appropriate for the particular nominal voltage of the circuit under test. For small installations (i.e. nominally rated up to 100 A per line conductor), a portable hand-held instrument for measuring loop impedance may be used and the measured value can be used to determine the prospective fault current I_{pf} (i.e. $I_{pf} = U_n/Z$). For the larger installations, a more sophisticated instrument will be required. As with all testing and measurement taking, there is an element of danger to the operator of the instrument (and bystanders), so it is essential that tests are all undertaken by competent persons who have had experience in the field and are fully versed in the necessary safe systems of working.

In designing an installation it is essential for the designer to make an assessment of the fault levels with which protective devices will have to cope. For the most part, measurement of prospective fault current will not be an option and, consequently, calculations will need to be undertaken. To estimate fault current, all that is required (applying Ohm's law) is a value for the source voltage and a value for the loop impedance at the point in question. The former is usually quite simply a matter of assuming that the source voltage will be the supply nominal voltage and will remain constant (i.e. an infinite busbar) under fault conditions. The latter parameter is more difficult to calculate unless all the constituent parts of the loop are known, if not in terms of impedance, then in terms of length, cross-sectional area, conductor material and cable construction, from which a deduction can be made about the resistance and inductive reactance of the constituent parts.

In making such calculations, there is an enormous scope for error from different sources. For example, resistance is a function of temperature (which will rise under fault conditions), whereas inductive reactance is not affected in the same way. With the very high currents which will flow under fault conditions, inductive reactance will be difficult to predict with any degree of certainty, and this parameter may be influenced by many unknowns, such as, for example, the actual inductive reactance of switchboards, busbar trunking and so on. Resistances of joints may also vary under fault currents. The calculations, therefore, need to be treated with caution, and it will often be found by testing at a later stage that the calculated value of prospective fault current was more pessimistic than the measured value indicates. This, of course, is welcomed in consideration of short-circuit capacities of protective devices and the effects of electromagnetic forces, but less welcomed with regard to the disconnection times (the lower the fault current, the longer the disconnection with fuses) and to the let-through energy, which is likely to be greater at lower fault currents.

For selection of an appropriate fault current protective device, Regulation 533.3 calls for account to be taken of the maximum fault current and also, where the device is to provide protection against fault current only (and not against overload current), the minimum fault current. The highest fault current will occur when the fault loop impedance is at its lowest value. This will be the case where a fault is considered immediately downstream of the protective device in question and when the conductors upstream are at the lowest temperature likely to be encountered during service. Unless better information and knowledge are available, this temperature will normally be taken as 20 °C and it will be found that cable manufacturers publish cable data (for conductor resistances) at this temperature. In cases where a lower temperature is encountered (e.g. in a refrigerated building) or where a higher temperature is known (e.g. boiler plant room), the given resistance data need correction to allow for these different temperatures. The maximum value of fault current is crucially important in the consideration of short-circuit capacities and the effects of electromagnetic forces.

The minimum value of fault current will occur when conductors are at their normal operating temperature (e.g. 70 °C for copper–PVC conductors carrying rated current) and when a fault occurs at the most remote end of the circuit in question. With certain exceptions, the conductor temperature for calculation purposes is the average of the initial and final temperatures given in Table 43.1 of BS 7671. If the conductors upstream of the protective device and circuit under consideration are much larger (say, at least twice) than the downstream conductors, then the fault current will have less effect in increasing their temperature and they can generally be assumed to maintain their normal operating temperature. BS 7671 allows this same assumption to be made for the purposes of protection against electric shock under earth fault conditions (Appendix 14 and the notes to Tables 41.2 - 41.4 refer).

In the following calculations, a simplistic method has been adopted that should be appropriate for most design appraisals of fault level. There may be times where a more rigorous approach will be required; in such cases, reference to BS EN 60909 *Short-circuit calculation in three-phase a.c. systems* will be essential. The cited examples assume that the voltage will be constant throughout the fault (infinite busbar), and this will be acceptable where the supply is derived from the public supply network. For other sources, such as generators (normal and standby supplies), it may be necessary to consult the generator manufacturer in order to define the transient characteristics of the set under fault conditions.

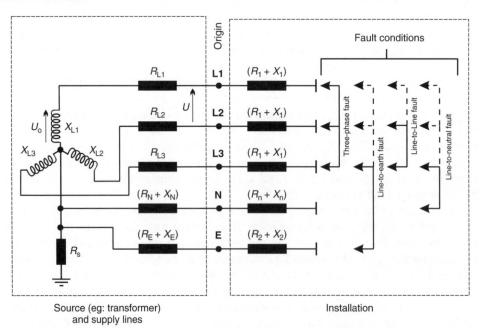

Figure 7.4   TN-S system: examples of the various fault conditions

It should be borne in mind that where items of large equipment that capable of energy storage are part of the installation, these too may be a source (in parallel with the normal supply source) capable of feeding a fault. This may be the case, for example, where a large motor instantaneously converts to a generator in the event of a fault (and loss of supply voltage). This energy source, therefore, needs to be taken into account in the assessment of prospective fault current.

Figure 7.4 shows a TN-S system and identifies examples of the various fault conditions, each being dealt with in the following text. Equivalent circuits for short-circuit fault conditions are the same irrespective of system type and earthing arrangements, whereas earth-fault equivalent circuits do differ. In the figure, the symbols given have the following meanings:

R_{L1} and X_{L1}	the resistive and reactance components respectively of impedance of the
R_{L2} and X_{L2}	supply source (transformer) and supply lines up to the origin of the installation
R_{L3} and X_{L3}	(L1 = line 1; L2 = line 2; L3 = line 3)
R_N and X_N	the resistive and reactance components respectively of impedance of the neutral supply conductor
R_E and X_E	the resistive and reactance components respectively of the supply protective conductor from the MET to the source star/neutral point
R_1 and X_1	the resistive and reactance components respectively of the installation line conductor(s)
R_n and X_n	the resistive and reactance components respectively of the installation neutral conductor
R_2 and X_2	the resistive and reactance components respectively of the installation cpc(s)
R_S	the resistance of the supply source earthing electrode
U_o	the source nominal voltage to Earth
U	the source nominal voltage line-to-line.

Resistive and reactive components of impedance cannot be added together arithmetically since they are in quadrature (i.e. displaced by 90°). The addition must be made vectorially, as shown in the 'impedance triangle' given in Figure 7.5. The triangle is right-angled, so that the hypotenuse (impedance) is the square root of the sum of the resistance squared plus the reactance squared, as given in Equation (7.9). The angle ϕ is the angular displacement between impedance and resistance and $\cos \phi$ is the power factor.

$$Z = \sqrt{R^2 + X^2} \tag{7.9}$$

7.5.2   Calculation of inductive reactance

Because of space limitations, it is not the intention here to reiterate the examination of inductive reactance calculations, which are dealt with admirably in the IEE Guidance Note No 6: *Protection against overcurrent*, but rather to draw the designer's attention to the need to be aware of the complexities in the more rigorous analysis in determining fault current with some greater degree of accuracy.

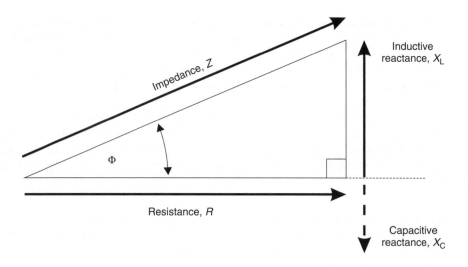

Figure 7.5 Impedance triangle

Inductive reactance is a function of the mutual inductance between conductors carrying current and, for steady-state conditions, is only present in a.c. systems where the instantaneous current is continuously changing in magnitude. When considering fault current, it is normally more advantageous to consider each circuit component or conductor separately (e.g. line, neutral and protective conductor), but the inductive reactance is a function of all the conductors combined in close proximity. For example, the reactance of two conductors close together will not be the same as three similar conductors in trefoil or when the two conductors are a distance apart. Additionally, the reactance will be different for the various fault conditions, and IEE Guidance Note No 6: *Protection against overcurrent* addresses the various cases.

7.5.3   Evaluation of k for different temperatures

Data for the k values for the more commonly used conducting materials (copper and aluminium) with their insulating materials (e.g. PVC, rubber, etc.) are given in Table 43.1 of BS 7671. This table also lists the assumed initial temperatures and the limiting final temperature appropriate to the particular type of insulation. The data given for k values are only valid for fault current flow duration of up to 5 s. Where faults take longer to clear, the cable manufacturer should be consulted to clarify whether or not the cable is protected for a fault of this longer duration. Where greater accuracy is required than that obtained by the use of the adiabatic equation, the calculation should be carried out as laid down in BS 7454:1991 (IEC 60949) *Method of calculation of thermally permissible short-circuit currents, taking into account non-adiabatic heating effects*.

The factor k, used in the adiabatic equation of Regulation 434.5.2, is based on the resistivity of the conductor, the temperature coefficient and the heat capacity of the conductor material. Its value is also dependent on the initial temperature and final temperature of the conductor, the latter being dictated by the type of insulation. The factor is derived

Table 7.4 Material data for evaluating k

Material	$B(°C)$	$Q_c(10^{-3}$ J $°C^{-1}$ mm$^{-3})$	$Q_{20}(10^{-6}$ Ω mm)	$C = [Q_c(B+20)/Q_{20}]^{0.5}$
Copper	234.5	3.45	17.241	226
Aluminium	228.0	2.50	28.264	148
Lead	230.0	1.45	214.000	41
Steel	202.0	3.80	138.000	78

from Equation (7.10), and Table 7.4 provides the necessary data for its use:

$$k = \sqrt{\frac{Q_c(B+20)}{Q_{20}}} \times \ln\left(1 + \frac{\theta_f - \theta_i}{B + \theta_i}\right) \tag{7.10}$$

where Q_c(J $°C^{-1}$ mm^{-3}) is the volumetric heat capacity of the conductor material, B ($°C$) is the reciprocal of temperature coefficient of resistivity at 0 $°C$ for the conductor, Q_{20} (Ω mm) is the electrical resistivity of conductor material at 20 $°C$, θ_i ($°C$) is the initial temperature of conductor, θ_f ($°C$) is the final temperature of conductor, and ln indicates logarithm to the base e.

Equation (7.10) can be much simplified by making the substitution given in Equation (7.11) for C (a constant for a particular conductor), which results in the formula given in Equation (7.12):

$$C = \sqrt{\frac{Q_c(B+20)}{Q_{20}}} \tag{7.11}$$

$$k = C\sqrt{\ln\left(1 + \frac{\theta_f - \theta_i}{B + \theta_i}\right)} \tag{7.12}$$

For example, if we take a copper conductor insulated with a 90 $°C$ thermosetting polymer with an initial temperature θ_i of 90 $°C$ and a final temperature of 250 $°C$ (as given in Table 43.1 of BS 7671) and apply these data together with those contained in Table 7.4, we get a k value, given in Equation (7.13) as 143, which is in agreement with that given in BS 7671:

$$k = 226\sqrt{\ln\left(1 + \frac{250 - 90}{234.5 + 90}\right)}$$

$$= 226\sqrt{\ln 1.493} \tag{7.13}$$

$$= 226 \times 0.633 = 143$$

There may be occasions where the designer has selected a conductor size with a current-carrying capacity much in excess of the design load current, perhaps because of volt-drop considerations. In such a case, the initial conductor temperature will be less than that assumed in Table 43A of BS 7671; consequently, a higher k coefficient will apply. Take, for example, a similar conductor to that given above and at an initial temperature of, say, 50 °C. By applying the above equation we get a k coefficient of 165:

$$k = 226\sqrt{\ln\left(1 + \frac{250 - 50}{234.5 + 50}\right)}$$

$$= 226\sqrt{\ln 1.703} \tag{7.14}$$

$$= 226 \times 0.730 = 165$$

This calculation is really only worthwhile where there is a substantial difference between the initial temperature given in Table 43A of BS 7671 and the actual initial temperature, and its application will be very limited. Where used, the average value of conductor temperature under fault conditions should also be recalculated.

7.5.4 Calculation of impedance of steel enclosures

The calculation of the impedance of steel enclosures requires special consideration in a.c. systems, because the magnetic effects of steel are significant and cannot be ignored. In the case of steel conduit, the relationship of these magnetic effects is not linear and it is necessary, in order to simplify matters, to treat fault currents up to 100 A and those above 100 A quite separately and assume linearity in these two ranges. Both the resistive and reactance components of impedance are affected. As with steel conduit, the impedance of steel trunking is also affected by magnetic flux linkages. However, the corrective procedure in the case of trunking is simplified by virtue of the general acceptance that trunking is not acceptable as a protective conductor for circuits with a fault current exceeding 100 A. Though not a common practice, it ought to be appreciated that a protective conductor or complementary protective conductor should not be run outside, and in close proximity to, steel trunking and steel conduit. The magnetic screening effects will increase the impedance of that conductor to a point where it ceases to become an effective protective conductor.

Perhaps not always readily appreciated, the metallic sheath or armouring of a cable (e.g. steel-wire armour) is a conductor enclosure. As such it is subjected to similar considerations with regard to the magnetic effects and the resultant changes to resistance and reactance. In the use of aluminium-wire armoured single-core cables, adjustment is also required to the resistance value. Generally speaking, such cables are bonded to earth at both ends of their run (two-point bonding); consequently, earth fault current, in a three-wire system, will flow in the three parallel paths provided by the armouring of each of the single-core cables. The adjustments for resistance and reactance for MICCs are much simpler because the magnetic effects are much less or nonexistent. IEE Guidance Note No 6: *Protection against overcurrent* gives detailed methods for making such adjustments to resistance and reactance where steel enclosures are involved in the wiring system.

7.5.5 Resistance and inductive reactance values

The values for resistance of most conductors may be obtained from BS EN 60228 *Specification for conductors in insulated cables and cords*. Alternatively, resistance and reactance values may be acquired by reference to the volt-drop tables in Appendix 4 of BS 7671. It is important to note, however, that the values so gained by these methods should be treated as the 'base' values and corrected to take account of the particular fault condition envisaged and the configuration of the conductors (e.g. trefoil or flat formation). IEE Guidance Note No 6: *Protection against overcurrent* addresses this subject in great depth.

To illustrate by (an unlikely) example how the 'base' resistance and reactance values may be obtained from the volt-drop tables of Appendix 4, let us consider a circuit of two 240 mm^2 single-core cables to BS 6004, of length 200 m, run in conduit (reference method B) with many draw-in boxes. From Table 4D1B of BS 7671, the resistance ($r = 0.195$ mV A^{-1} m^{-1}), inductive reactance ($x = 0.260$ mV A^{-1} m^{-1}) and impedance ($z = 0.330$ mV A^{-1} m^{-1}) are given. To obtain the values for the 200 m run, the given data are simply multiplied by 200 and, to give ohms rather than milliohms, divided by 1000. The values become $R = 0.039$ Ω, $X = 0.052$ Ω and $Z = 0.066$ Ω. In the case of data provided in the tables for three-phase, the volt-drop values, for each line conductor, are $\sqrt{3}r$, $\sqrt{3}x$ and $\sqrt{3}z$. These should be divided by $\sqrt{3}$ to obtain the resistance, reactance and impedance values (and multiplied by the run length) for conductors.

It is also important to note that the volt-drop data given in Appendix 4 of BS 7671 for single-core cables relate to axial spacing in the range of one or two cable diameters and when spaced at greater distances (i.e. more than one cable diameter between cables) the loop reactance is increased and due allowance should be made in this respect.

7.5.6 Temperature adjustments to resistance values

The values of resistances obtained from the volt-drop tables (as detailed in Section 7.5.5 above) need to be corrected for temperature. The correcting factor is different for faults occurring when the circuits are not loaded from those developing when the conductors are at their maximum permissible working temperature. The correction factors are given in Table 7.5 and should only be applied to the resistances and not to the reactances, as these are not temperature dependent.

Table 7.5 Resistance/temperature correction factors for conductors on full-load and no-load

Ref	Conductor on full-load (correction to working temperature)			Conductors without load (correction to 20 °C)	
	Conductor working temperature (°C)	Final temperature (°C)	Correction factor	Conductor maximum working temperature (°C)	Correction factor
A	60	200	1.24/1.00[a]	60	0.86
B	70	160	1.15/1.00[a]	70	0.83
C	70	140	1.12/1.00[a]	–	–
D	85	160	1.12/1.00[a]	85	0.79
E	85	140	1.09/1.00[a]	–	–
F	85	220	1.21/1.00[a]	–	–
G	90	250	1.25/1.00[a]	90	0.78

[a]For the purpose of determining the contribution of the circuit conductors to the earth fault loop impedance Z_s for protection against electric shock.

7.5.7 Line-to-neutral short-circuits

In a short-circuit between a line conductor and the neutral conductor, the fault current I_F is limited only by the impedance of the short-circuit loop. Referring to Figure 7.6, the short-circuit current at the origin of the installation is given by Equation (7.15). For short-circuits elsewhere, Equation (7.16) is appropriate.

$$I_F = \frac{U_o}{(R_L + R_N) + j(X_s + X_L + X_N)}$$

$$= \frac{U_o}{\sqrt{(R_L + R_N)^2 + (X_s + X_L + X_N)^2}} \tag{7.15}$$

$$I_F = \frac{U_o}{\sqrt{(R_L + R_N + R_1 + R_n)^2 + (X_s + X_L + X_N + X_1 + X_n)^2}} \tag{7.16}$$

where R_L is the resistance of the supply line conductor, R_N is the resistance of the supply neutral conductor, X_s is the source inductive reactance (note: the source resistance is, in this instance, considered to be small compared with the inductive reactance and, therefore, is neglected), X_L is the inductive reactance of the supply line conductor, X_N is the inductive reactance of the supply neutral conductor, R_1 is the resistance of the installation line conductor (where appropriate, further identified by reference to the particular line number or colour), R_n is the resistance of the installation neutral conductor, X_1 is the inductive reactance of the installation line conductor, (where appropriate, further identified

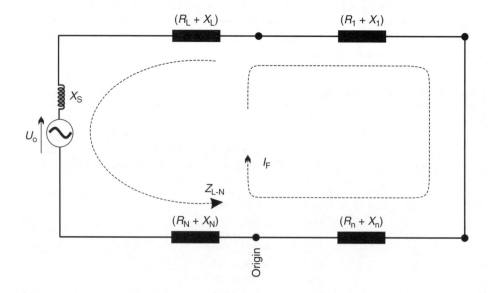

Figure 7.6 Line-to-neutral fault equivalent circuit

by reference to the particular line number or colour) and X_n is the inductive reactance of the installation neutral conductor.

For the smaller single-phase (up to 100 A, 230 V) installation supplied from the public network, the source line–neutral impedance Z_{L-N} (see Figure 7.6) may, generally speaking, be taken as a maximum of 0.8 Ω for a TN-S system supply (the maximum Z_{L-N} may be less, depending on the particular supply characteristics) and 0.35 Ω for a TN-C-S supply, although confirmation from the ED will always be necessary. The minimum value can be assumed to be 14 mΩ where the prospective fault current is given as 16 kA (230 V/16 000 A = 0.014 Ω). This range of values should be taken into account.

It may be difficult to obtain the separate values for resistance and reactance of the source and supply lines, and for the smaller installation, at any rate, and the Z_{L-N} quoted by the ED may have to be taken as wholly resistive. This would, for circuits up to 100 A (or up to 35 mm^2), tend to produce slightly optimistic results in the assessment of fault current. The formula for the calculation is

$$I_F = \frac{U_o}{Z_{L-N} + R_1 + R_n} \tag{7.17}$$

For circuits above 100 A, the resistance and reactance terms of both the source and the installation circuit should be taken into account.

7.5.8   Line-to-line short-circuits

Line-to-line short-circuit currents are less than three-phase fault currents. Referring to the equivalent circuit in Figure 7.7, the general formula for calculating the prospective line-line short-circuit current is given in Equation (7.18). For assessment at the origin of the installation, the simplified formula is given in Equation (7.19).

$$
\begin{aligned}
I_F &= \frac{\sqrt{3}U_o}{(R_{L1} + R_{L2} + R_{1(L1)} + R_{1(L2)}) + j(X_{S1} + X_{S2} + X_{L1} + X_{L2} + X_{1(L1)} + X_{1(L2)})} \\
&= \frac{\sqrt{3}U_o}{\sqrt{(R_{L1} + R_{L2} + R_{1(L1)} + R_{1(L2)})^2 + (X_{S1} + X_{S2} + X_{L1} + X_{L2} + X_{1(L1)} + X_{1(L2)})^2}}
\end{aligned}
\tag{7.18}
$$

$$
\begin{aligned}
I_F &= \frac{\sqrt{3}U_o}{\sqrt{(R_{L1} + R_{L2})^2 + (X_{S1} + X_{S2} + X_{L1} + X_{L2})^2}} \\
&= \frac{\sqrt{3}U_o}{Z_{L-L}}
\end{aligned}
\tag{7.19}
$$

where Z_{L-L} is the external line-to-line impedance.

Where circuit conductors are not greater than 35 mm^2 the reactance terms are largely insignificant and may be neglected, simplifying the formula further, as given in Equation (7.20). If the installation circuit line conductors are of the same resistance, as is usually the case, then a further simplification is possible, as shown in Equation (7.21), giving a

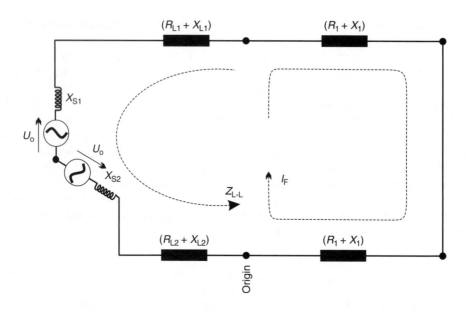

Figure 7.7   Line-to-line fault equivalent circuit

somewhat optimistic assessment of fault current, since the addition is arithmetic and not vectorial.

$$I_F = \frac{\sqrt{3}U_o}{Z_{L-L} + R_{1(L1)} + R_{1(L2)}} \tag{7.20}$$

$$I_F = \frac{\sqrt{3}U_o}{Z_{L-L} + 2R_1} \tag{7.21}$$

7.5.9   Three-phase short-circuit

Referring to Figure 7.8, the general formula for three-phase fault current is given in Equation (7.22) and for faults at the origin of an installation Equation (7.23) modifies the formula:

$$
\begin{aligned}
I_F &= \frac{U_o}{(R_{L1} + R_1) + j(X_{S1} + X_{L1} + X_1)} \\
&= \frac{U_o}{\sqrt{(R_{L1} + R_1)^2 + (X_{S1} + X_{L1} + X_1)^2}}
\end{aligned} \tag{7.22}
$$

$$I_F = \frac{U_o}{R_{L1} + j(X_{S1} + X_{L1})} = \frac{U_o}{\sqrt{R_{L1}^2 + X_{L1}^2}} \tag{7.23}$$

For supplies up to 100 A, where the inductive reactances are negligible, a further simplification to Equation (7.22) may be made, as given in Equation (7.24), again giving

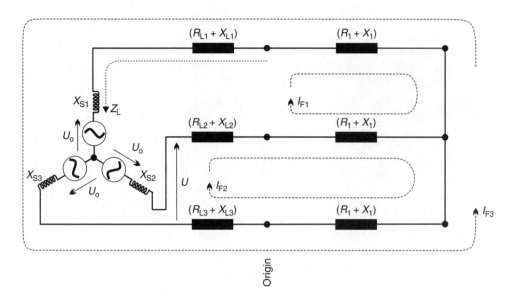

Figure 7.8 Three-phase fault equivalent circuit

a somewhat optimistic assessment of fault current magnitude:

$$I_F = \frac{U_o}{\sqrt{(R_{L1} + R_1)^2 + (X_{S1} + X_{L1})^2}} = \frac{U_o}{Z_L + R_1} \tag{7.24}$$

As a matter of interest, the relationship between line–earth voltage U_o and the line–line voltage U is shown in Figure 7.9. By way of explanation of the figure, the three phases are equi-separated by 120° (i.e. $3 \times 120° = 360°$). Looking to the bisected triangle CNE, the angle $\theta = 60°$ and the length E–B is given by $\sin \theta$, which is $\sin 60°$ (which equals $\sqrt{3}/2$). If length N–B represents the line-earth voltage U_o and length C–B depicts the line-to-line voltage U, then it can be seen that U is portrayed by length C–B (twice the length E–B), which equals $\sqrt{3}U_o$.

7.5.10 Line-to-earth faults

Evaluation of line–earth loop impedance is essential not only because of the consideration of fault protection (touch voltage magnitude and disconnection times), but also from the standpoint of selection of line conductor(s) in terms of energy-withstand of equipment and the assessment of required short-circuit capacities of associated protective devices.

Prospective earth fault currents in TN systems (TN-C, TN-C-S and TN-S) are generally much higher than those in TT, principally because the former systems employ a metallic link between the installation and the source, whereas part of the loop in the TT system consists of the soil and associated electrodes.

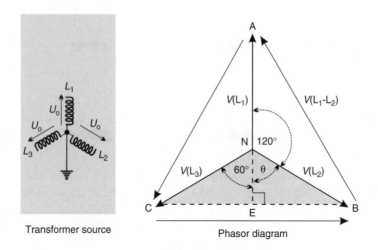

Figure 7.9 Relationship between line–neutral and line–line voltages

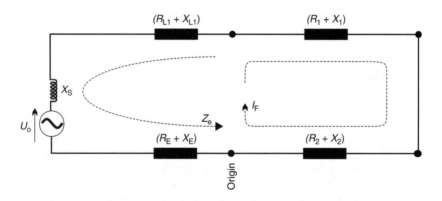

Figure 7.10 Line-to-earth fault equivalent circuit

The general formulae for calculating prospective earth fault current are given in Equations (7.25) and (7.26), and Equation (7.27) should be used when considering fault current at the origin of the installation. Figure 7.10 shows a line-to-earth fault equivalent circuit.

$$I_F = \frac{U_o}{(R_{L1} + R_E + R_1 + R_2) = j(X_s + X_E + X_1 + X_2)} \tag{7.25}$$

$$I_F = \frac{U_o}{\sqrt{(R_{L1} + R_E + R_1 + R_2)^2 + (X_s + X_E + X_1 + X_2)^2}} \tag{7.26}$$

$$I_F = \frac{U_o}{\sqrt{(R_{L1} + R_E)^2 + X_s^2}} = \frac{U_o}{Z_e} \tag{7.27}$$

With circuits up to 100 A (or cross-sectional area $\leq$35 mm^2) the reactance of the conductors of that circuit is small and, for all practical purposes, may be ignored. The formula set out in Equation (7.25) can thus be simplified to that given in Equation (7.28). The assumption that these reactances may be neglected is restricted to where the line and protective conductors are run together in close proximity; this assumption cannot be applied where this is not so or where the two conductors are separated by ferromagnetic material, such as steel conduit or steel trunking.

$$I_F = \frac{U_o}{\sqrt{(R_{L1} + R_E + R_1 + R_2)^2 + X_s^2}} \tag{7.28}$$

Where supplies are derived from the public LV supply network, the external line–earth loop impedance may be obtained on request from the ED. For other supplies, including generator supplies, reference to the ED and source equipment manufacturer will be necessary (see also Section 7.5.11 of this Guide).

7.5.11   Fault current at the origin of an installation

Irrespective of whether the supply is to be obtained from the public network or private generating plant, the designer will need to know the fault level at the origin of the installation as required by Regulation 313.1. In the case of public network supplies, the Energy Networks Association has published data for prospective fault current in their *Engineering Recommendations P25/1* for the smaller installations of 100 A single-phase and for *P26* for three-phase services.

For the 100 A single-phase public network supply, a prospective fault current of 16 kA is normally quoted by the ED, but this is often referring to the 'tee off' point in the road or pavement, and the linking supply cable to the installation will often reduce this to a much lower value. The reduction in fault level will depend on the cross-sectional area and length of the service cable, which is shown graphically in Figure 7.11. Providing the overcurrent protective devices near the origin are capable of coping with 16 kA, there should be no further need to consider the attenuation of the fault level in the linking supply cable. Consumer units complying with BS EN 60439-3 incorporating Annex ZA of June 2006 have a conditional rating and may be used on supplies with a fault level not exceeding 16 kA without difficulty. In areas of high population density, consideration and application of the attenuation of the supply service cable should be avoided owing to the continual changes to the network. For three-phase systems, higher fault levels can be anticipated. It is normally considered necessary to consult the ED, at an early stage of design, to obtain values of fault level and external loop impedance.

Where supplies are taken at HV and LV supplies are derived from a privately owned transformer, the fault level and impedances should be obtained from the HV ED and the transformer manufacturer. These fault levels are often expressed in megavolt-amperes (MVA) and the derivation of prospective three-phase short-circuit current and ohmic impedance values is given in Equations (7.29) to (7.31):

$$\text{Fault level} = \frac{3V_L I_F}{10^6} = \frac{\sqrt{3}V_{L-L} I_F}{10^6} \tag{7.29}$$

$$I_F = \frac{\text{Fault level (MVA)} \times 10^6}{3V_L} \qquad (7.30)$$

$$Z_L = \frac{3V_L^2}{\text{Fault level (MVA)} \times 10^6} \qquad (7.31)$$

where Fault level (MVA) is the fault level in megavolt-amperes, V_L (V) is the rated line-to-neutral (or line-to-earth) voltage, V_{L-L} (V) is the rated line-to-line voltage, I_F (A) is the prospective three-phase short-circuit current and $Z_L(\Omega)$ is the ohmic impedance from the star point of the source to the point of the fault. (For example, for a fault on the secondary side of the transformer, Z_L would include the line impedance of the HV source and HV network, the impedance of the transformer and the line impedance of the circuit on the secondary side of the transformer up to the point of fault – all referred to the transformer secondary side.)

Transformer impedances are normally expressed in per-unit or percentage terms, which, when used with fault levels (MVA), makes life easier, and avoids complications, when working with different voltages, as would be the case on the primary and secondary sides of a transformer. The impedance in per-unit terms is

$$Z_{pu} = \frac{Z_L \times \text{Transformer rating}}{V_L^2} \qquad (7.32)$$

In carrying out such calculations, the transformer manufacturer should always be consulted with regard to fault level and impedances, but Table 7.6 gives typical values for per-unit impedances for some small 11/0.400 kV delta–star ('Dy 11') transformers.

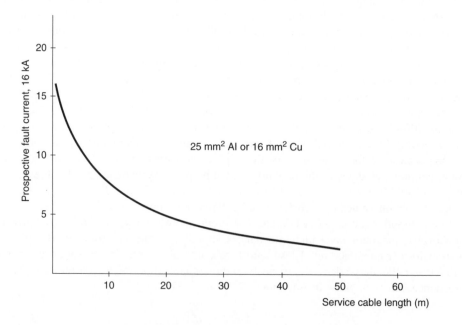

Figure 7.11   An example of attenuation of prospective fault current in service cable

Table 7.6 Typical per-unit impedance values for 11/0.400 kV delta–star transformer

Power rating (kVA)	Impedance[a]	
	p.u.	%
100–250	0.042	4.2
251–500	0.044	4.4
501–750	0.047	4.7
751–1000	0.050	5.0
1000–1500	0.053	5.3
1501–2000	0.056	5.6

[a]Transformer impedances are predominantly inductive reactances (X) and in most cases the resistive component may be ignored.

The fault level of generators is again normally expressed in MVA and the impedance declared in per-unit or percentage terms. To illustrate the process by example, take a three-phase 15 MW, 0.8 lagging power factor, operating at 11 kV and with a line impedance of 1.59 Ω. The line-to-Earth voltage V_L is given by Equation (7.33) and the three-phase short-circuit current I_F by Equation (7.34):

$$V_L = \frac{V_{L-L}}{\sqrt{3}} = \frac{11000}{\sqrt{3}} = 6350 \text{ V} \tag{7.33}$$

$$I_F = \frac{V_L}{Z_L} = \frac{6350}{1.59} = 3994 \text{ A} \tag{7.34}$$

The fault level of this generator, the apparent power rating S and the per-unit (and percentage) impedance Z_{pu} are as stated in Equations (7.35), (7.36) and (7.37) respectively. The active power P referred to in Equation (7.36) is the rated power of the generator (i.e. 15 MW).

Fault level $= 3V_L I_F$

$$= 3 \times 6350 \times 3994 = 76.1 \times 10^6 \text{ VA} = 76.1 \text{ MVA} \tag{7.35}$$

$$S = \frac{P}{\text{Power factor}} = \frac{15 \times 10^6}{0.8} = 18.75 \times 10^6 \text{ VA} = 18.75 \text{ MVA} \tag{7.36}$$

$$Z_{pu} = \frac{\text{Power rating}}{3V_L^2} \times Z_L$$
$$= \frac{18.75 \times 10^6}{3 \times 6350^2} \times 1.59 = 0.246 \text{ (p.u.)} \tag{7.37}$$

As previously mentioned, the use of per-unit (p.u.) terms is particularly advantageous in calculating fault levels in systems where different voltages are present, as would be the case where a system employs 33 kV, 11 kV and 400 V voltages. It is beyond the scope of this Guide to address this topic fully in the way it deserves, other than to draw attention to some of the considerations. A more rigorous analysis, involving a positive, negative and zero sequence current study, may be required to be undertaken in all but the very small systems. Where more than one source can supply a fault, all such sources need consideration in the fault calculations, as would the short-circuit transient fault currents of generators. It should not be forgotten that a motor that is running prior to a fault may also contribute significantly to the fault level (release of stored energy). The designer should consult BS EN 60909: *Short-circuit current calculation in three-phase systems* for further guidance.

7.6   Characteristics of protective devices

In every case, except as stated later, Regulation 434.5.1 requires that the rated short-circuit breaking capacity of each fault current protective device is not less than the maximum prospective fault current at the point where the device is installed. Fault current, as mentioned earlier, embraces both short-circuit and earth fault currents. Whilst the currents are normally of the same order of magnitude, it is worth remembering that earth fault currents can in some cases be greater than short-circuit current at the same point in the circuit. Take, for example, the case of the remotely located switchboard fed with mineral-insulated copper-sheathed distribution (or, if you prefer, sub-mains) cables. The cpc (cable copper sheath) is of substantially lower impedance than that of the associated live conductors, resulting in a higher earth fault current than a short-circuit current at the same point in the circuit. There are other cases where, for different reasons, the earth fault current may exceed the short-circuit current by a considerable amount, and this cannot be ignored.

The rated short-circuit breaking capacity of protective devices is addressed in the appropriate British or Harmonized Standard, and this feature is normally distinctively marked on each device. In some cases this is given in coded form using letters, as shown for some of the devices in Table 11.5 of Chapter 11 of this Guide.

The exception to the rule (Regulation 434.5.1) is that a protective device may have a lower rated short-circuit breaking capacity than the maximum prospective fault current at its point of installation provided that there is another fault current protective device (or devices) on the supply side that has the necessary rated short-circuit breaking capacity. In these circumstances, the characteristics of the devices must be coordinated so that the let-through energy of the devices does not exceed the capabilities of the device in question. In other words, the device in question is capable of withstanding, without damage, the let-through energy of the upstream device.

The general requirement for protection against fault current, given in Regulation 434.5.2, is that disconnection time under fault condition must not be greater than the time required to raise the conductor to its limiting temperature. The disconnection time t is limited, therefore, to that given in the adiabatic equation and repeated in Equation (7.38):

$$t = \frac{k^2 S^2}{I_F^2} \qquad\qquad (7.38)$$

where t is the time in which the temperature of the live conductors will be raised to the maximum limiting temperature, S (mm^2) is the conductor cross-section area, k is the particular factor for the conductor (from Table 43.1 for common materials) and I_F (A) is the fault current (rms for a.c.).

It will be seen that Equation (7.39) is obtained by transposing the elements of Equation (7.38); on one side is the energy-withstand of the cable k^2S^2 and on the other side is the let-through energy $I_F^2 t$ of the protective device:

$$k^2 S^2 = I_F^2 t \tag{7.39}$$

For very short disconnection times (less than 0.1 s) for current-limiting protective devices, Regulation 434.5.2 requires that the energy-withstand of the cable k^2S^2 is greater than the let-through energy $I^2 t$ of the protective device as quoted in the applicable British or Harmonized Standard or by the manufacturer. In other words, Equation (7.40) must be satisfied:

$$I^2 t < k^2 S^2 \tag{7.40}$$

The 'quoted' let-through energy $I^2 t$ in Equation (7.40) may exceed the product of the rms fault current squared I_F^2 and the short disconnection time t. This is because a short disconnection time can be associated with a fault close to a generator or transformer. In such an inductive circuit, the instantaneous fault current is likely to be asymmetrical if the fault occurs at a time when the instantaneous voltage is near a point of zero magnitude in its cycle.

Where a device is intended to provide protection against fault current and also meets the requirements for overload protection (Regulation 435.1), and it has a rated short-circuit breaking capacity not less than the maximum prospective fault current at its point of installation, then it may be assumed that the downstream live conductors are protected against fault current for radial circuits. For conductors in parallel and for noncurrent-limiting circuit-breakers, this assumption should be checked (Regulation 435.1) to confirm its validity by the use of Equation (7.39) or, for very short disconnection times, Equation (7.40).

It is also wise to check the validity of the assumption for final circuits where an RCD that is providing additional protection against electric shock in accordance with Regulation 415.1.1 is also relied on for fault protection in accordance with Regulation 411.3.2.1. In such circumstances, protection against fault current (Equation (7.39) or (7.40)) may be the limiting factor on the circuit length.

Table 43.1 of BS 7671 gives two values for k in some instances. The lower of the two should be used for cables of cross-sectional area exceeding 300 mm^2. Knowing the k value and the cross-sectional area S of a particular conductor, it is fairly straightforward to plot the 'adiabatic line' on to the time/current characteristics for the more common overcurrent protective devices given in Appendix 3 of BS 7671. The procedure involves only the identification of two points and the connection of those two points with a straight line. The first step is to convert Equation (7.40) to the $\log_{10}$ scale to match the log–log scale given in the Appendix 3 time/current characteristics, as shown in Equation (7.41). Further rearrangement, given in Equation (7.42), makes the current I the subject, and this enables further calculations to be made simpler:

$$2 \log I + \log t = 2(\log k + \log S) \tag{7.41}$$

$$\log I = \log k + \log S - 0.5 \log t \tag{7.42}$$

Both k and S are constants for a particular conductor, so it can be seen from Equation (7.42) that the log–log representation of $k^2 S^2$ is a straight line with a gradient of -0.5; hence, it is possible to plot the adiabatic line knowing only one value. However, the two points will be calculated to reinforce the method. From the foregoing equations we obtain

$$I = \frac{kS}{t^{0.5}} \tag{7.43}$$

For a disconnection time of 0.01 s (10^{-2} s), Equation (7.43) yields $I = 10\ kS$. For a disconnection time t of 1 s (10^0), $I = kS$. To illustrate by example, take a 70 °C copper–PVC conductor of cross-sectional area of 25 mm^2 and with a k value (from Table 43.1 of BS 7671) of 115. At $t = 10^{-2}$ s, $I = 10(115 \times 25) = 28750$ A and at $t = 10^0$, $I = 115 \times 25 = 2875$ A, which gives the two points for plotting. It is worth noting that increasing the time from 0.01 s to 1 s (by a factor of 100) has the result of reducing the current that can be tolerated by a factor of 10. This, when related to the log–log scale, is a gradient of -0.5, as mentioned previously.

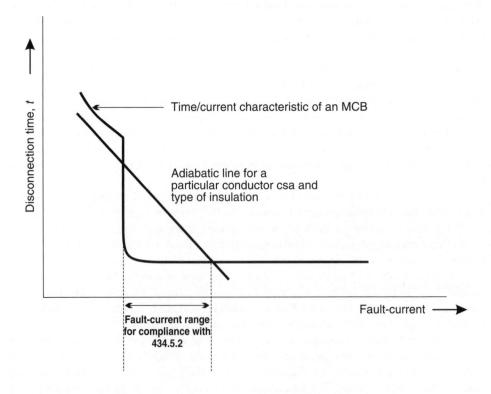

Figure 7.12 Range of fault current restriction of MCBs

It is important to recognize that, in the case of MCBs, the adiabatic line intersects the time/current characteristics in two places, as shown in Figure 7.12. This implies that, for compliance with Regulation 434.5.2, only a certain range of prospective fault current can be tolerated, as indicated in the figure.

It should be noted that the time/current characteristics in Appendix 3 of BS 7671, unlike that in Figure 7.12, are 'cropped' at a time of 0.1 s. This reflects the requirement that, for disconnection times shorter than 0.1 s, the value of let-through energy for use in fault current protection calculations should be obtained from the applicable British or Harmonized Standard or from the manufacturer, as mentioned above in relation to Equation (7.40).

7.7   Overcurrent protection of conductors in parallel

Regulation 433.4.1 stipulates that, where parallel conductors are installed and these share the current equally and are protected by a single device, the effective I_z is the sum of the current-carrying capacities of the paralleled conductors.

Regulation 433.4 requires that, except for ring final circuits, there must not be any branch circuits or devices for isolation and switching in any of the parallel conductors when protected by a single overcurrent protective device.

The situation where the currents in parallel conductors is unequal is covered in Regulation 433.4.2, which then requires that the requirements for overload protection of each of the parallel conductors must be considered separately.

Unequal sharing of the current between parallel conductors is likely where more than two conductors are run in parallel. In the absence of more precise information, a 'rule of thumb' is to allow only, say, 75% of the individual current-carrying capacity of three or more conductors in parallel to contribute to I_z. For example, a four-conductor line could be regarded as having a total I_z of three times the current-carrying capacity of the individual conductors. The assumption of equal load current sharing may not be valid for some cabling arrangements and configurations. IEE Guidance Note No 6: *Protection against overcurrent* addresses the issue of where current equality or otherwise may be expected. When considering current-carrying capacity and the correction factors for grouping of a paralleled conductor circuit, each conductor should be regarded as a 'circuit'.

Even when the above requirements are met, further consideration is required in addressing all the locations where a fault may occur and the resultant fault current sharing of conductors. One possible solution to any difficult cases may be to employ overcurrent protection in each 'leg' arranged to disconnect both 'legs' under fault conditions.

Figure 7.13 shows the line and cpcs in parallel configuration and Figure 7.14 illustrates the equivalent circuit for the arrangement. The symbols used in these figures and the following equations have the meanings attributed below:

U_0	the nominal voltage to Earth (V)
Z_a	the impedance per unit length of each conductor in the line 'circuit' (Ω m^{-1})
Z_b	the impedance per unit length of each conductor in the protective conductor 'circuit' (Ω m^{-1})
Z_A	the effective impedance in the line 'circuit' up to the fault (Ω)

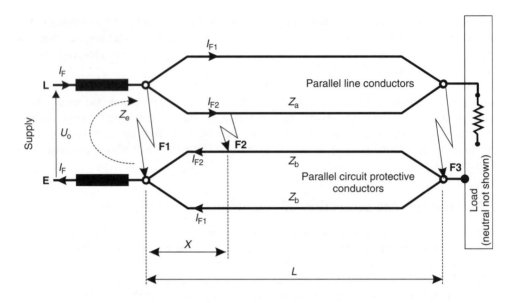

Figure 7.13 Earth faults in parallel conductors

Z_B	the effective impedance in the protective conductor 'circuit' up to the fault (Ω)
Z_e	the line–earth loop impedance upstream of the origin of the circuit (Ω)
I_F	the total earth fault current (A)
I_{F1}	the earth fault current in one leg (the longer path to the earth fault) of the parallel circuit conductors (A)
I_{F2}	the earth fault current in one leg (the shorter path to the earth fault) of the parallel circuit conductors (A)
L	overall length of parallel circuit (m)
x	distance from origin of circuit to location of earth fault (m)
F1, F2, F3	three positions where an earth fault (between one line conductor 'leg' and one cpc 'leg') is considered.

It can be shown that the contribution Z_A made by the parallel line conductors to the total line–earth loop impedance for an earth fault developing anywhere in the circuit is

$$Z_A = Z_a \left(\frac{2Lx - x^2}{2L} \right) \tag{7.44}$$

A similar equation can be obtained for Z_B, and Equation (7.45) is an expression for the total effective line–earth loop impedance Z_{eff} up to the point of the earth fault:

$$Z_{eff} = Z_e + \left[(Z_a + Z_b) \left(\frac{2Lx - x^2}{2L} \right) \right] \tag{7.45}$$

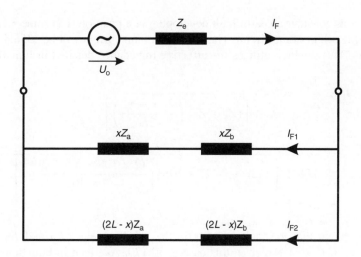

Figure 7.14   Earth faults in parallel conductors: equivalent circuit

For an earth fault at the most remote end of the circuit, shown as F3 in Figure 7.13, Equation (7.45) can be simplified by substituting L for x, as given in Equation (7.46):

$$Z_{\text{eff}} = Z_e + \left[\frac{L(Z_a + Z_b)}{2} \right] \tag{7.46}$$

To illustrate the different magnitudes of fault currents, let us consider a parallel circuit with the following parameters:

$$Z_e = 0.08 \ \Omega, \ Z_a = 0.28 \ \text{m}\Omega \ \text{m}^{-1}, \quad Z_b = 0.41 \ \text{m}\Omega \ \text{m}^{-1}, \quad L = 100 \ \text{m}, \quad U_o = 230 \ \text{V}$$

First, let us consider an earth fault (F1) at the origin of the parallel circuit where $x = 0$. The earth fault current I_F will be 2875 A, as given in Equation (7.47), and the parallel conductor currents I_{F1} and I_{F2} will be zero:

$$I_F = \frac{U_o}{Z_e} = \frac{230}{0.08} = 2875 \ \text{A} \tag{7.47}$$

Let us now consider an earth fault (F3) at the remote end of the parallel circuit where $x = L$. Making use of Equation (7.46) and substituting values we get a total earth fault current I_F of 2009 A, as given in Equation (7.48). The fault currents in the two 'legs', I_{F1} and I_{F2}, will, at 1004.5 A, be equal in magnitude.

$$I_F = \frac{U_o}{Z_{\text{eff}}} = \frac{U_o}{Z_e + \dfrac{L(Z_a + Z_b)}{2}}$$

$$= \frac{230}{0.08 + \dfrac{100(0.28 + 0.41) \times 10^{-3}}{2}} = 2009 \ \text{A} \tag{7.48}$$

Finally, let us consider an earth fault developing at a position (F2) some 30 m (i.e. $x = 30$ m) from the origin of the circuit. The total earth fault current will be 2356.5 A, as given in Equation (7.49) – which utilizes the effective impedance detailed in Equation (7.45):

$$I_F = \frac{U_o}{Z_{eff}} = \frac{230}{Z_e + \left[(Z_a + Z_b)\left(\dfrac{2Lx - x^2}{2L}\right)\right]}$$

$$= \frac{230}{0.08 + \left\{(0.28 + 0.41) \times 10^{-3}\left[\dfrac{(2 \times 100 \times 30) - 30^2}{2 \times 100}\right]\right\}} \tag{7.49}$$

$$= 2356.5 \text{ A}$$

It can be shown that the fault currents in the parallel 'legs', I_{F2} and I_{F1}, will be in the ratio of $x : (2L - x)$, which in this case is 30:170. As given in Equations (7.50) and (7.51) respectively, I_{F1} is 353.5 A and I_{F2} is 2003.0 A:

$$I_{F1} = I_F\left(\frac{x}{2L}\right) = 2356.5\left(\frac{30}{200}\right) = 353.5 \text{ A} \tag{7.50}$$

$$I_{F2} = I_F\left(\frac{2L + x}{2L}\right) = 2356.5\left(\frac{200 - 30}{200}\right) = 2003.0 \text{ A} \tag{7.51}$$

The foregoing example illustrates the inequality of earth fault currents in the two legs of the parallel circuit other than at the remote end. As x goes to zero, I_{F2} will tend to I_F (2875 A); this fault condition, when considered on no-load, will represent the maximum earth fault current which can flow in each 'leg'. The minimum earth fault current will occur when an earth fault occurs at the end under load conditions. Appropriate adjustments for resistances relating to the different temperatures may be necessary (see Section 7.5.6 of this Guide).

7.8   Coordination of overload and fault current protection

Regulation 435.2 calls for the let-through energy of a fault current protective device to be such that no damage is sustained by a separate downstream overload protective device. It is necessary, therefore, to consider the electrical characteristics of each protective device to establish that no damage will occur to downstream equipment when under fault conditions.

The concept of let-through energy is often best understood when presented in graphical form as in Figure 11.5 of this Guide. In most cases, coordination will involve only ensuring that the energy-withstand of conductors k^2S^2 is not less than the let-through energy I^2t of the overcurrent device protecting the circuit. For motor circuits, coordination may be provided in accordance with BS EN 60947-4-1 *Specification for motor starters for voltages up to and including 1000 V a.c. and 1200 V d.c.* (see also Section 7.3.2 of this Guide).

7.9   Protection according to the nature of circuits and distribution systems

The requirements of this section of BS 7671 (431.1) are categorized in terms of line conductors and neutral conductors as they relate to the different system types (e.g. TT, TN-S, TN-C-S, etc.).

Except where Regulation 431.1.2 applies (mentioned below), Regulation 431.1.1 calls for overcurrent detection to be provided for each line conductor, with the detector being arranged to disconnect that line under overcurrent conditions. It is not required to disconnect the other live conductors (other line conductors and/or the neutral conductor) of the circuit unless, by not doing so, danger would be caused. This would be the case where, for example, the circuit related to a three-phase load such as a motor and here disconnection of all the three phases would need to be employed or other appropriate precautions would need to be taken to remove the potential for danger.

Where the installation forms part of a TN system or a TT system (Regulation 431.1.2) and a circuit is supplied between line conductors, and the neutral conductor is not distributed, overcurrent detection need not be provided for one of the line conductors provided the two following conditions are met:

- the circuit is provided with differential protection on the supply side which would cause disconnection, in the event of unbalanced loads, of *all* line conductors;
- the neutral conductor is not distributed from an artificial point on the load side of the differential protection.

The requirements for neutral conductors of TN systems and TT systems are addressed by Regulation 431.2.1. This regulation requires that, in situations where the neutral conductor is of at least equivalent cross-sectional area to that of the associated line conductor(s), overcurrent protection (detection and disconnection) is not necessary for the neutral conductor. In cases where the cross-sectional area of the neutral conductor is less than that of the associated line conductor(s), overcurrent detection is required for the neutral conductor, appropriate to the cross-sectional area of that conductor, as is a means of disconnection of the associated line conductors (but not necessarily the neutral conductor). In either case, if the current in the neutral conductor is expected to exceed that in the line conductors, then the requirement of Regulation 431.2.3, relating to harmonics, applies. This regulation calls for overcurrent detection to be provided for the neutral conductor of a multiphase circuit, if, due to harmonic content in line conductor currents, the current in the neutral conductor is expected to exceed the current-carrying capacity of that conductor. The detection is required to cause the disconnection of the line conductors, but not necessarily the neutral conductor.

The overcurrent detection requirements just described for the neutral conductor of a TN or TT system also apply to a PEN conductor, except that a means of automatic disconnection must not be provided for a PEN conductor, because of its protective function (Regulation 431.2.1).

In the case of an IT system, Regulation 431.2.2 requires that the neutral conductor shall not be distributed except where one of the following requirements (1), (2) or (3) is met:

1. Overcurrent detection is provided for the neutral conductor of every circuit. The detection must cause the disconnection of all the live conductors of the corresponding circuit, including the neutral conductor.
2. The neutral conductor is effectively protected against fault current by a protective device installed on the supply side, such as at the origin of the installation, in accordance with Regulation 434.5.
3. The circuit is protected by an RCD with a rated residual operating current $I_{\Delta n}$ not exceeding 0.2 times the current-carrying capacity of the corresponding neutral conductor. The RCD must disconnect all the live conductors of the circuit, including the neutral conductor, and must have adequate breaking capacity in all poles.

7.10   Protection against undervoltage

Undervoltage or the loss of voltage (loss of supply) can be a potential danger. Equipment having had no precautions taken against this phenomenon may be damaged by, for example, the drawing of excess current, as in the case of a three-phase motor drawing current from only two phases. Additionally, there is the real danger of equipment starting unexpectedly on restoration of supply or improved voltage. Section 445 of BS 7671 deals with these dangers and stipulates precautions necessary to prevent them.

Regulation 445.1.1 makes clear that, where a motor circuit is concerned, Regulation 552.1.3 calls for every motor to be equipped with a means to prevent automatic restarting on restoration of supply or restoration of full voltage after a period in which voltage has been reduced or lost. This applies only to situations where unexpected restarting of motors could cause danger. It would not necessarily apply to a number of motors being started sequentially (to minimize the effects of simultaneous starting currents on maximum demand metering), provided suitable precautions had been taken to obviate any associated risks.

As also stated in Regulation 445.1.1, if current-using equipment is likely to be damaged by a reduction in, or loss of, voltage, then either suitable precautions need to be taken to obviate the risk *or* confirmation obtained from the person responsible for the operation and maintenance of the equipment that the likely damage is an acceptable risk. Acceptable risks would not include those associated with danger to persons or livestock from shock, fire, burns or injury from mechanical movement.

As Regulation 445.1.2 confirms, the device for protection against undervoltage may incorporate a time delay (time delay from sensing undervoltage to the time when device operates) provided the equipment so protected is not likely to be damaged during the period of the delay. Regulation 445.1.3 makes further demands in that any time-delay mechanism should not adversely affect the operation of any control or protective device (e.g. emergency switch or overcurrent device). Furthermore, as Regulation 445.1.5 requires, an undervoltage protective device must not automatically reclose unless such reclosure is unlikely to cause danger. Finally, all such devices, as called for in Regulation 445.1.4, are required to be suitable for the equipment they protect; therefore, reference will be needed to the appropriate BS or CENELEC standard to ascertain the specific requirements in this respect.

7.11 Protection against overvoltage

7.11.1 General

Sections 442 and 443 of BS 7671 include requirements for protection of LV installation against overvoltages from two distinct types of cause:

- earth faults in the HV system or faults in the LV system (Section 442);
- lightning strikes on the supply distribution system (overvoltages of atmospheric origin), or due to switching equipment forming part of the LV installation (Section 443).

7.11.2 Temporary overvoltages due to earth faults in the high-voltage system or faults in the low-voltage system

When a fault occurs between the HV (typically 11 kV) system and Earth in a transformer substation supplying an LV installation, the following two types of overvoltage may affect the installation for the duration of the fault.

- a power frequency fault voltage U_f between exposed-conductive-parts and Earth;
- power frequency stress voltages (U_1 and U_2) between the line conductor and the exposed-conductive-parts of equipment, in the substation and the installation respectively.

The maximum permitted magnitude and duration of U_f and the maximum permitted magnitudes and durations of U_1 and U_2 are given in Regulations 442.2.1 and 442.2.2 respectively. However, to predict the actual magnitude and durations of U_1 and U_2 in order to check compliance with these regulations requires the use of certain detailed information about the HV and LV systems, as explained in Regulation 442.2.

Fortunately, as stated in Regulation 442.2.3, the requirements of Regulations 442.2.1 and 442.2.2 are deemed to be met if the installation receives its supply at LV from a system for distribution of electricity to the public, in which case the designer of the LV installation does not need to predict the actual magnitudes and durations of U_f, U_1 and U_2. This deemed-to-comply status assumes that the public electricity supply distribution system is appropriately designed and constructed, as is the case in Great Britain, where EDs meet the relevant requirements of the ESQCR by applying the Distribution Code. For an installation that receives its LV supply from a privately owned distribution transformer supplied at HV by an ED in accordance with the ESQCR, it is reasonable to extend the deemed-to-comply status conferred by Regulation 442.2.3 to that installation, provided that the installation designer ensures that the requirements of the Distribution Code are complied with in the design and construction of the privately owned substation.

Regulations 442.3, 442.4 and 442.5 require consideration to be given to the stress voltages that would occur in an installation in the event of loss of the neutral conductor in a TN or TT system, an earth fault in an IT system with distributed neutral, or a short-circuit between a line conductor and a neutral conductor. In practice, there is usually little that the installation designer and constructor can do to meet the requirements of these regulations beyond selecting equipment with appropriate insulation voltage ratings, such as 600/1000 V cables for an installation of nominal voltage 230/400 V.

7.11.3   Overvoltages of atmospheric origin or due to switching

Where an installation is supplied by or includes LV overhead lines, and the anticipated thunderstorm activity in the area exceeds 25 thunderstorm days per year, Regulation 443.2.3 requires additional protection to be provided against overvoltages caused by lightning strikes, such as by means of surge protective devices.

Fortunately, as indicated in Regulation 443.2.2, such additional protection is generally considered to be unnecessary in the UK, where the number of thunderstorm days per year is between five and ten. However, the omission of additional protection is conditional on the impulse-voltage-withstand of equipment in the installation being in accordance with Table 44.3 of BS 7671.

Table 44.3 sets out the required minimum impulse withstand voltage (in kilovolts) for various categories of equipment. Examples of equipment within each category are given in Table 44.4 of BS 7671. If such equipment conforms with the relevant BS EN product standards, then the impulse-withstand voltage requirements can be assumed to have been met. In the UK, where products conforming to the relevant BS EN product standards are used, compliance with Section 443 can generally be assumed.

As an alternative to the use of the thunderstorm days per year criteria in Regulations 443.2.2 and 443.2.3, Regulation 443.2.4 gives a risk assessment method that may be used to assess the need for surge protection.

8

Isolation and switching

8.1 General

It is worth noting that the isolation and switching arrangements make a major contribution to the overall safety of an installation. *The Electricity at Work Regulations 1989* (Regulation 12) calls for the function of isolation to be a secure operation. Whilst the installation designer and constructor cannot be held responsible for the use of the installation, it is incumbent on them both to confirm that satisfactory provision is made for installed isolating devices so that they are accessible and, where necessary, securable. Isolation and switching cannot be properly provided for without first addressing the nature of the installation, its use and its operational procedural requirements.

The phrase 'isolation and switching' used throughout BS 7671 and elsewhere has significance and it is essential to appreciate that an isolating device may be something other than a switching device (e.g. a plug and socket). Hence the use of this phrase to embrace all the aspects which would not be covered by the term 'switching'. Isolation and switching embraces four distinct concepts each with its own particular requirements for safety, namely: (i) isolation, (ii) switching off for mechanical maintenance, (iii) emergency switching and (iv) functional switching. Figure 8.1 sets out these measures and identifies their purpose and the personnel who are intended to operate them.

Means of isolation is required for every installation whereas switching off for mechanical maintenance and emergency switching and emergency stopping will only be required where the type of installation and its usage demand such provision. The designer should consult, at an early stage, with the installation user regarding the intended utilization of the installation and the level of skills, capabilities and competency of the persons involved in its usage. Other considerations needed to be addressed at an early stage include the assessment of all the external influences (e.g. water, dust, mechanical vibration and impact). This may well affect the designer's selection of types of devices and their positioning in order to expedite ready access for operation, inspection and testing, and maintenance of devices, as required by Regulation 513.1.

In many instances a number of items for isolation and switching will be located together in an allocated switchroom. In this case the designer should provide adequate space in

---

A Practical Guide to The Wiring Regulations: 17th Edition IEE Wiring Regulations (BS 7671:2008)
Fourth Edition Geoffrey Stokes and John Bradley
© Geoffrey Stokes and John Bradley. Published by John Wiley & Sons, Ltd

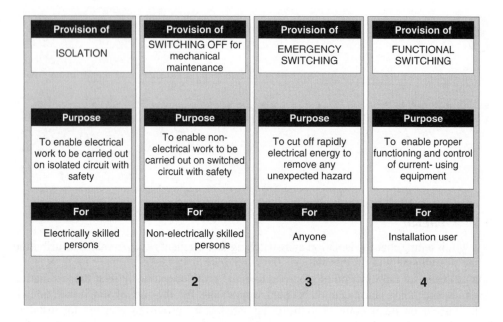

Figure 8.1   Isolation and switching: the four concepts

the switchroom for electrically skilled persons to work on the equipment (switchboards and the like) without impediment and in safety. The installation owner/user should be made aware that switchrooms are not provided for the storage of paint, cardboard boxes, brooms and the like and should be kept free of such material at all times. When siting switchgear (e.g. distribution boards) the designer must confirm that access can be gained for maintenance without difficulty and that adequate working space is provided.

It is important to note that a switch is essentially an *on-load* mechanical device capable of making, carrying and breaking load current. It may also be capable of carrying, within specified limits (current magnitude and duration), overload current and in some cases it may be capable of making (but not breaking) fault current. Significantly, semiconductor devices (e.g. thyristors) are not considered to be switches as far as BS 7671 is concerned though, of course, their use as functional switching and control is not precluded.

An isolator is defined as 'a mechanical switching device which, in the open position, complies with the requirements specified for the isolation function'. An isolator, which is otherwise known as a disconnector, can be, and often is, an off-load device and as such access to it must be restricted to electrically skilled persons. An isolator provides for a specified contact clearance distance (allowing for creepage, etc.) when the device is in the 'open' or 'OFF' position (see BS EN 60947-3). Isolating switches (switch disconnecters) complying with BS EN 60947-3 may also serve as isolating devices and, because these devices are capable of switching load current, they may be operated by electrically unskilled personnel for non-isolation purposes. Devices for isolation include:

- isolator (disconnector) – off-load device,
- isolating switch (switch-disconnector) – on-load device,

- circuit-breaker – on-load device,
- fuse-switch – on-load device,
- switchfuse – on-load device,
- fuse – off-load device,
- solid link – off-load device,
- plug and socket-outlet – off-load device.

The use of a plug and socket-outlet for isolation purposes, as permitted by Table 53.2 of BS 7671, is restricted to circumstances where it is located in close proximity to the equipment being isolated and where the electrically skilled person has the plug and/or socket-outlet under their control at all material times.

The purpose of providing isolation is to cut off the supply from the installation, or part of it, for the safety of electrically skilled persons in carrying out modification and/or electrical maintenance to the affected circuit(s). It is not a function of isolation to provide safety to non-electrically skilled personnel engaged in non-electrical maintenance, this being the subject of the provision of devices for 'switching off for mechanical maintenance'.

The fundamental requirements for isolation and switching are addressed in Regulations 132.15.1 and 132.15.2. These regulations call for effective means, located in a readily accessible position, to be provided for the cutting off of all voltage from every installation and, as is necessary to prevent or remove danger, to cut off all voltage from every circuit and from all electrical equipment. The latter regulation calls for every *fixed* electric motor to be provided with an efficient means of switching off, again located in a readily accessible position, to which access may be gained without danger.

Table 8.1 summarizes the relevant Sections and Regulations covering the various aspects of isolation and switching.

The number and siting of isolating devices needs very careful consideration by the designer, who would necessarily take into account the installation usage and operational requirements. A means of isolation is required in the main switch but, on some larger commercial and industrial installations, it may be necessary or desirable to have a main switch for a number of separate installations (e.g. security systems or fire alarms or standby supplies). Where such an arrangement is adopted, identification as to purpose and operational procedures is of crucial importance so that no misunderstanding or ambiguity can exist in the design purpose and limitations of operation of such devices. As Regulation 537.2.1.2 demands, provision needs to be made to obviate the risks associated with the inadvertent or unintentional energization and re-energization of the isolated circuit. The designer should also be mindful of the installation user's responsibility in effecting safe systems of work (e.g. permit to work) in the provision of such devices. Again, the designer's careful attention to the detailed requirements in this respect is essential and the following options should be considered:

- Locking of switchroom which houses the isolating device(s). This would necessitate the person carrying out the work on the isolated circuit holding the only key to the switchroom. This procedure may be inconvenient and, in some circumstances, may lead to danger where, for example, access by others to the switchroom for prompt disconnection may be delayed but, on the other hand, having more than one key presents obvious dangers.

Table 8.1 Summary of regulations (in addition to the fundamental requirements) for isolation and switching

Aspect	Protective measure	Application of protective measure	Protective devices
Isolation and switching generally	Chapter 53 - General	Chapter 53 - General	Chapter 53 - General
	Section 537	Section 537	Table 53.2 Section 537
Isolation	Chapter 53 Regulation Group 537.2	Chapter 53 Regulations 537.2.1.1 to 537.2.1.7	Chapter 53 Table 53.2
			Regulations 537.2.2.1 to 537.2.2.6
Switching off for mechanical maintenance	Chapter 53 Regulation Group 537.3	Chapter 53 Regulations 537.1.1.1 and 537.3.1.2	Chapter 53 Table 53.2
			Regulations 537.3.2.1 to 537.3.2.6
Emergency switching	Chapter 53 Regulation Group 537.4	Chapter 53 Regulations 537.4.1 to 537.4.1.4	Chapter 53 Table 53.2
			Regulations 537.4.2.1 to 537.4.2.8
Functional switching	Chapter 53 Regulation Group 537.5	Chapter 53 Regulations 537.5.1.1 to 537.5.1.4	Chapter 53 Table 53.2
			Regulations 537.5.2.1 to 537.5.2.3

- Padlocking or other means of locking provided for the individual isolating device – sole key available only to person working on isolated circuit(s).
- Appropriate warning notices put on devices only accessible to electrically skilled personnel – notices used on their own are not without risk as there is the possibility of the warning being ignored.
- Locking of switchgear enclosure cover – sole key available only to person working on isolated circuit(s) or a locking system for each person working on a circuit.
- Secure closure of switchgear enclosure cover together with warning notice – access only by means of a tool.

Where fuses and solid-links are used for isolation purposes, special care is needed to ensure that these are not withdrawn or reinstated under load, or indeed, fault conditions. Links removed should be retained by the person working on the isolated circuit and replacements must not be readily available to others who may unwittingly create a serious hazard by re-energization of the circuit being worked upon.

As called for in Regulation 537.2.1.7, provision must be made for disconnecting the neutral conductor from the supply to facilitate testing procedures. Whether this facility is given in the form of a joint or disconnectable link, it must be in a readily accessible position, and be mechanically adequate to reliably maintain continuity throughout the lifetime of the installation.

8.2   Main switch

Regulations 537.1.4 and 537.1.6 call for a main linked switch or linked circuit-breaker to be provided at or near the origin of the installation. The origin is defined, in Part 2 of BS 7671, as 'the position at which electrical energy is delivered to an installation'. In many, if not most, cases the origin of the installation will be at the load-side terminals of the supply meter(s). In practice, it would generally be necessary to install a main switch (or main circuit-breaker) adjacent to the meter(s) (say, within two or three metres). Where installations are supplied by more than one source, these requirements apply equally to each source and, in this case, notices will be required to warn of the dangers in that the installation may have more than one point of main isolation. Alternatively, the main switches (or main circuit-breakers) are to be provided with a suitable interlocking system (electrical and/or mechanical) to obviate the inherent dangers associated with installations which need isolation from more than one source.

The main switch or circuit-breaker must interrupt all the line conductors and, with certain exceptions, the neutral conductor, of the supply. In a TT system or an IT system, the neutral conductor must be interrupted in all cases. In a TN-S or a TN-C-S system, the neutral conductor need not be interrupted if this conductor can be regarded as being reliably connected to Earth through a suitably low impedance (Regulation 537.1.2 refers), and provided the main switch is not intended for operation by ordinary persons on a single-phase supply, such as in a household or similar installation (Regulation 537.1.4 refers). The neutral conductor of a low voltage supply given in accordance with the *Electricity Safety, Quality and Continuity Regulations 2002* can be regarded as being reliably connected with Earth through a suitably low impedance. However, outside of England, Scotland and Wales, confirmation should be sought from the distributor that the supply conforms to the *ESQCR* requirements in this respect (Section 114 of BS 7671 refers).

Table 8.2 summarizes the requirements relating to the main linked switch or linked circuit-breaker. Figure 8.2 illustrates the necessary isolation of supply poles, for a.c systems, relating to the different system types and earthing arrangements.

Regulation 543.3.4 precludes the use of a switching device to break a protective conductor, and Regulation 537.1.2 precludes the isolation or switching of a PEN conductor (see also Regulation 537.5.1.4). There is an exception to this rule: where an installation is supplied by more than source, one of which needs a separate means of earthing independent of the other, then it is permissible, under certain conditions, to break the protective conductor, as indicated in Regulation 537.1.5. This exception, for example, caters for the circumstances of parallel generators feeding an installation. It may be necessary, on occasions, to switch the conductor connecting the neutral point of a source to earth, to prevent winding damage otherwise resulting from excessive (third harmonic) circulating currents. In the case of a generator running in parallel with the public supply network, the ED

Table 8.2 Principal requirements for main switch or main circuit-breaker

Requirement	Regulation
Located at, or near, the origin of the installation	537.1.4
If more than one source, main switch required for each source	537.1.6
If more than one source, interlocking and/or warning notice required	537.1.6
Capable of providing isolation	537.1.4
Capable of switching supply *on load*	537.1.4
Identified as to purpose (appropriately labelled)	514.1.1
Interrupt both live conductors on a single-phase supply (main switch for ordinary persons' use)	537.1.4

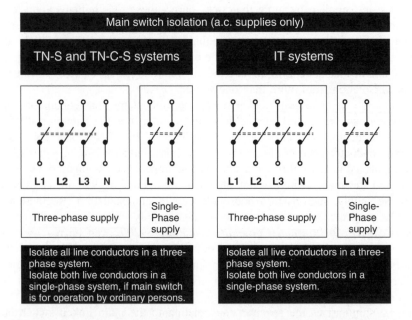

Figure 8.2 Pole isolation requirements for the main switch depending on system type

must be consulted before such measures are implemented. Any such device for switching a protective conductor would also need to interrupt the associated live conductors at substantially the same time.

Although the main switch or main circuit-breaker of an installation provides a means of switching and breaking the supply on load for every circuit of the installation (including final circuits), these requirements will, where required for safety reasons, need to be applied as necessary to parts (or sections) of the installation as well as for individual final circuits.

Although Regulation 537.1.4 requires that the main linked switch or linked circuit-breaker shall be situated as near as practicable to the origin of the installation, it is unlikely to be practicable to position this device actually *at* the origin. Almost inevitably, therefore, there will be a certain length of meter 'tails', however short, between the origin and the supply terminals of the main switch (or main circuit-breaker) that cannot be isolated by this device and for which the only means that could be used for isolation is likely to be the switchgear that the ED generally provides at the intake position, which generally consists of a fused cut-out. This is acceptable provided that the ED has expressly confirmed the use of their equipment for this purpose. However, in any event, a main switch or main circuit-breaker will be required at the downstream end of the meter 'tails'.

Regulation Group 537.3 also calls for electrical switching for mechanical maintenance for certain parts of the installation, and Regulation 537.1.1 and Table 53.2 recognize that all the functions envisaged for isolation and switching may be performed by a common device provided that it meets all the various requirements for all the functions.

Before moving on to the other aspects of isolation and switching, it is worth noting that some requirements are common to all. Whilst it is important not to confuse the requirements, or the different concepts, it should not be overlooked that some devices will meet all the various requirements. It is not uncommon for devices to serve more than one function, as in the case of the main switch, which can act both as a means of isolation and as a means of emergency switching (and possibly, in some cases, functional switching). Similarly, a cord-operated switch can perform as a switch for mechanical maintenance (except two-way and pull-cord switches without contact indication) provided it is under the supervision of the person carrying out the work, as well as its functional role in switching the load. Take, too, the example of the milling machine with its own combined isolator and starter which can serve not only the functional requirements but can also provide for isolation, emergency switching (if equipped with no-volt release as required by Regulation 552.1.3) and meet adequately the needs for switching for mechanical maintenance if adjacent to the equipment it serves. Finally, the latch-OFF-type push button may provide emergency switching and a device for switching off for mechanical maintenance (but see Figure 8.3) and, in some cases, also meet the functional requirements, provided, of course, that all the requirements are met for each function. Machine limit switches relate only to functional switching and should not be regarded as meeting the requirements for other aspects of isolation and switching.

8.3   Isolation

The purpose of providing isolation is so that live parts (normally energized) can be safely isolated (separated) from their source(s) so that work may be carried out on, or near, those parts without the risk of electric shock to electrically skilled persons. It should be noted that isolation is an *off-load* function.

Regulation 537.2.1.1 requires that every circuit must be capable of being isolated, though in achieving this objective it is permissible to provide isolation to a group of circuits providing due consideration is taken of the operational requirements and service conditions. Clearly, to provide group isolation to circuits, many of which may need to be in constant use, is going to be highly inconvenient to the operator and is unlikely to satisfy the requirements of BS 7671. One undesirable consequence of grouped-circuit isolation is the temptation for work to be carried out with the circuit still live, and for

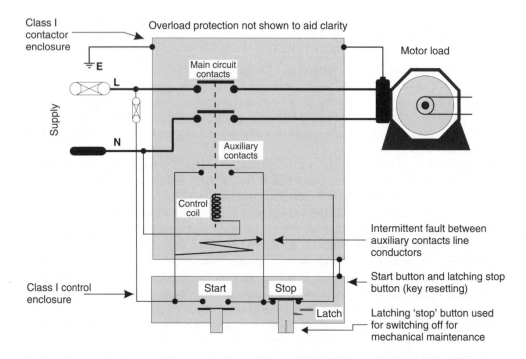

Figure 8.3 An example of an unsatisfactory arrangement for switching for mechanical maintenance

this reason, if none other, the designer would be ill advised to skimp on the number of isolating devices.

Similarly, if such devices are not located conveniently there will be the temptation to circumvent the isolation procedures in an effort to save time in total disregard of the safety implications.

As indicated by Regulation 537.2.1.1, where the system is TN-S or TN-C-S, isolation of all line conductors (not necessarily the neutral conductor) is required, whereas isolation of all live conductors (lines and neutral) is needed on TT or IT systems. Table 8.3 outlines the isolation requirements relating to the different system types and Table 8.4 provides a summary of where isolation is required (see also Figure 8.2 for pole isolation in the main switch).

As called for in Regulation 537.2.1.2, precautions need to taken to prevent equipment from being inadvertently or unintentionally energized. Where an item of equipment or an enclosure contains more than one circuit each with separate isolation (i.e. not isolated by a single device), Regulation 537.2.1.3 requires that the equipment or enclosure be accompanied by a durable notice (see Figure 9.13 for example) on or near the equipment to alert persons of the dangers before gaining access to live parts. The notice should give advice on the need to isolate all devices and indicate where the isolation devices are located, to enable complete isolation to be performed. Alternatively, a suitable interlocking arrangement, ensuring isolation of all circuits, is to be provided which will prevent access to live parts.

Table 8.3 Isolation requirements relating to system types

System type	Number of phases	Isolation required for
TN-C	Single-phase	Line only
	Three-phase	Three lines only
TN-C-S	Single-phase	Line only
	Three-phase	Three lines only
TN-S	Single-phase	Line only
	Three-phase	Three lines only
TT	Single-phase	Line and neutral
	Three-phase	Three lines and neutral
IT	Single-phase	Line and neutral
	Three-phase	Three lines and neutral

Table 8.4 Equipment and circuits where isolation is required

Isolation required	Comments
At origin of every installation	Applies to every source (or tariff), usually provided in the form of main switch or main circuit-breaker located as near as possible to the origin of installation
For every circuit or group of circuits	Group isolation only if service conditions permit. Where isolating device is remote from the equipment it serves, it must be securable in the *OFF* position
For every discharge lighting installation operating at HV	Isolation to be provided in LV circuit. Where isolating device is remote from the equipment it serves, it must be securable in the *OFF* position
For every distribution board (for electrical servicing, maintenance and testing and inspection)	Where isolating device is remote from the equipment it serves, it must be securable in the *OFF* position
For every switchboard (for electrical servicing, maintenance and testing and inspection)	Any isolating device used in conjunction with a circuit-breaker must be interlocked. Where isolating device is remote from the equipment it serves, it must be securable in the *OFF* position

As required by Regulation 537.2.1.4 for situations where inductive and capacitive loads or equipment are involved, suitable arrangements need to be taken to discharge any stored energy ($0.5\,Cv^2$ and/or $0.5\,Li^2$) upon isolation, such as by automatic or manual discharge to earth. Generally speaking, LV installations do not present too many problems in this respect, but particular care is needed in consideration of isolation of equipment which contains capacitors (e.g. fluorescent luminaires, power factor correction equipment, etc.) and other components capable of storing energy.

All isolating devices need clear identification as to their purpose, and Regulation 537.2.2.6 calls for clear and durable marking to this end (label to identify circuit(s) and equipment served). See also Chapter 9 of this Guide.

In situations where isolating devices are used in conjunction with a circuit-breaker, to isolate main switchgear, the isolator must be interlocked with the circuit-breaker or, alternatively, be so positioned or safeguarded so that it may be opened only by skilled persons (Regulation 537.2.2.4 refers).

Where an isolating device is located remotely from the equipment it serves, it is necessary, for compliance with Regulation 537.2.1.5, for the device to be provided with a means of securing it in the open (OFF) position. This can be achieved by, for example, a locking arrangement on an MCB, a locked distribution board with only one key or with a removable key or handle which is one of a kind. Most manufacturers have developed and now market locking devices for switchgear, including MCBs, which require the use of a tool for release or, if the padlocking type, a key to unlock.

For installation of electric discharge lighting operating at open-circuit voltages exceeding LV (e.g. more than 1000 V a.c.), Regulation 537.2.1.6 requires isolation to be provided in at least one of the following ways for every self-contained luminaire or every circuit supply luminaires:

- self-contained luminaire with interlock arranged to automatically disconnect the supply *before* access to live parts can be gained (the interlock being additional to the switch normally used for controlling the circuit), and/or
- local effective means of isolation of the circuit from the supply (the isolation being additional to the switch normally used for controlling the circuit), and/or
- be provided with a lockable switch or switch with removable handle provided that the switch, when remote, is lockable in the open position (Regulation 537.2.1.5).

Regulation 555.1.3 makes additional demands for isolation of step-up transformers, in that such isolation must be in all live supply conductors, which includes, by definition, the neutral conductor, irrespective of system type and earthing arrangements. Similarly, electrode boilers warrant additional requirements (Regulation 554.1.2) in that all electrodes, including the neutral, require isolation.

To ensure there is sufficient isolating distance between contacts (in the absence of international agreement, normally not less than 3 mm in the open position), devices for use for isolation must be selected to be in accordance with the types' product standards listed as being suitable for this purpose in Table 53.2 of BS 7671. Regulation 537.2.2.2 requires that the contacts of a device for isolation are either visible or *clearly* and *reliably* indicated, with such indication only occurring *after* the specified isolating distance has been obtained in each and every pole. Semiconductor devices are precluded from being used as isolating devices by Regulation 537.2.2.1, as are items of equipment of overvoltage category II or I to BS EN 60664-1 (that is, equipment not intended to be used as part of a fixed installation – Table 44.4 of BS 7671 gives examples).

Multiple isolation, as Regulation 537.2.2.5 confirms, should preferably be effected by a single device isolating all the necessary poles (as would be required for a three-phase, three-wire circuit) or by a number of single-pole devices (e.g. fuse-links) mounted together (as would be adequate for a three-phase, four-wire circuit with single-phase loads).

Regulation 537.2.2.4 requires that the link, where it is used for the neutral conductor connection at an isolator, must meet at least one of the following requirements:

Table 8.5 Principal requirements for isolating devices

Requirement	Regulation
Non-automatic device (i.e. manually operated)	573.1
Contacts visible or reliably indicated	573.2.2.2
Adequate contact separation (3 mm)	Table 53.2 isolating device
To incorporate means of discharge of inductive and capacitive charges, where appropriate	537.2.1.4
Not susceptible to the effects of vibration or impact	537.2.2.3 and 537.2.2.4
Lockable, if remote	537.2.1.5
Identified as to purpose (appropriately labelled)	537.2.2.6
When used in conjunction with circuit-breaker, interlocking and/or safeguarding required	537.2.2.6
Semiconductor devices not acceptable	537.2.2.1
Equipment of overvoltage category II or I not acceptable	537.2.2.1

- the link to be accessible only to skilled persons;
- the link may only be removed by the use of a tool.

Regulation 537.2.2.3 calls for isolating devices to be such that unintentional reclosure cannot occur (i.e. constructed or installed so that vibration or mechanical impact cannot reclose the device). Furthermore, provision is also required for the isolating device to be unsusceptible to inadvertent or unauthorized operation. This, in practice, means that suitable arrangements should be made for locking facilities, where necessary.

Table 8.5 summarizes the requirements for isolating devices and Table 8.6 gives examples of acceptable and unacceptable isolating devices.

It is important in the design and construction of isolation facilities that consideration is given to the intended use of the installation and the persons likely to operate such devices. In most cases the operator will be a 'skilled person' and/or an 'instructed person' under direct supervision (see definitions), and it would normally be assumed that only such persons would be permitted to operate these devices (unless, of course, the devices served other purposes). The designer should also be aware that the installation user will be required to have satisfactory procedural systems to effect safe isolation (e.g. safe systems of work, permits-to-work) and these should be borne in mind in the electrical design. Such procedures may, for example, include:

- opening up of isolating device;
- checking that all poles are fully open;
- proving dead (by testing);
- earthing of isolated live parts (usually associated with HV isolation, but applies equally to LV system where a voltage to Earth may exist on the isolated live parts by virtue of, for example, capacitive and/or inductive loads);

Table 8.6 Examples of acceptable and unacceptable devices for isolation

Device	Principal characteristic	Restriction on use as isolator
Isolator (disconnector)	Off-load device	Only to be used where load current is zero or negligible
Isolating switch (switch-disconnector)	Suitable for making and breaking load current	None
MCB	Suitable for making and breaking load current	None, providing the necessary contact separation is afforded by the particular device. Refer to manufacturers for advice
MCCB	Suitable for making and breaking load current	None, providing the necessary contact separation is afforded by the particular device. Refer to manufacturers for advice
Fuses	Suitable for off-load operation only	Generally acceptable for off-load isolation except for circumstances where the neutral also requires isolation (e.g. TT systems). *Note.* Whilst a fuse may be capable of making and breaking rated load current, it would be considered dangerous to replace a fuse on a faulted circuit
Links (both line and neutral)	Suitable for off-load operation only	Generally acceptable for off-load isolation. *Note.* Whilst a link may be capable of making and breaking rated load current, it would be considered dangerous to replace a link on a faulted circuit
Fused connection unit	Suitable for off-load operation only	Generally acceptable for off-load isolation of current-using equipment connected to it, except where neutral requires isolation (e.g. TT systems). · This general acceptance presupposes that precautions are taken to remove the possibility of inadvertent fuse replacement
Plugs and sockets and similar devices	May or may not be suitable for making and breaking load current	Refer to manufacturers for advice
Semiconductor devices(e.g. thyristors and triacs)	Deemed not to provide the necessary degree of electrical separation	Not to be used for isolation purposes
Micro-gap switches (e.g. plate switches)	Deemed not to provide the necessary degree of contact separation	Not to be used for isolation purposes

- provision for safe and ready disconnection of isolated circuit (sometimes used for short-term isolation);
- padlocking facility or other means of securing for the prevention of danger to *all* persons working on the circuit(s);
- use of warning notices.

The installation designer may find it beneficial to hold discussions with the installation user at the early stages of design to establish the proposed procedural safety system intended, so that due account of this may be implemented into the electrical design. Whilst BS 7671 does not address this aspect, the designer would need to be mindful of the installation user's duties and responsibilities under the *Health and Safety at Work etc. Act 1974* and more particularly the *Electricity at Work Regulations 1989*. The designer should make themselves aware of the user's intended procedures for isolation of equipment, which may well take the form of 'permit-to-work' procedures. Such a procedure may commence with a check that the circuit(s) can be 'isolated without inconvenience or danger' followed by isolation, taking care that the load current is zero if the isolating device is an off-load mechanism. A check should now be made that all poles of the isolator are open and the downstream circuit conductors are proved 'dead' by application of a voltage-detection test instrument (e.g. test lamp). The functionality of the instrument that is to be used for testing the circuit should be checked both before and after the test, such as on a proving unit or live circuit. The isolating device now requires securing in the open position, followed by the provision of warning notices and safety instructions, as appropriate. On completion of the work, a check should be made that all enclosure covers and barriers removed for the purpose of carrying out the work are replaced and secured in place. The isolator can now be closed, with warning notices and safety instructions removed, to enable the final recommissioning of the circuit(s) concerned.

8.4   Switching off for mechanical maintenance

Whilst there is no definition of 'switching off for mechanical maintenance', it is widely accepted to mean the electrical disconnection of equipment to facilitate work of a non-electrical nature to be carried out safely. 'Mechanical maintenance' is defined and relates to the refurbishment and replacement and cleaning of lamps and other non-electrical parts of equipment and electrically driven or electrically controlled plant and machinery. Since switching off for mechanical maintenance is not isolation of live parts, the cleaning and replacement of lamps (with live parts accessible) needs special consideration by the person responsible for the installation maintenance, who should be mindful of the statutory requirements of the *Electricity at Work Regulations 1989*.

Devices for 'switching off for mechanical maintenance' should not be used where the maintenance work involves the access to live parts even though the predominant task is mechanical related. Any such work where live parts are exposed to touch, or are in close proximity, must be subjected to the more onerous requirements for isolation.

Mechanical maintenance and the need to switch off for carrying out such work relates to where electrically unskilled persons perform the necessary tasks without the risk of physical injury; these requirements, therefore, are not addressing isolation of live parts, on which electrically skilled persons will work, but *normal non-electrical maintenance*.

The dangers associated with mechanical maintenance are those involving burns or injury from mechanical movement or from electromagnetic radiation, as envisaged by Regulation 537.3.1.1. Switching off for mechanical maintenance, therefore, is to protect non-electrically skilled persons from injury which may result from contact with electrically heated equipment or moving parts of electrically activated machinery. Conventionally, replacement and cleaning of lamps is considered to be mechanical maintenance for these purposes (though it is accepted that this may involve limited access to live parts, as in the case of lampholders), as is cleaning of non-electrical parts of equipment, plant and machinery. For electrically powered equipment within the scope of BS EN 60204 *Safety of machinery*, Regulation 537.3.1 specifies that the requirements for switching off for mechanical maintenance of that standard shall apply. For example, Part 1 of BS EN 60204, *General requirements*, gives requirements for the selection and erection of devices for switching off for the prevention of unexpected start-up. In cases where very large plant is involved, where it can be expected that persons undertaking non-electrical maintenance are required to work inside the plant, consideration should be given to providing isolation, rather than switching, for mechanical maintenance to obviate the risks of malfunction or misuse of the switching device.

There is a need to identify the switching device clearly, as to its purpose, and a need to locate it in a suitably accessible position (Regulation 537.3.2.4). If located at some position remote from the equipment it serves, then precautions need to be taken (Regulation 537.3.1.2) to avoid inadvertent or unintentional closure or reclosure of the switching device. This may, for example, be achieved by the use of a suitable locking device. Where the switching device is under the control of the person undertaking the maintenance work and is adjacent to the equipment, locking arrangements may be unnecessary.

For items of equipment needing mechanical maintenance, the switching device must be in the main circuit supplying the equipment (Regulation 537.3.2.1), except that interruption of a control circuit of a drive is permitted instead if a condition equivalent to the direct interruption of the main supply is provided. The control circuit may be switched provided the British or Harmonized Standard for the equipment permits such a method or, alternatively, an equal degree of safety is achieved by some other means, such as mechanical restrainers. Special consideration is needed if the switch is not provided in the main circuit and the control circuit is switched instead; reflection on the effects of a fault on the control circuit (possibly causing the main circuit to be re-energized) will be paramount in assessing the degree of safety. Figure 8.3 illustrates an example where switching for mechanical maintenance, provided by a latching-type control circuit 'stop' button (with provision for resetting by key), is less than satisfactory. Consider an intermittent earth fault in the control circuit at the location shown in the figure. The intermittent fault between the auxiliary contact's line conductors may cause the energization of the coil, which in turn would cause the motor contacts to 'make', starting the motor. With the auxiliary contacts now 'made', the motor would continue to run, leading to serious hazards for the maintenance worker operating on the machinery driven by the motor.

The device (or control switch) must be manually operated (i.e. non-automatic) and its contact gaps must be visible externally or a clear and reliable indication of the contact position must be given by some other means (Regulation 537.3.2.2), the 'OFF' position indication only occurring after the contacts have opened sufficiently on each pole. The device must have the capability to interrupt the full load current of the circuit(s) concerned

Table 8.7 Principal requirements for devices for switching off for mechanical maintenance

Requirement	Regulation
Operated by hand	537.3.2.2
Suitably located	537.3.2.4
Capable of switching load current (off-load device unsuitable)	537.3.2.5
Contacts visible or reliably indicated	537.3.2.2
Lockable, if remote	537.3.1.2
Not susceptible to the effects of vibration and impact	537.3.2.3
Identified as to purpose (appropriately labelled)	537.3.2.4
Preferably to be inserted in the main supply to equipment	537.3.2.1
Permissible to insert in control circuit if BS or CENELEC standard permits or if equal degree of safety is provided	537.3.2.1
Plug and socket-outlet up to 16 A	537.3.2.6

(Regulation 537.3.2.5) and must be selected and/or erected so that unintentional reclosure does not occur due to, for example, vibration or mechanical impact.

Table 8.7 gives a summary of the essential requirements for a device for switching off for mechanical maintenance.

The need for switching off for mechanical maintenance must be assessed individually for each item of electrical equipment, bearing in mind all factors associated with its maintenance. It is essential that such switching is provided for:

- electric-motor-activated machines;
- equipment with electrically heated surfaces which may be touched (e.g. lamps, heaters, cooking appliances);
- electromagnetic equipment where there is a risk of an accident.

Table 8.8 provides examples of devices for switching off for mechanical maintenance.

Where a higher degree of hazard is evident, the more onerous requirements relating to isolation should be considered together with suitable warning notices provided for added security and safety in performing such operations.

8.5 Emergency switching and other forms of switching for safety

8.5.1 General

The requirements for emergency switching must be applied to all equipment or parts of the installation which could otherwise present a potential danger. In such instances, if cutting off the electricity is not rapid, then danger may occur or perhaps further danger would not be prevented (Regulation 537.4.1.1). The means of emergency switching adopted must be such that an additional hazard is not introduced and interference with the operation does not occur (Regulation 547.4.1.4). This will involve the installation designer taking

Table 8.8   Examples of devices for switching off for mechanical maintenance

Devices for switching off for mechanical maintenance	Comment/restrictions, if any
Cord-operated switch (pull-switch)	Only if it incorporates reliable contact indication (e.g. flag indication). Must be under the continuous control of person carrying out maintenance
MCBs	If remote, to be locked *OFF*
MCCBs	If remote, to be locked *OFF*
Devices also providing isolation	If remote, to be locked *OFF*
Stop-lock switch	If remote, the stop-lock circuit must have priority and cannot be over-ridden by starting circuit or starting method
Stop-latch switch (emergency stop button)	Must be under the continuous control of person carrying out maintenance and must provide equal degree of safety as if device is located in main circuit (see also Figure 8.3)
Control switch operating contactor	If remote, to be locked *OFF*
Plug and socket or similar devices, such as a luminaire supporting coupler	Must be adjacent to equipment and under the continuous control of person carrying out maintenance

account of the effects of the emergency switching of plant and machinery which may as a result create additional hazards. Regulation 537.4.1.3 demands that an emergency switching arrangement is such that one single action will interrupt the appropriate supply. This implicitly means that such devices must be uncomplicated in their switching action. Not only must the device act promptly, but the effect on the equipment must also be rapid. This may necessarily also require provision for emergency stopping (see Section 8.6 of this Guide). Generally speaking, emergency switching devices will be of the type that needs resetting after the event and that would require a deliberate act to re-energize the circuit concerned.

Emergency switching is intended, in general, for use by *any* person in the event of an emergency in order to prevent or remove a hazard. Therefore, not later than at handover, the installation designer should make the installation user aware of the locations, purposes and functions of the emergency switching arrangements and devices installed. The hazards envisaged by BS 7671 include, for example, electric shock, burns, fire and injury from mechanical movement. Depending on the particular installation, there may well be other dangers which will necessitate the provision of emergency switching measures to electrical equipment. The facility for emergency switching is required anywhere where hazards may be reasonably envisaged from events such as fires, electric shock, explosions or accident. Such switching is required not only to safeguard persons, but also to protect livestock and, where appropriate, property.

Emergency switching should not be confused with isolation (a different concept), and there will be many circumstances where emergency switching will not be required. However, emergency switching will be required for any equipment or any part of an installation where, by virtue of the particular characteristics of the equipment, it may be necessary to disconnect rapidly to prevent or remove a hazard or danger. Having commented that emergency switching and isolation are different in concept, the requirements for pole switching, depending on the system type and earthing arrangements, are the same (see Figure 8.2).

Installations embodying electrically activated (powered or controlled) machines are particularly susceptible to this type of danger and would, therefore, need emergency switching facilities. Such machines would include individually operated machines, as well as, for example, complex production lines comprising a number of drives operating simultaneously and which may need overall emergency switching in addition to individual switching. Escalators and conveyor systems are prime examples requiring emergency switching. For electrically powered equipment within the scope of BS EN 60204 *Safety of machinery*, Regulation 537.4.1 specifies that the requirements for emergency switching of that standard shall apply. For example, Part 1 of BS EN 60204, *General requirements*, gives requirements for the selection and erection of devices for emergency switching. Other examples would include electric cookers and other catering equipment likely to present hazards from fire and explosion. In all cases, the emergency switching device(s) should be easily accessible and, as a rule of thumb, should not normally be more than 2 m from the equipment it serves, and the normal pathway to the device must be unimpeded by obstructions, dangerous moving parts and heat sources.

Emergency switching devices must (Regulation 537.4.2.1) be capable of interrupting the supply on load (i.e. the full load current), and in the case of motors must be adequate to switch the much greater current associated with stalled-rotor conditions. The arrangement for switching must (Regulation 537.4.2.2) consist of:

- a single device (cutting off all poles of the circuit to the equipment), or
- a combination of a number of items of equipment capable of being operated by a single action (e.g. contactor controlled by emergency stop push) (see also Section 8.6 in this Guide: Emergency stopping).

Regulation 537.4.1.3 calls for all devices to act as directly as is possible on the circuit supply conductors. A plug and a socket-outlet or similar device should *not* be selected for the function of emergency switching (Regulation 537.4.2.8 refers). The handle, push button or other means of operation of an emergency switching device must be clearly identified, preferably by the colour red with a contrasting background (Regulation 537.4.2.4). The device must (Regulation 537.4.2.7) be clearly identified as to its purpose and, with account being taken of intended use of the premises, access to it must not be impeded by the emergency conditions that are envisaged (Regulation 537.4.2.5). Where there is a risk that, by inappropriate operation of the device, an additional danger may be created, arrangements need to be made to restrict access to it to skilled and instructed persons.

The emergency switching requirements apply to all equipment where a hazard may be expected in normal use. The device for such purposes must be located in a readily accessible position such that the operator is not put in danger in the act of operation. Emergency switching of a number of appliances in the same room is not precluded by the requirements, and neither is the incorporation of the device into the appliance it serves, where appropriate.

Regulation 537.4.2.3 requires that, where a circuit-breaker or contactor is used which is operated remotely (e.g. from a remote emergency push button), the arrangement is to be such as to open the circuit on *de-energization* of the coil, or that another technique of suitable reliability is used. The implication here is that emergency switching must not rely on energization of a coil which will, in circumstances of power failure to the control circuit, be ineffective.

The handle or push button of the device, as required by Regulation 537.4.2.5, must also be readily accessible and in the vicinity in which the anticipated hazard may occur and, where appropriate, at other positions from where the danger may be removed (including remotely). The push button must, as required by Regulation 537.4.2.6, be the latching 'OFF' type and the arrangement must be such that re-energization does not occur on release of the push button or handle. In other words, a separate action is required for re-energization; for example, by a key switch (these requirements, however, do not apply in cases where the emergency switching device and the means of re-energization are under the control of the same person). Table 8.9 gives a summary of the requirements for emergency switching devices.

For the more complex industrial installation where process machinery is involved, consideration should be given to the consequential effects of emergency switching off of one machine on other machines in the line process. Some form of sequential emergency switching (and emergency stopping) may be appropriate. The designer and the person responsible for process safety should consult in order to achieve the highest degree of safety in the workplace.

In, for example, electrical test room situations, emergency switching may be referred to as 'emergency tripping'; both terms should not be confused with 'emergency stopping',

Table 8.9   Principal requirements for devices for general emergency switching

Requirement	Regulation
Where practical, non-automatic device	537.4.2.3
Switching of all the live conductors (except the neutral where excused by Regulation 537.1.2) is required if there is a risk of electric shock	537.4.1.2
Capable of switching load current	537.4.2.1
In the case of motors, must be capable of switching stalled-rotor current	537.4.2.1
To be single-action device	537.4.1.3
Identified as to purpose (appropriately labelled)	537.4.2.7
Device to act as directly as possible on circuit concerned	537.4.1.3
Suitably located	537.4.2.5
If device is used to operate contactor controlling the circuit, switching is required to *de-energize* coil, or another technique of suitable reliability must be used	537.4.2.3
If push button, must be the latching off type	537.4.2.6
Plug and socket-outlet not acceptable	537.4.2.8
Operating knob, lever or button, preferably colour *RED*	537.4.2.4

which refers to the arresting (stopping) of electrically activated mechanical movements, whereas emergency switching refers only to rapid disconnection of the electricity supply to equipment. In rooms used by students or trainees, it may be appropriate to increase the number of emergency switching devices.

8.5.2 *The firefighter's switch*

The firefighter's switch is a particular form of switching for safety (and of isolation) and, as Regulation 537.6.1 requires, must be provided for LV circuits supplying all exterior installations operating at a voltage exceeding LV (more than 1000 V a.c. or 1500 V d.c. between conductors, or 600 V a.c. or 900 V d.c. between conductor and Earth) and for all interior HV discharge lighting installations. 'Exterior installations' include covered markets, shopping arcades and shopping malls, though a temporary installation of such equipment for exhibitions would not be required to meet these requirements. There is also a dispensation for portable discharge lighting luminaires up to 100 W, provided they are fed by a readily accessible socket-outlet.

Regulation 537.6.3 demands that the provision of firefighter's switches shall also comply with the requirements of the local Fire Authority, which may differ in one area from another. It is important, therefore, for the designer to consult the Fire Authority at an early stage to establish its particular needs.

Where a number of such exterior installations exist which require provision of a firefighter's switch, the preferable method is to control all the installations by a single switch as envisaged by Regulation 537.6.2. Similarly, a separate single device is also required for all interior HV installations. Where a number of circuits require switching by a firefighter's switch, this may be effected by, for example, switches with multipole switching or by a contactor controlled by a firefighter's switch. This requirement may be difficult to achieve in practice where, for example, a number of items of equipment need switching through a single device and where they may be fed from separate supplies. In such cases, the Fire Authority must be consulted in an effort to resolve these practical difficulties.

The principal features of a firefighter's switch and its mounting positions are summarized in Tables 8.10 and 8.11 respectively, and Figure 8.4 illustrates these requirements pictorially.

8.6 Emergency stopping

Emergency stopping, as opposed to emergency switching, provides for the arresting of mechanical movement of electrically activated equipment and in most cases, of necessity, involves the rapid disconnection of the supply to the equipment (e.g. electric motor). Where necessary, further initiation of a stopping mechanism (e.g. *de-energization* of a magnetic clutch or *energization* of a magnetic or electrically operated brake is also required).

The requirements for emergency switching should be applied generally to emergency stopping, which must, in any event, be provided where there is a risk from mechanical movement of electrically actuated equipment. Such emergency stopping devices must be readily accessible, and in those cases where there is more than one means of starting a machine, interlocking or other precautions are required so as to obviate the risks of unexpected restarting.

Table 8.10 Principal requirements for the firefighter's switch

Requirement	Regulation
Be coloured *RED*	537.6.4
Be marked or labelled *firefighter's switch* – see (Figure 9.19)	537.6.4
Marking or labelling not less than 36 point.	537.6.4
Minimum dimension of marking or labelling – 150 mm × 100 mm	537.6.4
ON and *OFF* positions clearly indicated	537.6.4
ON and *OFF* positions clearly visible to firefighter on ground or other standing	537.6.4
Be constructed so that switch cannot be inadvertently switched *ON*	537.6.4
Be constructed so as to facilitate operation by firefighter (pole operated)	537.6.4

Table 8.11 Principal requirements for the positioning of the firefighter's switch

For emergency switching of exterior installations	For emergency switching of interior installations
Fixed on outside of building in a conspicuous position adjacent to controlled equipment or affix notice to, or near, switch indicating where the controlled equipment is located	Fixed on outside of building in a conspicuous position near main entrance or other position agreed with the Fire Authority
Switch to be readily accessible (unimpeded access)	Switch to be readily accessible (unimpeded access)
Mounted at not more than 2.75 m from floor or hard-standing beneath the switch	Mounted at not more than 2.75 m from floor or hard-standing beneath the switch
Where more than one device, each to be identified clearly as to its purpose (labelled)	Where more than one device, each to be identified clearly as to its purpose (labelled)
Switch and positioning to meet *all* the requirements of the Fire Authority.	Switch and positioning to meet *all* the requirements of the Fire Authority

In certain cases, retention of supply will be needed to, for example, effect an electric braking facility (e.g. electric crane), and Regulation 537.4.2.2 would not preclude this arrangement where it is necessary.

8.7 Functional switching

BS 7671 defines *functional switching* as an 'operation intended to switch "on" or "off" or vary the supply of electrical energy to all or part of an installation for normal operating purposes'. Regulation 537.5.1.1 calls for a functional switch to be provided for each part of a circuit which may require to be operated independently of any other part or circuit. In other words functional switching can be described as switching necessary for the everyday control and functioning of electrical equipment.

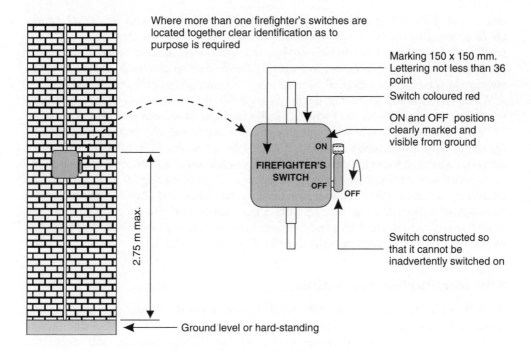

Where more than one firefighter's switches are located together clear identification as to purpose is required

Marking 150 x 150 mm. Lettering not less than 36 point

Switch coloured red

ON and OFF positions clearly marked and visible from ground

ON

FIREFIGHTER'S SWITCH

OFF

OFF

2.75 m max.

Switch constructed so that it cannot be inadvertently switched on

Ground level or hard-standing

Figure 8.4 Requirements for a firefighter's switch

Whilst Regulation 537.5.1.2 does not require a functional switch to control all live conductors (e.g. line and neutral), it does stipulate, by reference to Regulation 530.3.2, that such switching must not be located in the neutral conductor alone. Although Regulation 537.5.1.3 recognizes that a functional switching device may control a number of items of equipment which are intended to operate simultaneously, it does require that all current-using equipment be controlled by a suitable functional switching device. Examples of functional switching devices are easily recognized and would include:

- micro-gap switches (e.g. plate switches, grid switches and switches integral to switched socket-outlets);
- contactors and relays;
- electronic devices (e.g. thyristors, transistors, triacs);
- change-over devices;
- machine limit switches;
- devices serving other functions (e.g. isolating devices).

Functional switches are not precluded for use for other purposes (e.g. switching off for mechanical maintenance) provided they meet all the particular requirements for all the functions they are required to perform. In the case of appliances, the device for functional switching can be, and often is, an integral part of the appliance.

Off-load isolators, also known as disconnectors, must not be used for functional switching, and neither must fuses and links be used for such a purpose (see Regulation 537.5.2.3);

and, as called for in Regulation 537.5.2.1, devices for functional switching must be capable of undertaking the most demanding duty intended by the design. Regulation 537.5.2.2 recognizes that functional switching devices may control the current without necessarily the opening and closing contacts, as might be the case of an electronic device in which the current waveform is modified by employing waveform chopping techniques.

Where a functional switching device is used for the change-over from one source to another, as might be the case where standby facilities have been provided, Regulation 537.5.1.4 calls for the device to switch *all* live conductors (lines and neutral). Additionally, the device must not be such as to connect more than one source in parallel, unless parallel operation was intended in the design, and that disconnection of a PEN conductor must only be provided where the design expressly requires such switching (e.g. to prevent unwanted circulating currents). Particular attention must be paid to control circuitry, inasmuch as a malfunction is likely to occur as a result of a fault arising between the control circuit and other live conductors, protective and functional conductors, exposed-conductive-parts and extraneous-conductive-parts, as envisaged by Regulation 537.5.3.

8.8   Identification and notices

As with all switchgear, devices for switching for safety need clear and unambiguous identification as required by Section 514 of BS 7671 (see also Chapter 9 of this Guide). Switchgear for isolation, switching off for mechanical maintenance, and emergency switching are also subject to the specific identification requirements, such as in Regulation 537.2.2.6 (isolation) and 537.6.4 (firefighter's switching). Chapter 9 of this Guide provides examples of labels for identification of switchgear.

9

Equipment selection: common rules

9.1 General

Chapter 51 of BS 7671 deals with the common rules for selection and erection of equipment and is made up of five principal sections:

- compliance with standards;
- operational conditions and external influences;
- accessibility;
- identification and notices;
- prevention of mutual detrimental influence.

The requirements embodied in Chapter 51 are common rules and as such must be applied to every installation irrespective of its location and environment.

9.2 Compliance with standards

In terms of selection and erection of equipment, the fundamental rule, contained in Regulation 511.1, is that equipment must be constructed to an acceptable current Standard appropriate to the intended use of that equipment, as follows:

- British Standard (BS ...);
- Harmonized European (CENELEC) Standard (BS EN ...).

Where any of the above standards is inappropriate or inconvenient, equipment to a foreign national standard based on the corresponding IEC standard may be used provided the designer or specifier of the installation confirms that any differences between it and the appropriate British Standard or Harmonised European (CENELEC) Standard results in no less a degree of safety, as required by Regulation 511.2. BS 7671 recognizes

A Practical Guide to The Wiring Regulations: 17th Edition IEE Wiring Regulations (BS 7671:2008)
Fourth Edition Geoffrey Stokes and John Bradley
© Geoffrey Stokes and John Bradley. Published by John Wiley & Sons, Ltd

that equipment complying with a British Standard or European Standard is acceptable without further evidence of compliance with that standard (i.e. certification or approval of conformity) providing it is installed in a proper manner and without modification other than that acknowledged by the standard or recommended by the manufacturer. Where installed equipment is manufactured to a foreign standard, and its degree of safety has been assessed as acceptable, care must be taken to confirm its compatibility with other installation equipment.

Essentially, BS 7671 is concerned only with safety, and where reference is made to acceptable British and Harmonized Standards this relates only to the safety aspects of those standards. Occasionally, equipment may be subject to more than one standard; for example, one may refer to the safety considerations whilst another may refer to the performance specification. For compliance with BS 7671, only the relevant safety aspects need to be met. Generally, BS 7671 demands that equipment complies with a relevant and appropriate standard but does not go so far as to insist on certification that the requirements have been met, though the designer may well feel that this would be desirable.

Other equipment not complying with a British or Harmonized Standard is not ruled out provided the designer has verified, perhaps by the employment of a third-party certification body, that the degree of safety is not less than that provided by the British or Harmonized Standard relating to similar equipment.

Whilst BS 7671 does not primarily address the performance of the equipment, the designer would be well advised not to ignore the other facets of good installation design:

• equipment performance – reference to standard specifications;
• sound engineering principles and CPs;
• the correct functioning of equipment as specified by the user.

Equipment must be selected and erected so that it is fit for the purpose intended; and if there is any doubt in the designer's mind, the manufacturer's advice should be sought. Regulation 510.2 draws attention to the need for equipment to comply with all the relevant regulations in all parts of BS 7671. Appendix 1 of BS 7671 lists the British Standards to which specific reference is made in the various regulations.

9.3   Operational conditions, external influences and accessibility

The equipment must, of course, be fit for purpose and suitable for all the relevant operating conditions and for the external influences, as summarized in Table 9.1. Electrically driven machines warrant special consideration in this respect, in that motors must be matched with the mechanical load. The characteristics in terms of delivered power as a function of time are of critical importance, particularly under starting conditions. The designer would need to assess the characteristics of the mechanical load and the number of starting/stopping operations required and select a suitable motor. The demands of the motor operation will have a significant effect on the fixed wiring system and related equipment. Selection of such equipment will need to take into account not only the steady-state load, but also the dynamic behaviour, particularly under starting conditions. Motor starting techniques also need very careful consideration of the effects of such operations on other equipment within the installation and on the supply. Where large motors are concerned,

Table 9.1 Summary of the operational and other conditions to be considered in the selection of equipment

Aspect	Regulation	Requirement
Voltage	512.1.1	For TN and TT systems, equipment to be suitable for the nominal voltage to Earth U_o. Account to be taken of voltage variations. For IT systems, equipment to be suitable for the nominal voltage between line conductors. Account to be taken of voltage variations
Current	512.1.2	To be suitable for design current I_b. Account to be taken of capacitive and inductive effects. To be suitable for fault current for the duration of the fault (determined by the protective device characteristics). When connected to conductors operating at more than $70\,°C$, the current rating of the equipment may need to be reduced (manufacturer to be consulted)
Frequency	512.1.3	Where frequency affects the characteristics of the equipment, the rated frequency of the equipment must match that of the supply
Power	512.1.4	The power characteristics of the equipment to be suitable for the duty demanded
Compatibility	512.1.5	Equipment not to cause harmful effects to other equipment nor to the supply (including during load switching operations). Equipment producing third-harmonics needs special consideration
External influences (see Chapter 32, Section 522 and Appendix 5: Classification of external influences)	512.2 (132.5.1) 512.2.3	Equipment to be of the appropriate design suitable for the environmental conditions likely to be encountered. Where equipment does not inherently possess the characteristics suitable to the external influences, then suitable additional protection will be required; such protection is not to affect the operation of the equipment adversely. Where equipment is likely to be subjected to a number of different external influences, the degree of protection of the equipment is also required to take account of any mutual effect
Identification	514.1.2	All wiring to be arranged or marked so that it can be readily identified
Mutual detrimental influences	515.1 515.2	Equipment to be selected and erected so that no harmful influence takes place between electrical equipment and between electrical equipment and any nonelectrical service. Harmful influences may include electromagnetic radiation. Where it is expected that mutual detrimental influence may take place due to different types of current or voltage, the equipment must be effectively segregated.
Accessibility	513.1 (132.12)	Equipment to be installed so as to allow its operation, inspection and maintenance. Additionally, access to every connection is required. Where mounted in an enclosure or compartment, access for such purposes is not to be significantly adversely affected. This requirement does not apply to cable joints exempted under Regulation 526.3

consultation with the ED will be essential to determine the extent of the effects on the ED's network and the limits, if any, imposed on the starting current. Special consideration is also required in the selection of protective devices, switchgear and controlgear used in relation to loads which are highly inductive (e.g. motors, transformers, fluorescent loads) or capacitive (e.g. power factor correction equipment), as low power factors can seriously affect the equipment's ability to disconnect such loads. Consultation with the manufacturers of the equipment will be essential in many cases.

9.4   Identification and notices

Table 9.2 summarizes the rules for identification of conduit and conductors and Table 9.3 similarly outlines the requirements for notices, including those for identification of switchgear and controlgear. Figures 9.1–9.23 give examples of identification labels and warning notices to meet the requirements.

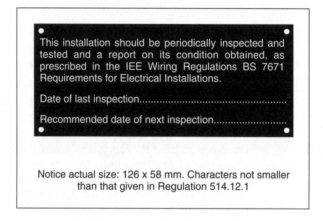

Figure 9.1   Periodic inspection notice

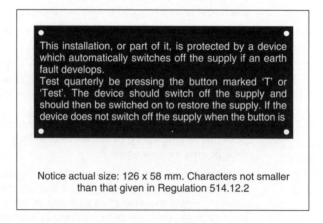

Figure 9.2   RCD notice

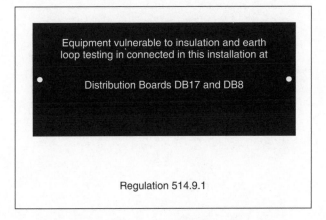

Figure 9.3 Equipment vulnerable to testing notice

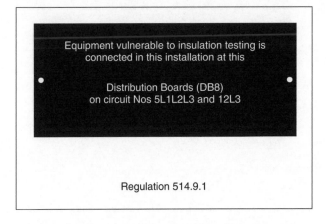

Figure 9.4 Equipment vulnerable to testing notice

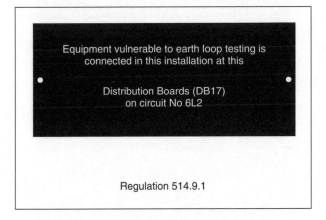

Figure 9.5 Equipment vulnerable to testing notice

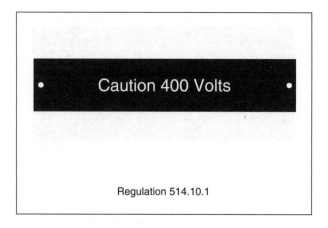

Figure 9.6   Caution 400 V notice

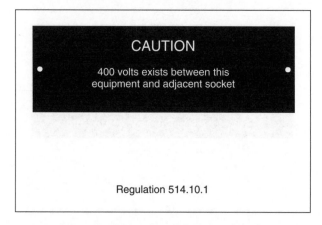

Figure 9.7   Caution 400 V (between equipment) notice

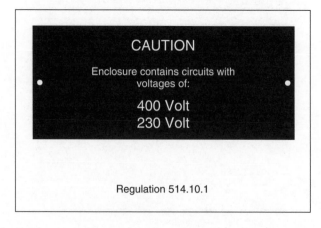

Figure 9.8   Different voltages notice

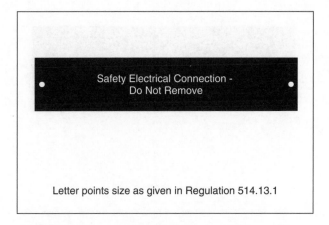

Figure 9.9 Earthing and bonding connection notice

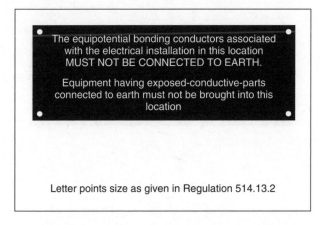

Figure 9.10 Warning notice for earth-free installations

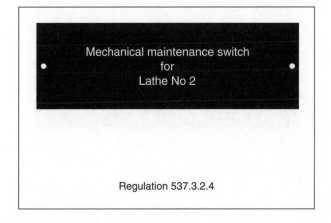

Figure 9.11 Mechanical switch notice

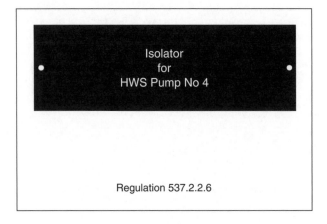

Figure 9.12   Isolator notice

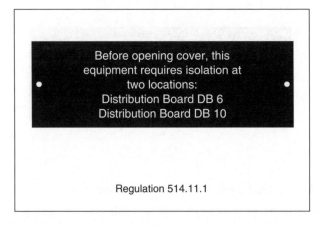

Figure 9.13   Isolation required at more than one point notice

Figure 9.14   Nonstandard colours notice

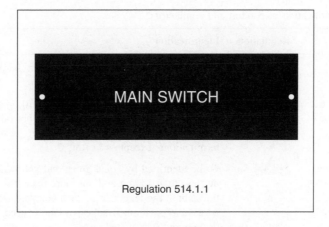

Figure 9.15   Main switch notice

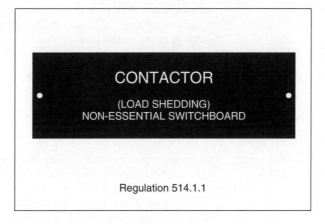

Figure 9.16   Contactor designation notice

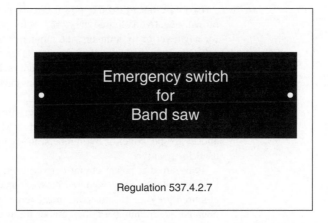

Figure 9.17   Emergency switch notice

Table 9.2 Identification of conduit and conductors[a]

Ref.	Item	Regulation	Identification
A	Conduit	514.2.1	Coloured orange where it is required to be distinguished from other service pipes
B	Protective conductor – generally	514.4.2	Coloured green–yellow (70:30 or 30:70). This colour combination not to be used for any other purpose. Single-core green–yellow cables not to be overmarked at terminations, except as in Ref. D
C	Protective conductor – bare conductor and busbar	514.4.2	To be identified by equal green and yellow stripes between not less than 15 mm and not more than 100 mm either throughout the length or in each compartment and at each accessible position. If adhesive tape is used, it must be bi-coloured (green–yellow)
D	PEN conductor – generally (when insulated)	514.4.3	Either green-and-yellow throughout its length with, in addition, blue markings at the terminations, or blue throughout its length with, in addition, green-and-yellow markings at the terminations
E	Functional earth	514.4	Telecommunications functional earth: coloured cream
F	Cores of cables everywhere	514.4.5	The colour green on its own must not be used
G	Cores of cables used as live conductors	514.4 and 514.5	Single-phase power circuit: coloured brown for line and blue for neutral; or L for line and N for neutral; or by numbers, the number 0 being reserved for neutral. Three-phase power circuit: lines coloured brown, black and grey as appropriate, and blue for neutral; or L1, L2 and L3 (as applicable) for lines, and N for neutral; or by numbers, the number 0 being reserved for neutral. Two-wire unearthed d.c. power circuit: positive coloured brown, negative coloured grey; or L+ for positive and L- for negative; or by numbers, the number 0 not being used. Two-wire negative earthed d.c. power circuit: positive coloured brown, negative coloured blue; or L+ for positive and M for negative; or by numbers, the number 0 being used for negative. Two-wire positive earthed d.c. power circuit: positive coloured blue, negative coloured grey; or M for positive and L- for negative; or by numbers, the number 0 being used for positive. Two-wire d.c. power circuit derived from a three-wire system: outer positive coloured brown, outer negative coloured grey; or L+ for outer positive and L- for outer negative; or by numbers, the number 0 not being used.

Table 9.2   *(continued)*

Ref.	Item	Regulation	Identification
			Three-wire d.c. power circuit: positive coloured brown, mid-wire coloured blue and negative coloured grey; or L+ for positive, M for mid-wire and L- for negative; or by numbers, the number 0 being used for the mid-wire. (Only the mid-wire may be earthed)
			Control circuits, extra-low voltage and other applications: line conductor coloured brown, black, red, orange, yellow, violet, grey, white, pink or turquoise, and neutral conductor or mid-wire coloured blue. Alternatively, L for line conductor and M for neutral or mid-wire; or by numbers, the number 0 being used for the neutral or mid-wire
H	Bare *live* conductors	514.4.6	Identification by the use of coloured tapes, discs or sleeves or by painting – colours as for Ref. G
I	Cables buried in the ground	522.8.10	Cables to be marked by covers or suitably marked by marking tape where not enclosed in ducts or conduit

[a]Identification of cores of cables is required at least at terminations, but preferably throughout their length.

In the most part, BS 7671 does not prescribe the size of labels and warning notices; obviously, however, they need to be adequately proportioned so that lettering is easily legible. There are, however, four instances where the minimum letter sizes (point sizes) are required: the periodic inspection notice (see Figure 9.1), the RCD test notice (see Figure 9.2) and the earthing and bonding connection notices (see Figures 9.9 and 9.10).

Cable core identification (colour coding) in BS 7671 has not changed from the 16th Edition of the Wiring Regulations (BS 7671:2001), including amendment number 2, 2004.

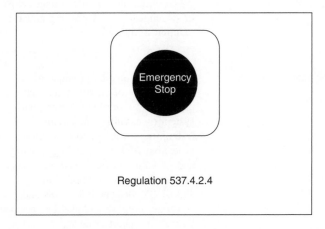

Figure 9.18   Emergency stop notice

Table 9.3 Warning notices and identification of switchgear and controlgear

Ref.	Item	Regulation	Identification and warning notices	See Figure
A	Live parts not capable of being isolated by a single device	514.11.1	Notice required to identify equipment that needs isolation at the appropriate locations and devices. The location of each isolating device must also be indicated	9.13
B	Isolating devices	537.2.2.6	Every device for isolation to be identified by durable marking to indicate relevant circuit(s)	9.12
C	Switches for mechanical maintenance	537.3.2.4	Every switch for the purpose of switching off for mechanical maintenance requires identification by durable marking	9.11
D	Voltages exceeding 230 V	514.10.1	Where a nominal voltage exceeding 230 V exists within an item of equipment and where such voltage would not be expected, a notice is required to warn skilled persons before access is gained to live parts. Not necessary for the label to be external, providing it is seen before access to live parts is gained.	9.6
			Where a nominal voltage exceeding 230 V exists between an item of equipment and another enclosure or accessory, a notice is required to warn skilled persons of the voltages	9.7
E			Where different nominal voltages exist in switchgear, etc., a warning notice is required to indicate the different voltages	9.8
F				
G	Diagrams, chart or table	514.9.1	Information must be provided to indicate: type of circuit, points served, number and size of conductors, wiring system type, vulnerable circuits to testing, the location and function of protective devices (including isolating and switching devices) and the method of compliance with Regulation 413-01-01. Distribution boards and consumer units should also be provided with such a chart or table	17.19(d)

Table 9.3   *(continued)*

Ref.	Item	Regulation	Identification and warning notices	See Figure
H	Equipment vulnerable to typical test	514.9.1	Where vulnerable equipment is connected to a circuit, a notice indicating this fact and a warning given that testing with common instruments (e.g. 500 V insulation test and earth loop impedance test) may damage the equipment, which should be tested separately. Ideally, a notice should be affixed at all points where tests are likely to be carried out and may be, for the smaller installation, in the form of a note appended to the circuit chart or diagram	9.3 9.4 9.5
I	RCDs	514.12.2	Wherever an RCD is installed, a notice is required drawing the user's attention to the need to test (with the mechanical test button) at intervals not exceeding 3 months	9.2
J	Periodic inspection and testing	514.12.1	On completion of an installation, a notice at or near the origin is required to alert the user to the need for periodic inspection and testing	9.1
K	Earthing and bonding connections	514.13.1	In an installation where ADS is used, a notice is required at every point of connection of an earthing conductor to an earth electrode and connection of a bonding conductor to extraneous-conductive-parts.	9.9
L	Locations with non-earthed bonding	514.13.2	In an installation where protection by earth-free local equipotential bonding or by electrical separation for the supply to more than one current-using item is employed, a notice is required warning amongst other things that such equipotential bonding must not be connected to Earth.	9.10
N	Nonstandard colours	514.14.1	Where an existing installation wired in the old cable colours is altered or extended such that it contains wiring to both the old colours and the harmonized colours, a notice is required giving warning of that fact	9.14

(continued overleaf)

Table 9.3   *(continued)*

Ref.	Item	Regulation	Identification and warning notices	See Figure
O	Dual supply	514.15.1	Where an installation has a generator used as an additional source in parallel with another source, a notice warning of the need to isolate both supplies before carrying out work on the installation is required at the locations given in Regulation 514.15.1	9.21
P	Earthing conductor connection	542.3.2	Label (as in Ref. K above) to be affixed to the connection of the earthing conductor to electrode or other means of earthing	9.9
Q	Switchgear – generally	514.1.1 514.8.1	Identified as to function and purpose. Additionally, if not in view of the operator and where this may be the cause of danger, then an indicator, complying with BS 4099 *Colour of indicator lights, push buttons, annunciators and digital readouts*, is required	9.15
R	Controlgear – generally	514.1.1	Identified as to function and purpose. Additionally, if not in view of the operator and where this may be the cause of danger, then an indicator, complying with BS 4099 *Colour of indicator lights, push buttons, annunciators and digital readouts*, is required	9.15
S	Protective device	514.8.1	Devices to be identified so that circuit protected is easily recognized (may be identified on circuit chart – see row H of this table)	–
T	Emergency switches	537.4.2.7	Devices for emergency switching to be durably marked	9.17
U	Emergency switches	537.4.2.4	Operating handle or push buttons of emergency switches to be clearly identified and preferably coloured red	9.18
V	Firefighter's switch	537.6.4	Firefighter's switches to be equipped with durable nameplate *Firefighter's switch*	9.19
W	Firefighter's switch location	537.6.3 (i)	Notice required to identify location of Firefighter's switch if not positioned on outside of building adjacent to the equipment it serves	9.20

Table 9.3 *(continued)*

Ref.	Item	Regulation	Identification and warning notices	See Figure
X	Firefighter's switch location	537.6.3 (iv)	Where more than one Firefighter's switch, identification is required as to the equipment each device serves	9.20
Y	Caravan extra-low voltage socket-outlet	721.55.2.2	Every socket-outlet supplied at extra-low voltage must have its voltage visibly marked	–
Z	Main isolating switch in caravan	721.537.2.1.1.1	Permanently fixed durable notice required adjacent to the main isolating switch in a caravan giving instruction for the electricity supply	9.23
ZA	Highway power temporary supply units	559.10.7.2	Durable label required on every temporary supply unit indicating maximum sustained current to be supplied by the unit	9.22

Where the protective measure of either earth-free local equipotential bonding or electrical separation for the supply to more than one item of current-using equipment is used, Regulations 418.2.5 and 418.3, respectively, require a warning notice as shown in Figure 9.10 to be fixed in a prominent position near every point of access to the location. Further such warning notices fixed along cable runs may be needed in some cases. It will be appreciated that a connection of an earth-free equipotential bonding conductor to Earth would destroy the protective provision for fault protection, as would bringing equipment having earthed exposed-conductive-parts into the location.

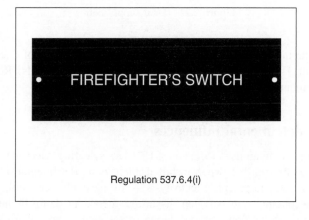

Figure 9.19 Firefighter's switch notice

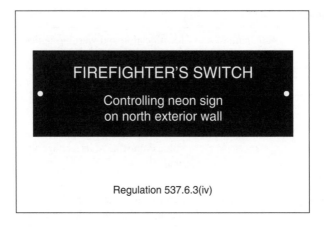

Figure 9.20   Remote fireman's switch notice

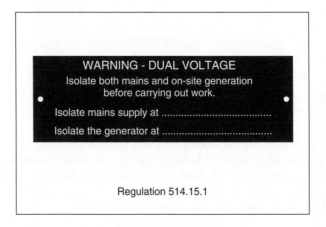

Figure 9.21   Dual supply notice

Where pictographic signs are used to give warning of the risks of electric shock they must accord with The Health and Safety (Safety Signs and Signals) Regulations 1996 (Statutory Instrument 1996 No. 341).

9.5   Mutual detrimental influences

As demanded in Regulations 515.1 and 515.2, in selecting electrical equipment the designer needs to confirm that, when installed, there is no detrimental influence between the various items of equipment. Electrolysis is of particular concern, since this may occur at an unacceptable rate between dissimilar metals in damp or humid conditions and earth leakage current flow can accelerate metal degradation. Corrosion, too, needs special consideration, and enclosures in damp environments should be selected from

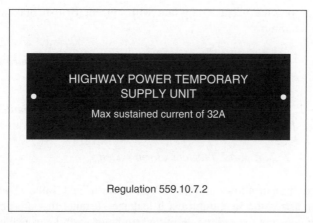

Figure 9.22   Highway power temporary supply unit notice

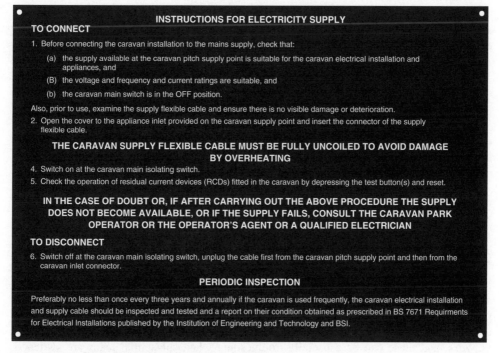

Figure 9.23   Caravan supply instructional notice

corrosion-resistant materials. Ferrous-based metals are particularly prone to corrosion where moisture or water is present and, where nonferrous materials are not available for selection, the use of a protective coating should be considered, taking care to avoid any adverse effects to the operation of the equipment so protected.

Another important aspect is the effect of different electrical systems on each other. LV installations complying with BS 7671 would not normally have adverse effects on other electrical systems (or vice versa) compliant with current British Standards and British Standards CPs, including:

- BS 6701 Parts 1 and 2 *Code of practice for installation of apparatus intended for connection to certain telecommunication systems*;
- BS 5266 Parts 1 and 3 *Emergency lighting*;
- BS 5839 Parts 1, 2, 3 and 4 *Fire detection and alarm systems for buildings*;
- BS 6739 *Code of practice for instrumentation in process control systems: installation design and practice*;
- BS 4737 Parts 1, 2, 3, 4 and 5 *Intruder alarm systems*.

Most electrical equipment allows current to leak to earth, and Table J1 of IEE Guidance Note No. 1 summarizes the maximum earth leakage currents that are permitted. Again, equipment which does not exceed the permitted limits of earth leakage current should not have any adverse effect on other equipment and on other systems.

Regulations 515.3.1 and 515.3.2 deal with electromagnetic compatibility in two important respects:

- Electrical equipment must be selected so that their immunity levels are such so as to be capable of coping with the likely electromagnetic influences in normal use from other equipment. In assessing the immunity levels, due account should be taken of the intended level of continuity of service of the application. In the absence of requirements of a particular standard for appropriate immunity levels, Regulation 515.3.1 requires that reference is made to BS EN 50 082 (now replaced by BS EN 61000).
- Emission levels of electromagnetic influences of equipment must be such that the influences cannot cause unacceptable interference with other electrical equipment either by conduction or propagation through air. Where necessary, the effects of emissions must be minimized. In the absence of requirements of a particular standard for appropriate low emission levels, Regulation 515.3.3 requires that reference is made to BS EN 50 081 (now replaced by BS EN 61000).

9.6   Compatibility

It is essential when selecting equipment that the consequential effects of the operation of such equipment on other equipment and other systems are carefully considered. Major problems can occur in the switching (on and off) of large loads, particularly if these are inductive (e.g. motors). Switching of any load produces higher than normal transient voltages, and these can be especially troublesome when substantial loads are switched. The effects of switching of large loads should not be overlooked with respect to the supply, which may transmit voltage transients (spikes) to other consumers on the network.

Information technology equipment, such as computers, data transmission facilities, electronic office equipment and point-of-sales apparatus, is particularly prone to interference from high voltage spikes. Careful attention to circumventing potential problems will be required. Consideration of information technology equipment is beyond the scope of this book, but guidance is available in the form of the *Applications Manual: Information Technology and Buildings* published by CIBSE.

9.7 Operation and maintenance manual

Guidance relating to the Operation and Maintenance Manual is beyond the scope of this Guide, but the requirements and advice are readily available in the form of Section 6 of the Health and Safety at Work etc. Act 1974, BS 4884 *Specification for technical manuals* and BS 4940 *Recommendations for the presentation of technical information about products and services in the construction industry*.

10

Wiring systems

10.1 Wiring systems

10.1.1 Wiring systems: general

There are numerous types of wiring system available to the designer nowadays, and much will depend on the designer's preferences and on the particular type of installation, its use and its environment, and also on any economic constraints. It is important, to afford compliance with BS 7671, to select a type of wiring system which is recognized as complying with the appropriate standard. Table 10.1 identifies wiring systems and cables specifically mentioned in BS 7671 and cross-refers to regulation numbers and Appendix 4 tables (of BS 7671) as, and where, appropriate. Table 10.2 lists some less-familiar cables. Cables manufactured to other recognized standards, provided they meet an equivalent degree of safety, are not precluded, but it is for the designer to confirm that such equivalence exists. Where necessary, it will be for the designer to obtain a 'Certificate of Compliance' from, for example, a third-party certification body, but such a certificate is not required by BS 7671.

Table 10.3 lists some of the currently available cables with low emission of smoke and corrosive gases when subjected to fire conditions. These cables should be considered for locations where there is a high degree of public access where dense and rapid formation of smoke and fumes may induce panic or injury. Similarly, it is desirable that escape routes and the like, which should not ideally be impeded by smoke and fumes, should be evaluated with a view to using this type of cable (see BS EN 60702, BS 6724, BS 6883 and BS 7211 for further information on such cables). These low-emission cables, where used, should not be installed in nonmetallic trunking and conduit unless they, too, are of the appropriate materials.

As called for in Regulation 521.10.1, excepting protective conductors, fixed-wiring nonsheathed cables must be enclosed in ducting or conduit or in a trunking system that provides at least the degree of protection IP4X or IPXXD and in which the cover can only be removed by means of a tool or deliberate action. Regulation 521.10.2 requires bare live (line and neutral) conductors to be installed on insulators. Regulation 521.9.3

A Practical Guide to The Wiring Regulations: 17th Edition IEE Wiring Regulations (BS 7671:2008)
Fourth Edition Geoffrey Stokes and John Bradley
© Geoffrey Stokes and John Bradley. Published by John Wiley & Sons, Ltd

Table 10.1 Common wiring systems: standards

Ref	Wiring system component	Regulation or Appendix 4 Table	British Standards
	Conduit, trunking and busbar systems		
A	Busbar trunking systems and powertrack systems	521.4	BS EN 60439-2
B	Conduit systems	521.6	BS EN 61386 series
C	Cable trunking and cable ducting systems	521.6	Appropriate part of the BS EN 50085 series
D	Cable tray and cable ladder systems	521.6	BS EN 61537
	Cables		
Copper			
F	Single-core 70 °C thermoplastic with and without sheath, nonarmoured	4D1A and 4D1B	BS 6004, BS 6231, BS 6346
G	Multicore 70 °C thermoplastic with and without sheath, nonarmoured	4D2A and 4D2B	BS 6004, BS 6346
H	Single-core 70 °C thermoplastic, armoured	4D3A and 4D3B	BS 6346
M	Multicore 70 °C thermoplastic, armoured	4D4A and 4D4B	BS 6346
I	Flat 70 °C thermoplastic with protective conductor, sheathed	4D5	BS 6004
J	Single-core 90 °C thermosetting with and without sheath, nonarmoured	4E1A and 4E1B	BS 5467, BS 7211
K	Multicore 90 °C thermosetting, nonarmoured	4E2A and 4E2B	BS 5467
L	Single-core 90 °C thermosetting, nonmagnetic armoured	4E3A and 4E3B	BS 5467, BS 6724
M	60 °C rubber-insulated flexible cables (not cords)	4F1A and 4F1B	BS 6007
N	85 °C and 150 °C rubber-insulated flexible cables	4F2A and 4F2B	BS 6007
O	Flexible cords	4F3A and 4F3B	BS 6141, BS 6500
P	Mineral insulated (copper conductors and sheath)	4G1A and 4G1B 4G2A and 4G2B	BS EN 60702-1
Aluminium			
Q	Single-core 70 °C thermoplastic (aluminium conductors) with and without sheath, nonarmoured	4H1A and 4H1B	BS 6004, BS 6346
R	Multicore 70 °C thermoplastic (aluminium conductors), nonarmoured	4H2A and 4H2B	BS 6004, BS 6346

Table 10.1 *(continued)*

Ref	Wiring system component	Regulation or Appendix 4 Table	British Standards
S	Single-core (aluminium conductors), nonmagnetic armoured, 70 °C thermoplastic insulated	4H3A and 4H3B	BS 6346
T	Multicore (aluminium conductors), armoured, 70 °C thermoplastic insulated	4H4A and 4H4B	BS 6346
U	Single-core 90 °C thermosetting (aluminium conductors), nonarmoured	4J1A and 4J1B	BS 5467, BS 6724
V	Multicore 90 °C thermosetting (aluminium conductors), nonarmoured	4J2A and 4J2B	BS 5467
W	Single-core 90 °C thermosetting (aluminium conductors), nonmagnetic armoured	4J3A and 4J3B	BS 5467, BS 6724
X	Multicore 90 °C thermosetting (aluminium conductors), armoured	4J4A and 4J4B	BS 5467, BS 6724

Table 10.2 Some less familiar cables

Ref	Cable type	British Standard	Possible utilization
A	Impregnated paper-insulated aluminium-sheathed (CONSAC)	BS 5593	PME power supplies
B	PVC-insulated cables for switchgear	BS 6231	Switchgear and panel wiring. See also BS 7671, Appendix 4, Table 4D1A
C	Rubber-insulated flexible	BS 6708	For use in mines and quarries
D	Flexible cables	BS EN 50214	Lift installations

Table 10.3 Cables with low emission of smoke and corrosive gases under fire conditions

Ref	Cable type	British Standard
A	Nonsheathed single-core thermosetting insulated	BS 7211
B	Armoured thermosetting insulated	BS 6724
C	Elastomer insulated (types C and D only), armoured and nonarmoured	BS 6883
D	Mineral insulated (except those served PVC overall)	BS EN 60702

allows flexible cables and flexible cords to be used as fixed wiring provided that their installation complies with the generality of BS 7671.

10.1.2   Fire performance of wiring systems

The toxic effects of PVC under fire conditions have long been recognized as a problem with regard to the effective escape of persons from buildings and the like. With the introduction of low smoke, halogen-free (LSHF) cables, the designer should seriously consider their use in locations where smoke and fumes may have disastrous effects on people trying to escape from a fire even in its early stages, especially where persons are likely to be panicked by the suggestion of toxic fumes. Equally, locations which house expensive computer and other electronic equipment should be considered for this treatment to prevent damage to sensitive components by toxic gaseous substances (e.g. hydrochloric acid produced by burning PVC).

Table 10.4, developed from BS 6387:1994 *Specification for performance requirements for cables required to maintain circuit integrity under fire conditions*, gives a summary of the fire test methods and related performance. Three categories are listed:

- tests related to resistance to fire alone;
- tests related to resistance to fire with water spray;
- tests related to resistance to fire with mechanical shock.

For example, a cable with the designation letters AWX indicates that it has met the requirements laid down by the standard for a 650 °C, 3 h fire resistance test with water spray and mechanical shock. The CWZ cable has met the most onerous performance requirements of a 3 h, 950 °C test with water and mechanical shock and it is this category which designers may wish to consider for the most demanding tasks of fire survival and circuit integrity. Fire alarm bell circuits may be one such situation where cables with superior integrity and fire resistance may be worthy of consideration in this respect.

Table 10.4   Fire performance requirements (BS 6387)

Resistance to fire alone			Resistance to fire with water spray	Resistance to fire with mechanical shock	
Test duration (min)	Test temp. (°C)	Designation letter		Designation letter	Test temp. (°C)
180	650	A	W	X	650
180	750	B	W	Y	750
180	950	C	W	Z	950
20	950	D	W	Z	950

10.2   External influences

10.2.1   External influences: general

Section 522 of BS 7671 deals with the many external influences to which equipment, including wiring systems, may be exposed. Tables 10.5, 10.6 and 10.8 catalogue the various factors to be taken into account in the selection of equipment for a particular environment. These tables should be taken only as a guide, and reference to BS 7671 will be needed in ascertaining the precise requirements. The designation letters (e.g. AE) given in Table 10.5 are taken from Appendix 5 (see also Chapter 32) of BS 7671, but it is important to recognize that there is not a direct correlation between these letters and the International Protection (IP) Code given in BS EN 60 529, except in the case of the letters AD (water).

The classification of external influences and their coding has been developed internationally for IEC Publication 60364 (the international wiring rules). Each code is given two letters and a number. The first letter indicates whether the code relates to environmental conditions (A), utilization (B) or the construction of the building (C). The second letter signifies the particular external influence (e.g. E in the case of solid foreign bodies). The number gives the degree or extent of the influence (e.g. 4 for dust); see Table 10.10 of this Guide.

Table 10.10 provides a summary of the International Protection (IP) Code, given in BS EN 60 529, and should be used only as a guide; reference should be made to the standard for the precise wording used. The index gives information as to the resistance of equipment to the ingress of both solid foreign bodies and human body parts, and the ingress of water.

The designer's main task in this respect is first to establish and assess all the external influences in all the distinct areas of the installation and its environment. Where more than one external influence is present in any area, the designer will need to take account of them all in matching equipment, including wiring systems, to the external influences (i.e. to the most onerous conditions).

Having assessed the external influences by observation, enquiry and previous experience, the designer has then to confirm that all equipment selected and methods of installation meet the necessary requirements. With regard to the International Protection (IP) Code, it is helpful to remember that the first numeral (see Table 10.10) relates to the resistance to solid objects in terms of (a) protection of persons from contact and (b) protection of equipment from solid bodies. The second numeral refers to the protection of equipment from the ingress of water. When equipment is selected correctly, protection against ingress of solids and water will be achieved, as will the necessary degree of protection against electric shock from the hazard of contact with a live part.

Where BS 7671 demands a particular degree of protection and quotes, for example, an IP index such as IPX4, the X in place of the first numeral indicates that protection against the ingress of solid foreign bodies is not important in the context of the particular consideration and can be any numeral (0 to 6). If the equipment is electrical, then the X should be replaced by a minimum of 2 to give the necessary degree of protection against contact with a live part. Therefore, if IPX4 is required by a particular regulation and this equipment houses electrical parts, then the necessary degree of protection would become

Table 10.5 List of external influences

Environmental considerations

Ref		Ambient temperature (AA) (°C)	Humidity (AB)	Altitude (m) (AC)	Water (AD)	Foreign bodies (AE)
1	Range and designation letters	AA1: −60 to +5 AA2: −40 to +5 AA3: −25 to +5 AA4: −5 to +40 AA5: +5 to +40 AA6: +5 to +60 AA7: −25 to +55 AA8: −50 to +40	AB	AC1: ≤2000 AC2: >2000	AD1: Negligible AD2: Drops AD3: Sprays AD4: Splashes AD5: Jets AD6: Waves AD7: Immersion AD8: Submersion	AE1: Negligible AE2: Small AE3: Very small AE4: Light dust AE5: Moderate dust AE6: Heavy dust

Ref		Corrosion (AF)	Impact (AG)	Vibration (AH)	Other mechanical stresses (AJ)	Flora (AK)
2	Range and designation letters	AF1: Negligible AF2: Atmospheric AF3: Intermittent AF4: Continuous	AG1: Low AG2: Medium AG3: High	AH1: Low AH2: Medium AH3: High		AK1: No hazard AK2: Hazard

Ref		Fauna (AL)	Electromagnetic (AM)	Solar (AN)	Seismic (AP)
3	Range and designation letters	AL1: No hazard AL2: Hazard	AM1: Level AM2: Signalling voltages AM3: Voltage amplitude variations AM4: Voltage unbalance AM5: Power frequency variations AM6: Induced low-frequency voltage AM7: D.c. in a.c. network AM8: Radiated magnetic fields	AN1: Negligible AN2: Medium AN2: High	AP1: Negligible AP2: Low AP3: Medium AP4: High

AM9: Electric fields
AM21: High frequency, etc.
AM22: Conducted ... nano
AM23: Conducted ... micro
AM24: Conducted oscillatory
AM25: Radiated HF
AM31: Electrostatic discharges
AM41: Ionization

Air movement (AR)
AR1: Low
AR2: Medium
AR3: High

Wind (AS)
AS1: Low
AS2: Medium
AS3: High

Contact with earth (BC)
BC1: None
BC2: Low
BC3: Frequent
BC4: Continuous

Evacuation (BD)
BD1: Normal
BD2: Difficult
BD3: Crowded
BD4: Difficult and crowded

Materials (BE)
BE1: No risk.
BE2: Fire risk
BE3: Explosion risk
BE4: Contamination risk

	Lightning (AQ)
4 Range and designation letters	AQ1: Negligible
	AQ2: Indirect
	AQ3: Direct

Utilization considerations

	Capability (BA)	**Resistance (BB)**
4 Range and designation letters	BA1: Ordinary	
	BA2: Children	
	BA3: Handicapped	
	BA4: Instructed	
	BA5: Skilled	

Buildings

	Materials (CA)	**Structure (CB)**
5 Range and designation letters	CA1: Non-combustible	CB1: Negligible risk
	CA2: Combustible	CB2: Fire propagation
		CB3: Structural movement
		CB4: Flexible

Table 10.6 External influences: temperature, water, solid bodies and corrosion

Influence	Class'n	Regulation	Main effect(s)	Likely location(s)	Attention
Ambient temperature	AA	522.1.1 and 522.1.2	Adverse effect, particularly on some cables, in use and in handling. Handling in low temperatures can cause damage to some cables (e.g. PVC)	Anywhere where temperatures are at variance (high and low) to normal and, in particular, where equipment is positioned close to heat sources. Floor heating systems. Cold rooms, etc.	Check temperature limits for in use and handling – use alternative equipment if necessary. Also, if ambient temperature is significantly different from the 30 °C, tabulated current-carrying capacities, given in Appendix 4 of the Wiring Regulations for 30 °C, will need modification by application of correction factor from Tables 4B 1 and 4B 2. Avoid use of PVC cables in very low temperatures
Heat sources	–	522.2.1 and 522.2.2	Equipment degradation and in particular to wiring systems (PVC can suffer softening at temperatures approaching 115 °C, leading to decomposition and the emission of corrosive gases)	In direct sunlight and anywhere where heat is generated	Placed away from source, shield, reinforced insulation, derate equipment and/or select alternative equipment
		522.2.2		In luminaires where cables are in close proximity to heat sources (lamps and controlgear)	Cables to be suitable for highest temperature encountered or sleeved with material which is suitably temperature rated
Water	AD	522.3.1	Equipment degradation by corrosion and electrolytic action and, in the case of waves, mechanical damage	E.g. car wash plants, swimming pools, kitchens, exterior locations	Select appropriate IP rating for equipment

High humidity	AB	522.3.2		Collection of condensation in equipment in locations such as saunas, etc.	Provide drain holes where draining can be effective
Waves	AD	522.3.3		E.g. marinas and water treatment works	Provide mechanical protection (against impact, vibration and other stresses)
Solid foreign bodies,	AE	522.4.1	Equipment degradation and overheating	Locations generally	Select appropriate IP rating for equipment
including dust	AE4 AE5 AE6	522.4.2	Dust, etc. accumulations leading to potential explosion risks	E.g. sawmills, quarries and cement works	Select appropriate IP rating for equipment
Corrosive and polluting substances	AF	522.5.1	Equipment corrosion and deterioration	E.g. chemical works and waste disposal works	Select equipment made from materials resistant to such substances or otherwise provide protection to equipment
	AF	522.5.2	Equipment degradation by erosion	Damp and/or polluting atmospheres, generally	Avoid liability to produce electrolytic action from different metals coming into contact with each other
	AF	522.5.3	Mutual or individual equipment hazardous degradation	Damp and/or polluting atmospheres, generally	Ensure, as far as is practicable, that equipment is not placed in contact with other equipment

Table 10.7 External influences: mechanical impact, mechanical stresses and vibration

Influence	Class'n	Regulation	Main effect(s)	Likely location(s)	Attention
Mechanical impact	AG	522.6.1	Unacceptable and potentially hazardous damage	Generally	Select equipment suitable for location, bearing in mind operational conditions
	AG 2 and AG 3	522.6.2	Unacceptable and potentially hazardous damage from medium- or high-severity impact	Generally	Select wiring system suitable for such impact or relocate equipment or provide additional mechanical protection
	AJ	522.8.10	Underground cable damage	Generally	Underground cable to be of the earthed armoured or metal sheath type (except cables in conduit or ducts). All cables to be marked by cable covers or marker tape
	AG	522.6.4	Damage to wiring systems in floors	Generally	Select wiring system capable of withstanding any likely damage or provide additional mechanical protection
	AG	522.6.5	Damage to wiring systems in floor joists and ceiling battens	Generally	Leave at least 50 mm between cable and top, or bottom as appropriate, of joist or batten or use cable with earthed metallic covering (e.g. MICC) or earthed metallic conduit or trunking or mechanical protection sufficient to prevent penetration by nails, screws and the like
	AG	522.6.6	Damage to concealed wiring systems in walls	Generally	Install cables more than 50 mm behind finished wall surface, or vertically/horizontally from an electrical point, accessory or switchgear or where cables can be reasonably expected (i.e. see Figure 10.1), and cables to be provided with additional protection by an RCD in compliance with 415.1.1 unless the installation intended to be under supervision of a skilled or instructed

		522.6.7		Generally	person. Alternatively, cables to have earthed metallic covering (e.g. MICC), earthed steel conduit or trunking, or thick steel plate protection
	AG	522.6.8	Damage to concealed wiring systems in walls or partitions with an internal construction that includes metallic parts (except nails screws and the like), except where the installation intended to be under supervision of a skilled or instructed person	Generally	Use cable with earthed metallic covering (e.g. MICC) or earthed metallic conduit or trunking or mechanical protection sufficient to prevent penetration by nails, screws and the like. Alternatively, cables to be provided with additional protection by an RCD in compliance with 415.1.1
Vibration	AH 2 and AH 3	522.7.1	Stressed or loosening of connections	Structures and equipment liable to medium- and high-severity vibration	Select equipment, including wiring system (e.g. flexible cables) and support systems capable of withstanding the level of vibration
	AH	522.7.2	Fixed installation of suspended equipment, such as luminaires	Suspended equipment liable to vibration	Use cables with flexible cores
Mechanical stresses	AJ	522.8.1	Damage to insulation, sheaths and connections of cables	Generally, in installation, use and maintenance	Select suitable equipment and/or take adequate precautions

(continued overleaf)

Table 10.7 (*continued*)

Influence	Class'n	Regulation	Main effect(s)	Likely location(s)	Attention
		522.8.2	Cable damage during installation	Generally, in inst., use and maintenance	If laid in the structure, conduit or ducting to be completed before circuit cables are drawn in, except for pre-wired conduit assembly specifically designed for the installation
		522.8.3	Cable suffering damage during installation and subsequently due to inadequate bend radii	Generally, in inst., use and maintenance	Carefully follow manufacturer's advice and official guidance publications
		522.8.4	Unsupported cable suffering damage under own weight	Generally, in inst., use and maintenance	Provide adequate cable supports – consult cable manufacturer, if necessary. Provide adequate terminations (e.g. cable glands)
		522.8.5	Damage to terminations and connections	Generally, in inst., use and maintenance	
		522.8.6	Damage caused by excessive tensile and torsional stresses to flexible cables	Generally, in inst., use and maintenance	Select suitable equipment and provide adequate cable supports

Table 10.8   Notches and drilled holes in timber joists

	Position of notch/hole[a]	Max. depth of notch	Max. diameter of hole[b]	Min. spacing between holes
Notch	Between 0.07 and 0.25 times joist span form the joist supports	0.125 times the depth of joist	–	–
Hole	Between 0.25 and 0.4 times joist span form the joist supports	–	0.25 times the depth of joist	Spacing not less than three hole diameters (centre to centre)

[a]No notches or holes permitted in roof rafters except where the rafter is birdsmouthed. Depth not to exceed 0.33 times the depth of the rafter.
[b]Holes to be drilled in the neutral axis.

IP24 because of the need to protect persons from contact with live parts. Where a product is claimed to meet the degree of protection of, say, IP20 or IP40, this means that there is no assessment of the degree of protection against the ingress of water.

10.2.2   Temperature

Ambient temperature is an important consideration for the designer, in that it affects the type of wiring system to be selected. In this assessment, the highest and the lowest temperatures are important considerations. The highest is important because it is necessary to confirm that the maximum operating and final temperatures of the cable are not exceeded and to assess the current-carrying capacities of the conductors (tabulated values of current-carrying capacities given in Appendix 4 are based on an ambient temperature of 30 °C). Where the ambient temperature(s) are not 30 °C, application of the rating factors given in Tables 4B1 and 4B2 of BS 7671 will provide the corrected current-carrying capacities for the particular temperature condition under consideration. At lower temperatures some cables may suffer damage, particularly during installation, and it is always advisable when installing wiring systems to check with the manufacturers as to the performance of their products under different temperature conditions.

For most installations, 30 °C may reasonably be taken as the norm, though in most occupied buildings 25 °C may be nearer the mark. Cables in loft spaces may well be subjected to an ambient temperature of 30 °C (solar gain) during the summer months. However, whilst 30 °C may be taken as the general ambient temperature, care should be taken where equipment is sited close to heat sources (e.g. radiant heaters, unlagged water heaters). PVC insulation installed in high temperatures will soften, and when subjected to temperatures approaching 115 °C it will severely soften with the ultimate consequence of decomposition and the emission of corrosive products which may attack adjacent material. Accessories, too, may be unsuitable for the temperatures reached, and the designer should check with the standard for the particular equipment to confirm that it will be suitable for the highest temperatures likely to be encountered before a selection is made (see also Table 10.6 of this Guide).

10.2.3   Water

The terms 'drip-proof', 'splash-proof', etc., though widely used, are not defined in any standard. The degrees of protection against the ingress of water (and solid bodies) claimed for products is that pertaining in their 'installed' condition. It is important to recognize that, in order to maintain this protection during and after the product is installed, careful installation is paramount. Holes for cable entry or fixing should not be drilled unless they are necessary and care is taken to fill and seal them as necessary. All wiring system accessories, such as cable glands, should also maintain the same degree of protection.

Water can be in the form of drops (such as rain) and moisture (such as in damp, humid conditions or fine spray). Where equipment, such as cable sheaths and armouring and metallic conduit and metallic trunking, is exposed to the weather or to damp conditions indoors, it and its fixings must be of the corrosion-resistant type (e.g. galvanized) or with a corrosion-resistant finish (e.g. plastic covering). Metallic materials must not be in contact with dissimilar metals where electrolytic action is liable to erode one or other or both of the metals concerned. Where there is the likelihood of a wiring system being susceptible to the ingress of moisture, holes must be left for safe drainage of any build-up of condensation or other aqueous deposits.

Particular care is needed in selection of equipment with regard to terminations, including joints in cables. Where insulating material is used to seal enclosures, such sealing must be capable of withstanding the normal temperature range without losing its ability to provide a proper seal. Ducting, cable ducts and trunking must be sealed so that ingress of water is prevented or such enclosures should be located so that such ingress is not possible. Where equipment is liable to be affected by waves (e.g. marinas), special mechanical protection may be necessary (see Sectio 10.2.6 and Table 10.6 of this Guide).

10.2.4   Solid foreign bodies

Protection against solid bodies (including dust) is required either to protect persons against coming into contact with live parts or to prevent entry of other bodies, such as tools, into the equipment. Particular attention should be paid to prevent dust (e.g. cement) and other debris being deposited into flush conduit system during installation. This may be done by the use of temporary conduit box lids, stop-ends, etc., thus preventing blockages later when drawing in cables with the consequential potential for insulation and/or sheath damage.

Dust, too, can present problems to installations in service and can lead to overheating of equipment and eventually to being the cause of fire and, in some extreme cases, to a potentially explosive situation (see also Table 10.6 of this Guide).

10.2.5   Corrosive and polluting substances

There are a number of corrosive and/or polluting substances commonly in use in building works which may, if not properly considered in the context of the electrical installation, lead to erosion or other degradation of equipment, particular wiring systems. Lime, cement, plaster and materials containing magnesium chloride can cause problems with wiring systems in direct contact with these materials. This can often be avoided by the use of protective coatings applied to the wiring systems during erection or their separation from the offending materials by the use of some noncorrosive media (e.g. plastic capping). The corrosive reactions are usually intensified by the presence of water or damp conditions, and special care is needed in the selection of equipment and its fixings in these conditions.

Flora and mould growth is another worrying problem, as is the placing of wiring systems touching oak and other acidic woods. Again, very careful consideration should be given to these materials when selecting wiring materials, and physical separation by the use of noncorrosive covering should be considered. Some manufacturers can supply insulating backing materials for use in such circumstances.

As with damp situations, the designer should avoid contact with dissimilar metals which may otherwise be degraded by electrolytic action. Unplated or bare aluminium in contact with copper or brass (or any other alloy with a copper content) should be avoided. The use of plated materials minimizes electrolytic action, and, where necessary, metal materials should be placed in enclosures which prevent the ingress of moisture or be otherwise encapsulated in suitable material (e.g. bitumen).

In addition to all the many possible corrosive effects between the electrical equipment and building construction materials, another important consideration is the effect of the environment in which the installation is to be installed and used. Inhospitable environments can have very serious adverse effects on equipment, including conductors (insulation, sheaths and armouring) and enclosures. In all cases of concern the designer should seek the advice of the product manufacturer to establish its performance under the environmental conditions perceived. Hydrocarbons (e.g. petroleum products, creosote) may attack rubber and PVC, and plasticizers have been known to migrate from polystyrene (a common heat insulator) to PVC (i.e. mutual detrimental influence). Where there is doubt about the environmental conditions, the designer would need to consult an expert in the particular environmental conditions before deciding on the selection of equipment (see also Table 10.6 of this Guide).

10.2.6 Mechanical damage: general

Equipment (switchgear, accessories and wiring systems) must generally be protected against impact, vibration and other mechanical stresses, such as abrasion. Where, for example, the utilization of the building is such that equipment is likely to suffer impact, the equipment must be capable of withstanding that impact or, alternatively, the equipment must be protected at least in the area subject to the impact, without rendering it unsafe or unserviceable. By far the best solution, if practicable, is to locate equipment in positions where it is not susceptible to mechanical damage. Socket-outlets, for example, can by rule of thumb be positioned about 150 mm above finished floor level or above a work surface where the threat of mechanical damage by furniture and household tools is likely to be minimal. However, the requirements of local or national building regulations, such as Part M (Access and use of buildings) of the Building Regulations in England and Wales, may mean that socket-outlets (and other accessories) have to be installed at a greater height above the floor to assist in their use by disabled persons.

Care must also be taken of wiring systems where they pass through holes in walls and in metalwork where precautions need to be taken to prevent damage to the cables by abrasion against sharp edges This can usually be minimized, in the case of holes in walls, by proper supports of cables or properly formed ducts and, in the case of abrasion by metalwork, by brass/plastic bushes and grommets.

Table 10.7 summarizes the requirements relating to mechanical damage from impact, vibration and other mechanical stresses.

10.2.7 Mechanical damage: concealed and buried cables

Where nonmetallic sheathed or armoured cables (e.g. PVC/PVC) are run on walls prior to plastering, they are sometimes encased in metal or plastic channelling; this method serves as a useful means of fixing cables and provides some measure of protection against the plasterer's trowel. However, it does not provide mechanical protection sufficient to prevent penetration by nails, screws and the like as envisaged by indent (iv) of Regulation 522.6.6. Indent (v) of Regulation 522.6.6 gives a method of installing such cables in recognized zones of 150 mm width, i.e. 150 mm from the ceiling horizontally, 150 mm from the corner between two adjoining vertical walls in both directions, or vertically and horizontally from an accessory or electrical point. Figure 10.1 graphically illustrates these zones. Except where the cables are in an installation intended to be under the supervision of a skilled or instructed person, such as in certain commercial premises, Regulation 522.6.7 requires the cables to be provided with additional protection by an RCD having the characteristics specified in Regulation 415.1.1 (see Section 5.7 of this Guide). Where it is not possible to locate such cables in the prescribed zones of Regulation 522.6.6(v), local mechanical protection will be required; this can be achieved, for example, by earthed steel conduit drops.

Additional protection by an RCD having the characteristics specified in Regulation 415.1.1 is also required to nonmetallic sheathed cables (e.g. PVC/PVC) concealed, at whatever depth from the surface, in a wall or partition having an internal construction that includes metallic parts other than fixings such as nails and screws (Regulation 522.6.8 refers). However, this requirement does not apply if the installation is intended to be under the supervision of a skilled or instructed person.

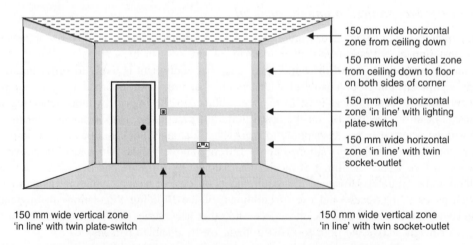

150 mm wide horizontal zone from ceiling down

150 mm wide vertical zone from ceiling down to floor on both sides of corner

150 mm wide horizontal zone 'in line' with lighting plate-switch

150 mm wide horizontal zone 'in line' with twin socket-outlet

150 mm wide vertical zone 'in line' with twin plate-switch

150 mm wide vertical zone 'in line' with twin socket-outlet

Wiring systems not within the designated zones must be not less than 50 mm from the surface or be adequately protected against mechanical damage by earthed metal or by the use of earthed metallic sheathed cables (e.g. MICC).

Figure 10.1   'Safe' zones for mechanically unprotected nonmetallic wiring systems

Wiring systems are frequently installed in roof spaces, buried in inter-floor spaces and in partition walls, and special care is needed in terms of locating these and providing additional mechanical protection where needed. Generally, cables need mechanical protection everywhere where they are likely to be penetrated by screws, nails and the like or to be of a type incorporating an earthed metallic covering (e.g. MICC) or enclosed in earthed steel conduit. Cables buried at more than 50 mm from the surface of a wall are not generally considered to be a potential risk (except as envisaged in Regulation 522.6.8, mentioned above, which requires measures to be taken to prevent metallic parts, such as a frame, in the internal construction of a wall or partition from becoming live, even where the concealed cables are at a depth of more than 50 mm from the surface).

Modern building methods sometimes employ narrow (e.g. 50 mm) partitioning boards, and these can present problems for the installer in providing mechanical protection. There are the options of using either MICC or steel conduit, but in many cases these wiring systems will be impracticable. An alternative approach would be to restrict all cables to the 'safe' zones (see Figure 10.1) and provide accessory points on both sides of the partition. This would indicate to others that cables may have been installed vertically or horizontally from these points. However, this is not necessary where the location of an accessory, point or switchgear on one side of a wall or partition of 100 mm thickness or less can be determined from the reverse side, as the 'safe' zone(s) created by the point, accessory or switchgear is deemed by Regulation 522.6.6 to extend to the reverse side.

Wiring systems (except earthing conductors – see Chapter 12 of this Guide) installed beneath the ground must be installed in ducts or pipes. Alternatively, they must be of the armoured or metal-sheathed type (e.g. SWA cable) which must be earthed. Cables in ducts and pipes located in the ground should preferably also be of one of these types. Underground cables must be laid deep enough (usually taken as a minimum of 600 mm, in the absence of precise knowledge) so that they are unlikely to sustain mechanical damage from hand tools (e.g. spades, forks and picks). They must be identified by cable marking tiles or covers or by suitable marking tape, for compliance with Regulation 522.8.10. Figure 10.2 shows a typical section of a wiring system buried in the ground. Additionally, cable routes and their depth should be recorded on the installation's 'as-fitted' drawings for future reference.

Figure 10.3 shows four different methods of complying with the requirements of BS 7671 for cables taken through joists in intermediate floor spaces. Table 10.8 gives guidance with regard to forming notches and drilling holes in timber joists.

10.2.8   Damage by fauna, flora and mould growth

Just two Regulations, 522.9.1 and 522.10.1, deal with the external influences of fauna, flora and mould growth. These call for suitable precautions to be taken against the effects of conditions likely to cause a hazard (AK2 in the case of flora and mould growth and AL2 in the case of fauna).

The term 'fauna' includes such animals as rats, mice and glis-glis. In locations where these may be present, care should be taken in selecting the wiring system or making provision for additional protection. Mechanically unprotected cables, such as PVC and rubber, are liable to damage by gnawing by rodents, particularly if the cables block their

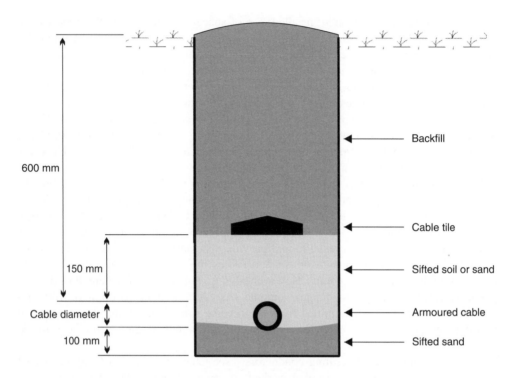

600 mm

150 mm

Cable diameter

100 mm

Backfill

Cable tile

Sifted soil or sand

Armoured cable

Sifted sand

Figure 10.2   A typical section of a buried wiring system

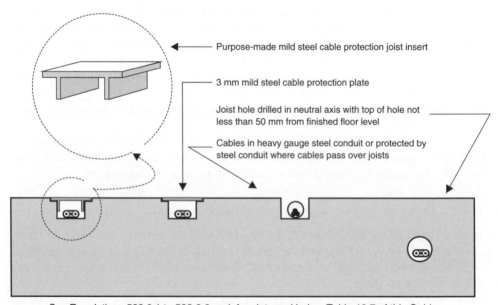

Purpose-made mild steel cable protection joist insert

3 mm mild steel cable protection plate

Joist hole drilled in neutral axis with top of hole not less than 50 mm from finished floor level

Cables in heavy gauge steel conduit or protected by steel conduit where cables pass over joists

See Regulations 522.6.1 to 522.6.6 and, for slots and holes, Table 10.7 of this Guide

Figure 10.3   Cables through timber joists: four methods of mechanical protection

established run or what they consider to be a potential run. Depending on the extent of the problem, wiring systems should be selected to minimize the potential damage by such creatures, and metal conduit and trunking systems and MICC would lend themselves to such applications.

10.2.9 Building design considerations

Regulations 522.15.1 and 522.15.2 call for precautions to be taken where there is the possibility of building structural movement. These precautions would need to take account of relative movement of the building and cables and their supports arranged so as not to permit the conductors to be excessively stressed. More importantly, the systems of protection (e.g. overcurrent protection and fault protection) must be capable of tolerating any structural movement without impairing their operation in any way. Where the structure is considered to be flexible and/or unstable and, therefore, is subject to external influence CB4 (flexible structure), flexible wiring systems (i.e. flexible cables) must be employed in order that relative movement can be accommodated without damage to the electrical systems. Structural movement can, for example be expected in towers, masts, bridges and the like, and the installation designer should consult with the structural engineer where the structural stability is in question.

10.2.10 Solar radiation

Regulation 522.11.1 calls for cables affected by direct sunlight to be resistant to damage by ultraviolet light. PVC cables to BS 6346 and synthetic rubber (but not natural rubber) probably afford the best performance for outside locations not in the shade. Alternatively, general cables may be used provided that they they are shielded from the sun, but it is essential that the shielding material does not restrict ventilation of the cable.

Another important consideration of direct sunlight is the solar gain, which effectively increases the ambient temperature. In the absence of better data, it is generally considered prudent to allow a 20 °C increase in ambient temperature from solar gain; this would translate to a decrease of about 7–13% in current-carrying capacity of the cables, depending on the particular wiring system. In other words, as a rule of thumb, the required current-carrying capacity of cables in direct sunlight without shielding should be increased by approximately 10%.

Table 10.9 summarizes the requirements relating to solar radiation and also includes information concerning flora and fauna and building design considerations.

10.3 Proximity to other services: general

Regulations contained in Section 528 of BS 7671 set out the requirements for precautionary measures to be taken where wiring systems are in close proximity to other electrical services, or to communications cables or to nonelectrical services. Regulation 528.1 deals with wiring systems interrelated with other electrical systems, Regulation 528.2 deals with wiring systems interrelated with communications cables, and Regulation Group 528.3 sets out the requirements concerning the proximity of wiring systems to nonelectrical systems (e.g. water, oil, gas, steam, chemical and other services). Table 10.10 gives the International Protection (IP) code.

Table 10.9 Other external influences

Influence	Class'n	Regulation	Main effect(s)	Likely location(s)	Attention
Flora and mould growth	AK2	522.9.1	Equipment degradation by vegetation, etc.	Agricultural and horticultural and exterior locations generally	Select suitable equipment and/or adopt special preventative measures
Fauna	AL2	522.10.1	Equipment degradation by animals (e.g. rats, mice)		Select suitable wiring system or adopt special protective measures, such as by mechanical characteristics of wiring system, location selected or additional local or general protection from mechanical damage
Solar radiation	AN2	522.11.1	Damage to wiring system by significant exposure to direct sunlight (including ultraviolet). Temperatures approaching 115 °C will have detrimental effects on natural rubber	Generally on exterior locations	Select suitable wiring system capable of withstanding the effects of direct sunlight or provide suitable shielding Some cable sheaths have ultraviolet stabilizers Cables subjected to solar gain will need their current-carrying capacity derated (as a general rule, the ambient temperature of 50 °C may be used)
Building design	CB3	522.15.1	Damage to wiring system and/or protective measures due to relative structural movement	Building in which structural movement occurs	Select suitable wiring system capable of withstanding the effects of structural movement and/or take precautions to prevent damage due to excessive mechanical stresses
	CB4	522.15.2	Damage to wiring system and/or protective measures due to movement of structure	E.g. unstable and flexible structures	

Table 10.10 International Protection (IP) code[a]

		First numeral Degree of protection		Second numeral Degree of protection
0	A[b] B[b]	No special protection No special protection	0	No special protection.
1	A B	Protection against accidental or inadvertent contact by large surface of body (e.g. hand) but not protected against deliberate access. Protection against the ingress of objects greater than 50 mm diameter.	1	Protection provides for no harmful effects from water dripping on to the enclosure.
2	A B	Protection against contact of the ingress of the standard finger (see Standard). Protection against the ingress of objects greater than 12 mm diameter and of length greater than 80 mm.	2	Protection provides for no harmful effects from water dripping on to the enclosure when tilted up to 15° from the vertical (sometimes referred to as 'drip proof').
3	A B	Protection against the ingress of objects such as tools and wires of greater than 2.5 mm thick. Protection against the ingress of objects such as tools and wires of greater than 2.5 mm thick.	3	Protection provides for no harmful effects from rain falling at any angle from the vertical (sometimes referred to as 'rain proof').
4	A B	Protection against the ingress of wires and strips more than 1 mm thick. Protection against the ingress of solid objects of more than 1 mm thick.	4	Protection provides for no harmful effects from water splashed from any direction (sometimes referred to as 'splash proof').
5	A B	Complete protection against contact assured. Complete protection against harmful deposits of dust assured (dust may enter but not in sufficient amounts to interfere with operation).	5	Protection provides for no harmful effects from water projected by a nozzle from any direction, under defined conditions (sometimes referred to as 'jet proof').
6	A B	Complete protection against contact assured. Complete protection against ingress of dust assured (sometimes referred to as 'dust tight').	6	Protection provides against ingress of water from heavy seas or power jets, under defined conditions. Protection against conditions on ships (sometimes referred to as 'watertight equipment').
			7	Protection provides against ingress of water from immersion, under defined conditions of pressure and time.
			8	Protection provides against ingress of water from indefinite immersion, under specified pressure.

[a]The table has been developed from BS EN 60529 (see the standard for precise wording). The two numerals in the table may be followed by up to one additional letter (A, B, C or D), denoting the degree of protection against access to hazardous parts, and one supplementary letter (the most common of which are H, for high-voltage apparatus, M and S, relating to ingress of water, and W, relating to specified weather conditions).

[b]'A' represents protection of persons from contact with live or moving parts. 'B' represents protection of equipment against solid bodies.

10.3.1 Proximity of electrical wiring systems to other electrical systems

The requirements of Regulation 528.1 relating to the proximity of wiring systems to other electrical services are concerned only with electrical safety. Segregation requirements considered necessary for other reasons, such as for reasons associated with a safety service, are the remit of the appropriate BS CP and/or Chapter 56 of BS 7671.

The requirements as far as electrical safety is concerned are set out in Table 10.11 and Figure 10.4.

10.3.2 Proximity of electrical wiring systems to communications cables

Regulation 528.2 deals with the proximity of underground electrical wiring systems to underground telecommunications cables. The regulation requires a clearance of at least 100 mm between these services, including at any points where they cross. Alternatively, a fire-resistant partition must be provided between the cables, formed of materials such as bricks or fire-retardant conduit.

The requirements of Regulation 528.2 relate only to electrical safety. They do not deal with avoiding electromagnetic or electrostatic interference to telecommunications circuits caused by electrical wiring systems, for which measures may have to be taken. BS 6701 deals with segregation requirements for telecommunications, and some information is given Table 10.12 of this Guide. The BS EN 50174 series of standards gives segregation requirements for information technology cables.

Band I and Band II circuits enclosed in the same trunking compartment, every cable insulated for the highest voltage present.

Band I and Band II circuits enclosed in the same conduit system, every cable insulated for the highest voltage present.

Band I and Band II circuits enclosed in the same conduit or trunking system, each Band I conductor of multicare cable separated from Band II circuit by an earthed metallic screen with current-carrying capacity equal to that of the largest Band II conductor.

Band I and Band II circuits enclosed in the same trunking or conduit system, earch Band I conductor of multicare cable separated from Band II circuit by an earthed metallic screen with current-carrying capacity equal to that of the largest Band II conductor.

Band I and Band II circuits enclosed in the same conduit or trunking system, each Band I conductor of multicare cable or cord individually or collectively insulated for highest voltage present in the Band II circuit.

For contols and outlets for Band I and Band II mounted in a common box, accessory or black, the cables and connections of the different voltage Bands must be segregated by an effective partition, which, if metal, must be earthed.

Figure 10.4 Examples of circuit segregation (Band I and Band II circuits)

Table 10.11   Summary of requirements for circuits at different voltage bands[a]

Ref	Regulation no.	Circuit in close proximity		Admissible	Conditions
A	528.1	Voltage exceeding LV	Band I (e.g. extra-LV) or Band II (e.g. LV)	No	None
B				Yes	Every cable or conductor insulated for the highest voltage present
C				Yes	Each conductor of multicore cable is insulated for the highest voltage present
D				Yes	The cables are insulated for their system voltage and contained in separate compartments of ducting or trunking system, or in separate conduit, trunking or ducting systems
E				Yes	The cables are installed on a cable tray with having a partition that provides physical separation
F		Band II (e.g. LV)	Band I (e.g. extra-LV)	No	None
G				Yes	Every cable or conductor insulated for the highest voltage present
H				Yes	Each conductor of multicore cable is insulated for the highest voltage present
I				Yes	The cables are insulated for their system voltage and contained in separate compartments of ducting or trunking system, or in separate conduit, trunking or ducting systems
J				Yes	The cables are installed on a cable tray with having a partition that provides physical separation
K				Yes	In a multicore cable, cores of Band I circuits are separated from cores of Band II circuits by an earthed metallic screen with current-carrying capacity equivalent to that of largest core of Band II circuits

[a]For SELV and PELV systems, the requirements of Regulation Group 414.4 apply. See Section 5.6 of this Guide.

Table 10.12 Minimum separation distances between LV a.c. cables and telecommunications cables

Ref	Voltage range to earth U_o (V)	Location	Restriction(s)	Minimum distance (or other method) (mm)
A	$50 < U_o \geq 600$	Interior – generally	None	50
B			LV cables enclosed in conduit in earthed metallic conduit	Nil
C			LV cables enclosed in conduit in earthed metallic trunking	Nil
D			LV cables are the mineral insulated type	Nil
E			LV cables are the armoured type	Nil
F			LV cables sharing the cable tray	50
G		Interior – crossovers	Where LV cables and telecoms cables, of necessity, crossover	Additional site-applied insulation required at cross-overs (except where either cable is armoured)
H	$600 < U_o \geq$ 900	Interior – generally	None	150
I			Nonflammable, non-conductive and rigid divider to maintain separation between the two sets of cables	50
J		Exterior – generally	None	150
K			Nonflammable, non-conductive and rigid divider to maintain separation between the two sets of cables	50

10.3.3   Proximity of electrical wiring systems to nonelectrical systems

Regulations 528.3.1–528.3.5 deal with wiring systems which are in proximity to non-electrical systems and where there may be a consequential potential hazard. If so, they must be provided with additional suitable protection so that deleterious effects on the electrical system, both by virtue of its proximity and all likely operations associated with the electrical and other systems, are avoided. Generally speaking, wiring systems should be located, where practicable, away from condensation, water, steam, smoke and fumes, compressed air systems and oil systems. Where this is not possible or desirable, preventative measures need to be taken to obviate the risks of mutual detrimental effects on the various systems, unless the systems are under constant effective supervision.

Where, because of the proximity of the systems, shielding of the electrical system is necessary, account must be taken of the effects of this shielding on the electrical system. In particular, the current-carrying capacities of wiring systems may be reduced owing to inhibited heat dissipation from the conductors due to the presence of the shield.

General installation wiring systems must not be run in lift shafts and hoist shafts unless they form part of the lift installation or hoist (as defined in the BS EN 81-1 series), as called for in Regulation 528.3.5.

10.4 Methods of installation of cables

10.4.1 General

There are only three regulations (521.1 to 521.3) in Chapter 52 of BS 7671 dealing with the various methods of installation of wiring systems based on cables, and most of the related information is embodied in Appendix 4: *Current-carrying capacity and voltage drop for cables and flexible cords*.

The tabulated values given in Appendix 4 of BS 7671 for current-carrying capacities and for voltage drop are based upon an ambient temperature of 30 °C and a frequency of 49–61 Hz. Where these parameters are different, the tabulated values of current-carrying capacities will need correction by applying rating factors given in Tables 4B1 and 4B2 of Appendix 4 in the case of temperature, and where the frequency is significantly outside the range given by applying a correction factor of $f_1/50$ (where f_1 is the actual frequency of the system) to the reactance component of the voltage-drop tabulated values given.

Table 10.13 lists the 11 Reference Methods and gives examples of the 54 Installation Methods as they relate to current-carrying capacities. The Reference Methods are those methods of installation for which the current-carrying capacity has been determined by test or calculation in order to cover all of the Installation Methods. It would impracticable to establish current ratings individually for every installation method by calculation or test and publish these, since this would result in the same current rating for many Installation Methods.

10.4.2 Current-carrying capacities, cross-sectional area of conductors and conductor operating temperatures

Section 523 of BS 7671 deals with operational aspects as they relate to current-carrying capacities of cables. The factors which affect the determination of the current-carrying capacity of a cable are:

- conductor material;
- conductor insulating material;
- the method of wiring system installation (see Table 10.13);
- grouping of conductors with other conductors.

The basic requirement, embodied in Regulation 523.1, is that the maximum permitted normal operating temperature of the conductor t_p is not more than the tabulated values given in Table 52.1 of BS 7671 (see Table 10.15 of this Guide). To achieve compliance with this requirement, tabulated values of current-carrying capacities I_t given in Appendix 4 of BS 7671 are to be used as a starting basis for calculating the effective current-carrying

Table 10.13 Methods of installation of wiring systems

Reference method	Examples of commonly used installation methods	
	Installation Method no.	Description
A	1, 2	Nonsheathed cables (Installation Method 1) or multicore cable (Installation Method 2) in conduit embedded in a wall consisting of an outer weatherproof skin, thermal insulation and an inner skin of wood or wood-like material having a thermal conductance not less than $10 \, W/m^2$ K. The conduit is in contact with the inner skin
	3	Multicore cable embedded directly in a wall consisting of an outer weatherproof skin, thermal insulation and an inner skin of wood or wood-like material having a thermal conductance not less than $10 \, W/m^2$ K. The cable is in contact with the inner skin
B	4, 5	Nonsheathed cables (Installation Method 4) or multicore cable (Installation Method 5) in conduit mounted on a wooden or masonry wall, such that the gap between the conduit and the surface is less than 0.3 times the conduit diameter
	6, 8	Nonsheathed cables (Installation Method 6) or multicore cable (Installation Method 8) in trunking run horizontally on a wooden or masonry wall
	7, 9	Nonsheathed cables (Installation Method 7) or multicore cable (Installation Method 9) in trunking run vertically on a wooden or masonry wall
C	20	Single-core or multicore cable mounted on a wooden wall so that the gap between the cable and the surface is less than 0.3 times the cable diameter. The term 'masonry' is taken not to include thermally insulating materials
	30	Single-core or multicore cables on unperforated cable tray
D	70	Multicore unarmoured cable drawn into a 100 mm diameter plastic, earthenware or metallic duct laid in direct contact with soil having a thermal resistivity of 2.5 K m/W and at a depth of 0.8 m
	72, 73	Sheathed armoured cable direct in ground having a thermal resistivity of not less than 2.5 K m/W, either without added mechanical protection (Installation Method 72) or with added mechanical protection, such as cable covers (Installation Method 73).

Table 10.13 (*continued*)

Reference method	Examples of commonly used installation methods	
	Installation Method no.	Description
E, F and G	31 to 35	Single-core or multicore cable in free air, such as on perforated cable tray (Installation Method 31), brackets or a wire mesh tray (Installation Method 32), supports such as cable cleats (Installation Method 33), cable ladder (Installation Method 34), or suspended from or incorporating a support wire or harness (Installation Method 35). The cable is supported such that the total heat dissipation is not impeded and care is taken that natural air convection is not impeded (e.g. a clearance is maintained between a cable and any adjacent surface of at least 0.3 times the cable external diameter for multicore cables or 1.0 times the cable diameter for single-core cables). Heating due to solar radiation and other sources must be taken into account
100 and 101	100, 101	Flat twin and earth cable to BS 6004 Table 8 clipped direct to a wooden joist above a plasterboard ceiling with a U value not less than $0.1 \, \text{W/(m}^2 \, \text{K)}$. For Installation Method 100 the thermal insulation does not exceed 100 mm in thickness. For Installation Method 101, the thermal insulation exceeds 100 mm thickness
102 and 103	102, 103	Flat twin and earth cable to BS 6004 Table 8 in a stud wall with thermal insulation with a U value not less than $0.1 \, \text{W/(m}^2 \, \text{K)}$. For Installation Method 102 the cable is touching the inner surface of the wall. For Installation Method 103 the cable is not touching the inner surface of the wall

capacity I_z of the installed cable conductor under the particular installation conditions. Extra care must be taken where the conductor operates at a temperature greater than 70 °C (e.g. 90 °C or 105 °C) to confirm that the equipment in which it terminates is not adversely affected by this higher temperature. In these circumstances the designer should consult the appropriate standard and/or the equipment manufacturer to confirm that, as a consequence of using the cable with a higher operating temperature, the current rating of the connected equipment is adequate, bearing in mind that it may need to be reduced because of the higher operating temperature (see Regulation 512.1.2).

To determine the effective current-carrying capacity I_z of a particular cable (the maximum current that can be carried in specified conditions without the conductors exceeding the applicable temperature in Table 52.1 of BS 7671 (see Table 10.15 of this Guide)), Equation (10.1) should be used:

$$I_z = I_t \times C_a \times C_g \times C_i \tag{10.1}$$

where I_t is the tabulated value of current-carrying capacities given in Appendix 4 of BS 7671, C_a is the rating factor for ambient temperature, C_g is the rating factor for grouping of conductors and C_i is the rating factor for thermal insulation.

By way of example is a circuit in which the tabulated value of current-carrying capacity for the cable $I_t = 50$ A, the temperature correction factor $C_a = 0.87$ (cable with 70 °C thermoplastic run through room of ambient temperature of 40 °C – see Table 4B1 of Appendix 4 of BS 7671), $C_g = 0.79$ (circuit installed single layer on a wall and grouped with two other circuits – see Table 4C1 of Appendix 4 of BS 7671) and $C_i = 0.78$ (cable surrounded by thermal insulation for a length of 100 mm – see Table 52.2 of BS 7671). For the purpose of this example, the grouping, insulation and ambient temperature conditions occur together; if they occurred at different parts of the circuit, the worst-case correction factor would be taken to calculate I_z. The calculation is as given in Equation (10.2).

$$I_z = I_t \times C_a \times C_g \times C_i$$
$$= 50 \times 0.87 \times 0.79 \times 0.78 = 26.8\text{A} \qquad (10.2)$$

Establishing the normal steady-state current I_b of a connected load of known power P (W) and power factor ($\cos \phi$) operating at a nominal line to Earth voltage U_o is, for a single-phase load, given in Equation (10.3) and that for a three-phase load is as set out in Equations (10.4) and (10.5) for star and delta connected loads respectively:

$$I_b \text{ (A)} = \frac{P}{U_o \times \cos \phi} \qquad (10.3)$$

$$I_b \text{ (A)} = \frac{P}{3 \times U_o \times \cos \phi} \qquad (10.4)$$

$$I_b \text{ (A)} = \frac{P}{\sqrt{3} \times U_n \times \cos \phi} \qquad (10.5)$$

Power factor can be determined from the relationship given in Equation (10.6), where S (VA) is the apparent power:

$$\text{power factor} = \cos \phi = \frac{P}{S} \qquad (10.6)$$

To apply the above formula it is necessary to know both the power rating (in terms of watts) and the apparent power S (VA), and these are usually readily available from the current-using equipment manufacturer, as is a rating for the power factor for most equipment.

In the case of discharge lighting, the assessment of the design current I_b is not a straightforward matter, in that an allowance has to be made for control gear losses and low-order harmonics. In the absence of detailed information, a first-order approximation of I_b is given in Equation (10.7). This assumes a corrected power factor of not less than 0.85 lagging (phase angle $\phi \leq -31.8°$). Where luminaires are not power factor corrected they will typically have a power factor of between 0.5 leading and 0.3 lagging (ϕ between $+60°$ and $+72.5°$) and this equation may not be valid.

$$I_b = \frac{1.8 \times P}{U_n} \tag{10.7}$$

where P (W) is the sum of the lamp rated wattage and U_n (V) is the nominal voltage of circuit.

The next step is to determine the ambient temperature and, hence, the ambient temperature rating factor C_a. This may be assessed by a suitable thermometer in the case where the building is already operational, but consultation with other professionals involved in the design process will be inevitable where the building is itself in the design stage. Where the ambient temperature(s) have been assessed, the rating factor is obtained from Table 4B1 or 4B2 of Appendix 4 of BS 7671 as appropriate. Where values are not given in these tables, special further consideration is required to establish an appropriate rating factor (e.g. where cables are affected by solar radiation or other heat sources – see Section 10.2.10 of this chapter). Figure 10.5 shows the typical relationship between temperature rating factor (and, hence, current-carrying capacity) and the ambient temperature.

Reference to Tables 4C1–4C5 of Appendix 4 of BS 7671 will be necessary to obtain the rating factor for grouping C_g for the particular wiring system selected and the installation method(s) to be employed.

Regulation 523.7 calls for cables preferably to be run in such a way as to avoid thermal insulating materials. Where this is not practicable, the regulation calls for the cross-sectional area of the cable to be selected to meet the requirements of Chapter 43 of BS 7671 for protection against overcurrent. In other words, the cross-sectional area must be chosen so that the cable's effective current-carrying capacity, taking account of contact with thermally insulating material (as well as the other particular installation conditions affecting the cable), will be sufficient so that the temperature of the conductors will not exceed the applicable value given in Table 52.1 of BS 7671. The regulation also requires account to be taken of the nature of the load (e.g. cyclic, see Section 11.3

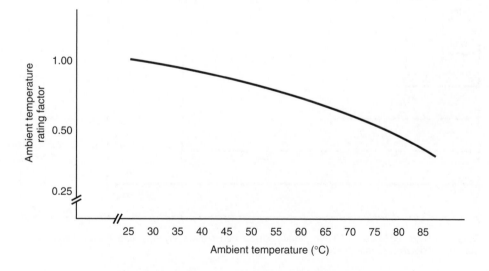

Figure 10.5 Typical temperature correction curve

of this Guide) and diversity (see Section 4.2 of this Guide). For certain methods of installing cables in contact with thermally insulating material, current-carrying capacities are tabulated in Appendix 4 of BS 7671. This is true for Installation Methods 1, 2 and 3, which correspond to Reference Method A, and Installation Methods 100 to 103, which correspond to Reference Methods 100 to 103 respectively (see Table 10.13 of this Guide for more details). Where a cable (up to 10 mm^2) is totally surrounded by thermally insulating material (e.g. in a cavity wall insulated with vermiculite granules or the like) having a thermal conductivity greater than 0.04 W/(K m), the appropriate factor given in Table 52.2 of BS 7671 should be used if the length of the affected cable is up to 400 mm. Where this length is 500 mm or more, the factor may be taken as 0.50 unless better, more precise information is to hand. In either case, the factor (from Table 52.2 or 0.5, as applicable) should be applied as a multiplier to the current-carrying capacity of the cable clipped direct to a surface and open (Reference Method C). For cables greater than 10 mm^2, routes through thermal insulation should be avoided unless suitable authoritative advice is available in respect of derating of current-carrying capacities. It should be noted, too, that where cable penetrations through walls, etc. are sealed (for compliance with Regulations 527.2.4–527.2.6) with material that is thermally insulating, the current-carrying capacities of the cables should be reduced accordingly. Figure 10.6 gives a pictorial representation of the derating factor C_i to be applied.

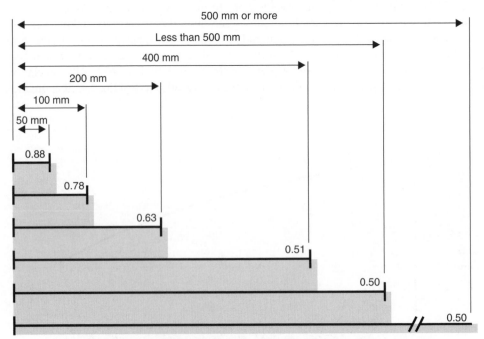

Data for a cable, up to csa of 10 mm^2, totally surrounded in thermally insulating material having a thermal conductivity greater than 0.04 mW/Km.

Figure 10.6   Derating of a cable when installed in thermally insulating material

Table 10.14 Current-carrying capacities of two-core and cpc copper cables to BS 6004 at ambient temperature of 30 °C

Ref	Csa[a] (mm²)	Installation Method 20 (Reference Method C)	Current-carrying capacity (A)				
			Length cables run through thermal insulation				
			50 mm	100 mm	200 mm	400 mm	500 mm
			Derating factor				
			0.88	0.78	0.63	0.51	0.50
A	1.0	16.00	14.08	12.48	10.08	8.16	8.00
B	1.5	20.00	17.60	15.60	12.60	10.20	10.00
C	2.5	27.00	23.76	21.06	17.01	13.77	13.50
D	4.0	37.00	32.56	28.86	23.31	18.87	18.50
E	6.0	47.00	41.36	36.66	29.61	23.97	23.50
F	10.0	64.00	56.32	49.92	40.32	32.64	32.00
G	16.0	85.00	74.80	66.30	53.55	43.35	42.50

[a]Cross-sectional area.

Table 10.14 gives data in terms of current-carrying capacities at an ambient temperature of 30 °C using Installation Method 20 (Reference Method C) and those when the derating factors are applied for two-core and cpc copper cables to BS 6004.

The rating factor C_c for the overcurrent protective device is unity unless the device is a semi-enclosed fuse to BS 3036 (rewirable fuse), when a factor of 0.725 must be used, or unless the cable installation method is 'in a duct in the ground' or 'buried direct', when a factor of 0.9 must be used, or unless the device is a semi-enclosed fuse to BS 3036 *and* the cable installation method is 'in a duct in the ground' or 'buried direct', when a factor of 0.653 must be used (i.e. 0.725 × 0.9). The rating factor C_c does not affect effective current-carrying capacity I_z of a cable (item 5.4 of Appendix 4 of BS 7671 refers) and, hence, it does not appear in Equation (10.1) of this Guide. However, C_c is used as a devisor where it is desired to calculate the lowest acceptable value of tabulated current-carrying capacity I_t for a cable to be used in a circuit, as in the case of Equations 1 to 4 in item 5 of Appendix 4 of BS 7671.

Excepting ring final circuits, where a circuit comprises two or more cables in parallel, the routes of the cables should be identical, as should the cable construction, cross-sectional area, disposition and length, so that load current may be shared equally in conductors (as required by Regulation 523.8). Such cases need very careful consideration with regard to fault current (wherever the fault may occur) and its effect on each conductor to confirm that automatic disconnection will occur, for protection against overcurrent and electric shock under single fault conditions where appropriate (see also Section 7.7 of this Guide).

Where cable connections are made directly to bare conductors or bare busbars, the designer needs to establish that the cable insulation is not adversely affected by the operating temperature of the busbars. This may be considerably higher than the temperature the conductor insulation can tolerate without damage. Where the cable insulation will not cope with this temperature, the insulation should be stripped back where safe to do so, or replaced by site-applied insulation of suitable temperature rating.

Regulation 523.5 calls for cables of different temperature rating run in the same enclosure to be assessed at the lower or lowest of the temperature ratings. For example, if two cables run in a conduit, one having a rating of 70 °C and the other 90 °C, then they should be assessed as both having a maximum normal operating temperature of 70 °C as far as current-carrying capacity is concerned.

The maximum conductor operating temperatures are given in Table 52.1 (Regulation 523.1) of BS 7671 and are reproduced in somewhat different format in Table 10.15. It is important to note that it is the insulating material, not the conductor, which determines both the maximum permitted conductor operating temperature and the final fault temperature limit.

Regulation 524.1 lays down the rules for the minimum cross-sectional area of each conductor of an a.c. circuit or a d.c. circuit, except as provided for extra-low voltage lighting installations according to Regulation 559.11.5.2. Table 52.3 of BS 7671 stipulates the minimum cross-sectional areas in terms of nonsheathed and sheathed cables, bare conductors and flexible connections with nonsheathed and sheathed cables. For copper conductors either insulated or within cables, a minimum of 1 mm^2 cross-sectional area is required for power and lighting circuits. For the same use, an aluminium conductor would need to be not less than 16 mm^2, which rules it out for most practical applications. Where bare conductors are used for power circuits, a minimum of 10 mm^2 is required for copper and 16 mm^2 for aluminium. Where aluminium conductors are used, the terminating connectors must be tested and approved for their specific use.

Signalling and control circuits are subject to a minimum cross-sectional area of 0.5 mm^2 for copper cables and insulated conductors and to 4 mm^2 if the conductors are bare, except that, for copper cables, the minimum cross-sectional area is 0.1 mm^2 if the circuit is for signalling control for electronic equipment. The requirements relate more to mechanical strength than to current-carrying capacity.

Flexible final connections to specific appliances must be copper of minimum cross-sectional area as specified in the appropriate standard. For other appliances, the minimum cross-sectional area is 0.5 mm^2 copper. The minimum sizes needed to meet the requirements of Section 524 will, of course, be overridden where consideration of the current-carrying capacity of the conductor demands a larger cross-sectional area.

Regulations 524.2 and 524.3 lay down the requirements for the minimum cross-sectional area of the neutral conductor of a circuit. As given in Regulation 524.2, there are three classes of circuit where the neutral conductor, if any, is required to have a cross-sectional area not less than that of the line conductor. The first of these is single-phase two-wire circuits. The second is polyphase circuits (e.g. three-phase) and single-phase three-wire circuits (split-phase), where the cross-sectional area of the line conductor does not exceed 16 mm^2 for copper or 25 mm^2 for aluminium. The third is three-phase circuits where the total harmonic distrortion due to third harmonics or multiples thereof is greater than 15% of the fundamental line current, as referred to in Regulation 523.6.3. A reduced neutral conductor is permitted, by Regulation 524.3, in a polyphase circuit (e.g. three-phase) where the cross-sectional area of each line conductor exceeds 16 mm^2 for copper or 25 mm^2 for aluminium, provided the three conditions stated in that regulation are fulfilled simultaneously. The first condition is that the expected maximum current in the

Table 10.15 Maximum conductor operating temperatures

Ref	Conductor and insulation materials	Csa (mm^2)	Conductor operating temp. limit (°C)	Limiting final operating temp. (°C)	BS 7671 Appendix 4 Table
Copper					
A	Copper with 70 °C PVC (general purpose)	<300	70	160	4D1,4D2, 4D3, 4D4 and 4D5
B	Copper with 70 °C PVC (general purpose)	≥ 300	70	140	
C	Copper with 90 °C thermosetting	All	90	250	4E1, 4E2, 4E3 and 4E4 4F1
D	Copper with 60 °C thermosetting	All	60	200	
Aluminium					
E	Aluminium with 70 °C PVC (general purpose)	< 300	70	160	4H1, 4H2 4H3 and 4H4
F	Aluminium with 70 °C PVC (general purpose)	≥ 300	70	140	
G	Aluminium with 90 °C thermosetting	All	90	250	4J1, 4J2 4J3 and 4J4
H	Aluminium with 60 °C thermosetting	All	60	200	
MICC					
I	Copper conductors and sheathed mineral insulated. Exposed to touch (bare and plastic covered)	All	70	160	4G1
J	Copper conductors and sheathed mineral insulated. Bare but not exposed to touch nor in contact with combustible materials	All	105	250	4G2

neutral conductor in normal service, inclusive of any harmonics, does not exceed the current-carrying capacity of that conductor. The second condition is that the neutral conductor is protected against overcurrent in accordance with Regulation 431.2 (see Section 7.9 of this Guide). The third condition is that the cross-sectional area of the reduced neutral conductor is not less than 16 mm^2 for copper or 25 mm^2 for aluminium. However, in respect of this third condition, when determining the size required for the reduced neutral conductor, account must be taken of the requirement of Regulation 523.6.3 for situations where the neutral conductor carries current without a corresponding reduction in the load of the line conductors (i.e. due to harmonic currents). Where a reduced cross-sectional area is considered, the normal neutral current should, of course, be taken into account and the conductor sized accordingly. Current in the neutral conductor of the three-phase circuit as a result of harmonic currents in the line conductor can be expected where low-order harmonics are generated by, for example, information technology switch-mode power supplies. These supplies typically employ rectifiers with the d.c. output connected to capacitors which then operate a switch-mode regulator. In three-phase circuits supplying such equipment, the low-order harmonic currents generated can produce a neutral current exceeding those in the associated line conductors. It is known that a neutral current up to 1.73 times ($\sqrt{3}$) of that of the line conductors can exist. When contemplating the installation of such equipment, it is advisable to obtain from the manufacturers the necessary information relating to load current in terms of root-mean-square (rms) and its harmonic components. It is important to recognize that third-order harmonics add arithmetically in the neutral conductor (in three-phase circuits).

Some motors, other inductive loads and electronic equipment are prone to large inrush currents, and this should be taken into account when considering current-carrying capacities. Particular care is required when evaluating an intermittent load subject to frequent starting and stopping and where there may be substantial cumulative effects of these currents. The designer would be well advised to consult the manufacturer of such equipment and to obtain a load profile and other characteristics.

10.4.3   Voltage drop

The requirements concerning voltage drop, set out in Regulations 525.1–525.3, relate only to the safety issues of electrical equipment performance. They do not address the other operational requirements, which may include, for example, efficiency. It may be that compliance with BS 7671 in this respect is not the only consideration in assessing voltage drop: some equipment may operate at less than its optimum efficiency at voltages permitted by BS 7671. For example, some forms of lighting sources may have considerably reduced lifespan and/or efficiency at voltages other than those prescribed by the manufacturer. In some cases, the operational considerations may place more stringent limits on voltage drop than the specified requirements. Some cases have been reported where a reduction of 5% of nominal voltage reduces the efficiency of particular lamps by as much as 20%. When one considers that a distributor is permitted a tolerance of +6% to −10% of the nominal declared voltage, it can be seen that the designer may find difficulty in providing supplies to such circuits if optimum efficiency is to be achieved.

From the safety standpoint, the basic requirement, embodied in Regulations 525.1 and 525.2, is that the voltage supplied to the current-using equipment provides for safe operation. This may be ascertained by reference to the relevant standard where that standard has addressed the safe functioning requirements (e.g. BS EN 60335). Regulation 525.3 provides a 'deemed to comply' status provided the voltage drop from the origin of the installation (supply point) to the terminals of all the current-using equipment or to socket-outlets is not greater than that stated in Appendix 12 of BS 7671. For an LV installation supplied directly from a public LV distribution system, the limiting voltage drop stated in Appendix 12, expressed with respect to the nominal voltage, is 3% for lighting equipment and 5% for other uses. At these limits, the voltage at the current-using equipment may be 91% or 89% of the nominal voltage respectively, i.e. 230 V $- (6 + 3)\% = 209.3$ V or 230 V $- (6 + 5)\% = 204.7$ V.

As permitted by Regulation 525.4, a greater voltage drop than prescribed in Appendix 12 may be permitted in the case of motors under starting conditions and other current-using equipment having high inrush current. However, the designer will need to confirm that such voltage drops as may occur will not impair the *safe functioning* of the equipment and that the voltages delivered are within the limits set out in the appropriate standard and/or the manufacturer's recommendations.

Although the actual voltage of the supply at the origin may not vary much from its nominal value, the designer should consider this as a bonus and not rely on it staying at a particular value. The public distribution network must be regarded as dynamic, continually under change and development. Because of this, the supply may be connected to different points on the network during the lifetime of the installation. The supply voltages may change from time to time, though the nominal voltage remains the same. The designer should assume the philosophy of 'plan for the worst and hope for the best'. For the purpose of calculating the voltage at the terminals of the current-using equipment, the voltage at the origin should be taken as the nominal voltage less 6% in the case of supplies derived from the public network.

The tabulated figures of voltage drop given in Appendix 4 of BS 7671 relate to the voltage dropped per metre per ampere of load current and assume that the conductors are at their maximum permitted operating temperature t_p. They apply only to frequencies in the range 49–61 Hz. Where higher frequencies are involved, the values of voltage drop are increased and allowance should be made in this respect. For cables of cross-sectional area of 16 mm^2 or less, the inductive reactance component of impedance may be ignored. For cables of greater cross-sectional area than 16 mm^2, this component becomes more significant and cannot be neglected. Where values are tabulated for single-core armoured cables, the tables in BS 7671 assume that the armour is bonded to earth at *both* ends of the run. Where this is not the case, further consideration in this respect is required.

The method of assessing voltage drop where conductors are at their maximum permitted normal operating temperature t_p is given by

$$V_d \text{ (V)} = \frac{\text{mV per metre (tabulated)} \times I_b(\text{A}) \times \text{length(m)}}{1000} \tag{10.8}$$

Values given in Appendix 4 of BS 7671 for three-phase circuits assume balanced conditions and refer to line-to-line voltage. Where values are given for cables of cross-sectional area greater than 16 mm^2 they are presented in terms of their resistive component, $(\text{mV}/(\text{A m}))_r$ and their reactive component $(\text{mV}/(\text{A m}))_x$. The voltage drop per metre, therefore, is as given by

$$V_d(\text{mV}) \text{ per metre} = I_b\sqrt{V_{d(r)}^2 + V_{d(x)}^2} \qquad (10.9)$$

where I_b (A) is the design current, $V_{d(r)}$ (mV) is the resistive component of voltage drop per metre and $V_{d(x)}$ (mV) is the reactive component of voltage drop per metre.

When using the tables giving the voltage drop values without further consideration of the actual cable operating temperature and using only the assumed maximum permitted operating temperature t_p, the resulting calculated voltage drops may be pessimistically high. Cables will operate at less than their maximum permitted normal operating temperature if the actual load current is less than the current-carrying capacity of the cable, as so often is the case. Equation (10.10) may be used where the protective device is other than a BS 3036 semi-enclosed (rewirable) fuse, to establish the correction factor for the operating temperature of the conductor C_t. Hence, a more accurate assessment of voltage drop may be made. It can be used only where the ambient temperature is not less than 30 °C.

$$C_t = \frac{(230 + t_p) - [(t_p - 30)(C_a^2 C_g^2 - I_b^2/I_t^2)]}{230 + t_p} \qquad (10.10)$$

where C_t is the rating factor for the operating temperature of the conductor, t_p (°C) is the maximum permitted operating temperature, C_a is the rating factor for ambient temperature, C_g is the rating factor for grouping, I_b (A) is the design current and I_t (A) is the current tabulated in Appendix 4 of BS 7671 for single circuits for a particular installation method in an ambient temperature of 30 °C.

Having established a value for C_t, this is then applied as a multiplying factor only to the tabulated resistive components of the tabulated voltage drops, $(\text{mV/(A m)})_r$, because the reactive component is unaffected by temperature. For the larger conductor, where the ratio of reactance to resistance is more than 3:1, this correction factor C_t is not worth considering for all practical purposes. Where there is no correction required for both ambient temperature and grouping, Equation (10.10) is simplified to that given in Equation (10.11) for PVC-insulated cables (with a maximum permitted operating temperature of 70 °C) and to that given in Equation (10.12) for 90 °C cables:

$$C_t = \frac{13}{15} + \frac{2}{15}\left(\frac{I_b}{I_t}\right)^2 \qquad (10.11)$$

$$C_t = \frac{13}{16} + \frac{3}{16}\left(\frac{I_b}{I_t}\right)^2 \qquad (10.12)$$

A further correction is possible for power factor, in that multiplying the tabulated $(\text{mV/(A m)})_r$ by cos ϕ (power factor) and the tabulated $(\text{mV/(A m)})_x$ by sin ϕ gives a more accurate voltage-drop figure.

10.4.4 Grouping

The question of rating factors for grouping is one which has vexed the industry for many years. The basic theory underlying the derating of current-carrying capacities of grouped cables is that the dissipation of heat generated in the cable when current flows ($I^2 R$ losses) is inhibited or retarded when other cables are in close proximity and possibly generating heat themselves.

Rating factors C_g for groups of circuits of single-core cables and one or more multicore cables are given in Tables 4C1–4C5 of Appendix 4 of BS 7671. All factors for groups of more than one are less than unity and are to be applied to the tabulated current-carrying capacity of single circuit cables given in Tables 4D1–4D5, 4E1–4E4, 4F1–4F3, 4G1–4G2, 4H1–4H4, and 4J1–4J4 of Appendix 4. Where appropriate, different values are given, depending on whether the cables are touching or spaced apart by one cable diameter.

The Appendix 4 tables assume that the cables so grouped are all the same size and equally loaded, which is seldom the case in practice. Where circuits are known to carry up to only 30% of their grouped current-carrying capacity (i.e. $0.3C_g I_t$), they may be overlooked when assessing the correction factor for the remainder of the grouped circuits (Regulation 523.5 refers). For example, if a circuit cable had an 'ungrouped' current-carrying capacity I_t of, say, 10 A and the appropriate grouping correction factor C_g was 0.60 and, consequently, the effective current-carrying capacity was 6 A, and the actual load current was known not to exceed 1.8 A, then this circuit could be ignored from the point of view of contributing to the number of grouped circuits. This example may apply, for instance, where the cross-sectional area of conductors of a long run has been determined from voltage-drop considerations rather than current-carrying capacity.

Cables with different conductor operating temperatures must be assessed on the basis of the lower or lowest temperature, as required by Regulation 523.5.

For convenience, Table 10.16 gives the C_g rating factors for grouping of circuits (not conductors). Row A gives values for PVC/PVC-insulated and -sheathed cables clipped direct to a nonmetallic surface (Installation Method 1). Row B gives those relating to similar cables clipped direct in single layer formation. Row C provides figures for nonsheathed single-core cables in conduit and Row D gives data for similar cables in trunking. These values are from Table 4C1 of Appendix 4 of BS 7671. The rating factors need not be considered where the horizontal clearances between cables exceed twice the cable diameter. Where no data are provided, it has to be assumed that such grouping factors have not been evaluated. In any event, grouping circuits in large numbers can often prove to be uneconomic because of the high degree of derating, and the designer would, in these circumstances, wish to avoid the problem by increasing the number of separate runs of circuits.

10.5   Resistances of copper conductors

For use in connection with determining the value of earth fault loop impedance (Z_s) of a circuit, Table 10.17 gives the resistance values of copper conductors at 20 °C and the values $R_1 + R_2$ with and without the multiplying factors relevant to the type of insulation (making allowance for the temperature rise due to load current).

Table 10.16 Correction factors C_g for grouping for PVC/PVC cables and PVC-insulated cables enclosed in conduit and trunking

Ref	Wiring system	Installation method[a]	Correction factor C_g										
			Number of circuits[b]										
			2	3	4	5	6	7	8	9	12	16	20
A	PVC/PVC cables bunched and clipped direct to non-metallic surface	20	0.80	0.70	0.65	0.60	0.57	0.54	0.52	0.50	0.45	0.41	0.38
B	PVC/PVC cables clipped direct to non-metallic surface-single layer	20	0.85	0.79	0.75	0.73	0.72	0.72	0.71	0.70	–	–	–
C	PVC cables in metallic and non-metallic conduit	4	0.80	0.70	0.65	0.60	0.57	0.54	0.52	0.50	0.45	0.41	0.38
D	PVC cables in metallic and non-metallic trunking	6 and 7	0.80	0.70	0.65	0.60	0.57	0.54	0.52	0.50	0.45	0.41	0.38

[a]Installation method of BS 7671 Appendix 4.
[b]Number of circuits or multicore cables.

10.6 Electrical connections

Section 526 of BS 7671 deals with electrical connections between conductors and equipment and with their accessibility. Electrical connections, wherever they occur, must be selected and installed so that they perform their function throughout the lifetime of the installation. Account should be taken of the environment in which the connections are located. Particular care is needed to avoid corrosion by external influences and by electrolytic action of dissimilar metals in close contact, especially in damp conditions. It goes without saying that all connections must provide durable and reliable electrical continuity and be sufficiently mechanically robust to cope with all the likely external influences (e.g. vibration). Good workmanship is always an essential part of effecting sound connections. Floating connectors in enclosures are best avoided, and the use of fixed connector blocks provides for a much tidier, well-ordered job and stands a better chance of being maintained properly in the future. Connections which are not properly made are the cause initially of

Table 10.17 Resistances of copper conductors[a]

Ref	Conductor resistances at 20 °C					70 °C thermoplastic (PVC)		85 °C thermosetting		90 °C thermosetting	
	Line conductor		Cpc		$R_1 + R_2$	Cpc as core of cable (Table 54.3) $1.20 \times (R_1 + R_2)$	Cpc *not* as core of cable (Table 54.2) $(1.20 \times R_1) + (1.04 \times R_2)$	Cpc as core of cable (Table 54.3) $1.26 \times (R_1 + R_2)$	Cpc *not* as core of cable (Table 54.2) $(1.26 \times R_1) + (1.04 \times R_2)$	Cpc as core of cable (Table 54.3) $1.28 \times (R_1 + R_2)$	Cpc *not* as core of cable (Table 54.2) $(1.28 \times R_1) + (1.04 \times R_2)$
	mm²	mΩ/m	mm²	mΩ/m	mΩ/m	mΩ/m	mΩ/m	mΩ/m	mΩ/m	mΩ/m	mΩ/m
A	1.0	18.100	1.0	18.100	36.200	43.440	40.544	45.612	41.630	46.336	41.992
B	1.5	12.100	1.0	18.100	30.200	36.240	33.344	38.052	34.070	38.656	34.312
C	1.5	12.100	1.5	12.100	24.200	29.040	27.104	30.492	27.830	30.976	28.072
D	2.5	7.410	1.0	18.100	25.510	30.612	27.716	32.143	28.161	32.653	28.309
E	2.5	7.410	1.5	12.100	19.510	23.412	21.476	24.583	21.921	24.973	22.069
F	2.5	7.410	2.5	7.410	14.820	17.784	16.598	18.673	17.043	18.970	17.191
G	4.0	4.610	1.5	12.100	16.710	20.052	18.116	21.055	18.393	21.389	18.485
H	4.0	4.610	2.5	7.410	12.020	14.424	13.238	15.145	13.515	15.386	13.607
I	4.0	4.610	4.0	4.610	9.220	11.064	10.326	11.617	10.603	11.802	10.695
J	6.0	3.080	2.5	7.410	10.490	12.588	11.402	13.217	11.587	13.427	11.649
K	6.0	3.080	4.0	4.610	7.690	9.228	8.490	9.689	8.675	9.843	8.737
L	6.0	3.080	6.0	3.080	6.160	7.392	6.899	7.762	7.084	7.885	7.146
M	10.0	1.830	4.0	4.610	6.440	7.728	6.990	8.114	7.100	8.243	7.137
N	10.0	1.830	6.0	3.080	4.910	5.892	5.399	6.187	5.509	6.285	5.546
O	10.0	1.830	10.0	1.830	3.660	4.392	4.099	4.612	4.209	4.685	4.246
P	16.0	1.150	6.0	3.080	4.230	5.076	4.583	5.330	4.652	5.414	4.675
Q	16.0	1.150	10.0	1.830	2.980	3.576	3.283	3.755	3.352	3.814	3.375
R	16.0	1.150	16.0	1.150	2.300	2.760	2.576	2.898	2.645	2.944	2.668
S	25.0	0.727	10.0	1.830	2.557	3.068	2.776	3.222	2.819	3.273	2.834
T	25.0	0.727	16.0	1.150	1.877	2.252	2.068	2.365	2.112	2.403	2.127
U	25.0	0.727	25.0	0.727	1.454	1.745	1.628	1.832	1.672	1.861	1.687
V	35.0	0.524	16.0	1.150	1.674	2.009	1.825	2.109	1.856	2.143	1.867
W	35.0	0.524	25.0	0.727	1.251	1.501	1.385	1.576	1.416	1.601	1.427
X	35.0	0.524	35.0	0.524	1.048	1.258	1.174	1.320	1.205	1.341	1.216

[a]The appropriate temperature corrected values of R_1 and R_2 should always be used in design calculations.

high-resistance joints leading, through continual expansion and contraction due to varying load conditions, to loose joints and eventually to the risk of fire. Great care is needed in selecting connections where they interface with conductors of the higher normal operating temperature (e.g. 90 °C and 105 °C), which are in increased usage nowadays. Connections should be selected so that they are able to cope with these higher temperatures without any adverse effects either on the connection or the conductor insulation.

Where enclosures and boxes are used for making connections, they should be selected so as to afford ample space for the connections and the conductors, and the designer should not forget the practical difficulties of terminating the larger conductors, especially solid aluminium cores. At terminations of MICC, the correct sleeving and other sleeving should always be used to match the temperature rating of the seal.

Termination glands for rubber- and plastic-insulated cables must comply with BS 6121. Where the armouring forms part or all of the cpc it is inadvisable to rely on detachable plates of equipment as a permanent and reliable path for earth fault current. Detachable plates are often covered with 'non-conductive' finishes and are usually secured by a few very small screws, which are unlikely to sustain earth fault current adequately. In such circumstances, it should be considered essential to make the protective conductor connection via a gland tag, nut, bolt and washers with a crimped protective conductor connecting the tag to the earthing terminal in the equipment, repeating the procedure for the outgoing cable (see Figure 10.7).

Regulations 526.1–526.9 set out the requirements for such connections and their accessibility, and these are summarized in Table 10.18.

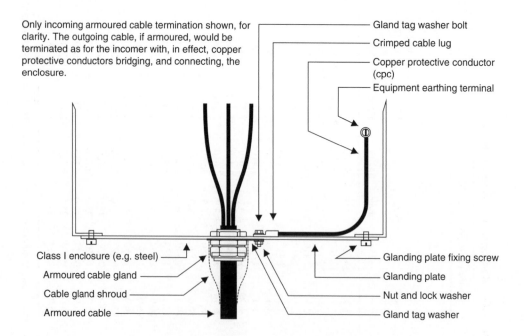

Only incoming armoured cable termination shown, for clarity. The outgoing cable, if armoured, would be terminated as for the incomer with, in effect, copper protective conductors bridging, and connecting, the enclosure.

Gland tag washer bolt

Crimped cable lug

Copper protective conductor (cpc)

Equipment earthing terminal

Class I enclosure (e.g. steel)

Armoured cable gland

Cable gland shroud

Armoured cable

Glanding plate fixing screw

Glanding plate

Nut and lock washer

Gland tag washer

Figure 10.7   Termination of armoured cables

Table 10.18 Electrical connections and their accessibility

Ref	Regulation	Requirement	Comment
A	526.1	Durable electrical continuity	
B	526.1	Adequate mechanical strength and connection locking arrangements, where necessary	Adequate mechanical strength against, for example, impact and vibration. Use of locking nuts on threaded connectors may be necessary where subject to vibration and/or thermal cycling
C	526.2	Suitable for the conductor material	Avoid metals where there is likely to be electrolytic action
D	526.2	Suitable for the conductor insulating material	Consider the temperatures of the connection under all conditions and that of the insulating material
E	526.2	Suitable for the conductor cross-sectional area and shape and the number of conductor strands	Care must be exercised with connections of multistrand and shaped conductors
F	526.2	Suitable for the number of conductors to be connected together	Connector must be capable of accepting all conductors without modification of the conductors (no cutting out strands)
G	526.2	If soldered connections are used they must be capable of withstanding the temperature attained under normal and fault conditions	Assess the temperature reached under the conditions stated and select suitable jointing material
H	526.2	If soldered connections are used they must be selected taking account of creep, mechanical stresses and temperature rise in service	

(continued overleaf)

Table 10.18 *(continued)*

Ref	Regulation	Requirement	Comment
I	526.7	Connections in enclosures must have adequate mechanical and protection against all the relevant external influences	External influences include water, temperature, dust, corrosive substances, impact, vibration, flora and fauna, and solar radiation
J	526.5	Connections of live and PEN conductors must be in enclosures consisting of one or more of the following: • accessories complying with the appropriate standard; • equipment enclosure complying with the appropriate standard; • an enclosure formed partly by building material which when tested to BS 476, Part 4 can be considered to be noncombustible	Enclosures to BS 4662 and BS 5733 are also suitable for this purpose
K	526.9	Sheathed cables from which the sheath has been removed and unsheathed cables where they emerge from conduit, trunking and ducting must be terminated in one of the above enclosures (Ref J)	Cores of sheathed cables which have a portion of their sheath removed for termination must be contained in the enclosure, i.e. sheath must go right into enclosure and leave no core exposed
L	526.3	Except as stated below (Ref M), all connections and joints must be accessible for maintenance, inspection and testing	

M	526.3	Connections and joints which need not be accessible are: • joints designed to be buried in the ground; • encapsulated joints; • compound filled joints; • cold tail connection of a heating system (e.g. pipe-tracing elements, ceiling and floor embedded elements); • welded joint; • soldered joint; • brazed joint; • mechanical joint by compression tool • joint forming part of an item of equipment complying with the relevant British Standard, and not intended to be accessible (such as a joint within a circuit–breaker, between parts thereof)	Compression joints to be certified as complying with BS 4579 or BS EN 61238 and the manufacturer's instructions to be adhered to particularly in the use of suitable tools and methods
N	526.8.1 to 526.8.3	Connections involving multiwire, fine wire and very fine wire conductors must meet the following requirements: • be made in suitable terminals or the conductor ends must be suitably treated; • soldering (tinning) of whole conductor end is not permitted where screw terminals are used; • soldered (tinned) conductor ends are not permitted at connection or junction points subject to relative movement between the soldered and nonsoldered parts of the conductor	To avoid inappropriate separation or spreading of individual wires To avoid the connection becoming loose due to creep To prevent the conductor fracturing due to vibration at the interface between the soldered and nonsoldered parts of the conductor

10.7   Cable supports and cable management systems

10.7.1   General

The regulatory requirements for cable supports are embodied in Section 522 of BS 7671 and in particular in Regulation 522.8.5, which calls for supports to cables so that there is no undue mechanical strain on conductors, including those at termination points. The spacings of such supports will be a matter of judgement and will depend on the location and whether cables will eventually be seen or hidden (e.g. in an intermediate floor space).

BS 7671 really only considers the safety aspects of support systems, and additional supports will very often be needed to make the finished installation look aesthetically pleasing. Where installations are subjected to vibration at medium or high severity and/or higher risk of mechanical impact, additional supports and mechanical protection will be needed, as necessary. With all the many cable management systems available on the market, there is now no excuse for untidy or unsafe cable installations.

10.7.2   Maximum cable support spacings

Table 10.19 gives suggested maximum distances between supports for accessible and inaccessible cables for general application for both vertical and horizontal runs (dimensions are given in both metric and imperial for those still reluctantly 'going metric inch-by-inch'); these distances will need modification for special installations and circumstances.

10.7.3   Overhead cables between buildings

Table 10.20 gives suggested minimum heights and maximum spans for overhead cables linking buildings on a site. They should only be applied to the general case and not to special cases, including agricultural and horticultural locations, yacht marinas and the like. These will need individual consideration and the cables may need to be much higher above the ground. The amount of sag depends on the cable weight and the tension to which it is subjected and, of course, the distance between supports. The point at which the maximum sag occurs will be at the centre of the run if the suspension points are at the same height, but will be off centre towards the lower suspension point if the heights are different.

10.7.4   Supports for conduits

Supports for conduit come in a number of different types and are suitable for different environments. Table 10.21 gives suggested maximum spacing for fixing of conduit, but further supports will be necessary in many circumstances to enhance the appearance, especially on surface installations. Figure 10.8 gives some examples of the types of conduit support system available.

10.7.5   Minimum bending radii of cables

For compliance with Regulation 522.8.3, it is important when installing cables that the radius of bends is such as not to cause damage either at installation stage or indeed subsequently. Table 10.22 gives the minimum bending radii for some common types of

Table 10.19 Maximum cable support spacings for armoured, nonarmoured and MICCs

Ref	Cable diameter ϕ_C[a]		Example[c]	Vertical[d]		Horizontal[e]	
	mm	inch[b]		mm	inch[b]	mm	inch[b]
Nonarmoured cables							
A	up to 9	≤0.35	1 to 10 mm² BS 6004 (6181Y) PVC/PVC 1-core; 1 mm² BS 6004 6242Y PVC/PVC 2-core/cpc	400	16	250	10
B	9–14	0.39–0.55	16 to 35 mm² BS 6004 (6181Y) PVC/PVC 1-core; 1.5 to 4 mm² BS 6004 (6242Y) PVC/PVC 2-core/cpc; 1 and 1.5 mm² BS 6004 (6243Y) PVC/PVC 3-core/cpc	400	16	300	12
C	15–19	0.59–0.75	6 to 10 mm² BS 6004 (6242Y) PVC/PVC 2-core/cpc. 2.5 and 4 mm² BS 6004 (6243Y) PVC/PVC 3-core/cpc	450	18	350	14
D	20 to 40	0.79–1.57	16 mm² 6242Y PVC/PVC 2-core/cpc. 6 to16 mm² BS 6004 (6243Y) PVC/PVC 3-core/cpc	550	22	400	16
Armoured cables							
E	9–14	0.39–0.55	1.5 and 2.5 mm² PVC BS 6346 SWA 2-core; 1.55 mm² PVC BS 6346 SWA 3-core; 1.5 mm² PVC BS 6346 SWA 4-core	450	18	350	14
F	15–19	0.59–0.75	4 and 6 mm² PVC BS 6346 SWA 2-core; 2.5 and 6 mm² PVC BS 6346 SWA 3-core; 2.5 to 4 mm² PVC BS 6346 SWA 4-core	550	22	400	16
G	20–40	0.79–1.57	10 and 16 mm² PVC BS 6346 SWA 2-core; 6 to 16 mm² PVC BS 6346 SWA 4-core; 10 and 16 mm² PVC BS 6346 SWA 3-core	600	24	450	18
MICCs							
H	up to 9	≤0.35	E.g. Light duty 500 V – 4-core 2.5 mm² – 4L1 (not served)	800	32	600	24
I	9–14	0.39–0.55	E.g. Light duty 500 V – 7-core 2.5 mm² – 7L2.5 (not served); E.g. Heavy duty 750 V – 2-core 10 mm² – 2H10 (not served)	1200	48	900	36
J	15–19	0.59–0.75	E.g. Heavy duty 750 V – 2-core 25 mm² – 2H25 (not served); E.g. Heavy duty 750 V – 4-core 16 mm² – 4H16 (not served)	2000	80	1500	60

[a]Where the cable is of the flat type, the major axis is taken a s the cable diameter. For larger cable diameters consult the manufacturer.
[b]Dimensions in inches are approximate.
[c]Examples cited, in terms of popular BCA's references, are not comprehensive and serve only to provide comparison.
[d]Vertical spacings relate to runs up to 30° from the vertical axis.
[e]Horizontal spacings relate to runs up to 60° from the horizontal axis.

Table 10.20 Minimum heights and maximum spans of cables linking buildings

Ref	Type of wiring system	Minimum cable height above ground level[a]						Maximum span[a]	
		Where accessible to vehicle traffic		At vehicle crossing		Where inaccessible to vehicle traffic			
		m	ft[b]	m	ft[b]	m	ft[b]	m	ft[b]
A	Plastics-sheathed cables without intermediate support	5.20	17.10	5.80	19.10	3.50	11.50	3.00	9.80
B	Plastics-sheathed cables enclosed in galvanized conduit of diameter not less than 20 mm and not jointed in span	5.20	17.10	5.80	19.10	3.00	9.90	3.00	9.80
C	Plastics-sheathed cables with separate catenary wire	5.20	17.10	5.80	19.10	3.50	11.50	No limit	
D	Oil resisting, flame retardant (hofr) without intermediate support	5.20	17.10	5.80	19.10	3.50	11.50	3.00	9.80
E	Oil resisting, flame retardant (hofr) enclosed in galvanized conduit of diameter not less than 20 mm and not jointed in span	5.20	17.10	5.80	19.10	3.00	9.90	3.00	9.80
F	Oil resisting, flame retardant (hofr) with separate catenary wire	5.20	17.10	5.80	19.10	3.50	11.50	No limit	
G	Covered or bare overhead lines supported only at the ends of the span	5.20	17.10	5.80	19.10	3.50	11.50	30.00	98.00
H	Overhead cable with integral catenary wire	5.20	17.10	5.80	19.10	3.50	11.50	Not exceeding manufacturer's guidance	
I	Covered or bare overhead lines meeting the requirements of the Electricity Safety, Quality and Continuity Regulations 2002, as amended	5.20	17.10	5.80	19.10	3.50	11.50	No limit	

[a]Suspension point heights must be chosen to allow for the inevitable sag between supports.
[b]Dimensions in feet are approximate.

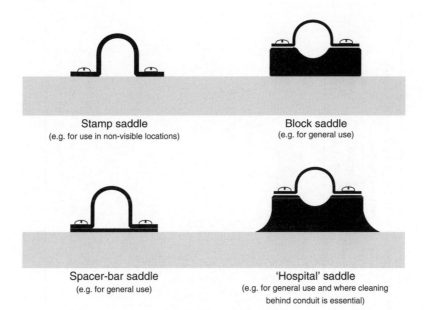

Figure 10.8   Typical conduit support systems

Table 10.21   Maximum spacing for supports for conduits

Ref	Conduit outside diameter (mm)	Maximum distance between supports					
		Steel conduit		Rigid non-metallic		Flexible conduit	
		m	ft in[a]	m	ft in[a]	m	ft in[a]
Horizontal runs							
A	16	0.75	2′5$\frac{1}{2}$″	0.75	2′5$\frac{1}{2}$″	0.30	0′11$\frac{3}{4}$″
B	20	1.75	5′ 9″	1.50	4′ 11″	0.40	1′ 3$\frac{3}{4}$″
C	25	1.75	5′ 9$\frac{3}{4}$″	1.50	4′ 11″	0.40	1′ 3$\frac{3}{4}$″
D	32	2.00	6′ 6$\frac{3}{4}$″	1.75	5′ 9$\frac{3}{4}$′	0.60	1′ 11$\frac{1}{2}$″
E	38	2.00	6′ 6$\frac{3}{4}$″	1.75	5′ 9$\frac{3}{4}$″	0.60	1′ 11$\frac{1}{2}$″
F	50	2.25	7′ 4$\frac{1}{2}$″	2.00	6′ 6$\frac{3}{4}$″	0.80	2′ 7$\frac{1}{2}$″
Vertical runs							
G	16	1.00	3′ 3$\frac{1}{4}$″	1.00	3′ 3$\frac{1}{4}$″	0.50	1′ 7$\frac{1}{2}$″
H	20	2.00	6′ 6$\frac{3}{4}$″	1.75	5′ 9$\frac{3}{4}$″	0.60	1′ 11$\frac{1}{2}$″
I	25	2.00	6′ 6$\frac{3}{4}$″	1.75	5′ 9$\frac{3}{4}$″	0.60	1′ 11$\frac{1}{2}$″
J	32	2.25	7′ 4$\frac{1}{2}$″	2.00	6′ 6$\frac{3}{4}$″	0.80	2′ 7$\frac{1}{2}$″
K	38	2.25	7′ 4$\frac{1}{2}$″	2.00	6′ 6$\frac{3}{4}$″	0.80	2′ 7$\frac{1}{2}$″
L	50	2.50	8′ 2$\frac{1}{4}$″	2.00	6′ 6$\frac{3}{4}$″	1.00	3′ 3$\frac{1}{4}$″

[a]Imperial dimensions are approximate.

Table 10.22 Internal minimum bending radii for common wiring systems

Ref	Cable type	Conductor construction and material	Cable diameter[a] ϕ (mm)	Minimum internal radii factor n (radius $= n \times \phi$)	
				Single-strand conductor (solid)	Multistrand conductor
Nonarmoured					
A	Rubber PVC XPLE	Copper and aluminium circular and circular stranded conductors	$\phi \leq 10$	3	2
B			$10 < \phi \leq 25$	4	3
C			$25 < \phi$	6	6
D		Copper shaped conductors and solid aluminium conductors	All	8	8
Armoured					
E	Rubber PVC XPLE	Copper and aluminium circular and circular stranded conductors	All	6	
F		Copper shaped conductors and solid aluminium conductors	All	8	
MICC					
G	MICC – bare	Copper	All	6 but 3 if bend is not reworked	
H	MICC – with covering	Copper	All	6 but 3[b] if bend is not reworked	

[a] Where the cable is of the flat type, the major axis is the cable diameter.
[b] MICC diameter to be taken as overall diameter with covering.

cable. The radius dimension relates to the internal cable surface and is given in terms of the cable overall diameter.

10.7.6   Maximum cable trunking support spacings

Table 10.23 gives maximum support spacings for cable trunking in terms of trunking cross-sectional area and material for both vertical and horizontal runs. Additional supports may be required at changes of direction and other bends or, indeed, where the trunking is likely to be subjected to mechanical impact and vibration. The figures in the table do not apply to supports for busbar and lighting trunking, which should be carried out in accordance with the manufacturer's recommendations.

10.7.7   Other cable management systems

There are many other cable management systems available to the designer, and consideration should be given to their use for specific applications. Metallic and nonmetallic cable trays and ladders lend themselves to industrial installations and underfloor systems, and floor and dado trunking are often used in the commercial environment. Cornice and skirting trunking frequently have a place in the domestic installation, particularly for rewiring of solid-floor buildings. In all cases where these cable management systems are used, the requirements of BS 7671 should be observed and in particular those relating to grouping of cables, supports and segregation of different systems (e.g. power, computer data lines, fire alarms). Additionally, the recommendations of the manufacturers of such systems should be observed.

Table 10.23   Maximum spacings for cable trunking supports

Ref	Trunking csa (mm²)	Examples of trunking sizes[a]		Vertical		Horizontal	
		mm	inch[b]	m	ft in[b]	m	ft in[b]
Metallic trunking							
A	csa ≤ 700	25 × 25	1 × 1	1.00	3′ 3″	0.75	2′ 5″
B	700 < csa ⩽ 1500	38 × 38	1½ × 1½	1.50	4′ 11″	1.25	4′ 1″
C	1500 < csa ⩽ 2500	50 × 50	2 × 2	2.00	6′ 6″	1.75	5′ 8″
D	2500 < csa ⩽ 5000	100 × 50	4 × 2	3.00	9′ 10″	3.00	9′ 10″
E	5000 < csa	100 × 75	4 × 3	3.00	9′ 10″	3.00	9′ 10″
Nonmetallic trunking							
F	csa ⩽ 700	25 × 25	1 × 1	0.50	1′ 7″	0.50	1′ 7″
G	700 < csa ⩽ 1500	38 × 38	1½ × 1½	0.50	1′ 7″	0.50	1′ 7″
H	1500 < csa ⩽ 2500	50 × 50	2 × 2	1.25	4′ 1″	1.25	4′ 1″
I	2500 < csa ⩽ 5000	100 × 50	4 × 2	2.00	6′ 6″	1.50	4′ 11″
J	5000 < csa	100 × 75	4 × 3	2.00	6′ 6″	1.75	5′ 8″

[a]The examples are given in common sizes.
[b]Imperial dimensions are approximate.

10.8   Minimizing the risk of fire

Section 527 deals with the sealing of wiring system penetrations so as to prevent or retard the spread of fire. It calls for all such penetrations through walls, floors, roofs, ceilings, partitions and cavity barriers to be sealed so that the fire resistance stipulated for the element penetrated is met. In other words, the fire barrier properties of the structure holed by wiring systems should be maintained as if there had been no penetration. Where a wiring system has an internal cross-sectional area of not greater than 710 mm^2 (e.g. 32 mm conduit and 25 mm × 25 mm trunking) and is of the non-flame-propagating type, no further internal sealing is required, though external sealing is still required of course (see Figure 10.9). The sealing arrangement must be compatible with the wiring system with which it is in contact so that there is no deterioration of either.

The sealing material must also be resilient to all potential external influences (e.g. water, corrosive materials, mechanical stresses). It should allow for expansion and contraction due to the thermal effects of the wiring systems and changes in ambient temperature and should, in so doing, maintain its sealing characteristics intact. It should also be removable for future extensions to the wiring system, and the removal and replacement of the sealing material must not damage the wiring system.

Every sealing arrangement must be verified as complying with the manufacturer's instructions by visual inspection both in the erection stage and on completion and a record issued to confirm compliance. Where sealing arrangements have been dismantled

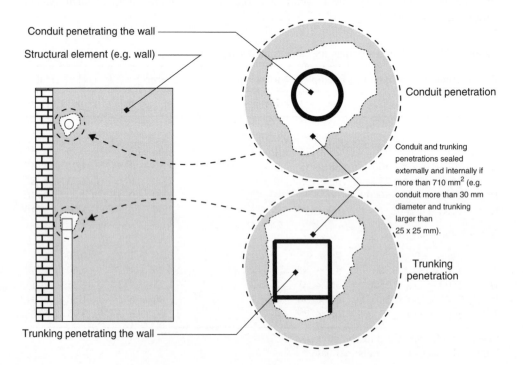

Figure 10.9   Sealing of wiring system penetrations

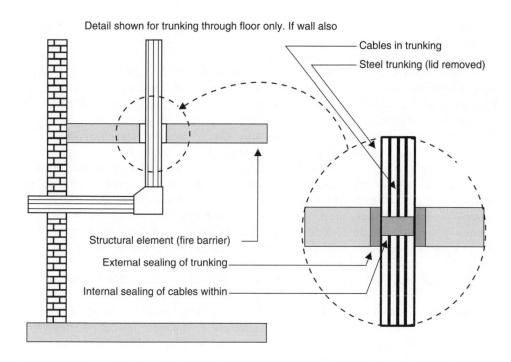

Detail shown for trunking through floor only. If wall also

Cables in trunking

Steel trunking (lid removed)

Structural element (fire barrier)

External sealing of trunking

Internal sealing of cables within

Figure 10.10   Trunking penetrating a structural floor and a wall

during alteration work, those arrangements should be resealed as soon as practicable so as to restore the fire resistance capabilities of the structure concerned. The designer should consider the use of wiring systems which have lower fire propagation characteristics (see Section 10.1.2) meeting the standard of performance set out in the relevant part of the BS EN 50266 series.

Figure 10.10 illustrates a typical arrangement of trunking penetrating a wall and a structural floor, which is intended to provide a fire barrier. Consultation with the person responsible for fire containment and with the structural engineer will almost certainly be inevitable for most installation designs.

Cables which do not comply with the flame propagation requirements of BS EN 60332-1-2 must be limited to the short length of connection of appliances to the fixed wiring. Such cable must not pass between fire-segregated compartments.

10.9   Electromagnetic and electromechanical effects

Electromagnetic effects of a.c. are of principal concern where single-core cables with steel wire or tape armouring are used, and Regulation 521.5.2 precludes the use of such cables for a.c. This regulation goes on to call for a.c. conductors enclosed in ferromagnetic materials (e.g. steel conduit and trunking) to have all line conductors, neutral, if any, and the cpc of a circuit to be contained within the enclosure. Conductors arranged in this manner will minimize the electromagnetic effects of current during both normal load conditions and fault conditions. Where the steel conduit or trunking serves as the cpc and is adequate, there is no further requirement in this respect regarding an additional cpc to be run in the conduit or trunking.

Conductors entering metal enclosures must not be separated by ferromagnetic material (e.g. steel) unless other precautions are taken to minimize the effects of eddy currents circulating in the steel producing small currents to flow in the line and neutral conductors. This may lead to an audible acoustic noise (induction hum) in extreme cases. Figure 10.11 graphically shows the eddy current effect.

Regulation 523.10 requires that the metallic sheaths and/or nonmagnetic armour of single-core cables be electrically bonded at both extreme ends of the run, sometimes known as solid bonding. Alternatively, for cables with conductors of more than 50 mm^2, the metallic sheaths or armouring may be bonded at one end only provided the sheaths or armouring are covered with a non-conducting outer sheath. When this 'single-point bonding' is employed, care must be taken to establish that the induced voltage from the armouring does not exceed 25 V with respect to the point at which the bonding has been effected. Other precautions may also be necessary to prevent corrosion by, for example, electrolytic action and damage to property, particularly when the conductors are subjected to fault conditions. Figure 10.12 illustrates the transformer effect of solid bonding.

Regulation 521.5.1 calls for every conductor to be of adequate strength to cope with electromechanical stresses. Conductors will tend to be subjected to mechanical forces during load conditions owing to expansion and contraction of conductors due to temperature changes. Changes in temperature may be associated with the ambient conditions and/or variations in load current. Stresses will occur by virtue of the difference in the coefficients of expansion of the conductors and that of the material with which it is in contact. Under

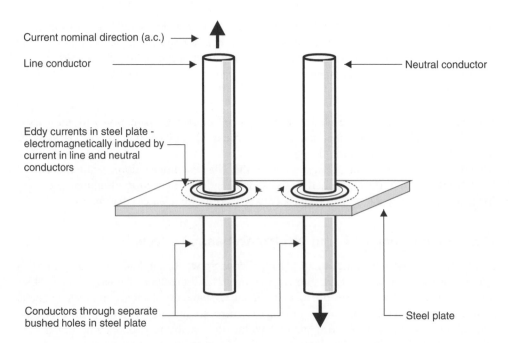

Figure 10.11   The eddy-current effect

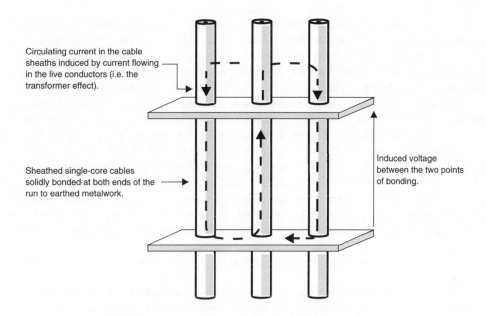

Circulating current in the cable sheaths induced by current flowing in the live conductors (i.e. the transformer effect).

Sheathed single-core cables solidly bonded at both ends of the run to earthed metalwork.

Induced voltage between the two points of bonding.

Figure 10.12   The transformer effect of 'solid bonding'

fault conditions, where currents may be of the order of magnitude of a 100 times the load current, serious distorting effects can occur on the cable. It is important for this reason, if for no other, that wiring systems are supported in accordance with BS 7671.

10.10   Conduit and trunking cable capacities

10.10.1   Conduit capacities

BS 7671 itself is not specific about the number of cables that are permitted to be installed in conduit systems, but there is a general requirement under Regulation 522.8 for cables to be installed so that they are not subjected to undue mechanical stresses.

The IEE Guidance Note *Selection and erection* gives, in its Appendix A, a method of determining the maximum number of cables permitted to be drawn into a conduit. Obviously, there are several factors which influence this number, including:

- the length of conduit run;
- the number of bends in the run;
- the radius of bends;
- accessibility points (adequate space for drawing in cables);
- proper fabrication of conduit system (e.g. conduit ends properly reamed);
- the construction of the cables (e.g. solid-single strand and multistrand);
- insulation rating (thickness);
- tolerance of conduit dimensions;
- conductor material (e.g. copper and aluminium);
- space for future extensions.

All these factors influence the number of cables which may be drawn into a conduit system. The method devised in Appendix A of the IEE Guidance Note is an attempt to provide useful advice to the designer in assessing the conduit sizes required for a particular number of cables (and circuits). In this method, PVC/copper cables have been allocated a term dependent on their cross-sectional area and the nature of the conduit run.

Cable and conduit terms for straight runs not exceeding 3 m are given Appendix A Table Al and Table A2 of the IEE Guidance Note *Selection and erection* respectively, and those for longer runs with and without bends are given in Tables A3 and A4. For convenience, these terms are summarized in Tables 10.24 and 10.25.

To use Tables 10.24 and 10.25, the designer first needs to assess the number and cross-sectional area of the PVC conductors and whether the intention is to use solid or stranded conductors (if there is an option for the particular sizes envisaged). The cable terms from Table 10.24 can then be identified and, where there are a number of different sizes, Equation (10.13) may be used to ascertain the sum of the cable terms:

$$T_t = (t_1 \times n_1) + (t_2 \times n_2) + (t_3 \times n_3) + \ldots + (t_n \times n_n) \tag{10.13}$$

where t_1 is the cable term for size n_1, t_2 is the cable term for size n_2, $t_3 =$ is the cable term for size n_3, t_n is the cable term for size n_n, n_1 is the quantity of cables of size 1, n_2 is the quantity of cables of size 2, n_3 is the quantity of cables of size 3, n_n is the quantity of cables of size n and T_t is the sum of all the cable terms.

To take a practical example, we have six 2.5 mm^2 (solid), twelve 1.5 mm^2 (stranded) and four 6 mm^2 cables to run 2.3 m in a conduit with three right-angled bends. Substituting cable term values obtained from Table 10.24 (rows reference A and B) for cable terms into Equation (10.13) we get the sum of terms for all the cables to be 958:

$$T_t = (39 \times 6) + (31 \times 12) + (88 \times 4) = 958 \tag{10.14}$$

We now know that the sum of terms for the cables is 958 and we need a value of conduit terms not less than this. From Table 10.25 we look for values in the row N3 which relates to a run of 2–2.5 m with three bends. We see for a 38 mm conduit the term is 900, which is less than we require. Therefore, the minimum conduit size that will be required is 50 mm with a conduit term of 1671.

Table 10.24   Terms for PVC/copper cables

Ref	Conduit runs	Conductor configuration	Terms							
			Conductor csa (mm^2)							
			1	1.5	2.5	4	6	10	16	25
A	Straight runs up to 3 m	Solid	22	27	39	–	–	–	–	–
B		Stranded	–	31	43	58	88	146	202	385
C	Straight runs over 3 m with and without bends	Solid	16	22	30	–	–	–	–	–
D		Stranded	–	22	30	43	58	105	145	217

Table 10.25   Terms for conduit

Ref	Conduit run length R (m)	Conduit term 16 mm	20 mm	25 mm	32 mm	38 mm	50 mm	63 mm
Straight runs with no bends								
A	$R \leq 3.0$	290	460	800	1400	1900	3500	5600
B	$3.0 < R \leq 3.5$	179	290	521	911	1275	2368	3826
C	$3.5 < R \leq 4.0$	177	286	514	900	1260	2340	3780
D	$4.0 < R \leq 4.5$	174	282	507	889	1244	2311	3733
E	$4.5 < R \leq 5.0$	171	278	500	878	1229	2282	3687
F	$5.0 < R \leq 6.0$	167	270	487	857	1199	2228	3599
G	$6.0 < R \leq 7.0$	162	263	475	837	1171	2176	3515
H	$7.0 < R \leq 8.0$	158	256	463	818	1145	2126	3435
I	$8.0 < R \leq 9.0$	154	250	452	800	1120	2080	3360
J	$9.0 < R \leq 10.0$	150	244	442	783	1096	2035	3288
Runs with one bend								
K1	$R \leq 1.0$	188	303	543	947	1325	2462	3977
L1	$1.0 < R \leq 1.5$	182	294	528	923	1292	2399	3876
M1	$1.5 < R \leq 2.0$	177	286	514	900	1264	2340	3780
N1	$2.0 < R \leq 2.5$	171	278	500	878	1229	2282	3687
O1	$2.5 < R \leq 3.0$	167	270	487	857	1199	2228	3599
P1	$3.0 < R \leq 3.5$	162	263	475	837	1171	2176	3515
Q1	$3.5 < R \leq 4.0$	158	256	463	818	1145	2126	3435
R1	$4.0 < R \leq 4.5$	154	250	452	800	1120	2080	3360
S1	$4.5 < R \leq 5.0$	150	244	442	783	1096	2035	3288
T1	$5.0 < R \leq 6.0$	143	233	422	750	1050	1950	3150
U1	$6.0 < R \leq 7.0$	136	222	404	720	1008	1872	3024
V1	$7.0 < R \leq 8.0$	130	213	388	692	968	1799	2906
W1	$8.0 < R \leq 9.0$	125	204	373	667	933	1734	2801
X1	$9.0 < R \leq 10.0$	120	196	358	643	900	1671	2700
Runs with two bends								
K2	$R \leq 1.0$	177	286	514	900	1260	2340	3780
L2	$1.0 < R \leq 1.5$	167	270	487	857	1199	2228	3599
M2	$1.5 < R \leq 2.0$	158	256	463	818	1145	2126	3435
N2	$2.0 < R \leq 2.5$	150	244	442	783	1096	2035	3288
O2	$2.5 < R \leq 3.0$	143	233	422	750	1050	1950	3150
P2	$3.0 < R \leq 3.5$	136	222	404	720	1008	1872	3024
Q2	$3.5 < R \leq 4.0$	130	213	388	692	968	1799	2906
R2	$4.0 < R \leq 4.5$	125	204	373	667	933	1734	2801
S2	$4.5 < R \leq 5.0$	120	196	358	643	900	1671	2700
T2	$5.0 < R \leq 6.0$	111	182	333	600	840	1560	2520
U2	$6.0 < R \leq 7.0$	103	169	311	563	788	1463	2364
V2	$7.0 < R \leq 8.0$	97	159	292	529	740	1375	2221
W2	$8.0 < R \leq 9.0$	91	149	275	500	700	1300	2100
X2	$9.0 < R \leq 10.0$	86	141	260	474	663	1232	1990

(*continued overleaf*)

Table 10.25 (*continued*)

Ref	Conduit run length R (m)	Conduit term 16 mm	20 mm	25 mm	32 mm	38 mm	50 mm	63 mm
Runs with three bends								
K3	$R \leq 1.0$	158	256	463	818	1145	2126	3435
L3	$1.0 < R \leq 1.5$	143	233	422	750	1050	1950	3150
M3	$1.5 < R \leq 2.0$	130	213	388	692	968	1799	2906
N3	$2.0 < R \leq 2.5$	120	196	358	643	900	1671	2700
03	$2.5 < R \leq 3.0$	111	182	333	600	840	1560	2520
P3	$3.0 < R \leq 3.5$	103	169	311	563	788	1463	2364
Q3	$3.5 < R \leq 4.0$	97	159	292	529	740	1375	2221
R3	$4.0 < R \leq 4.5$	91	149	275	500	700	1300	2100
S3	$4.5 < R \leq 5.0$	86	141	260	474	663	1232	1990
Runs with four bends								
K4	$R < 1.0$	130	213	388	692	968	1799	2906
L4	$1.0 < R \leq 1.5$	111	182	333	600	840	1560	2520
M4	$1.5 < R \leq 2.0$	97	159	292	529	740	1375	2221
N4	$2.0 < R \leq 2.5$	86	141	260	474	663	1232	1990

It may be, in some cases, more economical to run the cables in more than one conduit; in the foregoing example, had two conduits been used (instead of the one 50 mm conduit) and each contained three 2.5 mm^2 (solid), six 1.5 mm^2 (stranded) and two 6 mm^2 cables (assuming they can be divided in this way), we could have used two 32 mm conduits. This option may be easier to install, and the cable current-carrying 'capacities' would need less correction because of grouping.

10.10.2 Trunking capacities

In designing cable trunking systems it is important for the designer to select a size that will provide an adequate space factor and make an allowance for future extensions to the installation. The space factor is defined as the ratio of the sum of the cross-sectional areas of all the cables enclosed in the trunking to the effective internal cross-sectional area of the trunking. Where cables are not of circular section, their diameter is taken as the distance across their major axis.

As with cables in conduit, it is first necessary to assess the number and cross-sectional areas of cables which need enclosure in trunking and to determine their corresponding cable terms from Table 10.26, either from Row A or Row C dependent on whether trunking terms from Table 10.27 or Table 10.28 are to be used. Note that the cable terms for trunking are different from those where the cables are to be enclosed in conduit.

To take the practical example that we used for cables in conduit, we have six 2.5 mm^2 (solid), twelve 1.5 mm^2 (stranded) and four 6 mm^2 cables to run 2.3 m in trunking. The sum of the cross-sectional area cable terms T_s from the individual terms in Table 10.26 (Row A) is give by

$$T_s = (10.2 \times 6) + (8.1 \times 12) + (22.9 \times 4) = 250 \tag{10.15}$$

Table 10.26 Cable terms when installing in trunking

Ref		Cable terms for trunking[a]															
		1.5	1.5S	2.5	2.5S	4	6	10	16	25	35	50	70	95	120	150	240
A	CSA term[b]	8.1	7.1	11.4	10.2	15.2	22.9	36.3	–	–	–	–	–	–	–	–	–
B	BESA diameter[c]	3.2	3.0	3.8	3.6	4.4	5.4	6.8	8.0	–	–	–	–	–	–	–	–
C	BESA term[d]	9.6	8.6	13.9	11.9	18.1	22.9	36.3	50.3	75.4	95.0	132.7	176.7	227	284	346	552

[a]Cables marked with 'S' are solid conductors.
[b]CSA term to be used only with Table 10.27 data.
[c]British Electrical Systems Association (BESA) outside diameter of cable.
[d]BESA term to be used with Table 10.28 only.

Table 10.27   Trunking terms for use with cross-sectional area terms

Ref		Trunking dimensions (mm)										
A	Height	50	50	75	75	75	75	100	100	100	100	100
B	Depth	37.5	50	25	37.5	50	75	25	37.5	50	75	100
C	Terms[a,b]	767	1037	738	1146	1555	2371	993	1542	2091	3189	4252

[a]Terms only to be used with cross-sectional area cable terms from Table 10.26 Row A.
[b]Terms allow for 45% space factor and take into account the gauge of the trunking.

Having established the sum of all the cable cross-sectional area terms to be 250, we now have to find a trunking with a term not less than 250. We find, from Table 10.27, Row C, that 50 mm × 37.5 mm trunking provides a term of 767, which is more than adequate for enclosing the cables we proposed.

To take another example, we have eight 120 mm², twelve 95 mm² and four 150 mm² cables to enclose in trunking. From Table 10.26, Row C, we find that the cable BESA terms are 284, 227 and 346 respectively, which gives the sum of cable terms as 6380. To find a trunking size that will be adequate to enclose these cables we look to Table 10.28, Rows D and H, and find that the minimum size will be 150 mm × 100 mm with a trunking term of 6394.

It must be remembered that these terms provide for the necessary space factor of 45% but do not make any allowance for the inevitable future extensions to the wiring system. Neither do they take account of the bending radii of cables (see Table 10.22), which may well demand a larger size than that necessary to meet the space factor considerations. In making provision for additional future cables to be contained within the trunking, the designer should not overlook the need to allow for the added grouping in his assessment of the correction factors for the circuits considered initially. It may well be more attractive economically to have a number of trunking runs in order to reduce the effects of derating from grouping.

Where PVC cables are sheathed, a reduction in the number of cables will be necessary to allow for the increased cable overall diameter. Where cables of other types and construction are used, the manufacturer's advice relating to containment should be obtained.

Where the trunking is metal, irrespective of whether or not it is used as a cpc, all joints between lengths of trunking and between trunking and bends should be soundly made and be adequate to take the largest earth fault current reliably. Riveted joints should not be relied upon to cope with such fault currents, and copper links across joints with soundly bolted connections or other equally acceptable method should be used. Most trunking manufacturers can provide such links.

10.11   Maintainability

Section 529 of BS 7671 deals with the aspect of maintainability of an installation and calls for provision to be made for safe and adequate access to all equipment, including wiring systems. Regulation 341.1 requires that the person who will be responsible for maintenance should be consulted. Their capabilities will influence the designer's

Table 10.28 Trunking terms for use with BESA terms

Dimensions and BESA terms for trunking sizes 50 mm × 38 mm to 200 mm × 50 mm

Height	50	50	75	75	75	75	100	100	100	100	100	150	150	150	150	150	200	200
Depth	38	50	25	38	50	75	25	38	50	75	100	38	50	75	100	150	38	50
Gauge	1	1	1.2	1.2	1.2	1.2	1.2	1.2	1.2	1.2	1.4	1.6	1.6	1.2	1.2	1.6	1.6	1.6
BESA term	767	1037	738	1146	1555	2371	993	1542	2091	3189	4252	2999	3091	4743	6394	9697	3082	4145

Dimensions and BESA terms for trunking sizes 200 mm × 75 mm to 300 mm × 300 mm

Height	200	200	200	200	225	225	225	225	225	225	225	300	300	300	300	300	300	300	300
Depth	75	100	150	200	38	50	75	100	150	200	225	38	50	75	100	150	200	225	300
Gauge	1.6	1.6	1.6	1.6	1.6	1.6	1.6	1.6	1.6	1.6	1.6	1.6	1.6	1.6	1.6	1.6	1.6	1.6	2
BESA term	6359	8572	13 001	17 429	3474	4671	7167	9662	14 652	19 643	22 138	4648	6251	9590	12 929	19 607	26 285	29 624	39 428

decisions relating to the selection and erection of equipment for the installation in this respect. Appendix 5 of BS 7671 categorizes capabilities of persons as: BA1 – Ordinary, BA2 – Children, BA3 – Handicapped, BA4 – Instructed and BA5 – Skilled. Depending on the use of the building, the designer will need to decide what measures are required for maintenance to be carried out safely by the maintenance personnel involved.

Ideally, switchrooms should have adequate and safe access to all equipment from floor level, and use of the room for other purposes (e.g. storage, particularly of combustible materials) should be actively discouraged. Distribution boards, consumer units and the controlgear enclosures should be readily accessible from floor level, and live parts must be protected against inadvertent, unintentional or accidental contact. Care should be taken that all conductors are identified and located in an easily recognized sequence and that points of isolation of the equipment are indicated. Where there are live parts which cannot easily be isolated, they should be shrouded in insulating material and access denied. Overcurrent and other protective devices should be of the type which makes replacement by other types or rating difficult, and their functions and related information should be readily available.

Again ideally, luminaires which need removal for cleaning and the like should be fitted with luminaire supporting couplers and be so constructed to allow lamp replacement and cleaning to be carried out by unskilled personnel without potential danger. Where luminaires are at high level, consideration should be given to providing self-hoisting facilities to allow maintenance to be carried out at ground level.

An Operations and Maintenance Manual should be provided to the maintenance personnel, and guidance is given in this respect in BS 4940 *Technical information on construction products and services* and Section 6 of The Health and Safety at Work etc. Act 1974. Such a manual may, for example, include:

- the installation design;
- the installation as-fixed drawings, including circuit, protective devices and conductor identification;
- the Electrical Installation Certificate;
- the recommended installation reinspection intervals (including all different installations and, where necessary, different parts of installations);
- a schedule of maintenance intervals for all installed equipment;
- competence level required of personnel for each and every maintenance operation;
- the number of persons required for each and every maintenance operation;
- a schedule of life expectancy of installed equipment;
- manufacturers' technical specifications (includes, for example, accessories, luminaires, fire alarm and emergency lighting equipment);
- manufacturers' recommended maintenance procedures and intervals;
- isolation procedures;
- safe working procedures;
- all special instructions as to use and maintenance of equipment.

11

Switchgear, protective devices and other equipment

11.1 Switchgear and protective devices: general

As Regulation 511.1 clearly indicates, switchgear, including protective devices, must comply with the appropriate British or Harmonized Standard. Where this is not possible and the proposed switchgear does not meet the requirements of an appropriate British or Harmonized Standard, the onus is on the designer to determine that the equipment selected will provide a degree of safety not less than that inherent in similar equipment which does comply with an appropriate British Standard or Euronorm. In this respect it may be necessary to employ the services of a third-party certification body to assess the equipment and certify its degree of safety. Table 11.1 lists some common LV switchgear, including protective devices, together with their standard. The list should not be regarded as being exhaustive.

11.2 Switchgear and controlgear

11.2.1 Switchgear and controlgear: general

Chapter 53 deals in some detail with the regulatory requirements of switchgear for protection, isolation and for switching. Regulation Group 530.3 addresses the general and common requirements and in Regulation 530.3.2 calls for no fuse, unlinked switch and unlinked circuit-breaker to be inserted in the neutral conductor. This does not, however, preclude a *linked* switch or *linked* circuit-breaker interrupting the neutral conductor where the associated line conductor(s) are also interrupted at substantially the same time. In such cases, in a multiphase circuit (such as three-phase), the device must be of the type which does not break the neutral conductor before the break occurs in the associated line conductor(s) and does not make the line conductor(s) before the neutral (Regulation 530.3.1).

The current ratings of switchgear should be carefully assessed by checking that they match the characteristics of the load. It should not be assumed, for example, that the equipment will take its rated current continuously. If there is doubt, then the designer should

A Practical Guide to The Wiring Regulations: 17th Edition IEE Wiring Regulations (BS 7671:2008)
Fourth Edition Geoffrey Stokes and John Bradley
© Geoffrey Stokes and John Bradley. Published by John Wiley & Sons, Ltd

Table 11.1 Common LV switchgear, including protective devices, together with their standards

Ref	Equipment	Standard	Requirements/comments
Switchboards, distribution boards and consumer units			
A	LV switchgear and controlgear assemblies	BS EN 60439-1	Requirements for type-tested assemblies (TTAs) and partially type-tested assemblies (PTTAs)
B	LV switchgear and control-gear: switches, disconnectors, switch-disconnectors, and fuse combination units	BS EN 60 947-3	Requirements for defined switchgear
C	Busbar trunking systems	BS EN 60439-2	LV switchgear and controlgear assemblies. Particular requirements for busbar trunking systems (busways)
D	Fuseboards	BS EN 60439-3	Specification for LV switchgear and controlgear assemblies. Particular requirements for LV switchgear and controlgear assemblies intended to be installed in places where unskilled persons have access to their use. Distribution boards
E	Circuit-breaker boards.	BS EN 60439-3	
F	Consumer units	BS EN 60439-3 Including Annex ZA	
G	Distribution assemblies for construction and building sites	BS 4363	Requirements for six types of LV distribution assembly
		BS EN 60439-4	Requirements for LV assemblies
H	LV switchgear and controlgear: switches, disconnectors, switch- disconnectors and fuse combination units	BS EN 60439	Specification for LV switchgear and controlgear
Fuses and circuit-breakers			
I	HBC cartridge fuses	BS 1361 ≠IEC 269-1	General requirements
J	HBC cartridge fuses	BS 1362 ≠ IEC 269-1	General requirements

Table 11.1   *(continued)*

K	Semi-enclosed (rewirable) fuses	BS 3036	General requirements
L	MCB	BS EN 60 898	General requirements
M	MCCB	BS EN 60 947-2	General requirements
RCDs			
N	RCD	BS EN 61008	General requirements
		$\neq$ IEC 755	
O	Time-delayed RCD	BS EN 61009	General requirements
P	PRCD	BS 7071	General requirements
Q	SRCD	BS 7288	General requirements

[a]This listing is not exhaustive and serves only as a guide to the more commonly used equipment.
[b]The symbols # and $\neq$ signify an identical standard and a differing standard respectively.

confirm with the manufacturer and/or standard the actual circumstances under which the rating was assigned. In particular, the designer should bear in mind that consumer units and other distribution boards complying with BS EN 60439-3 are not usually designed to take their fully rated current continuously or for long periods. Where so employed they may be subjected to excessive temperature rise which may result in premature operation of protective devices (e.g. MCBs).

11.2.2   Switchgear and controlgear: forms of assembly

British Standard BS EN 60439-1:1999 addresses the specification for LV switchgear and controlgear assemblies. Its requirements include constructional considerations and allow for seven distinct forms of physical separation of circuits, which are shown graphically in Figure 11.1 and summarized in Table 11.2.

Depending on the circumstances, any of the forms of assembly may be employed. Very careful consideration is essential to establish the methods of operation and of maintenance of the equipment, with particular regard to personnel who will be required to do such work. Form 4b offers the highest level of safety and will usually be used where it is necessary for access to be gained without the need to isolate the whole assembly. The use of the other forms would require additional precautionary measures to be taken for compliance with the *Electricity at Work Regulations 1989*.

11.3   Selection of devices for overload and fault current protection: general

Regulation 533.2.1 demands that in the selection of devices for protection against overload the rated current of the protective device I_n must be chosen in accordance with the

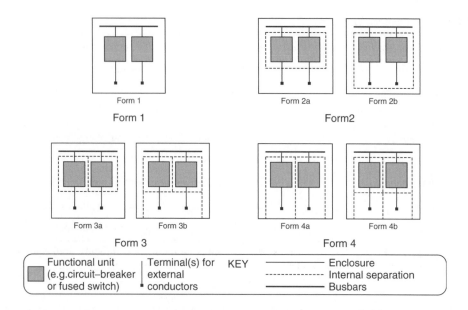

Figure 11.1   Forms of assembly of switchgear and controlgear

relationship given in Equations (11.1) and (11.2) (see Regulation Group 433.1):

$$I_n \leq I_b \leq I_z \tag{11.1}$$

$$I_2 \leq 1.45 I_z \tag{11.2}$$

where I_b is the circuit design current, I_z is the continuous service current-carrying capacity under the declared installation conditions, I_n is the rated current or current setting of the overload protective device and I_2 is the current that will cause the protective device to disconnect on overload

In every case, the peak load of the circuit must be taken into consideration. Assessment of the current-carrying capacity of conductors and the selection of the rated current of the protective devices must be made on the basis of the peak load. This regulation affords cyclic loads special consideration, inasmuch as the designer is entitled to treat such loads on the basis of a thermally equivalent load. In the absence of more precise information, one approach to establishing the relationship between a thermally equivalent load and a cyclic load is illustrated in Figure 11.2.

As can be seen in Figure 11.2, an example of a cyclic load is shown together with a thermally equivalent load based on energy (power × time) losses in the current-using equipment and circuit conductors, the areas under the 'curve' being equal. However, this is only one approach, and others may be more appropriate depending on the circumstances. In any event, it is necessary for the designer to establish that the overload protective devices will not operate unintentionally (i.e. prematurely) and to provide the necessary protection against overload.

Table 11.2 LV switchgear and controlgear assemblies form of assembly

Main criteria	Sub-criteria	Form	Restrictions	Comment
No internal separation		1	For use where assemblies can be isolated before gaining access and/or situated in areas restricted to authorized persons	Assembly may have exposed live parts when outer barriers removed (e.g. door opened)
Separation of busbars from the functional units	Terminals for external conductors not separated from busbars	2a	For use where assemblies can be isolated before gaining access and/or situated in areas restricted to authorized persons	Similar to Form 1, but internal separation of circuits minimizes the spread of the effects of faults
	Terminals for external conductors separated from busbars	2b		
Separation of busbars from the functional units and separation of all functional units from one another. Separation of terminals for external conductors from the functional units, but not from those of other functional units	Terminals for external conductors not separated from busbars	3a		Similar to Forms 1 and 2, except that circuits are separated from each other and from busbars
	Terminals for external conductors separated from busbars	3b		
Separation of busbars from all functional units and separation of all functional units from one another. Separation of terminals for external conductors associated with a functional unit from those of any other functional unit and the busbars	Terminals for external conductors in the same compartment as the associated functional unit	4a	Live parts to be isolated before gaining access to them, or equipped with additional internal barriers preventing unintentional contact	Form 4 affords internal separation of busbars from circuits, circuits from each other and outgoing terminals from busbars
	Terminals for external conductors not in the same compartment as the associated functional unit, but in individual, separate, enclosed protected spaces or compartments	4b		

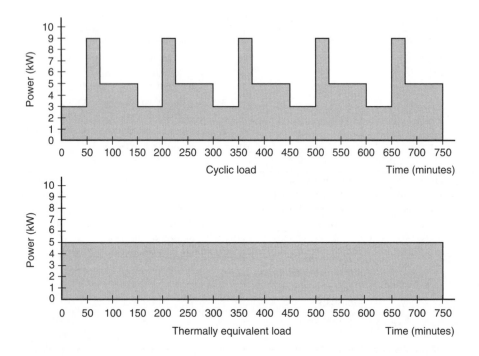

Figure 11.2   An example of a cyclic load and its thermally equivalent load

In the case of a normal motor circuit, the overload protective device often forms part of the starter controlgear and it is often positioned near to the motor. This is permitted by Regulation 433.2.2 where the stated conditions are met. In such cases it is not necessary to duplicate overload protection at the circuit origin, and the overcurrent protective device at this point may be selected on the basis of fault current protection only.

In fact, its rated current may be more than the circuit design current and, indeed, the current-carrying capacity of the wiring system. Figure 11.3 shows typical motor starting and running load profiles, together with the motor starter overload relay characteristic. The back-up fuse providing fault current protection only is also shown. It is worth noting that motor starting currents are much higher than steady-state conditions (running), and the initial inrush current may last for up to 10 s when direct-on-line starting methods are used, depending on the motor type and characteristics.

Regulation 533.3 demands that, in the consideration of fault current protective devices, for fault currents of up to 5 s duration, account is taken of the maximum fault current conditions and, where a device is used only for fault current protection, account is taken of the minimum fault current conditions. Generally speaking, the three-phase fault current (under no-load conditions) may be assumed to be the maximum fault current. However, it is important to recognize that earth fault currents may exceed line-to-line short-circuit currents (by up to approximately 5% in some cases). It is also vital to appreciate that the prospective fault current is attenuated (reduced) as the distance of the fault from the source is increased. Where circuits are of some considerable length, the fault levels will be significantly reduced at the remote end of the circuit. Selection of the overcurrent

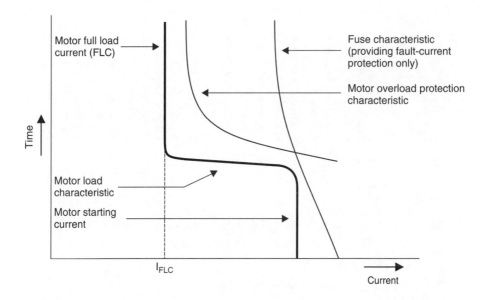

Figure 11.3   Typical characteristics of a motor load and associated protective devices

protective device would require these factors to be taken into consideration, particularly regarding disconnection time and the thermal capacity of the wiring system ($S^2 k^2$).

11.4   Overcurrent protective devices

Regulation 530.3.3 recognizes that a protective device may fulfil more than one function (protection against electric shock, protection against overcurrent and, thirdly, isolation). Where it does provide for more than one function, a protective device must meet *all* the requirements relating to every function. Table 11.3 provides a summary of protective devices and the functions they are commonly called upon to provide.

The choice of protective device will depend on many factors, including the system type and earthing arrangements, initial and maintenance costs and, not least, the designer's own preference. Table 11.4 summarizes the relative merits of the more common devices and lists, where appropriate, the main advantages and disadvantages. Table 11.5 details the rated short-circuit breaking capacities of the more common devices.

Where, in TN and TT systems, overcurrent devices are also used to provide protection against electric shock under single fault conditions (fault protection), Regulations 531.1.1 and 531.1.2 respectively call for the devices to be selected to meet the requirements of Chapter 41, which require the coordination of the characteristics of the devices and the line–earth loop impedance. In addition to providing automatic disconnection within the prescribed time limits, the devices must also interrupt the fault current sufficiently fast that it does not cause the line conductor or the cpc to exceed their maximum permissible final temperature. In assessing maximum conductor temperatures, the designer needs only to consider whether the adiabatic equation (given in Equation (11.3)) is satisfied for both line conductor and cpc (see Regulation 434.5.2 for live conductors and Regulation 543.1.3

Table 11.3 Common protection devices and the functions they are frequently called upon to perform

Ref	Protective device	Standard	Over-current	Protection against electric shock			Isolation and Fire switching		Comments
				Basic	Fault	Additional	Isolation	Switching	
A	MCB Type 1		✓	×	✓	×	✓	✓	See table footnotes b, c, d and e
B	MCB Type 2		✓	×	✓	×	✓	✓	See table footnotes b, c, d and e
C	MCB Type B		✓	×	✓	×	✓	✓	See table footnotes b, c, d and e
D	MCB Type 3		✓	×	✓	×	✓	✓	See table footnotes b, c, d and e
E	MCB Type C		✓	×	✓	×	✓	✓	See table footnotes b, c, d and e
F	MCB Type D	BS 3871 and BS EN 60 898	✓	×	✓	×	✓	✓	See table footnotes b, c, d and e. Limited scope for overcurrent and fault protection
G	MCB Type 4		✓	×	✓	×	✓	✓	See table footnotes b, c, d and e. Limited scope for overcurrent and fault protection

Table 11.3 *(continued)*

Ref	Protective device	Standard	Over-current	Protection against electric shock			Isolation and Fire switching			Comments
				Basic	Fault	Additional	Isolation	Switching	Fire	
H	MCCB	BS EN 60 947-2	✓	×	✓	×	✓	✓	×	See table footnotes b, c, d and e
I	HBC cartridge fuse	BS 88	✓	×	✓	×	✓	×	×	—
J	HBC cartridge fuse	BS 1361	✓	×	✓	×	✓	×	×	—
K	Cartridge fuse	BS 1362	✓	×	✓	×	✓	×	×	—
L	Semi-enclosed fuse (rewirable)	BS 3036	✓	×	✓	×	✓	×	×	—
M	RCD ($I_{\Delta n} \leq 30$ mA)	BS EN 61008-1	×	✓	✓	✓	✓	✓	✓	—
N	RCD ($I_{\Delta n} \geq 30$ mA)	BS EN 61008-1	×	×	✓	×	✓	✓	✓	Some measure of protection against fire
L	RCD ($I_{\Delta n} = 500$ mA)	BS EN 61008-1	×	×	✓	×	✓	✓	✓	Some measure of protection against fire

[a] ×: not commonly used or not permitted; ✓: commonly used.

[b] Suitable for isolation purposes provided the device affords the necessary contact separation and provides a reliable indication.

[c] If device is used as a remote isolating device, then it must be capable of securing in the open position.

[d] Device may be used for switching operations provided manufacturer agrees that it can be used for this purpose.

[e] Although device can provide overcurrent protection, it cannot directly protect against fire hazard.

Table 11.4 The relative merits of common overcurrent protective devices

Ref	Protective device	Standard	Initial cost	Maintenance costs	Advantages	Disadvantages
A	Semi-enclosed (rewirable) fuse	BS 3036	Low	Low	None other than costs	Low short-circuit breaking capacity. Cable current-carrying capacities C_c usually need to be increased. Prone to misuse particularly on replacement of fuse element. Tend to deteriorate with time. A possibility exists of fuse insertion on to a faulty circuit
B	Cartridge fuse	BS 1362	Low	Low		Low short-circuit breaking capacity. A possibility exists of fuse insertion on to a faulty circuit
C	HBC fuse	BS 88 Part 2	Medium	Medium	High short-circuit capacities capable of dealing with most fault levels encountered in practice. Less prone to misuse than semi-enclosed fuses	A possibility exists of fuse insertion on to a faulty circuit
D		BS 88 Part 6	Medium	Medium		

Table 11.4 (*continued*)

Ref	Protective device	Standard	Initial cost	Maintenance costs	Advantages	Disadvantages
E	HBC fuse	BS 1361 Type I	Medium	Medium	Fairly high short-circuit capacities capable of dealing with most fault levels encountered in practice. Less prone to misuse than semi-enclosed fuses	A possiblity exists of fuse insertion on to a faulty circuit
F		BS 1361 Type II	Medium	Medium	High short-circuit capacities capable of dealing with most fault levels encountered in practice. Less prone to misuse than semi-enclosed fuses	
G	MCB	BS 3871 and BS EN 60 898	Medium to high	Low	Generally provides for quicker disconnection. Short-circuit capacities available to deal with most fault levels up to 16kA. Less prone to misuse. The two types afford easy selection of various load characteristics	The need to maintain, and have ready, the necessary spares
H	MCCB	BS EN 60 947-2	High	Low	Generally provides for quicker disconnection. Short-circuit capacities available to deal with most fault levels. Adjustable overcurrent settings	

Table 11.5 Common overcurrent protective devices and their short-circuit capacities

Ref	Protective device	Standard	Short-circuit capacity	Nominal ratings	At nominal power factor	Comments
A	Semi-enclosed (rewirable) fuse	BS 3036	S1 – 1 kA	All ratings except 100 A	–	Short-circuit capacities marked on device
B			S2 – 2 kA	All ratings except 100 A		
C			S4 – 4 kA	30–100 A		
D	Cartridge fuse	BS 1362	6.0 kA	All	–	–
E		BS 88 Part 2	80 kA	All	–	For a.c. of 4.5 V or more
F			40 kA	All	–	For d.c. not exceeding 500 V
G	HRC fuse	BS 88 Part 6	–	–	–	Seek manufacturer's advice
H		BS 1361 Type I	16.5 kA	All	0.3	–
I		BS 1361 Type II	33.0 kA	All	0.3	–
J			M1 – 1 kA	All	0.85–0.90	Short-circuit capacities marked on device
K			M1.5 – 1.5 kA	All	0.80–0.85	
L	MCB	BS 3871 and BS EN 60 898	M2 – 2 kA	All	0.75–0.80	
M			M3 – 3 kA	All	0.75–0.80	
N			M4 – 4 kA	All	0.75–0.80	
O			M6 – 6 kA	All	0.75–0.80	
P			M9 – 9 kA	All	0.55–0.60	
Q			M16 – 16 kA	All		
R	MCCB	BS EN 60 947-2	–	–	–	Short-circuit capacities marked on device
S	Switchgear and controlgear assemblies	BS EN 60439	–	–	–	Seek manufacturer's advice

for protective conductors). The adiabatic equation featured in these two regulations is stated differently; for short-circuit considerations, the subject of the equation is time t (the maximum duration permitted in order that the conductor limiting temperature is not exceeded), and for earth fault analysis the minimum protective conductor cross-sectional area S is the subject. Equation (11.3) is in the form for calculating the time permitted to disconnect a short-circuit current:

$$t = \frac{k^2 S^2}{I^2} \tag{11.3}$$

where S (mm^2) is the cross-sectional area of the conductor, I (A) is the fault-current, t (s) is the duration of the fault (from occurrence to disconnection) and k is a factor for the conductor based on its resistivity, temperature coefficient and heat capacity.

In an IT system, the requirements in the event of a second fault where exposed-conductive-parts are connected together and the overcurrent device is used to provide protection against electric shock, the requirements relating to TN and TT systems in this respect must also be met (see Regulation 531.1.3).

There are three varieties of overcurrent protective device commonly used in LV installations, namely fuses, MCBs and MCCBs. These devices are also often used to provide protection against electric shock under fault conditions (fault protection) and for isolation and switching (see Table 11.3).

The time–current characteristics are given in Appendix 3 of BS 7671, together with accompanying tables to provide data on the magnitude of the fault current required to cause disconnection within the applicable time limit (e.g. 0.4 or 5 s). It should be borne in mind that, as mentioned earlier, it is the magnitude of the fault current that is the crucial factor in achieving the specified disconnection time. The limiting circuit impedance is dependent on the nominal voltage. Table 11.6 summarizes the fault current magnitudes required to effect automatic disconnection within 0.2, 0.4 1 and 5 s (the common maximum permissible disconnection times).

Overcurrent devices and their associated switchgear are commonly referred to by the number of poles in terms of switching and overcurrent detection in a.c. systems, and these are summarized in Figure 11.4. Note that overcurrent detection is normally provided in each line (L) conductor, but not in the neutral conductors, except where it is necessarily required (as shown in the DP device).

11.4.1 Fuses: general

There are principally four types of fuse in common usage in LV installations, namely the semi-enclosed (rewirable) fuses to BS 3036, HBC fuses to BS 88 and BS 1361 (commonly used in electricity distributor's cut outs) and the cartridge fuses to BS 1362 (commonly used in plug tops and fused connection units). All have their advantages and disadvantages, as summarized in Table 11.4; and for compliance with Regulation 533.1, an indication of their intended rated current must be provided on or adjacent to the fuse. Regulation 533.1.1.2 requires that where fuses are used that have fuse links likely to be removed or replaced by persons other than skilled or instructed persons, those fuses must meet the safety requirements of BS 88. The regulation goes on to require that such fuses must meet one of two requirements. Either they must have an indication of the type of

Table 11.6 Minimum fault current magnitudes (amperes) required for automatic disconnection within stated time limits for common overcurrent protective devices[a]

Minimum fault current magnitude (A)

Ref	Nom. rating (A)	Fuses to BS 88 gG Parts 2 and 6 Time (s)				Fuses to BS 1361 Time (s)				Fuses to BS 3036 Time (s)				Fuses to BS 1362 Time (s)		MCBs to BS 3871 and BS EN Time (s) 0.1 and 5					
		0.2	0.4	1	5	0.2	0.4	1	5	0.2	0.4	1	5	0.4	5	Type 1	Type 2	Type B	Types 3 and C	Type D	Type 4
A	5	–	–	–	–	25	22	18	14	32	24	24	13	–	–	20	35	25	50	100	250
B	6	31	27	23	17	–	–	–	–	–	–	–	–	–	–	24	42	30	60	120	300
C	10	51	45	39	31	–	–	–	–	–	–	–	–	–	–	40	70	50	100	200	500
D	13	–	–	–	–	–	–	–	–	–	–	–	–	95	60	–	–	–	–	–	–
E	15	–	–	–	–	80	70	60	46	125	90	90	43	–	–	60	105	75	150	300	750
F	16	95	85	72	55	–	–	–	–	–	–	–	–	–	–	64	112	80	160	320	800
G	20	150	130	110	79	155	135	110	82	180	130	130	60	–	–	80	140	100	200	400	1000
H	25	180	160	130	100	–	–	–	–	–	–	–	–	–	–	100	175	125	250	500	1250
I	30	–	–	–	–	240	200	170	125	300	210	140	87	–	–	120	210	150	300	600	1500
J	32	260	220	170	125	–	–	–	–	–	–	–	–	–	–	128	224	160	320	640	1600
K	40	–	–	240	170	–	–	–	–	–	–	–	–	–	–	160	280	200	400	800	2000
L	45	–	–	–	–	–	–	330	240	–	–	390	145	–	–	–	–	–	–	–	–
M	50	–	–	310	220	–	–	–	–	–	–	–	–	–	–	200	350	250	500	1000	2500
N	60	–	–	–	–	–	–	480	330	–	–	360	205	–	–	–	–	–	–	–	–

O	63	–	400	280	–	–	–	–	–	–	252	441	315	630	1260	3150
P	80	–	580	400	–	–	460	660	–	–	320	560	400	800	1600	4000
Q	100	–	790	550	–	–	630	–	1200	–	400	700	500	1000	2000	5000
R	125	–	1050	690	–	–	–	–	–	430	–	–	623	1263	2667	–
S	160	–	1400	900	–	–	–	–	–	–	–	–	–	–	–	–
T	200	–	1700	1200	–	–	–	–	–	–	–	–	–	–	–	–
U	315	–	–	1960	–	–	–	–	–	–	–	–	–	–	–	–
V	400	–	–	3000	–	–	–	–	–	–	–	–	–	–	–	–
W	450	–	–	3340	–	–	–	–	–	–	–	–	–	–	–	–
X	500	–	–	3980	–	–	–	–	–	–	–	–	–	–	–	–
Y	560	–	–	4430	–	–	–	–	–	–	–	–	–	–	–	–
Z	630	–	–	5100	–	–	–	–	–	–	–	–	–	–	–	–
A1	710	–	–	6090	–	–	–	–	–	–	–	–	–	–	–	–
B1	800	–	–	7530	–	–	–	–	–	–	–	–	–	–	–	–

[a]The current magnitudes given relate to devices manufactured to take full allowance of the operating range specified. Where manufacturers state that devices have been designed to operate within a narrower range, the current magnitudes may be modified appropriately. The entries denoted by a dash indicate that devices are not normally available or their use is impracticable. From the values of fault current data given, the appropriate Z_s may be calculated by application of the formula $Z_s = U_o / I_F$.

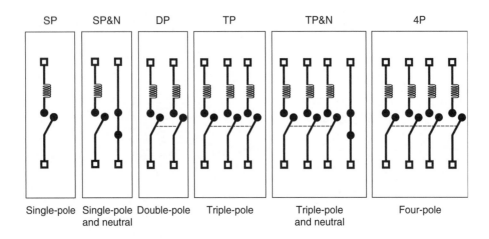

Figure 11.4   Identification of overcurrent switchgear in terms of the number of poles

fuse link intended to be used marked on or adjacent to the fuse, or they must be of a type whereby the fuse link cannot inadvertently be replaced by one having the same rated current but a higher fusing factor than intended.

11.4.2   Semi-enclosed fuses to BS 3036

The semi-enclosed fuse, more commonly known as a rewirable fuse, has enjoyed a long history of service in the United Kingdom. Despite its disadvantages, it is the most inexpensive device for overcurrent protection. Whilst most manufacturers have designed these fuses so that a fuse carrier of higher rated current cannot be inserted in the shielded circuit connection barrier, there is the real danger of inadvertent (and deliberate) replacement of the fuse element with a higher rating rendering the circuit unprotected against overload. As called for by Regulation 533.1.1.3, the preference is for cartridge-type fuse links, but where semi-enclosed fuses are used the fuse element wire must be selected to meet the sizes set out in Table 53.1 of BS 7671. A major disadvantage of the semi-enclosed fuse is its low rated short-circuit breaking capacity (1–4 kA – see Table 11.5 of this Guide), which restricts its use to locations where the fault level does not exceed these values or where back-up protection is provided by an upstream device that can break the prospective fault current. Another not inconsiderable drawback is the application of the rating factor C_c (0.725) to be applied as a multiplier to the tabulated current-carrying capacities of conductors resulting in an increase in conductor cross-sectional area of at least 38% and, in some cases, more.

11.4.3   High breaking capacity fuses to BS 88

HBC fuses to BS 88 are extensively used in industrial and commercial installations and, to a lesser extent, in the domestic field. Their high short-circuit capabilities make them suitable for most installations and make life much easier for the designer when considering fault levels throughout the installation. The principal variations are the BS 88-2.2 'gG' fuses for use by authorized persons and the BS 88-6 for use in 230/400 V industrial and commercial installations. Fuses are further identified as follows:

- 'gG' – (general purpose) fuse with full short-circuit breaking capacities for general application;
- 'gM' – fuse with full short-circuit breaking capacities for motor circuit protection application;
- 'aM' – fuse with partial-range short-circuit breaking capacities for motor circuit protection application.

Important considerations in design of circuits protected by these fuse links, particularly in regard to discrimination, are the pre-arcing energy $I^2t_{[pa]}$ and the total operating energy let-through $I^2t_{[t]}$; these data, and the operating characteristics, are readily available from the manufacturers.

Figure 11.5 shows graphically the cut-off and energy let-through of current-limiting overcurrent devices. The point A is where disconnection commences, the darkly shaded area 0At_1 represents the pre-arcing energy $I^2t_{[pa]}$, and the sum of both shaded areas 0At_2 portrays the total energy let-through $I^2t_{[t]}$ (units: A^2 s).

Figure 11.6 illustrates typical I^2t characteristics for a range of fuse links (rms symmetrical current 33 kA) on which the $I^2t_{[pa]}$ and the total energy let-through $I^2t_{[t]}$ are shown. Figures 11.7 and 11.8 show the typical cut-off characteristics for a range of HBC fuses and typical cut-off currents again for HBC fuses (for a particular prospective current) respectively.

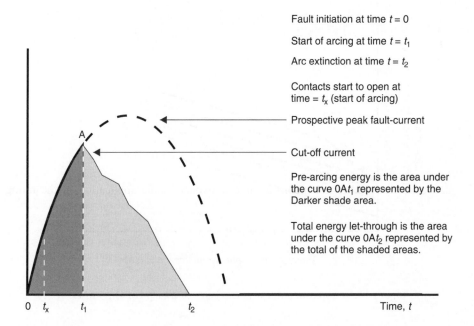

Figure 11.5   Cut-off current and energy let-through of a current-limiting overcurrent device

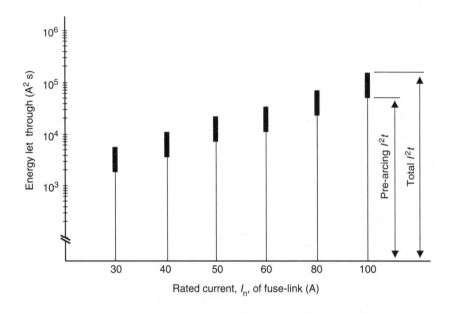

Figure 11.6   Typical I^2t characteristics of a range of fuses

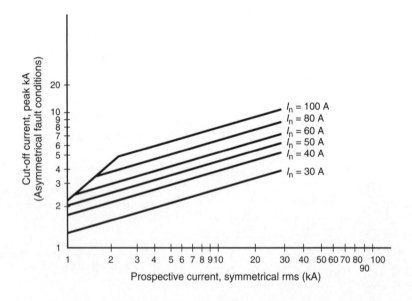

Figure 11.7   Typical cut-off characteristics of a range of HBC fuses

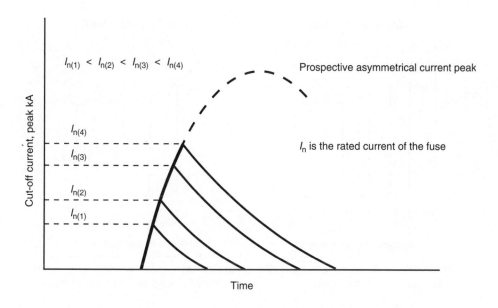

Figure 11.8   Typical cut-off currents for a range of HBC fuses, for a particular prospective current

11.4.4   High breaking capacity fuses to BS 1361

Fuses to BS 1361 are more commonly used in circuits for installations in domestic and similar premises and EDs make extensive use of them in their household service cut-outs. They are of lower rated short-circuit breaking capacity than BS 88 fuses, but at 16.5 kA they are adequate for most domestic applications. As with BS 88 fuses, operating data are freely available from the manufacturers.

11.4.5   Cartridge fuses to BS 1362

Cartridge fuses to BS 1362 are for general use primarily in the BS 1363 13 A fused plug and 13 A fused connection units. With a rated short-circuit breaking capacity of 6 kA they are capable of coping with fault levels available on final circuits in most cases, although this should be always be checked.

11.4.6   Miniature circuit-breakers to BS 3871 and BS EN 60 898

There are three important parameters to consider when selecting an MCB:

- the rated current;
- the type;
- the rated short-circuit breaking capacity.

The rated current is easily identified as the nearest available rating in amperes equal to, or more than, the design current I_b. The rated short-circuit breaking capacity, marked

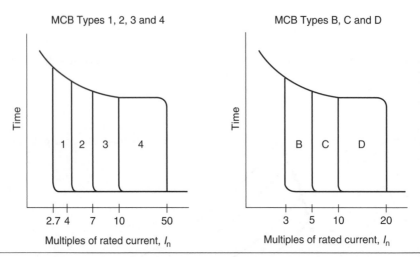

Figure 11.9   MCB characteristics

on the device, must be not less than the prospective fault current (short-circuit and earth fault current) unless a back-up device is employed upstream.

As illustrated in Figure 11.9, there are seven types of MCB currently in use in the industry, though BS 3871 has now been replaced by BS EN 60898. They vary in their operating characteristics, so that different current magnitudes are required to operate the devices instantaneously (defined as within 100 ms), although, within small variations, their overload characteristics are the same in all cases. The constructional standard (BS EN 60898) permits instantaneous operation to occur between gates (within a range), and this is defined for each type in terms of the rated current I_n. Figure 11.10 illustrates a typical operating characteristic of an MCB and Table 11.7 details the main features of the various types and the uses to which they are commonly put. When considering discrimination, the whole range for fault current operation needs to be taken into account, whereas for electric shock protection under fault conditions (fault protection) the worst-case value is all that it is necessary to consider. Limiting values of Z_s are based on these 'worst-case' values.

11.4.7   Moulded case circuit-breakers to BS EN 60947-2

MCCBs are commonly used as an alternative to HBC fuses for protecting circuits rated at 100 A or more. They are used extensively to protect against fault current and in some cases against overload. Their initial cost is high compared with switchgear incorporating HBC fuses, but they are often preferred in areas under the control of skilled persons where breakers with adjustable thermal–magnetic trip mechanisms may be used (on the higher rated units), which can aid discrimination. BS EN 60947 permits these devices to be calibrated at 20 °C or 40 °C, so it is important to know the temperature to which the device has been calibrated before making a selection. Where devices with adjustable settings are

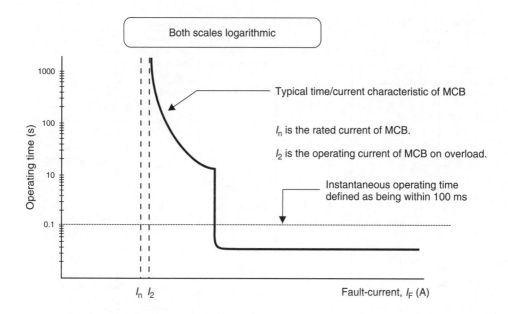

Both scales logarithmic

Typical time/current characteristic of MCB

I_n is the rated current of MCB.

I_2 is the operating current of MCB on overload.

Instantaneous operating time defined as being within 100 ms

Operating time (s)

1000

100

10

0.1

I_n I_2 Fault-current, I_F (A)

Figure 11.10   Typical operating characteristics of an MCB

employed in areas accessible only to authorized personnel, it will be beneficial if a notice, recording the settings, is affixed to or near the device. Adjustable fault current settings are often of the form of two distinct calibrated and marked levels (e.g. 2000 and 4000 A).

The rated short-circuit breaking capacities of MCCBs can be high (e.g. 150 kA), which, in circumstances of circuits with high fault levels, make them attractive to the designer. The time−current characteristics of such devices are published by the manufacturers, and the designer should consult with them in matching the device with the particular load and other relevant factors.

11.5   Residual current devices

11.5.1   Residual current devices: general

RCDs is the generic term for a range of devices, including those known as RCCBs. Also included in this term are the RCBO device, which incorporates an MCB with the RCD, the socket-outlet with RCD (SRCD) and the portable RCD (PRCD).

There are two principal types of operating characteristics of RCDs, namely the basic type and the polarized type, which are dealt with to some extent later. There are also RCDs with a time-delay operation, which is particularly useful in the pursuit of discrimination.

Regulations 531.2.1–531.5.1 set out the requirements for RCDs generally and Regulations 531.3.1, 531.4.1 and 531.5.1 call for the additional requirements to be met as they relate to TN, TT and IT systems respectively. Table 11.8 summarizes these various considerations.

In a fixed installation there are a number of instances where it is essential to employ an RCD, and these are summarized in Table 11.9.

Table 11.7   MCBs to BS 3871[a] and BS EN 60 898: main features

Ref	Type	Nom. rating I_n (A)	Overload characteristic[b]	Current band causing instantaneous operation I_{inst}	Current necessary for instantaneous operation	Common utilization
(i)	1	$I_n \leq 10$ $I_n > 10$	$1.50I_n$ $1.35I_n$	$2.7I_n$ to $4.0I_n$	$4I_n$	General circuits where load does not exhibit high inrush characteristics. Tungsten lighting loads
(ii)	B	All	$1.45I_n$	$3I_n$ to $5I_n$	$5I_n$	
(iii)	2	$I_n \leq 10$ $I_n > 10$	$1.50I_n$ $1.35I_n$	$4I_n$ to $7I_n$	$7I_n$	General circuits where load only exhibits moderate inrush characteristics. Large tungsten and fluorescent lighting loads
(iv)	C	All	$1.45I_n$	$5I_n$ to $10I_n$	$10I_n$	General circuits where load exhibits moderate to high inrush characteristics. E.g. motor loads, air conditioning plant, etc.
(v)	3	$I_n \leq 10$ $I_n > 10$	$1.50I_n$ $1.35I_n$	$7I_n$ to $10I_n$	$10I_n$	
(vi)	D	All	$1.45I_n$	$10I_n$ to $20I_n$	$20I_n$	General circuits where load exhibits high to harsh inrush characteristics. E.g. X-ray equipment, welding equipment, DOL motors, circuits with transformers, etc.
(vii)	4	$I_n \leq 10$ $I_n > 10$	$1.50I_n$ $1.35I_n$	$10I_n$ to $50I_n$	$50I_n$	

[a]BS 3871 was withdrawn on 1 July 1994.

[b]Current required for overload operation within 'conventional time' at reference temperature. Conventional time for Types 1, 2, 3 and 4 and for Types B, C and D up to $I_n \leq 63$ A is 1 h. For Types B, C and D of $I_n > 63$ A the conventional time is 2 h.

Table 11.8 Regulatory requirements for RCDs

Ref	Regulation	Essential requirements
A	531.2.1	RCD must disconnect all line conductors of the protected circuit at substantially the same time
B	531.2.2	All live conductors (line(s) and neutral) must pass through the transformer magnetic circuit (but not the protective conductor)
C	531.2.3	The residual operating current must be appropriate to the type of earthing system as required by Section 411
D	531.2.4	Account must be taken of earth leakage currents (which are not fault currents) so that the device selected is not prone to unwanted and/or unnecessary tripping. (Accumulative effects of leakage current need to be assessed, as do the effects of capacitors connected between live conductors and earth)
E	531.2.5	Where a circuit normally requiring a protective conductor is not so equipped, the RCD must not be considered to provide fault protection (even if $I_{\Delta n} \leq 30$ mA)
F	531.2.6	Where an RCD is energized or de-energized from an auxiliary source, it must be supervised, inspected and tested by a skilled or instructed person or such protection must fail to safety (i.e. protection is maintained even in the event of power failure to auxiliary source)
G	531.2.7	The RCD must be located so that it is not adversely affected by the presence of magnetic fields generated by other equipment. (Magnetic fields may have serious consequences on the transformer magnetic circuit of the RCD and may cause malfunction and may not fail to safety)
H	531.2.8	When used for fault protection in conjunction with, but separate from, an overcurrent device, it must be ascertained that the RCD is capable of withstanding without impairment all mechanical stresses arising from a fault on the downstream circuit(s)
I	531.2.9	Where two or more RCDs are connected in series, effective discrimination to prevent danger must be provided
J	531.2.10	An RCD that may be operated by a person who is not a skilled or instructed person must be designed and installed so that no alteration can be made to the setting or calibration of its rated residual operating current $I_{\Delta n}$ without a deliberate act using a tool or key, resulting in a visible indication of its setting or calibration
K	531.3.1	In TN systems where the requirements of Regulation 411.4.5 ($Z_s I_a \leq U_o$) cannot be met, an RCD may be used for fault protection. When so used, the exposed-conductive-parts must be connected to the MET or separate earth electrode. When a separate earth electrode is used, the RCD-protected circuit must be treated as part of a TT system and subjected to the requirements of Regulations 411.5.1–411.5.3.
L	531.4.1	In an installation forming part of a TT system and a sole RCD is used, this device must be positioned at the origin (front end) unless the equipment between the RCD and the origin is of Class II construction or equivalent insulation. See also Figure 11.11
M	531.5.1	In IT systems where disconnection of a first fault is not contemplated, the nonoperating residual current must not be less than the first fault circulating current

Table 11.9 Instances where RCDs will be needed, if appropriate

Ref	Locations	System type	Regulation	Requirement qualifications
A	Generally	All	411.3.3	Socket-outlets rated at 20 A or less for use by ordinary persons and intended for general use, except where for use under supervision of skilled or instructed persons or labelled/identified to supply a specific item of equipment ($I_{\Delta n} \leq 30$ mA)
B		All	411.3.3	Mobile equipment with a rated current not exceeding 32 A ($I_{\Delta n} \leq 30$ mA)
C			411.4.5 to 411.4.9	Where the line–earth loop impedance is too high to effect fault protection by use of an overcurrent protective device (e.g. MCB) and supplementary bonding is not provided
D		TT	411.5.2 to 411.5.4	Where the line–earth loop impedance is too high to effect fault protection by use of an overcurrent protective device (e.g. MCB) and supplementary bonding is not provided. Regulation 411.5.2 expresses a preference for RCDs in TT systems
E		IT	411.6.2 to 411.6.4	Where the line–earth loop impedance is too high to effect fault protection by use of an overcurrent protective device (e.g. MCB)
F		All	415.1.1 and 415.2.2	Where necessary to provide additional protection ($I_{\Delta n} \leq 30$ mA)
G		All	522.6.7 and 522.6.8	Cables concealed in a wall or partition at a depth of less than 50 mm from the surface, or, irrespective of depth, in a partition having an internal construction that includes metallic parts (except screws, etc), in an installation not intended to be under supervision of a skilled or instructed person ($I_{\Delta n} \leq 30$ mA). This does not apply to cables having protection complying with Regulation 522.6.7 or 522.6.8, as applicable (such as earthed steel conduit)

Table 11.9 *(continued)*

Ref	Locations	System type	Regulation	Requirement qualifications
H	Locations with risk of fire due to the nature of processed or stored materials	All	422.3.9	Wiring systems, except mineral insulated cables, busbar, trunking, powertrack systems, for protection against insulation faults. See Regulation 422.3.9 for details.
I	Electrode water heaters and boilers	All	554.1.4	For electrode water heaters and electrode boilers directly connected to the supply at a voltage exceeding extra-low, where specified in the regulation
J	Luminaires and lighting installations	All	559.10.3.2	Recommended for equipment such as lighting arrangements in places such as telephone kiosks, bus shelters and town plans ($I_{\Delta n} \leq 30$ mA)
K	Locations containing a bath or shower	All	701.411.3.3	All circuits of the location ($I_{\Delta n} \leq 30$ mA)
L	Swimming pools	All	702.410.3.4	Equipment in fountains, where protection is provided by ADS. $I_{\Delta n} \leq 30$ mA
			702.53	Certain equipment in zones 1 and 2, where protection is provided by ADS. $I_{\Delta n} \leq 30$ mA
			702.55.1	Electric heating units embedded in the floor, except where provided by SELV meeting specified requirements. $I_{\Delta n} \leq 30$ mA
M	Rooms containing a sauna cabin with a sauna heater	All	703.411.3.3	All circuits of the sauna ($I_{\Delta n} \leq 30$ mA). Such protection need not be provided for the sauna heater unless recommended by the manufacturer
N	Construction and demolition sites	All	704.410.3.10 and 704.411.3.2.1	Socket-outlets and circuits supplying hand-held equipment, where specified in these regulations
O	Agricultural and horticultural premises	All	705.411.1 and 705.422.7	Where specified in these regulations

(continued overleaf)

Table 11.9 *(continued)*

Ref	Locations	System type	Regulation	Requirement qualifications
P	Conductive locations with restricted movement	All	706.410.3.10	For the supply to fixed equipment where protection is Class II equipment or equivalent insulation ($I_{\Delta n} \leq 30$ mA)
Q	Caravan/camping parks and similar locations	All	708.553.1.13	Each caravan-pitch socket-outlet ($I_{\Delta n} \leq 30$ mA)
R	Marinas and similar locations	All	709.531.2	Individually for socket-outlets, and for circuits intended to supply houseboats, where protection is provided by ADS. $I_{\Delta n} \leq 30$ mA.
S	Exhibitions, shows and stands	All	711.410.3.4 and 711.411.3.3	Where specified in these regulations
T	Mobile or transportable units	All	717.415	Every socket-outlet intended to supply current-using equipment outside the unit, except socket-outlets protected by SELV, PELV or electrical separation ($I_{\Delta n} \leq 30$ mA)
U	Caravans and motor caravans	All	721.411.1	All circuits protected by automatic disconnection of supply ($I_{\Delta n} \leq 30$ mA)
V	Temporary installations for fairgrounds, amusement parks and circuses	All	740.410.3 and 740.415.1	Where specified in these regulations
W	Floor and ceiling heating systems	All	753.411.3.2 753.415.1	Floor and ceiling heating systems ($I_{\Delta n} \leq 30$ mA)

RCDs, in general, should be capable of withstanding the prospective fault current at the point of insertion into the circuit (see Regulation 434.5.1) unless they have back-up protection, in which the energy-withstand of the device must be more than match the energy let-through of the upstream device. It should also be borne in mind that when an RCD forms part of a consumer unit to BS EN 60439-3 (including Annex ZA), which itself has a 'conditional rating', the device may be used for prospective fault currents up to 16 kA.

Where RCDs meet all the prerequisites for an isolating device they may be used for isolation purposes. Where remote from the current-using device they serve they must be securable in the open (OFF) position for compliance with Regulation 537.2.1.5. It is vital to appreciate that RCDs are *not* devices suitable for providing protection against overcurrent.

In a domestic installation fed from a TT source and with 'front-end' protection provided by an RCD (also serving as a main switch, where suitable), it should be recognized that the RCD will only protect equipment that is downstream of it. For this reason, it is advisable to use a consumer unit of insulated construction with a wiring system which does not employ exposed-conductive-parts between it and the ED's meter. Where a metal-clad consumer unit is used, care should be taken to determine that the possibility of an earth fault occurring on the upstream side of the RCD is minimized or eliminated. One option is by the use of suitable protection of the incoming 'tails', as shown in Figure 11.11. It should not be forgotten that an earth fault developing on a metal-clad consumer unit would produce a touch voltage, with respect to 'true earth', on the consumer unit and all other conductive parts. This touch voltage would persist until the ED's fuse operated, and in

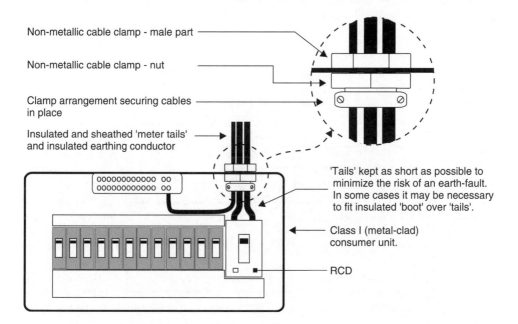

Figure 11.11   Class I consumer unit on a TT system

some cases may remain indefinitely. Incoming supply cables to the metal-clad unit should be insulated and sheathed, enter via an insulated bush and cable clamp, and should be kept as short possible within the enclosure so as to minimize the possibility of an earth fault.

11.5.2   Residual current devices: principles of operation

Figure 11.12 illustrates, in schematic single-phase form, the normal core-balancing RCD main components. The load is shown in fault condition with current allowed to pass via the protective conductor and earth electrode (e.g. separate electrode to the star/neutral of the supply transformer). In a normal healthy circuit, the line and neutral currents would be of the same magnitude and would produce equal and opposing magnetic flux in the common transformer core of the RCD, through which the conductors pass. When a fault occurs on the downstream line conductor, the fault current flows down the protective conductor. There is, therefore, an imbalance between the line and neutral conductors and, hence, the resultant magnetic flux operates the relay and disconnects the circuit. In many cases the search coil produces insufficient power to operate the tripping mechanism and some means of amplification will be necessary. This may be achieved in a number of ways, including the search coil output being fed into a solid-state amplifier.

It should be noted that the test button, when depressed, allows current to flow from the line conductor on the downstream side of the current transformer to the neutral on the upstream side, thereby creating an imbalance in the transformer and causing the device to operate. This test, therefore, only serves to confirm or otherwise that the device itself performs satisfactorily. It does *not* indicate that the means of protection against electric shock is adequate and effective, because the protective conductor(s) and the means of earthing are not included in the test.

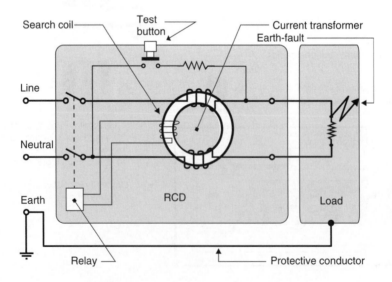

Figure 11.12   Principal components of an RCD

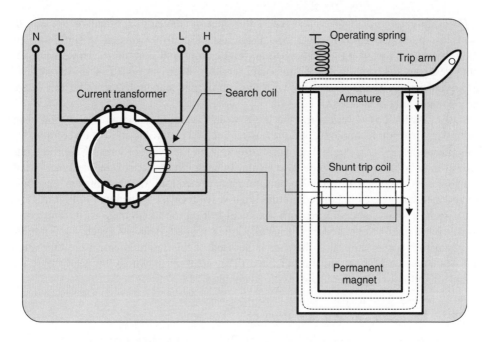

Figure 11.13   RCD polarized relay mechanism

Another method of amplification is to employ the polarized relay principle (illustrated in Figure 11.13), in which a permanent magnet is employed. The weak magnetic field set up by the magnet retains the armature (and tripping mechanism) under normal conditions. Under fault conditions, the output from the search coil 'de-energizes' the magnet sufficiently for the armature to be released (powered by the operating spring) and, thus, disconnection is effected.

Although Figures 11.12 and 11.13 show the single-phase system, the principle holds good for both three-phase three-wire and three-phase four-wire systems. This is because the phasor sum of currents in the live conductors always equals zero in a healthy circuit and only departs from zero when fault current flows elsewhere (e.g. a protective conductor or through a human body) to Earth.

As Table 11.1 shows, BS EN 61008 covers the constructional specification for RCDs, which are identified in two main groups or types of delay:

- without time-delay – type for general use;
- with time-delay – type S for selectivity.

Type S is useful in achieving discrimination with a downstream RCD in series. It has to be borne in mind that RCDs to BS EN 61008-1 permit the device to operate above $0.5I_{\Delta n}$ and that for an RCD of 30 mA sensitivity it may operate (trip) at a residual current above 15 mA. Using such a device at the 'front end' of an installation is generally undesirable, because the cumulative effects of leakage currents of the variously connected appliances may exceed this value. This arrangement is unlikely to afford compliance with Regulations 314.1 and 531.2.4.

The safety requirements embodied in the appropriate British Standard set out the limits for permitted earth leakage, and these are normally expressed in milliamps per kilovolt-ampere for stated conditions (e.g. temperature and humidity). These limits are quite generous for some items of equipment, and for others no limit is set. This makes it important for the designer to check with the appropriate standard(s) in order to evaluate the cumulative effects of earth leakage.

Many installations now incorporate loads which are used for control and monitoring and for information technology, and these often include electronic solid-state components. This phenomenon has increased the possibility that, in the event of an earth fault the fault current may be of a nonsinusoidal waveform. In such circumstances, the designer should consider using RCDs which include pulsating d.c. protection. Such waveforms may be of the rectified half-wave or chopped rectified half-wave forms, as illustrated in Figure 11.14.

Every RCD must be equipped with a manual test button to simulate earth fault conditions. This should be operated by the installation user at intervals not exceeding 3 months. It is imperative, therefore, that the user is advised of this requirement in the Operations and Maintenance Manual and further have their attention drawn to the need for testing by an affixed notice in compliance with Regulation 514.12.2.

11.5.3   Residual current monitors

As stated in the note to Regulation 411.1, an RCM is not a protective device. Its purpose is to monitor the residual current in a circuit and activate an alarm when this exceeds the

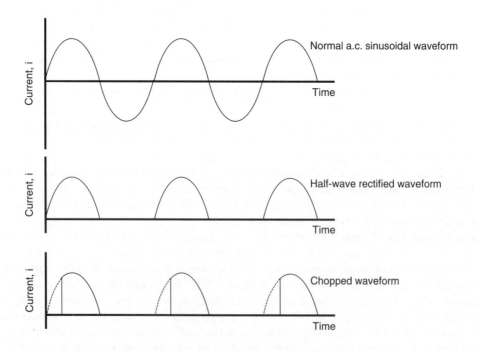

Figure 11.14   Examples of pulsating d.c. waveforms

operating value of the device. The alarm may be audible or audible and visual, but it is generally permitted by BS 7671 to cancel the audible alarm where a visual alarm is also provided.

Although RCMs are not required in the majority of installations, probably the most common application for them is in installations forming part of an IT system (see Section 5.3.6 of this Guide). Regulation Group 538.4, which specifies that RCMs shall comply with BS EN 62020, recommends that an RCM used in an IT system should be of the directionally discriminating type. This is so as to avoid operation of the alarm due to leakage currents caused by high leakage capacitances on the load side of the device, rather than due to earth fault current.

For the situation where an RCD is installed upstream of an RCM, Regulation 538.4.1 recommends that the rated operating current of the RCM does not exceed one-third of the rated residual operating current $I_{\Delta n}$ of the RCD. This is to increase the chances of the RCM giving a warning before the value of residual current is sufficient to operate the RCD.

Regulation 538.4.2 points out that an RCM may be installed to facilitate location of a first insulation fault in an IT system where interruption of such a fault is not required or permitted, and that, where used for this purpose, it is recommended that the RCM is installed at the beginning of the outgoing circuits.

11.6   Identification of overcurrent protective devices

Identification of equipment has to some extent been dealt with in Chapter 9 of this Guide. Section 533 of BS 7671 makes further demands for identification relating to overcurrent devices.

For *all* overcurrent devices (fuses and circuit-breakers), Regulation 533.1 demands that there must be an indication on (or adjacent to) the device of the rated current of the circuit it protects. Normally, this is achieved by the rated current being signified on the device itself, but in some cases it may be necessary to provide further notices.

Where semi-enclosed (rewirable) fuses are employed, the rated current and the size of the tinned copper fusewire need to be matched as given in Table 53.1 of BS 7671. It is not difficult to envisage that these devices might be prone to abuse and misuse. Regulation 533.1.1.2 calls for fuses, where they are likely to be replaced by other than skilled or instructed persons, to be marked or an indication provided near the device with the relevant type (e.g. British Standard number and any further necessary qualifications) or be of a type that cannot be replaced by a fuse of the intended rated current but of a higher fusing factor. In any event, the designer may consider it necessary to provide a comprehensive indication of the device's fault current capacity, rated current, type and standard specification reference.

As mentioned earlier in this chapter, where a circuit-breaker is equipped with adjustable fault current protection (and not subject to operation by other than a skilled or instructed person), a notice indicating the appropriate settings should be affixed on or near the device. These circuit-breakers must not be used where a person other than a skilled or instructed person may operate the device unless the settings can only be modified by means of a key or tool (to satisfy Regulation 533.1.2). Furthermore, the settings, if adjustable, must be visually indicated, but access to this information may involve the opening of the enclosure.

11.7 Discrimination

11.7.1 Discrimination: general

Regulation 536.2 demands that, where it is necessary to prevent danger or to facilitate proper functioning of the installation by selectivity (discrimination), the manufacturer's instructions shall be taken into account, so that the characteristics and settings of protective devices are such that discrimination is achieved. In terms of proper functioning of the installation, the consequences of faults should be restricted to the circuit concerned.

Discrimination relates to all faults, whether they be short-circuits (faults between live conductors) or line–earth faults. Consideration must be given to overcurrent devices and, where employed, RCDs where two or more such devices are configured in series. In simple terms, discrimination is achieved when a circuit under fault conditions causes only the protective device of the affected circuit to disconnect. Upstream protective devices remain unaffected, so that only the faulty circuit is disconnected from the supply. In order to carry out an analysis of discrimination it is necessary in many cases to know the fault levels (i.e. the prospective fault currents) at the various points of insertion of all the overcurrent protective devices. It is necessary to establish fault levels and the margins by which they are likely to change. For example, changes may occur because of modifications to the supply network. Discrimination should be designed to take account of this range.

Whilst it is not possible to consider all the possible combinations of protective devices in series, an attempt is made here to address the more common ones. In all cases, the designer should obtain the necessary operating characteristics from the manufacturers.

11.7.2 Discrimination: high breaking capacity–high breaking capacity fuses

In effecting discrimination between fuses in series, the principal factor is to establish that the total energy let-through $I^2t_{[t]}$ of the minor fuse (the downstream device) does not exceed the pre-arcing energy let-through $I^2t_{[pa]}$ of the major fuse (upstream device) (see Figure 11.6). It should be borne in mind that, at low fault levels, the difference between $I^2t_{[t]}$ and $I^2t_{[pa]}$ is very small and can be ignored for this purpose.

Two methods exist to determine whether or not discrimination will be achieved between fuses in series, namely that of visually correlating their time–current characteristics and that of using their I^2t characteristics. In order to decide which of these methods is the correct one to use in a particular case, it is important to establish the fault level at the various points in order to establish the approximate pre-arcing times and, in particular, whether they are more than or less than 20 ms. Where pre-arcing times exceed 20 ms, an assessment may be made using the time–current characteristics, but otherwise this should be made by assessing the I^2t characteristics.

Most manufacturers publish data on their fuses, including pre-arcing energy let-through and total energy let-through (A^2 s), from which discrimination can be assessed. In many cases, a table of fuse data is provided, for convenience, showing where discrimination will be achieved using various rated currents. Fuses complying with BS 88-2.2 and BS 88-6, for example, are deemed to discriminate with each other at fault currents up to 80 kA providing that a ratio of 1.6:1 rated currents between the major and minor fuse links is maintained.

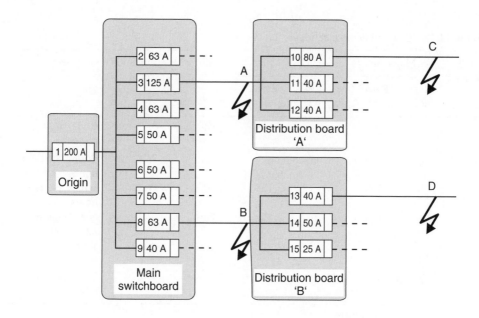

Figure 11.15   Small distribution layout using HBC fuses

Take a hypothetical small three-phase distribution example, shown schematically in Figure 11.15 (in single-line form). All overcurrent protective devices are fuses to BS 88-2.2 having the rated currents shown on the figure. Consider a fault occurring at each of the four points marked A, B, C and D in turn. Using the discrimination ratio of 1.6:1, the results are given in Table 11.10.

It can be seen from the above simplistic example that the designer needs to address discrimination, not just the loading in the selection of protective devices, in order to provide a safe installation. In the two cases shown in Table 11.10 where the ratios are less than 1.6:1, it may be that, in practice, discrimination may be achieved at certain fault levels. To be sure, it would be necessary to consider altering the ratio by, for example, replacing fuses No. 3 and No. 13 with fuses of rated currents 100 A and 35 A respectively, assuming, of course, that the circuit loading will permit.

11.7.3   Discrimination: miniature circuit-breakers–miniature circuit-breakers

Figure 11.16 illustrates a common occurrence of two MCBs in series. Both devices are Type B and the upstream unit, the major device, has been designated 'A' and the minor device is shown as 'B'. Since both devices are Type B, the instantaneous operation (tripping) will occur between three and five times their rated current. We need only consider the 'worst-case' instantaneous operating current (i.e. $5 \times I_n$) when evaluating fault protection, but we need to have regard to the range (three to five times I_n) within which instantaneous tripping occurs in addressing discrimination.

It is apparent from Figure 11.16 that there will be discrimination between the two MCBs under overload conditions occurring downstream of MCB 'B'. Under fault conditions

Table 11.10 Discrimination (see Figure 11.16)

Fault location	Major fuse		Minor fuse		Ratio[a]	Discrimination assured	Comments
	Ref no.	Rated current (A)	Ref no.	Rated current (A)			
A	1	200	3	125	1.60:1	Yes	Within the ratio limits
B	1	200	8	63	3.17:1	Yes	Within the ratio limits
C	1	200	3	125	1.60:1	Yes	Within the ratio limits
	3	125	10	80	1.56:1	No[b]	Not within the ratio limits
D	1	200	8	63	3.17:1	Yes	Within the ratio limits
	8	63	13	40	1.57:1	No[b]	Not within the ratio limits

[a]Ratios given are approximate.
[b]By consideration of the I^2t characteristics of the devices, it may be possible to show that discrimination can be achieved.

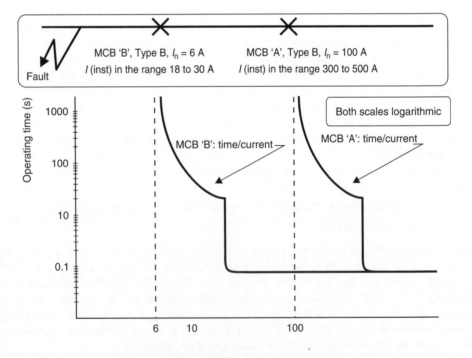

Figure 11.16 Discrimination between MCBs

(both short-circuit and earth fault) it is necessary to know the prospective short-circuit current and the prospective earth fault current in order to determine whether or not discrimination is achieved.

By way of a practical example, let MCB 'A' have a rated current of 100 A and MCB 'B' be rated at 6 A, the latter feeding a 230 V lighting circuit. Taking three cases, let us first consider the case with a prospective short-circuit current of 230 A (line–neutral impedance 1.0 Ω) and prospective earth fault current of 209 A ($Z_s \approx 1.1\Omega$). Clearly, we should take the higher value of prospective fault current represented here by the short-circuit fault current 230 A. It is easy to see that discrimination will be achieved because the prospective short-circuit current is less than three times the rated current (representing the lower end of the instantaneous tripping range) of MCB 'A'.

For the second case, consider a circuit with the same components but where the prospective earth fault current is 750 A ($Z_s \approx 0.31\ \Omega$). It is obvious that both MCBs will trip at this fault level; therefore, discrimination will not be obtained, with the consequence of losing power to all circuits fed by MCB 'A'. For the third and final case, let us attribute a prospective earth fault current of 328 A ($Z_s \approx 0.7\ \Omega$), where it will be evident that discrimination may be achieved (depending on the particular devices in question) but the designer could not claim discrimination with any degree of certainty and would wish to reconsider the distribution layout and amend it by, for example, inserting an intermediate device.

11.7.4 Discrimination: miniature circuit-breakers–fuse

MCBs and fuses are often used in series, and Figure 11.17 shows typical time–current characteristics of an MCB and a fuse. If selected correctly, then these devices will present little difficulty in achieving discrimination under overload conditions. However, it can be readily seen from the characteristic curves that discrimination will depend to a large extent on the prospective fault current magnitude. The figure illustrates that discrimination may be achieved for fault currents approaching $I_{F(m)}$. For fault currents at that value, discrimination will be doubtful, and there will be no discrimination for fault currents exceeding that value.

Figure 11.18 shows the time–current characteristics of MCBs rated at 16, 20 and 32 A together with that of a 13 A BS 1362 fuse (commonly used in 13 A plug tops to BS 1363 and fused connection units). Again, we can see that discrimination will depend on the prospective fault current. Taking the 32 A MCB and the fuse characteristic curves to illustrate the point, we can see that discrimination will not be achieved at fault levels above the current corresponding to Point X (about 120 A). It may be that, in the case of ring final circuits protected by a 32 A MCB, the lack of discrimination with the BS 1363 plug top fuse will not be an important consideration from the safety standpoint and the designer may feel that such a situation could be tolerated.

11.7.5 Discrimination: residual current devices

Where RCDs are used to protect against electric shock. Regulation 531.2.9 calls for discrimination between two or more such devices in series. The first, fundamentally important, point to recognize is that RCDs *do not* limit the prospective earth fault current I_F. For example, an RCD with rated residual operating (tripping) current $I_{\Delta n} = 100$ mA

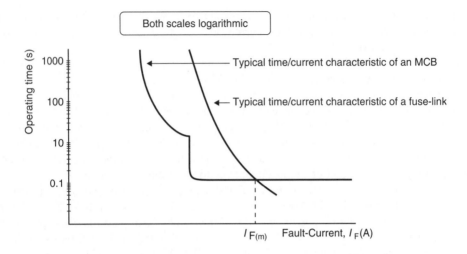

Figure 11.17 Typical time/current characteristics of an MCB and a fuse link

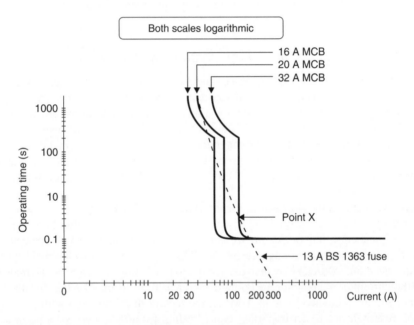

Figure 11.18 Time/current characteristics of MCBs (16, 20 and 32 A) and a BS 1363 fuse link

does not limit the prospective fault current to 100 mA. Prospective earth fault current is a function of the nominal voltage to earth U_o and the line−earth loop impedance Z_s, such that $I_F = U_o/Z_s$. For a fault current of 100 mA on a 230 V circuit, the impedance Z_s would be 2300 Ω $(230/(100 \times 10^{-3}))$. Even in a TT system it is unlikely that such a high line−earth loop impedance would be encountered; for this reason, attempting to

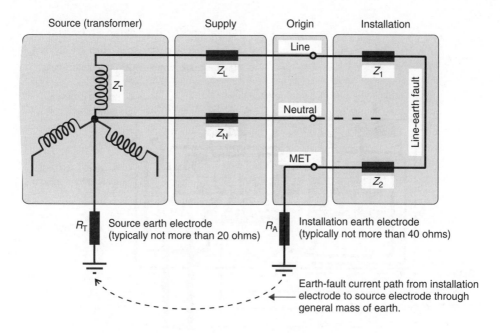

Figure 11.19   TT system circuit under earth fault conditions

discriminate on the basis of fault current magnitude is not a practical option. Figure 11.19 represents a circuit, in a TT system, under earth fault conditions.

Referring to Figure 11.19, the total line–earth loop impedance Z_s is given by

$$Z_s = Z_T + Z_L + Z_1 + Z_2 + R_A + R_T \tag{11.4}$$

where Z_s is the total line–earth loop impedance, Z_T is the transformer impedance, Z_L is the supply line cable impedance, Z_1 is the installation line cable impedance, Z_2 is the installation cpc impedance, R_A is the installation earth electrode and connecting protective conductor impedance and R_T is the source earth electrode and connecting protective conductor impedance.

The total line–earth loop impedance Z_s in a TT system is usually of the order of between 30 and 70 Ω, but in extreme cases it may be up to 200 Ω. The source earth electrode resistance can be up to 20 Ω and that of the installation electrodes is typically about 20–50 Ω. A line–earth loop impedance value of 46 Ω would produce a fault current of 5 A on a 230 V circuit (230/46).

Discrimination between RCDs can only really be accomplished using the time parameter. This can be achieved by the introduction of a time delay on the upstream RCD. In order to achieve discrimination between a time-delayed RCD upstream and a non-time-delayed (or shorter time-delayed) RCD downstream, it is essential that the downstream device operates (trips) before the time delay on the upstream device has expired. The operating times will depend to some extent on the earth fault current magnitude, but in practice it is normal to allow the time delay to be in excess of that which is strictly necessary to effect discrimination.

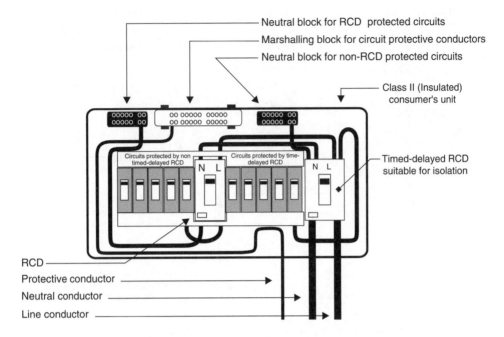

Figure 11.20   Split-load consumer unit (TT systems)

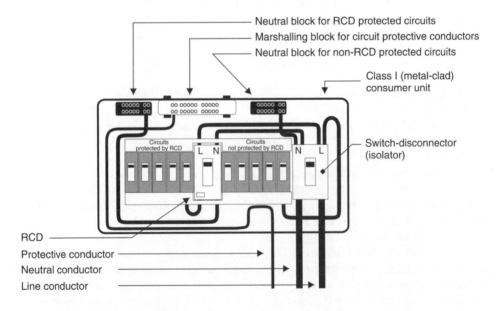

Figure 11.21   Split-load consumer unit (TN-S and TN-C-S systems)

In a domestic situation in a TT system (as in a TN system), it is normally necessary to provide additional protection to socket-outlets rated at 20 A or less that are for general use by an RCD ($I_{\Delta n} \leq 30$ mA) for compliance with Regulation 411.3.3. It would be undesirable to provide the 30 mA RCD protection at the 'front end', since it is likely that the cumulative effects of earth leakage currents of the various connected appliances and other factors would produce unwanted tripping, leaving the whole installation without power. A much more effective approach (but somewhat more expensive) would be to use, say, a 100 mA time-delayed RCD at the 'front end' with a 30 mA non-time-delayed RCD for additional protection to final circuits which require such protection. The arrangement is shown in Figure 11.20 and is currently available in the form of a 'split-load' consumer unit. A similar arrangement for a 'split-load' consumer unit for use on TN systems is illustrated in Figure 11.21.

11.8 Other equipment

11.8.1 Accessories

Section 553 of BS 7671 makes demands on the selection of accessories not included elsewhere in BS 7671, and these are summarized in Table 11.11.

11.8.2 Luminaires and lighting points

The regulatory requirements embodied in Section 559 of BS 7671 relating to luminaires and lighting installations are summarized in Table 11.12. Of particular interest is that Regulation 559.6.1.1 sets out the requirements for the connection of luminaires to the fixed wiring – see Ref. K in Table 11.12. Ceiling roses (for pendants) and batten-holders have their limited uses but LSCs will increasingly become commonplace, providing, as they do, an aesthetically pleasing and sound final connection with the added facility of enabling disconnection of the luminaire for cleaning purposes, etc. without disturbing the fixed installation. Decorative and domestic luminaire manufacturers should be encouraged to supply luminaires with a fitted integral LSC 'plug' (as some already do) so that it becomes standard practice for the fixed-wiring installer to terminate with an LSC 'socket'.

11.8.3 Heaters for liquids and other substances, including water

As with all equipment, water heaters must comply with the relevant British or Harmonized Standard. Regulations 554.1.1–554.1.7 that deal with electrode water heaters and boilers, whilst Regulation 554.2.1 specifically relates to heaters with immersed elements. Regulations 554.3.1–554.3.4 make demands in relation to water heaters having immersed and uninsulated elements. The requirements are summarized in Table 11.13. It is crucially important that the manufacturer's instructions and recommendations are followed when installing water heaters, particularly in the case of electrode heaters and boilers.

Table 11.11   Requirements relating to selection of plugs, socket-outlets and cable couplers

Ref	Accessory	British Standard	Requirements/comments	Regulation
A	Plugs and socket-outlets, except SELV		Must not be possible for any pin of a plug to make contact with a live part of its associated socket-outlet whilst any other pin is completely exposed nor with other types of socket-outlet within the same installation	553.1.1
B			Except for SELV and special circuits (see Row M), every plug and socket-outlet must be the nonreversible type	533.1.2
C			Socket-outlets for household or similar use must be of the shuttered type, preferably the 13 A to BS 1363	553.1.4
D		BS 1363	For 13 A fused two-pole and earth plug and socket-outlets (with BS 1362 fuses)	553.1.3
E		BS 546	For 2, 5, 15 and 30 A fused and nonfused two-pole and earth plug and socket-outlets (with BS 646 fuses)	
F		BS 196	For 5, 15 and 30 A fused and nonfused two-pole and earth plug and socket-outlets	
G		BS EN 60 309-2 (BS 4343)	For 16, 32, 63 and 125 A industrial-type plug and socket-outlets	
H			Wall mounted socket-outlets must be mounted at a sufficient height above finished floor or work surface to prevent mechanical damage to plug top flex by insertion of plug – usually taken as at least 150 mm above floor or work surface	553.1.6
I			Where for use with mobile equipment, socket-outlets to be positioned for equipment to be fed from adjacent socket-outlet taking into account the normal length of flexible cord provided on the equipment (usually taken as not more than 2 m)	553.1.7
J	Clock connector sockets		The connection of electric clocks may be other than plugs and socket-outlets listed above but must incorporate a fuse not exceeding 3 A to BS 646 or BS 1362. Nominal voltage not to exceed 250 V	

Table 11.11   *(continued)*

K	Shaver sockets	BS 3535	In bathrooms, shaver sockets must comply with this standard. Nominal voltage not to exceed 250 V	
L		BS 4573	In locations other than bathrooms, shaver socket must comply with this standard or BS EN 61558-2-5. Nominal voltage not to exceed 250 V	553.1.5
M	Special circuit socket-outlets		In a circuit having special characteristics where danger may otherwise occur or where it is necessary to distinguish it from other circuits in order to prevent danger, then it is permissible to use socket-outlets to standards not mentioned above. Nominal voltage not to exceed 250 V	
N	Cable couplers, except for SELV and a Class II circuit	BS 196, BS 6991, BS EN 61535, BS EN 60 309-2	Cable couplers to comply with one of these British Standards and must be of the nonreversible type and have provision for connection of a protective conductor	553.2.1
O			The cable coupler connector (not the plug) is connected to the cable from the supply	553.2.2

11.8.4   *Heating conductors and electric surface heating systems*

Regulations 554.4.1–554.4.4 set out the requirements relating to heating conductors and cables and Regulation 554.5.1 deals with electric surface heating systems, all of which are summarized in Table 11.14.

11.8.5   *Transformers*

There are only three regulations (555.1.1 to 555.1.3) relating to transformers as items of equipment, though their installation is, of course, subject to the generality of BS 7671. The first is concerned with autotransformers and requires that where such a transformer is employed and is connected to a circuit with a neutral conductor that conductor must be connected to the common terminal of the winding. Regulation 555.1.2 precludes the connection of *step-up* autotransformers to an IT system. In all cases where a step-up transformer is used, Regulation 555.1.3 calls for a linked switch to disconnect all live conductors (lines and neutral) simultaneously from the supply.

11.8.6   *Rotating machines*

Regulation 552.1.1 requires that all equipment associated with rotating machine circuits which carries starting, accelerating, braking and load currents must be suitable for such

Table 11.12 Requirements relating to luminaires and lighting installations

Ref	Accessory/ aspect	CAP	British Standard	Requirements	Regulation
A	Exclusions from scope of Section 559	All	BS 559 and BS EN 50107	HV signs supplied at LV (such neon tubes), and signs and luminous discharge installations supplied at a no-load voltage exceeding 1 kV and not exceeding 10 kV, are excluded from the scope of Section 559	559.1
B	Luminaires and lampholders	All		Luminaires to comply with relevant British Standard and be selected and erected to manufacturer's instructions	559.4.1
C		All		Luminaires without transformers or convertors but having extra-LV lamps in series are to be considered as LV equipment not extra-LV equipment, for the purpose of Section 559	559.4.2
D		All		If luminaires are installed in a pelmet, then there must be no adverse effects due to curtains or blinds	559.4.3
E		All		Luminaires to be adequately fixed by means able to support them, and in accordance with manufacturers' instructions. The fixing means for a pendant luminaire (such as the screw fixings supporting a ceiling rose) to be suitable for a mass of 5 kg, or more where needed. The weight of luminaires to be compatible with suspended ceiling or other supporting structure. Any cable/cords supporting luminaires from fixing means to be installed so there is no unsafe strain on conductors or terminations, etc.	559.6.1.5
F		SBC, BC	BS EN 61184	Type B15 and B22 lampholders to comply with BS EN 61184, rating T2	559.6.1.7

Table 11.12 *(continued)*

Ref	Accessory/ aspect	CAP	British Standard	Requirements	Regulation
G	Luminaires and lampholders	All		Only luminaires designed for the purpose are to be through wired. Temperature rating of through-wiring cables to be selected to suit information on luminaire or in manufacturer's instructions. Triple-pole disconnecting devices to be provided for groups of luminaires divided over three phases with only one neutral conductor	559.6.2.1, 559.6.2.2, 559.6.2.3
H		All		Thermal effects of radiant and convected energy to be taken into account when selecting and erecting luminaires	559.5.1
I				Lamp controlgear not to be used external to a luminaire unless marked as suitable for independent use according to the relevant standard	559.7
J				Compensating capacitors to comply with BS EN 61048 and be used with discharge resistors if exceeding 0.5 μF	559.8
K	Lighting points	All	BS 67, BS 6972, BS 7001, BS EN 60598, BS 1363-2, BS 546, BS EN 60309-2, BS 5733, BS 1363-4, BS EN 60670, BS 4662, IEC 61995-1	The connection at each fixed lighting point to be by a ceiling rose, LSC, batten lampholder or pendant set, directly connected luminaire designed for that purpose, socket-outlet, plug-in lighting distribution unit, 13 A fused connection unit, appropriate terminals enclosed in a box to specified British Standards, or a DCL	559.6.1.1
L		All		Ceiling roses and lampholders for filament lamps not to be operated at voltages normally exceeding 250 V	559.6.1.2

(continued overleaf)

Table 11.12 *(continued)*

Ref	Accessory/ aspect	CAP	British Standard	Requirements	Regulation
M		All		Unless specially designed for multiple pendants, ceiling roses must only be used for single pendants	559.6.1.3
N	Lighting points	All		LSCs not to be used for connecting equipment other than luminaires	559.6.1.4
O		SBC BC SES ES GES		Type B15, B22, E14, E27 and E40 lampholders with overcurrent protective device not exceeding 16 A	559.6.1.6
P		All	BS 3676/, BS EN 60669-1, BS EN 60669-2-1	Lighting points to be controlled by switches and/or automatic lighting control system	559.6.1.9
Q	Lighting track systems		BS EN 60570	Track systems for luminaires must comply with BS EN 60570	559.4.4
R	Outdoor lighting installations			Outdoor lighting installations do not include ED's equipment or temporary festoon lighting	559.3
S				Requirements for outdoor lighting installations, highway power supplies and street furniture. See Chapter 13 of this Guide for further information	559.10
T	Extra-low voltage lighting installations	All		The protective measure of FELV not to be used	559.11.1

Table 11.12 (*continued*)

Ref	Accessory/aspect	CAP	British Standard	Requirements	Regulation
U	Extra-low voltage	All		Extra-LV luminaires without protective conductor provisions to be installed only in SELV systems	559.11.2
V			BS EN 61558-2-6 BS EN 61347-2-2	Safety isolating transformers to comply with BS EN 61558-2-6 and either be of short-circuit proof type (both inherently and noninherently) or protected by a device complying with Regulation 599.11.4.2 on primary side. Electronic convertors to comply with BS EN 61347-2-2	559.11.3.1 559.11.3.2
W			BS EN 60598-2-23	If both circuit conductors are uninsulated then they must be protected by a device complying with Regulation 599.11.4.2 or the system must comply with BS EN 60598-2-23	559.11.4.1
X				Requirements for a device providing protection against risk of fire where both live conductors are uninsulated	559.11.4.2
Y				Metallic structural parts of building, such as pipes and parts of furniture, not to be used as live conductors	559.11.5.1
Z				Requirements for minimum conductor cross-sectional area (see regulation)	559.11.5.2
A1				Bare conductor may be used if nominal voltage is 25 V a.c or 60 V d.c. or less, subject to certain requirements (see regulation)	559.11.5.3
A2				In suspended systems, suspension devices, including conductors, for luminaires to be capable of supporting five times the mass of luminaires and lamps, and not less than 5 kg. Suspension system to be to walls and ceiling by insulated distance cleats and continuously accessible. Terminations and connections to meet certain requirements (see regulation)	559.11.6

Table 11.13 Requirements relating to heaters for liquids and other substances including water

Ref	Regulation	Requirements
Electrode water heaters and boilers		
A	554.1.1	To be used on a.c. systems only. Not for use on d.c.
B	554.1.2	Supply to be controlled by a linked circuit-breaker with overcurrent protection on each conductor connected to an electrode
C	554.1.3	Earthing generally in accordance with Chapter 54 of BS 7671
D	554.1.3	Heater or boiler shell to be connected to metallic armouring or sheath, if any, of supply cable
E	554.1.3	Protective conductor to be connected to shell and must meet the requirements of the adiabatic equation or Table 54.7 of BS 7671
F	554.1.4	RCD protection required for boilers and heaters operating on, and directly connected to, HV. RCD normally set at 10% of load (but may be increased to 15% if essential for stable operation of boiler or heater) and may incorporate a short time-delay mechanism to overcome imbalances of short duration
G	554.1.5	If three-phase LV, neutral must be connected to the shell in addition to protective conductor. Neutral current-carrying capacity no less than that of the largest line conductor
H	554.1.6	If single-phase LV, neutral must be connected to the shell in addition to protective conductor and to the supply neutral and earthing conductor
I	554.1.7	For heaters and boilers not piped to a water supply and not in contact with earth, and where the electrodes are shielded in insulating material so they cannot be touched whilst live, the linked circuit-breaker may be replaced by a phase conductor fuse and there is no need to connect the shell to the neutral
Heaters having immersed heating elements		
J	554.2.1	Automatic device required for preventing a dangerous rise in temperature
Water heaters with immersed and uninsulated elements		
K	554.3.1	Calls for requirements of the following regulations to be met but excludes electrode boilers and heaters
L	554.3.2	Metal parts in contact with water (excluding current-carrying parts, but including metal taps and covers) to be connected to the metallic water supply pipe to heater. This in turns needs connecting to the MET by means of an *independent* protective conductor

Table 11.13   *(continued)*

M	554.3.3	Permanent connection to supply required. Double-pole linked switch required either separate from heater and within easy reach or incorporated in heater. Plugs and socket-outlets not permitted for final connection. Where installed in a special location (e.g. bathroom), the requirements relating to the location must also be met
N	554.3.4	Installer to check that no single-pole device is fitted in the neutral conductor in any part of circuit between the heater and the installation origin

Table 11.14   Requirements relating to heating conductors and cables and electric surface heating systems

Ref	Regulation	Requirements
A	554.4.1	Heating conductors and cables passing through or in close proximity to material representing a fire hazard must be enclosed with material having ignitability characteristics 'P' as laid down in BS 476-12. Additionally, suitable and adequate mechanical protection must be provided
B	554.4.2	Cables laid direct in soil or in the building fabric must be capable of withstanding mechanical damage likely to be encountered
C	554.4.2	Cables laid direct in soil or in the building fabric must be of such construction as to be resistant to damage from moisture
D	554.4.2	Cables laid direct in soil or in the building fabric must be of such construction as to be resistant to damage from corrosion
E	554.4.3	Heating cables directly laid in soil, a road or in the building fabric must be completely embedded
F	554.4.3	Heating cables directly laid in soil, a road or in the building fabric must not suffer damage from normal relative movement
G	554.4.3	Heating cables directly laid in soil, a road or in the building fabric must comply fully with the manufacturer's instructions and recommendations
H	554.4.4	The load of each floor warming cable to be limited so that maximum conductor operating temperatures not to exceed the manufacturer's stated conductor temperature. Full consideration must also be given to other component parts, including seal and coverings and adjacent materials
I	554.5.1	Electric surface heating systems to comply with BS 6351 in respect of system design, installation and testing

currents and rated at not less than that given in the relevant British or Harmonized Standard. The effects of frequent starting and stopping have also to be taken into consideration, as do the consequential effects on the supply and the temperature rise on equipment, including conductors.

The ED should always be consulted where fairly large motors are concerned in order to establish the effects of such loads on the supply. The ED will be able to advise on the maximum 'voltage dip' that can be tolerated, and this will affect the choice of the method of starting rotating machines. Direct-on-line starting is commonly used for small motors and, where acceptable, provides for the most inexpensive method of starting. Other methods include star–delta starting for three-phase motors and the 'soft-start' thyristor starting techniques. These need to be considered together with the type of mechanical load characteristics and the ED's requirements. Lack of proper design coordination in this respect may result in 'voltage dips' on the supply which are unacceptable to the ED and to other users on the network.

Regulation 552.1.2 calls for motors of rating exceeding 370 W to be provided with overload protection. This requirement does not apply to equipment incorporating a motor which itself, as a whole, complies with the relevant British Standard. Unless there is a danger from so doing, Regulation 552.1.3 demands that every motor be provided with 'no-volt' release, which prevents the motor restarting on restoration of full supply voltage after stopping as a result of a reduction of supply voltage or a failure of the supply. This requirement does not preclude the use of automatic starting systems (e.g. on a process line) where adequate alternative measures have been taken to prevent danger from unexpected starting. Regulation 537.5.4.2 requires that, where a motor is equipped with reverse-current braking, precautions must be taken to prevent the motor running in the opposite direction after braking has been completed if there is a possibility of danger resulting from reverse running. Additionally, Regulation 537.5.4.3 requires that, where safety depends on the direction of rotation, provision must be made to prevent reverse operation due to any cause, including the loss of one phase on a multiphase motor. These last two regulations (537.5.4.2 and 537.5.4.3) apply only where the electrically powered equipment is within the scope of BS EN 60204. However, the scope of BS EN 60204 is wide ranging. It covers electrical, electronic and programmable electronic equipment and systems to machines not portable by hand while working, including a group of machines working together in a coordinated manner, with nominal supply voltages not exceeding 1000 V a.c. or 1500 V d.c.

12

Protective conductors, earthing and equipotential bonding

12.1 Protective conductors

12.1.1 Protective conductors: general

The requirements for protective conductors are embodied in Chapter 54, which sets out constraints in terms of minimum cross-sectional areas and acceptability of certain material and types, and Figure 12.1 lays out the various sections of Chapter 54 of BS 7671.

It is important to recognize that the term 'protective conductor' is generic and embraces the following particular protective conductors:

- earthing conductor;
- main protective bonding conductor;
- supplementary bonding conductor;
- cpc;
- combined protective and functional conductor.

The sizing requirements may differ depending on the function of the protective conductor. In all cases where the conductor has a covering, that covering must be coloured yellow/green (70%/30% or 30%/70%) to meet the requirements of Regulation 514.4.2. Any other colour (including green alone – Regulation 514.4.5) must not be used except for functional earthing conductors, where the colour cream must be used. Additionally, Regulation 514.3.2 requires every conductor to be identified at least at its terminations, though preference is expressed that it should be identified throughout its length. Where bare protective conductors are used, they should be identified by green/yellow, tape, disc or sleeving (Regulation 514.4.6).

Protective conductors may serve more than one function (e.g. main protective bonding conductor and cpc). They may serve the function of cpc to a number of circuits, as in the case of steel conduit enclosing a number of circuits. When so used, the protective

A Practical Guide to The Wiring Regulations: 17th Edition IEE Wiring Regulations (BS 7671:2008)
Fourth Edition Geoffrey Stokes and John Bradley
© Geoffrey Stokes and John Bradley. Published by John Wiley & Sons, Ltd

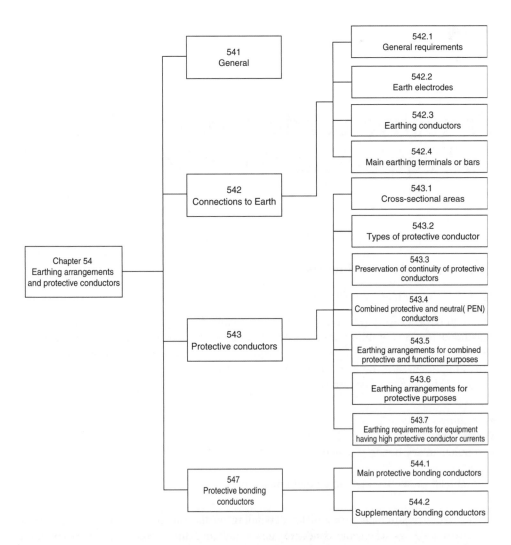

Figure 12.1   Arrangement of Chapter 54: earthing arrangements and protective conductors

conductor must fulfil the requirements relating to the most onerous for all the functions and/or circuits it is used to protect.

Where fault protection (formerly known in BS 7671:2001 as protection against indirect contact) is provided by an overcurrent protective device, Regulation 543.6.1, calls for the protective conductor to be incorporated in the same wiring system as associated live conductors, or at least in the immediate proximity of those live conductors.

12.1.2   Protective conductors: types

Section 543 of BS 7671 identifies the various types of protective conductors which are acceptable and those that are not. These requirements are summarized in Table 12.1.

Table 12.1 Acceptable and unacceptable types of protective conductor

Ref	Regulation	Type of protective conductor	Acceptable	Restrictions, if any/comments
A	543.2.2 543.2.3	Conductor in a cable	Yes	If cross-sectional area does not exceed 10 mm², to be copper
B		Single-core cable	Yes	
C		Insulated or uninsulated conductor in a common enclosure with live conductors	Yes	
D		Fixed insulated or uninsulated conductor	Yes	
E		Metal cable sheath, armouring or screen	Yes	See also Regulations 543.2.4 and 543.2.5
F		Metal conduit	Yes	Subject to a minimum cross-sectional area of 10 mm². See also Regulation 543.2.4
G		Metal enclosure (e.g. trunking).	Yes	Subject to a minimum cross-sectional area of 10 mm². See also Regulation 543.2.4
H		Electrically continuous conductor support system	Yes	Subject to a minimum cross-sectional area of 10 mm². See also Regulation 543.2.4
I	543.2.2, 543.2.6	Extraneous conductive part	Yes	Excepting gas and oil pipe lines, an extraneous conductive part may be used as a protective conductor. Subject to a minimum cross-sectional area of 10 mm². Must be protected against mechanical, chemical and electromagnetic deterioration. Electrical continuity must be assured. The cross-sectional area must meet the adiabatic equation (Regulation 543.1.3) or be verified by test to BS 5486 Part 1 provided always that precautions are taken against its removal and that the part has been considered for such use

(continued overleaf)

Table 12.1 *(continued)*

Ref	Regulation	Type of protective conductor	Acceptable	Restrictions, if any/comments
J	543.2.1	Gas pipes	No	Must never be used
K	543.2.1	Oil pipes	No	Must never be used
L	543.2.1	Flexible and pliable conduit	No	Must never be used
M	543.2.1	Support wires/flexible metallic parts	No	Must never be used
N	543.2.1	Constructional parts stressed in normal service	No	Must never be used
O	543.2.4	Metal enclosure of switchgear and controlgear	Yes	Must be protected against mechanical, chemical and electromagnetic deterioration. Electrical continuity must be assured. The cross-sectional area must meet the requirements of Regulation 543.1 for a protective conductor or be verified by test to BS EN 60439-1. The arrangement must provide for connection of other protective conductors at every predetermined point
P	543.2.4	Busbar trunking systems	Yes	
Q	543.2.5	Cable sheaths and other cable metal coverings including armouring, copper sheaths (MICC) and conduit	Yes	To be used as a protective conductor for the associated circuit. Must be protected against mechanical, chemical and electromagnetic deterioration. Electrical continuity must be assured. The cross-sectional area must meet the requirements of Regulation 543.1 for a protective conductor or be verified by test to BS EN 60439-1.

12.1.3   Protective conductors: thermal withstand

As with live conductors, the designer needs to consider the effects of earth fault current on protective conductors. It is important to recognize that the energy-withstand S^2k^2 of the protective conductor must not be less than the energy let-through I^2t of the protective device(s). This is satisfied if the adiabatic equation is met (see Equation (12.3)) or the cross-sectional area of the protective conductor is related to the line conductor cross-sectional area in accordance with Table 54.7 of BS 7671.

One could easily be forgiven for thinking that the cross-sectional area of the protective conductor is proportional to the fault current magnitude, i.e. the higher the fault current is, the larger the cross-sectional area is. This is not so, and it can be shown that in many circumstances a lesser cross-sectional area S will be acceptable with the increasing earth fault current. To illustrate by way of example, take a 100 A BS 88-2.2 HBC fuse and consider disconnection times of 0.1, 0.2, 0.4 and 5 s:

Fault current I_f (A)	Disconnection time (s)	Energy let-through I^2t (A^2 s)	Min. conductor S (mm^2) (with $k = 143$)
1400	0.1	196 000	3.1
1150	0.2	264 500	3.6
980	0.4	384 160	4.3
550	5.0	1 512 500	8.6

The example only serves to illustrate the point made and should not be taken to mean it is suggested that designing protective conductors to such close tolerances is advocated. Figure 12.2 shows graphically the relationship between fault current and the minimum

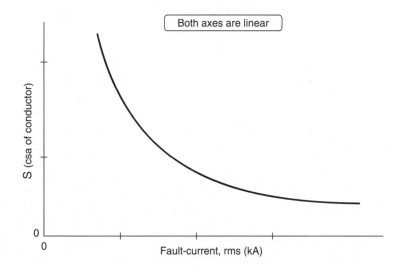

Figure 12.2   Relationship between fault current and the minimum protective conductor cross-sectional area for fuses

conductor cross-sectional area from the viewpoint of thermal protection where protection is provided by fuses. Where MCBs are employed, the curve is a horizontal straight line.

The earth fault current I_f for application of the adiabatic equation may be determined using

$$I_f = \frac{U_o}{Z_S} \tag{12.1}$$

where I_f is the Earth fault current in the circuit, U_o is the nominal line-to-Earth voltage and Z_s is the line-earth fault loop impedance in the circuit.

The value of Z_s used in Equation (12.1) should correspond to when the line and protective conductors of the circuit are at their normal operating temperature. If the conductors are at a lower temperature when their resistance is obtained, then account must be taken of the increase in their resistance with the increase in temperature due to load current. Table 10.17 of this Guide gives resistance values for copper line and protective conductors, making allowance for the temperature rise due to load current and relevant to the type of insulation.

The resistance of the line and protective conductors should be added to the external earth fault loop impedance Z_e to give Z_s, although for conductors larger than 16 mm^2 the inductive reactance should also be taken into account.

12.1.4 Protective conductors: sizes

The sizing of all protective conductors, except protective bonding conductors, relates primarily to thermal-withstand consideration of the conductor such that the insulating and/or adjacent materials do not exceed their final permissible temperatures under earth fault conditions. From such considerations, the sizing obtained should be regarded as the minimum. There may be cases where the cross-sectional area may need to be increased to reduce the protective conductor's contribution to the total line–earth loop impedance.

As stated in Regulation 543.1.1, the minimum cross-sectional area can be obtained in one of two ways: first, by reference to Table 54.7 of BS 7671; second, by application of the adiabatic equation in Regulation 543.1.3. For most designers, reference to Table 54.7 will be the easier of the two approaches, involving little calculation where the protective conductor is of the same material as the associated line conductor. Where this is not the case, the simple application of a multiplying factor of k_1/k_2 to the cross-sectional area S of the line conductor is all that is required.

The k factor takes account of the resistivity, temperature coefficient and heat capacity of the conductor, together with the initial and final admissible temperatures; k_1, the factor for the line conductor, is obtained from reference to Table 43.1 of BS 7671 and k_2, the factor for the protective conductor, is obtained from Table 54.2, 54.3, 54.4, 54.5 or 54.6 depending on the particular protective conductor's constructional aspects. By way of example, take a line conductor of 16 mm^2 copper and a steel protective conductor in the form of armouring of a 90 °C thermosetting cable. From Table 43.1 we get $k_1 = 143$ and

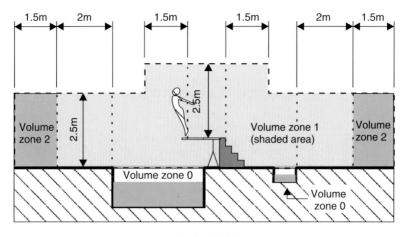

Section 'A-A'

Note: Installations outside volume zone 2 to comply with the generality of the Wring Regulations

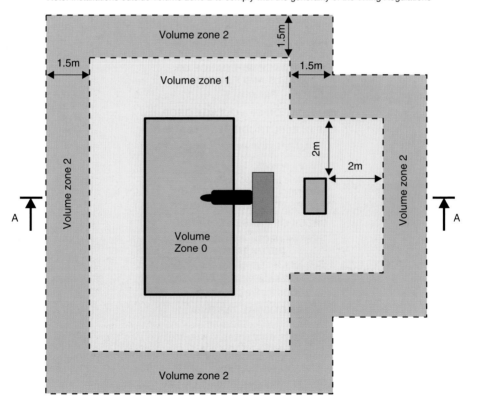

Plan

Plate 1: Swimming pool zones – section and plan. (See Page 400)

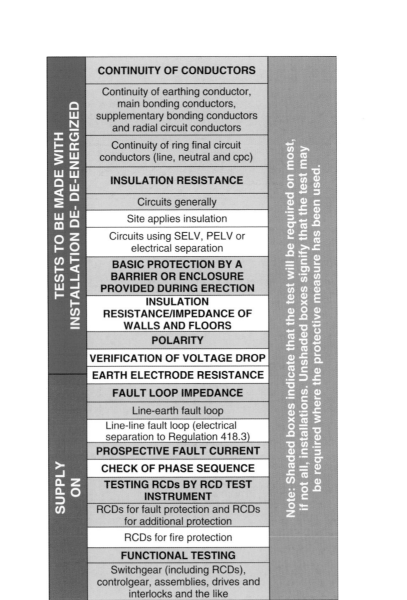

Plate 2: Sequence of tests for initial testing. (See Page 501)

This safety certificate is an important and valuable document which should be retained for future reference

ELECTRICAL INSTALLATION CERTIFICATE

Issued in accordance with *British Standard BS 7671- Requirements for Electrical Installations*

DETAILS OF THE CLIENT

Client / Address:

DETAILS OF THE INSTALLATION

The installation is:

Address:

New

Extent of the installation covered by this certificate:

An addition

An alteration

DESIGN

I/We, being the person(s) responsible for the design of the electrical installation (as indicated by my/our signature(s) below), particulars of which are described above, having exercised reasonable skill and care when carrying out the design, hereby CERTIFY that the design work for which I/we have been responsible is to the best of my/our knowledge and belief in accordance with BS 7671:   amended to   (date) except for the departures, if any, detailed as follows:

Details of departures from BS 7671, as amended (Regulations 120.3, 120.4):

The extent of liability of the signatory/signatories is limited to the work described above as the subject of this certificate.
For the **DESIGN** of the installation: ****** (*Where there is divided responsibility for the design*)

Signature	Date	Name (CAPITALS)		Designer 1
Signature	Date	Name (CAPITALS)	**	Designer 2

CONSTRUCTION

I/We, being the person(s) responsible for the construction of the electrical installation (as indicated by my/our signature below), particulars of which are described above, having exercised reasonable skill and care when carrying out the construction, hereby CERTIFY that the construction work for which I/we have been responsible is to the best of my/our knowledge and belief in accordance with BS 7671:   amended to   (date) except for the departures, if any, detailed as follows:

Details of departures from BS 7671, as amended (Regulations 120.3, 120.4):

The extent of liability of the signatory is limited to the work described above as the subject of this certificate.
For the **CONSTRUCTION** of the installation:

Signature	Date	Name (CAPITALS)	Constructor

INSPECTION AND TESTING

I/We, being the person(s) responsible for the inspection and testing of the electrical installation (as indicated by my/our signatures below), particulars of which are described above, having exercised reasonable skill and care when carrying out the inspection and testing, hereby CERTIFY that the work for which I/we have been responsible is to the best of my/our knowledge and belief in accordance with BS 7671:   amended to   (date) except for the departures, if any, detailed as follows:

Details of departures from BS 7671, as amended (Regulations 120.3, 120.4):

The extent of liability of the signatory/signatories is limited to the work described above as the subject of this certificate.
For the **INSPECTION AND TESTING** of the installation: Reviewed by †

Signature	Date	Signature	Date
Name (CAPITALS)	Inspector	Name (CAPITALS)	

DESIGN, CONSTRUCTION, INSPECTION AND TESTING *

* This box to be completed only where the design, construction, inspection and testing have been the responsibility of one person.

I, being the person responsible for the design, construction, inspection and testing of the electrical installation (as indicated by my signature below), particulars of which are described above, having exercised reasonable skill and care when carrying out the design, construction, inspection and testing, hereby CERTIFY that the said work for which I have been responsible is to the best of my knowledge and belief in accordance with BS 7671,   amended to   (date) except for the departures, if any, detailed as follows:

Details of departures from BS 7671, as amended (Regulations 120.3, 120.4):

The extent of liability of the signatory is limited to the work described above as the subject of this certificate.
For the **DESIGN**, the **CONSTRUCTION** and the **INSPECTION AND TESTING** of the installation.   Reviewed by †

Signature	Date	Signature	Date
Name (CAPITALS)		Name (CAPITALS)	

† The completed schedules of inspection and testing should preferably be reviewed by another competent person to confirm that the recorded results are consistent with electrical installation work conforming to the requirements of BS 7671

Page 1 of

This form is based on the model shown in Appendix 6 of BS 7671: 2008.
© Copyright The Electrical Safety Council (Jan 2008).

Please see the 'Notes for Recipients' on the reverse of this page.

ICM4/1

Plate 3: Electrical Installation Certificate – Page 1. Reproduced by kind permission of The Electrical Safety Council. (See Page 502)

Original (To the person ordering the work)

PARTICULARS OF THE ORGANISATION(S) RESPONSIBLE FOR THE ELECTRICAL INSTALLATION

DESIGN (1) Organisation †

Address:

Postcode

DESIGN (2) Organisation †

Address:

Postcode

† CONSTRUCTION Organisation

Address:

Postcode

INSPECTION AND TESTING Organisation †

Address:

Postcode

SUPPLY CHARACTERISTICS AND EARTHING ARRANGEMENTS
Tick boxes and enter details, as appropriate

✤ System Type(s)	✤ Number and Type of Live Conductors			Nature of Supply Parameters			✤ Characteristics of Primary Supply Overcurrent Protective Device(s)	
TN-S		a.c.	d.c.	Nominal voltage(s), $U^{(1)}$ / $U_0^{(1)}$	V	V		
TN-C-S	1-phase (2 wire)	1-phase (3 wire)	2 pole	Nominal frequency, $f^{(1)}$	Hz	Notes: (1) by enquiry	BS(EN)	
TN-C	2-phase (3 wire)		3-pole	Prospective fault current, I_{pf} $^{(2)(3)}$	kA	(2) by enquiry or by measurement	Type	
TT	3-phase (3 wire)	3-phase (4 wire)	other	External earth fault loop impedance, Z_e $^{(2)(3)}$	Ω	(3) where more than one supply, record the higher or highest values	Rated current	A
IT	Other	Please state		Number of supplies			Short-circuit capacity	kA

PARTICULARS OF INSTALLATION AT THE ORIGIN
Tick boxes and enter details, as appropriate

✤ Means of Earthing

		Details of Installation Earth Electrode (where applicable)
Distributor's facility:	Type: (eg rods), tape etc)	Location:
Installation earth electrode:	Electrode resistance, R_A: (Ω)	Method of measurement:

✤ Main Switch or Circuit-Breaker
* (applicable only where an RCD is suitable and is used as a main circuit-breaker)

			Maximum Demand (Load):		kVA / Amps *Delete as appropriate	**Protective measure(s) against electric shock:**		
Type: BS(EN)	Voltage rating	V				**Earthing and Protective Bonding Conductors**		
No of Poles	Rated current, I_n	A	**Earthing conductor**		**Main protective bonding conductors**		**Bonding of extraneous-conductive-parts (✓)**	
Supply conductors: material	RCD operating current, $I_{\Delta n}$ *	mA	Conductor material		Conductor material		Water service	Gas service
Supply conductors: csa	RCD operating time (at $I_{\Delta n}$) *	ms	Conductor csa	mm²	Conductor csa	mm²	Oil service	Structural steel
			Continuity check	(✓)	Continuity check	(✓)	Lightning protection	Other incoming service(s)

COMMENTS ON EXISTING INSTALLATION

In the case of an alteration or additions see Section 633

Note: Enter 'NONE' or, where appropriate, the page number(s) of additional page(s) of comments on the existing installation.

NEXT INSPECTION
§ *Enter interval in terms of years, months or weeks, as appropriate*

I/We, the designer(s), RECOMMEND that this installation is further inspected and tested after an interval of not more than §

† *Where the electrical contractor responsible for the construction of the electrical installation has also been responsible for the design **and** the inspection and testing of that installation, the 'Particulars of the Organisation Responsible for the Electrical Installation' may be recorded only in the section entitled 'CONSTRUCTION'.*

✤ *Where a number of sources are available to supply the installation, and where the data given for the primary source may differ from other sources, a separate sheet must be provided which identifies the relevant information relating to each additional source.*

Page 2 of _____

This form is based on the model shown in Appendix 6 of BS 7671: 2008.
© Copyright The Electrical Safety Council (Jan 2008).

Please see the 'Notes for Recipients' on the reverse of this page.

ICM4/3

Plate 4: Electrical Installation Certificate – Page 2. Reproduced by kind permission of The Electrical Safety Council. (See Page 504)

Original (To the person ordering the work)

SCHEDULE OF ITEMS INSPECTED

† *See note below*

PROTECTIVE MEASURES AGAINST ELECTRIC SHOCK

Basic and fault protection

Extra low voltage

☐ SELV   ☐ PELV

Double or reinforced insulation

☐ Double or Reinforced Insulation

Basic protection

☐ Insulation of live parts   ☐ Barriers or enclosures

☐ Obstacles **   ☐ Placing out of reach **

Fault protection

Automatic disconnection of supply

☐ Presence of earthing conductor

☐ Presence of circuit protective conductors

☐ Presence of main protective bonding conductors

☐ Presence of earthing arrangements for combined protective and functional purposes

☐ Presence of adequate arrangements for alternative source(s), where applicable

☐ FELV

☐ Choice and setting of protective and monitoring devices (for fault protection and/or overcurrent protection)

**Non-conducting location **

☐ Absence of protective conductors

**Earth-free equipotential bonding **

☐ Presence of earth-free equipotential bonding

Electrical separation

☐ For **one** item of current-using equipment

☐ For **more** than one item of current-using equipment **

Additional protection

☐ Presence of residual current device(s)

☐ Presence of supplementary bonding conductors

** *For use in controlled supervised/conditions only*

Prevention of mutual detrimental influence

☐ Proximity of non-electrical services and other influences

☐ Segregation of Band I and Band II circuits or Band II insulation used

☐ Segregation of Safety Circuits

Identification

☐ Presence of diagrams, instructions, circuit charts and similar information

☐ Presence of danger notices and other warning notices

☐ Labelling of protective devices, switches and terminals

☐ Identification of conductors

Cables and Conductors

☐ Selection of conductors for current carrying capacity and voltage drop

☐ Erection methods

☐ Routing of cables in prescribed zones

☐ Cables incorporating earthed armour or sheath or run in an earthed wiring system, or otherwise protected against nails, screws and the like

☐ Additional protection by 30mA RCD for cables concealed in walls (where required, in premises not under the supervision of skilled or instructed persons)

☐ Connection of conductors

☐ Presence of fire barriers, suitable seals and protection against thermal effects

General

☐ Presence and correct location of appropriate devices for isolation and switching

☐ Adequacy of access to switchgear and other equipment

☐ Particular protective measures for special installations and locations

☐ Connection of single-pole devices for protection or switching in line conductors only

☐ Correct connection of accessories and equipment

☐ Presence of undervoltage protective devices

☐ Selection of equipment and protective measures appropriate to external influences

☐ Selection of appropriate functional switching devices

SCHEDULE OF ITEMS TESTED

† *See note below*

☐ External earth fault loop impedance, Z_e

☐ Installation earth electrode resistance, R_A

☐ Continuity of protective conductors

☐ Continuity of ring final circuit conductors

☐ Insulation resistance between live conductors

☐ Insulation resistance between live conductors and Earth

☐ Protection by SELV, PELV or by electrical separation

☐ Basic protection by barrier or enclosure provided during erection

☐ Insulation of non-conducting floors or walls

☐ Polarity

☐ Earth fault loop impedance, Z_s

☐ Verification of phase sequence

☐ Operation of residual current devices

☐ Functional testing of assemblies

☐ Verification of voltage drop

SCHEDULE OF ADDITIONAL RECORDS* (See attached schedule)

Page No(s) _____

Note: *Additional page(s) must be identified by the Electrical Installation Certificate serial number and page number(s).*

† **All boxes must be completed.** *'✓' indicates that an inspection or a test was carried out and that the result was **satisfactory**. 'N/A' indicates that an inspection or test was **not applicable** to the particular installation.*

Page 3 of _____

* *Where the electrical work to which this certificate relates includes the installation of a fire alarm system and/or an emergency lighting system (or a part of such systems), this electrical safety certificate should be accompanied by the particular certificate(s) for the system(s).*

This form is based on the model shown in Appendix 6 of BS 7671: 2008.
© Copyright The Electrical Safety Council (Jan 2008).

ICM4/5

Plate 5: Electrical Installation Certificate – Page 3. Reproduced by kind permission of The Electrical Safety Council. (See Page 508)

Original (To the person ordering the work)

SCHEDULE OF CIRCUIT DETAILS
FOR THE INSTALLATION

TO BE COMPLETED IN EVERY CASE	TO BE COMPLETED ONLY IF THE DISTRIBUTION BOARD IS NOT CONNECTED DIRECTLY TO THE ORIGIN OF THE INSTALLATION*

Location of distribution board:	Supply to distribution board is from:		No of phases:	Nominal voltage:	V
	Overcurrent protective device for the distribution circuit:		Associated RCD (if any): BS(EN)		
Distribution board designation:	Type: BS(EN)	Rating:	A	RCD No of poles:	$I_{\Delta n}$	mA

CIRCUIT DETAILS

Circuit number and phase	Circuit designation	Type of wiring (see code below)	↑ Reference method	Number of points served	Circuit conductors: csa		Max. disconnection time permitted by BS 7671	Overcurrent protective devices					RCD	Maximum Z_s permitted by BS 7671
					Live (mm²)	cpc (mm²)	(s)	BS (EN)	Type No	Rating (A)	Short-circuit capacity (kA)		Operating current, $I_{\Delta n}$ (mA)	(Ω)

↑ See Table 4A2 of Appendix 4 of BS 7671: 2008

CODES FOR TYPE OF WIRING									
A	B	C	D	E	F	G	H	O (Other - please state)	
PVC/PVC cables	PVC cables in metallic conduit	PVC cables in non-metallic conduit	PVC cables in metallic trunking	PVC cables in non-metallic trunking	PVC/SWA cables	XLPE/SWA cables	Mineral-insulated cables		Page 4 of

* In such cases, details of the distribution (sub-main) circuit(s), together with the test results for the circuit(s), must also be provided on continuation schedules.

This form is based on the model shown in Appendix 6 of BS 7671: 2008.
© Copyright The Electrical Safety Council (Jan 2008).

See next page for Schedule of Test Results

ICM4/7

Plate 6: Electrical Installation Certificate – Page 4. Reproduced by kind permission of The Electrical Safety Council. (See Page 509)

SCHEDULE OF TEST RESULTS
FOR THE INSTALLATION

TO BE COMPLETED ONLY IF THE DISTRIBUTION BOARD IS NOT CONNECTED DIRECTLY TO THE ORIGIN OF THE INSTALLATION	Test instruments (serial numbers) used:	

Characteristics at this distribution board

Confirmation of supply polarity

★ See note below

Z_s Ω   Operating times of associated   At $I_{\Delta n}$ ms

I_{pf} kA   RCD (if any)   At $5I_{\Delta n}$ (if applicable)   ms

Test instruments (serial numbers) used:		
Earth fault loop impedance		RCD
Insulation resistance		Other
Continuity		Other

TEST RESULTS

Circuit number and phase	Circuit impedances (Ω)					Insulation resistance † Record lower or lowest value				Polarity	Maximum measured earth fault loop impedance, Z_s ★ See note below	RCD operating times	
	Ring final circuits only (measured end to end)			All circuits (At least one column to be completed)		Line/Line †	Line/Neutral †	Line/Earth †	Neutral/Earth			at $I_{\Delta n}$	at $5I_{\Delta n}$ (if applicable)
	r_1 (Line)	r_n (Neutral)	r_2 (cpc)	$R_1 + R_2$	R_2	(MΩ)	(MΩ)	(MΩ)	(MΩ)	(✓)	(Ω)	(ms)	(ms)

★   Note: Where the installation can be supplied by more than one source, such as a primary source (eg public supply) and a secondary source (eg standby generator), the higher or highest values must be recorded.

TESTED BY

Signature:		Position:		Page 5 of
Name: (CAPITALS)		Date of testing:		

This form is based on the model shown in Appendix 6 of BS 7671: 2008.
© Copyright The Electrical Safety Council (Jan 2008).

See previous page for Circuit Details

ICM4/9

Plate 7: Electrical Installation Certificate – Page 5. Reproduced by kind permission of The Electrical Safety Council. (See Page 511)

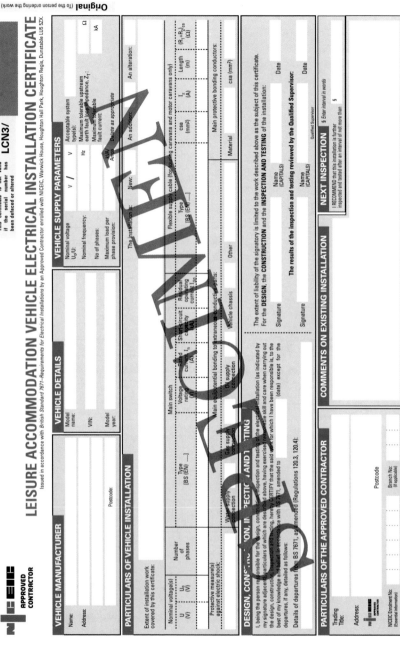

Plate 8: Leisure Accommodation Vehicle Electrical Installation Certificate – Page 1. Reproduced by kind permission of The Electrical Safety Council. (See Page 513)

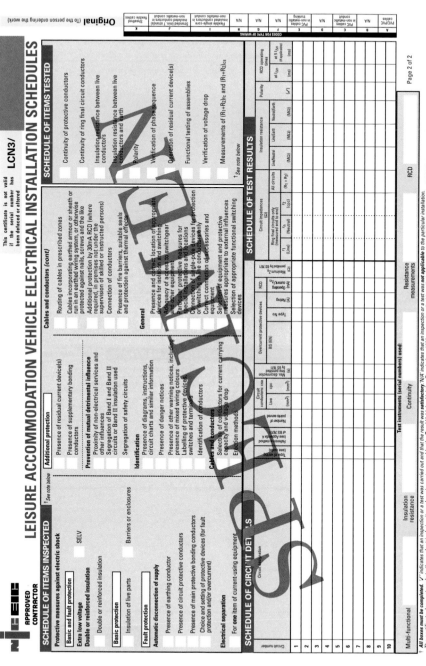

Plate 9: Leisure Accommodation Vehicle Electrical Installation Certificate – Page 2. Reproduced by kind permission of The Electrical Safety Council. (See Page 514)

IMM4

MINOR ELECTRICAL INSTALLATION WORKS CERTIFICATE

Issued in accordance with *British Standard BS 7671 - Requirements for Electrical Installations*

This safety certificate is an important and valuable document which should be retained for future reference

To be used only for minor electrical work which does not include the provision of a new circuit

PART 1: DETAILS OF THE MINOR WORKS

Client:

Date minor works completed:

Contract reference, if any:

Description of the minor works:

Details of departures, if any, from BS 7671 (as amended):

Location/address of the minor works:

Postcode

PART 2: DETAILS OF THE MODIFIED CIRCUIT

| System type and earthing arrangements: | TN-C-S | | TN-S | | TT | | TN-C | | IT | |

Protective measures against electric shock:

Overcurrent protective device for the modified circuit: BS(EN) | Type | Rating | A

Residual current device (if applicable): BS(EN) | Type | $I_{\Delta n}$ | mA

Details of wiring system used to modify the circuit: Type | Reference method | csa of live conductors | mm^2 | csa of cpc | mm^2

Where the protective measure against electric shock is ADS, insert maximum disconnection time permitted by BS 7671: | s | Maximum Z_s permitted by BS 7671 | Ω

Comments, if any, on existing installation:

PART 3: INSPECTION AND TESTING OF THE MODIFIED CIRCUIT AND RELATED PARTS

† *Essential inspections and tests*

† Confirmation that necessary inspections have been undertaken	(✓)	† Confirmation of the adequacy of earthing	(✓)
† Circuit resistance: $R_1 + R_2$	Ω or R_2 Ω	† Confirmation of the adequacy of protective bonding	(✓)
Insulation resistance: (* In a multi-phase circuit, record the lower or lowest value, as appropriate) Instrument Serial Nos	Line/Line* $M\Omega$	† Confirmation of correct polarity	(✓)
	Line/Neutral $M\Omega$	† Maximum measured earth fault loop impedance, Z_s	Ω
	Line/Earth* $M\Omega$	† RCD operating time at $I_{\Delta n}$ (if RCD fitted)	ms
	† Neutral/Earth $M\Omega$	RCD operating time at $5 I_{\Delta n}$, if applicable	ms

Agreed limitations, if any, on the inspection and testing:

PART 4: DECLARATION

I/We CERTIFY that the said works do not impair the safety of the existing installation, that the said works have been designed, constructed, inspected and tested in accordance with BS 7671   (IEE Wiring Regulations), amended to   and that the said works, to the best of my/our knowledge and belief, at the time of my/our inspection complied with BS 7671 except as detailed in Part 1.

Name (CAPITALS)

Signature

Position

Date

For and on behalf of (Trading title of electrical contractor)

Address and Postcode

This form is based on the model shown in Appendix 6 of BS 7671
© Copyright The Electrical Safety Council (Jan 2008).

Please see the 'Notes for Recipients' on the reverse of this page.

IMM4/1

Plate 10: Minor Electrical Installation Works Certificate. Reproduced by kind permission of The Electrical Safety Council. (See Page 515)

IPM4

PERIODIC INSPECTION REPORT
FOR AN ELECTRICAL INSTALLATION

Issued in accordance with *British Standard BS 7671- Requirements for Electrical Installations*

A. DETAILS OF THE CLIENT

Client:

Address:

B. PURPOSE OF THE REPORT

This Periodic Inspection Report must be used only for reporting on the condition of an existing installation.

Purpose for which this report is required:

C. DETAILS OF THE INSTALLATION

	Domestic	Commercial	Industrial

Occupier:

Description of premises:

Address:

Other (Please state)

Estimated age of the electrical installation: years

Postcode:

Evidence of alterations or additions   If yes, estimated age   years

Date of previous inspection:

Electrical Installation Certificate No or previous Periodic Inspection Report No:

Records of installation available: Records held by:

D. EXTENT OF THE INSTALLATION AND LIMITATIONS OF THE INSPECTION AND TESTING

Extent of the electrical installation covered by this report:

Agreed limitations (including the reasons), if any, on the inspection and testing:

This inspection has been carried out in accordance with BS 7671, as amended. Cables concealed within trunking and conduits, or cables and conduits concealed under floors, in inaccessible roof spaces and generally within the fabric of the building or underground, have not been visually inspected.

E. DECLARATION

I/We, being the person(s) responsible for the inspection and testing of the electrical installation (as indicated by my/our signatures below), particulars of which are described above (see C), having exercised reasonable skill and care when carrying out the inspection and testing, hereby declare that the information in this report, including the observations (see F) and the attached schedules (see H), provides an accurate assessment of the condition of the electrical installation taking into account the stated extent of the installation and the limitations of the inspection and testing (see D). I/We further declare that in my/our judgement, the said installation was overall in ✚   condition (see G) at the time the inspection was carried out, and that it should be further inspected as recommended (see I).

✚ *(Insert 'a satisfactory' or 'an unsatisfactory', as appropriate)*

INSPECTION, TESTING AND ASSESSMENT BY:

Signature:

Name: (CAPITALS)

Position:

Date:

REPORT REVIEWED AND CONFIRMED BY: † *See note below*

Signature:

Name: (CAPITALS)

Date:

†*The completed report should preferably be reviewed by another competent person to confirm that the declared overall condition of the electrical installation is consistent with the inspection and test results, and with the observations and recommendations for action (if any) made in the report.*

Page 1 of

This form is based on the model shown in Appendix 6 of BS 7671.
© Copyright The Electrical Safety Council (Jan 2008)

Please see the 'Notes for Recipients' on the reverse of this page

IPM4/1

Plate 11: Periodic Inspection Report – Page 1. Reproduced by kind permission of The Electrical Safety Council. (See Page 524)

Original (To the person ordering the work)

F. OBSERVATIONS AND RECOMMENDATIONS FOR ACTIONS TO BE TAKEN

Referring to the attached schedules of inspection and test results, and subject to the limitations at D:

There are no items adversely affecting electrical safety.

or

The following observations and recommendations are made.

Item No		Code †
1		

Note: If necessary, continue on additional pages(s), which must be identified by the Periodic Inspection Report date and page number(s).

† Where observations are made, the inspector will have entered one of the following codes against each observation to indicate the action (if any) recommended:-

1. 'requires urgent attention' or   2. 'requires improvement' or
3. 'requires further investigation' or   4. 'does not comply with BS 7671'

Please see the reverse of this page for guidance regarding the recommendations.

Urgent remedial work recommended for Items: Corrective action(s) recommended for Items:

G. SUMMARY OF THE INSPECTION

General condition of the installation:

Note: If necessary, continue on additional page(s), which must be identified by the Periodic Inspection Report date and page number(s).

Date(s) of the inspection: Overall assessment of the installation:

(Entry should read either **'Satisfactory'** or **'Unsatisfactory'**)

Page 2 of

Plate 12: Periodic Inspection Report – Page 2. Reproduced by kind permission of The Electrical Safety Council. (See Page 525)

IPM4

H. SCHEDULES AND ADDITIONAL PAGES

Schedule of Items Inspected and Schedules of Items Tested: Page No 4

Additional pages, including additional source(s) data sheets: Page No(s)

Schedule of Circuit Details for the Installation: Page No(s) 5

Schedule of Test Results for the Installation: Page No(s) 6

The pages identified here form an essential part of this report. The report is valid only if accompanied by all the schedules and additional pages identified above.

I. NEXT INSPECTION

I/We recommend that this installation is further inspected and tested after an interval of not more than

(Enter interval in terms of years, months or weeks, as appropriate)

provided that any items at F which have been attributed a Recommendation Code 1 *(requires urgent attention)* **and Code 2** *(requires improvement)* **are remedied without delay and as soon as possible respectively. Items which have been attributed a Recommendation Code 3 should be actioned as soon as practicable (see F).**

J. DETAILS OF ELECTRICAL CONTRACTOR

Trading Title:

Address:

Telephone number:

Fax number:

Postcode:

K. SUPPLY CHARACTERISTICS AND EARTHING ARRANGEMENTS

Tick boxes and enter details, as appropriate

System Type(s): TN-S, TN-C-S, TN-C, TT, IT

Number and Type of Live Conductors: a.c., d.c., 1-phase (2 wire), 1-phase (3 wire), 2 pole, 2-phase (3 wire), 3-pole, 3-phase (3 wire), 3-phase (4 wire), other, Other Please state

Nature of Supply Parameters: Nominal voltage(s) $U^{(1)}$ V, $U_0^{(1)}$ V; Nominal frequency $f^{(1)}$ Hz; Prospective fault current $I_{pf}^{(2)(3)}$ kA; External earth fault loop impedance $Z_e^{(3)(4)}$ Ω; Number of supplies

Notes: (1) by enquiry (2) by enquiry or by measurement (3) where more than one supply, record the higher or highest values (4) by measurement

Characteristics of Primary Supply Overcurrent Protective Device(s): BS(EN); Type; Rated current A; Short-circuit capacity kA

L. PARTICULARS OF INSTALLATION AT THE ORIGIN

Tick boxes and enter details, as appropriate

Means of Earthing: Distributor's facility; Installation earth electrode

Details of Installation Earth Electrode (where applicable): Type (eg rod(s), tape etc); Electrode resistance, R_A (Ω); Location; Method of measurement

Main Switch or Circuit-Breaker *(applicable only where an RCD is suitable and is used as a main circuit-breaker)*: Type BS(EN); No of Poles; Supply conductors material; Supply conductors csa mm²; Voltage rating V; Rated current, I_n A; RCD operating current, $I_{\Delta n}$ mA; RCD operating time (at $I_{\Delta n}$) ms

Maximum Demand (Load): kVA/Amps *Delete as appropriate

Protective measure(s) against electric shock:

Earthing and Protective Bonding Conductors:
Earthing conductor: Conductor material; Conductor csa mm²; Continuity check (✓)
Main protective bonding conductors: Conductor material; Conductor csa mm²; Continuity check (✓)
Bonding of extraneous-conductive-parts (✓): Water service, Gas service, Oil service, Structural steel, Lightning protection, Other incoming service(s)

Where a number of sources are available to supply the installation, and where the data given for the primary source may differ from other sources, a separate sheet must be provided which identifies the relevant information relating to each additional source.

Page 3 of

This form is based on the model shown in Appendix 6 of BS 7671.
© Copyright The Electrical Safety Council (Jan 2008)

Please see the 'Notes for Recipients' on the reverse of this page.

IPM4/5

Plate 13: Periodic Inspection Report – Page 3. Reproduced by kind permission of The Electrical Safety Council. (See Page 526)

SCHEDULE OF ITEMS INSPECTED † *See note below*

PROTECTIVE MEASURES AGAINST ELECTRIC SHOCK

Basic and fault protection

Extra low voltage

| | SELV | | PELV |

Double or reinforced insulation

| | Double or Reinforced Insulation |

Basic protection

| | Insulation of live parts | | Barriers or enclosures |
| | Obstacles ** | | Placing out of reach ** |

Fault protection

Automatic disconnection of supply

	Presence of earthing conductor
	Presence of circuit protective conductors
	Presence of protective bonding conductors
	Presence of earthing arrangements for combined protective and functional purposes
	Presence of adequate arrangements for alternative source(s), where applicable
	FELV
	Choice and setting of protective and monitoring devices (for fault protection and/or overcurrent protection)

Non-conducting location *

| | Absence of protective conductors |

Earth-free equipotential bonding *

| | Presence of earth-free equipotential bonding |

Electrical separation

| | For **one** item of current-using equipment |
| | For **more** than one item of current-using equipment** |

Additional protection

| | Presence of residual current device(s) |
| | Presence of supplementary bonding conductors |

** *For use in controlled supervised conditions only*

Prevention of mutual detrimental influence

	Proximity of non-electrical services and other influences
	Segregation of Band I and Band II circuits or Band II insulation used
	Segregation of Safety Circuits

Identification

	Presence of diagrams, instructions, circuit charts and similar information
	Presence of danger notices and other warning notices
	Labelling of protective devices, switches and terminals
	Identification of conductors

Cables and Conductors

	Selection of conductors for current carrying capacity and voltage drop
	Erection methods
	Routing of cables in prescribed zones
	Cables incorporating earthed armour or sheath or run in an earthed wiring system, or otherwise protected against nails, screws and the like
	Additional protection by 30mA RCD for cables concealed in walls (where required, in premises not under the supervision of skilled or instructed persons)
	Connection of conductors
	Presence of fire barriers, suitable seals and protection against thermal effects

General

	Presence and correct location of appropriate devices for isolation and switching
	Adequacy of access to switchgear and other equipment
	Particular protective measures for special installations and locations
	Connection of single-pole devices for protection or switching in line conductors only
	Correct connection of accessories and equipment
	Presence of undervoltage protective devices
	Selection of equipment and protective measures appropriate to external influences
	Selection of appropriate functional switching devices

SCHEDULE OF ITEMS TESTED † *See note below*

	External earth fault loop impedance, Z_e
	Installation earth electrode resistance, R_A
	Continuity of protective conductors
	Continuity of ring final circuit conductors
	Insulation resistance between live conductors
	Insulation resistance between live conductors and Earth
	Protection by SELV, PELV or by electrical separation

	Basic protection by barrier or enclosure provided during erection
	Insulation of non-conducting floors or walls
	Polarity
	Earth fault loop impedance, Z_s
	Verification of phase sequence
	Operation of residual current devices
	Functional testing of assemblies
	Verification of voltage drop

† *All boxes must be completed.*
'✓' *indicates that an inspection or a test was carried out and that the result was **satisfactory***
'✗' *indicates that an inspection or a test was carried out and that the result was **unsatisfactory***
'N/A' *indicates that an inspection or a test was **not applicable** to the particular installation*
'LIM' *indicates that, that exceptionally, a **limitation** agreed with the person ordering the work (as recorded in Section D) **prevented** the inspection or test being carried out.*

Page 4 of ▢

This form is based on the model shown in Appendix 6 of BS 7671
© Copyright The Electrical Safety Council (Jan 2008)

IPM4/7

Plate 14: Periodic Inspection Report – Page 4. Reproduced by kind permission of The Electrical Safety Council. (See Page 527)

SCHEDULE OF CIRCUIT DETAILS
FOR THE INSTALLATION

Original (To the person ordering the work)

TO BE COMPLETED IN EVERY CASE	TO BE COMPLETED ONLY IF THE DISTRIBUTION BOARD IS NOT CONNECTED DIRECTLY TO THE ORIGIN OF THE INSTALLATION*

Location of distribution board:

Supply to distribution board is from: No of phases: Nominal voltage: V

Overcurrent protective device for the distribution circuit: Associated RCD (if any): BS(EN)

Distribution board designation:

Type: BS(EN): Rating: A RCD No of poles: $I_{\Delta n}$ mA

CIRCUIT DETAILS

Circuit number and phase	Circuit designation	Type of wiring (see code below)	Reference method	Number of points served	Circuit conductors: csa		Max. disconnection time permitted by BS 7671 (s)	Overcurrent protective devices				RCD	Maximum Zs permitted by BS 7671
					Live (mm²)	cpc (mm²)		BS (EN)	Type No	Rating (A)	Short-circuit capacity (kA)	Operating current, $I_{\Delta n}$ (mA)	(Ω)

† See Table 4A2 of Appendix 4 of BS 7671.

CODES FOR TYPE OF WIRING									
A	B	C	D	E	F	G	H	O (Other - please state)	
PVC/PVC cables	PVC cables in metallic conduit	PVC cables in non-metallic conduit	PVC cables in metallic trunking	PVC cables in non-metallic trunking	PVC/SWA cables	XLPE/SWA cables	Mineral-insulated cables		

Page 5 of

* In such cases, details of the distribution (sub-main) circuit(s), together with the test results for the circuit(s), must also be provided.

This form is based on the model shown in Appendix 6 of BS 7671.

See next page for Schedule of Test Results

IPM4/9

Plate 15: Periodic Inspection Report – Page 5. Reproduced by kind permission of The Electrical Safety Council. (See Page 528)

Original (To the person ordering the work)

SCHEDULE OF TEST RESULTS
FOR THE INSTALLATION

TO BE COMPLETED ONLY IF THE DISTRIBUTION BOARD IS NOT CONNECTED DIRECTLY TO THE ORIGIN OF THE INSTALLATION	Test instruments (serial numbers) used:

Characteristics at this distribution board

Confirmation of supply polarity

★ See note below

Z_s Ω   Operating times of associated   At $I_{\Delta n}$ ms

I_{pf} kA   RCD (if any)   At $5I_{\Delta n}$ (if applicable)   ms

Earth fault loop impedance   RCD

Insulation resistance   Other

Continuity   Other

TEST RESULTS

Circuit number and phase	Circuit impedances (Ω)					Insulation resistance † Record lower or lowest value				Polarity	Maximum measured earth fault loop impedance, Z_s ★ See note below	RCD operating times	
	Ring final circuits only (measured end to end)			All circuits (At least one column to be completed)		Line/Line †	Line/Neutral †	Line/Earth †	Neutral/Earth			at $I_{\Delta n}$	at $5I_{\Delta n}$ (if applicable)
	r_1 (Line)	r_n (Neutral)	r_2 (cpc)	R_1+R_2	R_2	(MΩ)	(MΩ)	(MΩ)	(MΩ)	(✓)	(Ω)	(ms)	(ms)

★ Note: Where the installation can be supplied by more than one source, such as a primary source (eg public supply) and a secondary source (eg standby generator), the higher or highest values must be recorded.

TESTED BY

Signature: Position: Page 6 of ___

Name: (CAPITALS)   Date of testing:

This form is based on the model shown in Appendix 6 of BS 7671.
© Copyright The Electrical Safety Council (Jan 2008)

See previous page for Schedule of Circuit Details

IPM4/11

Plate 16: Periodic Inspection Report – Page 6. Reproduced by kind permission of The Electrical Safety Council. (See Page 529)

from Table 54.4 we get $k_2 = 46$, from which we can deduce the minimum cross-sectional area of the steel protective conductor as 50 mm^2:

$$S_p = S \times \frac{k_1}{k_2} = 16 \times \frac{143}{46} = 50 \text{mm}^2 \tag{12.2}$$

where S (mm^2) is the cross-sectional area of line conductor (copper), S_p (mm^2) is the cross-sectional area of protective conductor (steel), k_1 is the factor for the line conductor (copper) and k_2 is the factor for the protective conductor (steel).

The alternative method of obtaining the cross-sectional area of protective conductors by calculation very often results in smaller sizes than those given in Table 54.7, but it takes more of the designer's time unless computer facilities are used. The basic equation for such calculations is given in Regulation 543.1.3 and is as set out in Equation (12.3):

$$S_p \text{ (mm}^2) = I_F \frac{\sqrt{t}}{k} \tag{12.3}$$

where I_F (A) is the earth-fault rms current (with fault of negligible impedance) taking account of current limiting capabilities of the protective device (if any), t (s) is the operating time of protective device associated with the particular fault current magnitude and S_p (mm^2) is the minimum cross-sectional area of protective conductor.

If we transpose Equation (12.3) we can see the energy-withstand $S_p^2 k^2$ of the protective conductor must not be less than the energy let-through $I^2 t$ of the protective device, as indicated in Equation (12.4):

$$S_p k \geq I_F \sqrt{t} \equiv S_p^2 k^2 \geq I_F^2 t \tag{12.4}$$

Taking a practical example, an 80 A BS 88-2.2 fuse protects a 230 V circuit with live conductors of 25 mm^2 cross-sectional area where it is anticipated that the maximum total earth loop impedance Z_s will be 0.57 Ω and the protective conductor will be a separate copper conductor with 70 °C insulation and not bunched with other cables ($k = 143$ from Table 54.2). From this we can deduce that the fault current will be about 403.5 A (230 V/0.57 Ω) and from the time/characteristics for this device, we can see that the device will operate within 5 s. The minimum cross-sectional area of the protective conductor S_p will be 6.31 mm^2:

$$S_p = \frac{I_F \sqrt{t}}{k} = \frac{403.5 \times \sqrt{5}}{143} = 6.31 \text{ mm}^2 \tag{12.5}$$

Where the application of the adiabatic equation results in a nonstandard size, the cross-sectional area of the next larger standard size must be chosen as, 'clearly', to choose the next lower size would render the protective conductor thermally unprotected.

Table 12.2 rearranges the data given for k factors in Tables 54.2, 54.3, 54.4, 54.5 and 54.6 of BS 7671.

Table 12.2 Protective conductors: k factors

Ref	Protective conductor arrangement	Insulation/ covering	Chapter 54 Table	Temp. (°C) Initial	Final	Csa (mm²)	Copper	Aluminium	Steel	Lead
Protective conductors with insulation or covering										
A	Insulated conductor not part of cable not bunched with other cables. Separate bare conductor in contact with cable covering but not bunched (assumed initial temperature 30°C)	PVC – 70°C	54.2	30	160	≤300	143	95	52	–
B		PVC – 70°C		30	140	>300	133	88	52	–
C		PVC – 90°C		30	160	≤300	143	95	52	–
D		PVC – 90°C		30	140	>300	133	88	52	–
E		Thermosetting – 90°C		30	250	All	176	116	64	–
F	Insulated and noninsulated protective conductor incorporated in a cable or bunched with cables (assumed initial temperature 70°C)	PVC – 70°C	54.3	70	160	≤300	115	76	–	–
G		PVC – 70°C		70	140	>300	103	68	–	–
H		PVC – 90°C		90	160	≤300	100	66	–	–
I		PVC – 90°C		90	140	>300	86	57	–	–
J		Thermosetting – 90°C		90	250	All	143	94	–	–
K	Cable sheath or cable armouring protective conductor	PVC – 70°C	54.4	60	200	All	–	93	51	26
L		PVC – 90°C		80	200	All	–	85	46	23
M		Thermosetting – 90°C		80	200	All	–	85	46	23

Table 12.2 *(continued)*

Ref	Protective conductor arrangement	Insulation/covering	Chapter 54 Table	Temp. (°C) Initial	Temp. (°C) Final	Csa (mm²)	Copper	Aluminium	Steel	Lead
Protective conductors with insulation or covering										
N	Protective conductor formed by steel conduit, steel ducting and/or steel trunking enclosing cables listed. Where enclosed cables consist of different temperature ratings, the *k* values attributed to the lowest or lower temperatures must be used	PVC – 70°C	54.5	50	160	All	–	–	47	–
O		PVC – 90°C		60	160	All	–	–	44	–
P		Thermosetting – 90°C		60	250	All	–	–	58	–
Bare protective conductors										
Q	Visible conductors in restricted areas	Copper	54.6	30	500	All	–	–	82	–
R		Aluminium		30	300	All	–	125	–	–
S		Steel		30	500	All	228	–	–	–
T	Normal installation conditions	Copper		30	200	All	159	–	–	–
U		Aluminium		30	200	All	–	105	–	–
V		Steel		30	200	All	–	–	58	–
W	Where risk of fire exists	Copper		30	150	All	138	–	–	–
X		Aluminium		30	150	All	–	91	–	–
Y		Steel		30	150	All	–	–	50	–

12.1.5   Protective conductors: for combined protective and functional purposes

Where an earthing arrangement provides the means of fault protection and is also used for functional purposes, Regulation 543.5.1 demands that the requirements relating to the protective function must take precedence.

PEN conductors, providing both functions, may only be used (Regulation 543.4.2) if one of the following four conditions is met:

- the supply is obtained from a private generating plant, or
- the supply is obtained from a privately owned transformer arranged so that there is, excepting the earthing connection, no metallic connection between the installation and the public supply, or
- the supply is obtained from a privately owned convertor arranged so that there is, excepting the earthing connection, no metallic connection between the installation and the public supply, or
- the supply is obtained from a distributor with their express agreement and authorization for such a system to operate and the installation complies with all the prerequisite conditions for use with PEN conductors.

BS 7671 makes further demands relating to the conductor (summarized in Table 12.3) and requires that where the functions of neutral conductor and protective conductor are separated they must not be rejoined further downstream (see Figure 12.3).

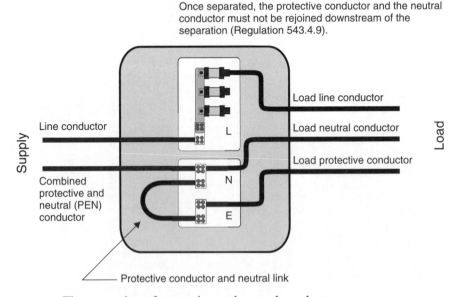

Figure 12.3   The separation of protective and neutral conductors

Table 12.3 Requirements for conductors for combined protective and neutral conductors

Ref	Regulation	Requirements
A	543.4.7 537.1.2	No means of isolation or switching permitted in the outer conductor of concentric cable
B	543.4.4	The outer conductor of a concentric cable only to be used as a conductor (protective and neutral) for that circuit and for no others. This does not preclude the use of multicore cables where, for example, additional cores are used for switching and control provided they are all associated with the same final circuit
C	543.4.6	Where joints are made in the outer PEN conductor of a cable, the continuity of that joint must be reinforced by a separate conductor linking the PEN conductor. The additional conductor must be of conductance required by Regulation 543.4.5. The additional conductor serves to supplement the continuity afforded by the normal sealing and clamping of the joint
D	543.4.8	Every joint in a PEN conductor, except as stated in Row E, must be insulated or have a covering suitable for the highest voltage
E	543.4.8	For cables complying with BS EN 60702-1 (MICC) and installed to meet the manufacturer's instructions, there are no additional regulatory requirements relating to insulation of joints
F	543.4.5	The conductance (measured at $20\,^{\circ}C$) of a single-core cable (outer conductor) must not be less than that of the internal conductor
G	543.4.5	The conductance (measured at $20\,^{\circ}C$) of the outer conductor of a multicore cable, serving a number of points in one final circuit, must not be less than that of one of the internal conductors
H	543.4.5	The conductance (measured at $20\,^{\circ}C$) of the outer conductor of a multicore cable, using paralleled internal conductors, must not be less than that of the parallel internal conductors
J	543.4.3	For the fixed installation and where there is no flexing, a copper conductor may be used as a PEN conductor provided it is of a cross-sectional area not less than $10\,mm^2$ and it is not supplied via an RCD
K	543.4.3	For the fixed installation and where there is no flexing, an aluminium conductor may be used as a PEN conductor provided it is of a cross-sectional area not less than $16\,mm^2$ and it is not supplied via an RCD
M	543.4.9	Where the protective and neutral functions are separated, the split conductors must be provided with separated terminals (or bars) for protective and neutral conductors with the PEN conductor connecting both terminals with a link of conductance satisfying Regulation 543.4.5. Once separated, the protective and neutral conductors must not be reconnected together (see Figure 12.3)

12.1.6   Protective conductors: electrical continuity

As stated in Regulation 543.3.1, all protective conductors must be protected against mechanical external influences such as impact and vibration and similarly protected against chemical reactions (e.g. between dissimilar metals). Additionally, account must be taken of electrodynamic effects. These effects may be severe where a high earth fault current flows and may put unacceptable mechanical forces on the protective conductor support system if it is less than adequate.

Unless the protective conductor is in the form of steel conduit or trunking or contained within a multicore cable, Regulation 543.3.2 calls for all conductors of cross-sectional area up to and including 6 mm^2 to have a covering (similar to the insulation provided on nonsheathed cables rated at not less than 450/750 V). Where the protective conductor is contained in a sheathed cable (e.g. PVC/PVC to BS 6004) and where the sheath is removed for termination purposes, insulated sleeving compliant with BS EN 60684 series must be used to sleeve the otherwise bare protective conductor (Regulation 543.3.2).

Generally speaking, no switching device is permitted to be installed in a protective conductor (Regulation 543.3.4), although mechanical joints with disconnectable links (removable only by use of a tool) are permitted for test purposes. However, there are exceptions to this rule, in that multipole linked switching and plug-in devices are permitted provided that the protective conductor is not interrupted before the associated live conductors and is made before the live conductors are re-established. Another important exemption relates to generator earthing and where parallel operation with other generators or with other sources is involved. A protective conductor linking the neutral point of a generator may be broken, if this is required for operational reasons, provided the switching arrangement is linked to the interruption of live conductors in the same manner as previously mentioned.

Where steel conduit and metallic-sheathed cables forming the protective conductor are terminated into an insulated enclosure, care must be taken to preserve the continuity of the conductor through the enclosure by, for example, the fitting of gland tag rings on the incoming and outgoing wiring system and providing an adequate conductor between the two. Of no less importance is the adequacy of metal enclosures to provide a reliable conductive path for earth fault current. Often, cables and conduits are terminated into switchgear via a steel plate bolted to the main enclosure by a few small bolts. With the application of paint and other non-conductive finishes to enclosures and plates, too much reliance must not be placed on the effectiveness of such connections; it is far better to preserve the earth fault path through the enclosure by means of gland tags and separate conductors.

12.1.7   Protective conductors: formed by steel conduit, trunking, etc

Whether or not steel conduit is used for the protective conductor, joints in that conduit must be mechanically sound and electrically continuous; this is possible only with fully tightened screwed couplers (sockets) and socket and bush unions. Slip conduit and conduit relying on pin grip sockets are not acceptable, since these joints cannot be relied upon to provide permanent and reliable connections either for the provision of a protective conductor or for the earthing of the steel conduit. As Regulation 543.3.3 requires, all joints in conduit, trunking, ducting and their support systems must meet the requirements of Regulation 526.3, which calls for joints (with certain exceptions) to be accessible.

Trunking should not be regarded as an effective cpc for circuits of nominal current exceeding 100 A unless it can be shown that it is capable of providing a permanent and reliable fault path for the prospective earth fault current.

Modern accessory and switchgear metal enclosures are often applied with paint or other non-conductive finishes which are not conductive to good electrical continuity between joints. These coatings must be removed where they are likely to affect the protective conductor or earthing continuity.

Where metal conduit and metal trunking are used as protective conductors (i.e. no separate cpc provided), it is necessary to equip the accessories with a protective conductor 'tail' to connect the earthing terminal of the accessory with the conduit or trunking system (Regulation 543.2.7). Although strictly not necessary for compliance with the Wiring Regulations, an earth 'tail' is considered good practice irrespective of whether the conduit provides the protective conductor or not. This good practice of providing an earth 'tail' from the accessory box (particularly where flush) to the accessory itself holds good for installations employing PVC/PVC cables (to BS 6004), although again not strictly necessary.

A separate enclosure for cables (as opposed to a composite cable) must not be used as a PEN conductor (Regulation 543.2.10) in an installation forming part of a TN-C system, and an exposed-conductive-part of equipment must not be used as a protective conductor for other circuits not associated with it unless fully compliant with Regulations 543.2.2–543.2.5.

12.1.8   Protective conductors: mineral-insulated cables

Mineral insulated cables to BS EN 60702-1 present little problem when considering the energy-withstand of the copper sheath and its use as a cpc. For multicore 500 V and 750 V grades of MICC, the effective sheath areas are in excess of their associated line conductor cross-sectional area. For single core the picture is different, as the effective sheath area of an individual single-core cable will not necessarily meet the cross-sectional area requirements of Table 54.7 of the regulations. However, single-core MICCs are normally run at least as a pair (single-phase circuit), or as a group of three or four (three-phase circuit), and in these arrangements the total sheath area for all the cables in the circuit will meet the cross-sectional area requirements of Table 54.7 if all the cables are of the same size and their sheaths are bonded together at both ends of the run.

Difficulty can sometimes be experienced in effecting and maintaining continuity of the protective conductor at the terminations of MICC into thin-walled accessories and other equipment. The use of 'earth-tail' pots is a useful way of achieving better continuity, as is the use of a zinc-plated lock-washer between the brass male bush and the enclosure. Additionally, internally threaded glands are available which allow the male bush to be screwed directly into the gland, dispensing with the conduit socket, and make a much more aesthetically pleasing installation finish on surface work.

12.1.9   Protective conductors of ring final circuits

In order to secure some added integrity of the cpc of ring final circuits, Regulation 543.2.9 calls for the protective conductor to be of the form of a ring with each 'leg' connected to the earthing terminal at the position from which it is supplied. This requirement does not apply, for obvious reasons, where a steel conduit or steel trunking or ducting wiring system provides the only cpc.

12.1.10 Protective conductors: armouring

When considering the use of cable armouring as a cpc the designer has two principal aspects to consider. First, the contribution that the armouring makes to the total line–earth loop impedance and, second, whether the protective conductor will be thermally protected. Where the contribution of impedance made by the armouring is too great to be acceptable, the designer may consider it necessary to provide a separate conductor connected in parallel with the armouring in order to reduce that contribution. Again, this may be a solution to the case where the armouring is not thermally protected (i.e. does not meet the adiabatic equation of Regulation 543.1.3). An alternative solution would involve the selection of a larger cable than would be dictated by current-carrying capacity considerations. However, the designer will find that, in most practical situations, steel armouring will be adequate for use as a cpc, but checks must be made to verify this of course.

Table 12.4 Cross-sectional areas of armour (round wires) of PVC cables to BS 6346 (all cables with stranded conductors)

Ref	Nominal csa of live conductors (mm^2)	Cross-sectional areas of armour (mm^2)												
		Copper live conductors										Aluminium live conductors		
		Power cables			Auxiliary cables								Power cables		
					Cores										
		2	3	4	5	7	10	12	19	27	37	48	2	3	4
A	1.5	15	16	17	20	39	45	70	78	90	–	–	–		
B	2.5	17	19	20	34	44	45	70	84	92	138	–	–	–	
C	4	21	23	35	40	72	72	84	128	144	163	–	–	–	
D	6	24	36	40	–	–	–	–	–	–	–	–	–	–	–
E	10	41	44	49	–	–	–	–	–	–	–	–	–	–	–
F	16	46	50	72	–	–	–	–	–	–	–	–	42	46	66
G	25	60	66	76	–	–	–	–	–	–	–	–	54	62	70
H	35	66	74	84	–	–	–	–	–	–	–	–	58	68	78
I	50	74	84	122	–	–	–	–	–	–	–	–	66	78	113
J	70	84	119	138	–	–	–	–	–	–	–	–	74	113	128
K	95	122	138	160	–	–	–	–	–	–	–	–	109	128	147
L	120	131	150	220	–	–	–	–	–	–	–	–	–	138	201
M	150	144	211	240	–	–	–	–	–	–	–	–	–	191	220
N	185	201	230	265	–	–	–	–	–	–	–	–	–	215	245
O	240	225	260	299	–	–	–	–	–	–	–	–	–	240	274
P	300	250	289	333	–	–	–	–	–	–	–	–	–	265	304
Q	400	279	319	467	–	–	–	–	–	–	–	–	–	–	–

Table 12.5 Cross-sectional areas of armour (round wires) of XLPE cables to BS 5467 and BS 6724 (LSHF)

Ref	Nominal csa of live conductors (mm²)	Cross-sectional areas of armour (mm²)					
		Power cables with stranded conductors					
		Copper live conductors			Aluminium live conductors		
		Cores					
		2	3	4	2	3	4
A	1.5	15	16	17	–	–	–
B	2.5	17	19	20	–	–	–
C	4	19	20	22	–	–	–
D	6	22	23	36	–	–	–
E	10	26	39	42	–	–	–
F	16	42	45	50	39	41	46
G	25	42	62	70	38	58	66
H	35	60	68	78	54	64	72
I	50	68	78	90	60	72	82
J	70	80	90	131	70	84	122
K	95	113	128	147	100	119	135
L	120	125	141	206	–	131	191
M	150	138	201	230	–	181	211
N	185	191	220	255	–	206	235
O	240	215	250	289	–	230	265
P	300	235	269	319	–	250	289

The story can be different when considering the use of the armouring as a main bonding conductor (as well as a cpc), and by application of Regulation 544.1.1 (see section 12.4.1 of this Guide) it will often be found that the armouring will not provide an adequate cross-sectional area for use as a main bonding conductor. It is important to recognize that the resistivity of aluminium is about 1.6 times that of copper and that of steel is of the order of eight times that of copper.

Cable manufacturers publish constructional and electrical data for all cables they produce, and Tables 12.4 and 12.5 summarize some of the information in the form of the cross-sectional area of round wire armouring of cables with stranded live conductors of both copper and aluminium to the specifications of BS 6346 (thermoplastic (PVC)), BS 5467 (thermosetting) and BS 6724 (thermosetting, having low emission of smoke and corrosive gases when affected by fire LSHF). Table 12.6 sets out the cross-sectional areas of steel conduit and steel trunking for some common sizes. Table 12.7 gives a correlation between the copper live conductors of armoured cables to BS 6346 and the equivalent copper (in terms of conductance) of the armouring to the nearest standard cross-sectional area below that calculated.

Table 12.8 summarizes, for 70 °C thermoplastic (PVC)-insulated armoured/metallic-sheathed cables, the relationship between the protective conductor and the line conductor referred to in Regulation 543.1.4 (Table 54.7) where they are of different metals. By

Table 12.6 Cross-sectional areas of steel conduit and steel trunking[a]

| | Steel conduit | | Steel trunking | |
| Nominal outside diameter (mm) | Csa (mm^2) | | Size (mm × mm) | Csa (mm^2) |
	Light gauge	Heavy gauge		
16	47	72	50 × 36.5	125
20	59	92	50 × 50	150
25	89	131	75 × 50	225
32	116	170	75 × 75	285
			100 × 50	260
			100 × 75	320
			100 × 100	440
			150 × 50	380
			150 × 75	450
			150 × 100	520
			150 × 150	750

[a]The cross-sectional areas are based on information in BS 4568 for conduit and BS 4678 for trunking. These cross-sectional areas should be regarded as a general guide only for conduit and trunking to Euronorm (EN) standards. These standards do not specify dimensions and therefore the manufacturer should always be consulted where accurate information on cross-sectional areas is required.

Table 12.7 Equivalent copper cross-sectional area of armouring for cables to BS 6346[a]

| Ref | Nominal csa of live conductors (mm^2) | Nominal csa of copper conductor providing equivalent conductance to that of the armouring (mm^2) | | |
		2-core	3-core	4-core
A	1.5	1.5	1.5	1.5
B	2.5	1.5	1.5	2.5
C	4	2.5	2.5	4
D	6	2.5	4	4
E	10	4	4	6
F	16	4	6	6
G	25	6	6	6
H	35	6	6	10
I	50	6	10	10
J	70	10	10	16
K	95	10	16	16
L	120	16	16 (Cu), 10 (Al)	25 (Cu), 16 (Al)
M	150	16	25	25
N	185	25	25 (Cu), 16 (Al)	25
O	240	25	25	35 (Cu), 25 (Al)
P	300	25	35	35
Q	400	25	35	50

[a]Unless otherwise stated, the data relate to cables having copper live conductors.

Table 12.8 Sheath or armour of a cable used as a circuit protective conductor: minimum cross-sectional area of S_p

Line conductor csa S (mm^2)	S_p (mm^2)			
	Sheath or armour			
	Copper	Steel	Aluminium	Lead
$S \leq 16$	S	$2.26S$	$1.24S$	$4.4S$
$16 < S \leq 35$	16	36	20	71
$S > 35$	$0.5S$	$1.13S$	$0.62S$	$2.2S$

referring back to Table 12.4, it can be seen that, for copper–PVC cables to BS 6346, all two-core cables up to 95 mm^2, all three-core cables up to 185 mm^2 and all four-core cables up to 240 mm^2 will have armouring of adequate energy-withstand, according to Table 54.7 of BS 7671. The same is true for the corresponding cables to BS 5467 and BS 6724 having copper conductors, if the conductors are operated at a temperature not exceeding 70 °C. Likewise, all 1.5, 2.5 and 4 mm^2 five-core to forty-eight-core auxiliary cable to BS 6346 will satisfy the armouring energy-withstand requirement. This table may be used in a similar manner to determine the energy-withstand capabilities of cables to the other standards listed.

As stated in Regulation 521.5.2, single-core armoured cables with steel armouring must not be used for a.c. circuits. Aluminium armoured cables, on the other hand, may be so used, but care is needed when the cables' protective conductors are connected to switchgear and the like to minimize the circulating currents.

Armoured cables always require correctly terminating to the equipment they serve irrespective of the constructional specification (i.e. Class I or Class II) of that equipment, not least in order to maintain the mechanical protection properties of the armoured cable. Appropriate cable glands to BS 6121 *Mechanical cable glands* or BS EN 50262 *Cable glands for electrical installations* should always be used together with gland earth tag washer, nut, bolt and washers. If there is any doubt about the equipment enclosure being reliable as a protective conductor, a connection (e.g. in a copper conductor) should be made between the bolt and the earthing terminal of the equipment. The use of BS 951 earthing and bonding clamps directly to the armouring should never be contemplated as a means of connection of the protective conductor as an alternative to proper glanding off. Such a method is unlikely to effect a reliable connection; deterioration can be expected over time as the armour wires are compressed into the insulating bedding, leaving the joint less than tight and, thus, providing a high resistance joint.

12.1.11 Protective conductors: 'clean' earths

The term 'clean' earth means different things to different people and it is important to establish the precise specification for such a facility. To this end the manufacturers of equipment for which a 'clean' earth is required should be consulted at an early stage. It is difficult to imagine a truly 'clean' earth, since all conductors are subject to electrical noise imported either from the supply or from adjacent electromagnetic fields, and the designer should obtain from the manufacturer the levels of noise which can and cannot be tolerated. Such an earth may be provided by a separate (usually a copper) cable from the equipment which utilizes it to the main earthing terminal of the installation. Alternatively, an additional core of a multicore cable may be used but, again, the manufacturer should be consulted with regard to induced voltages generated from the other cores of the cable.

Often, a 'clean' earth will provide the dual purpose of protective and functional conductor to the equipment concerned, but sometimes it only serves the functional objectives. In all cases the designer would need to consider the fault voltages that may arise between equipment connected to the 'clean' earth and other equipment and extraneous-conductive-parts connected to the normal ('unclean') protective conductors.

12.1.12 Protective conductors: proving and monitoring

There may be occasions where consideration of the integrity of the protective conductor may warrant earth fault monitoring and protective conductor monitoring (e.g. where flexible trailing leads supply mobile apparatus). When using such equipment it is important to confirm that the operating coil is connected in the pilot conductor and not the protective conductor (Regulation 543.3.5). Where such protection is provided, the installation of such equipment should comply with BS 4444 *Guide to electrical earth monitoring and protective conductor proving*. This CP addresses the principles of design, construction and application of such equipment. It includes earth monitoring units and earth proving units, and units with combined functions. When designing and installing such protection, reference to BS 4444 is essential. Figure 12.4 shows the basic circuit for a combined earthing proving and earth monitoring unit.

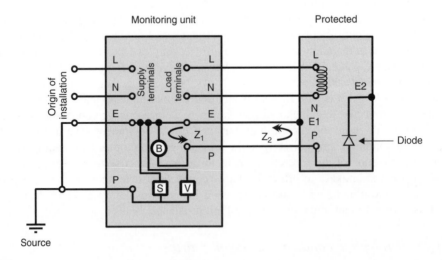

Figure 12.4 Outline circuit for a combined earth proving and earth monitoring unit. The key to the symbols used in the figure is as follows: L, line conductor; N, neutral conductor; E, earthing/protective conductor; E_2, pilot conductor earthing terminal; Z_1, source line-earth loop impedance; V, ELV source; B, balancing sensing network; P, pilot conductor terminal; E_1, equipment protective conductor terminal; S, sensing device; Z_2, impedance of loop (cpc and pilot conductor)

12.2   Earthing

12.2.1   Earthing: general

Earthing is defined as the act of connecting the exposed-conductive-parts of an installation to the MET of that installation. It should *not* be confused with protective equipotential bonding, which is different in concept in that it connects the extraneous-conductive-parts of the premises in which the installation is located to the MET.

Earthing serves two basic purposes:

- to limit the potential on current-carrying conductors with respect to the general mass of the Earth;
- to limit the potential on non-current-carrying conductors with respect to the general mass of the Earth.

In other words, the limiting of the potential on current-carrying conductors with respect to the general mass of the Earth effectively 'ties down' the live conductors to Earth or, if you prefer, to reference them to Earth potential. This is essential for the proper functioning (particularly under unbalanced load conditions) of the electrical system and is commonly referred to as 'system earthing'. Limiting the potential on normally non-current-carrying conductors with respect to the general mass of the Earth, on the other hand, relates to safety protective measures used, in the majority of cases, to protect humans and livestock against hazards (e.g. electric shock), and this is commonly known as 'equipment earthing'. Chapter 54 of BS 7671 deals with the earthing arrangements as they relate to installations and BS 7430 addresses earthing in the form of a CP.

Regulations 541.1 and 541.2 call for the means of earthing (and every protective conductor) to be selected and installed to afford compliance with BS 7671, and in particular with Chapter 54; and where the installation is subdivided, each part is required to comply.

12.2.2   Earthing: responsibilities

It is the consumer who is responsible for providing facilities for the earthing of their installation, not the ED. In practice, this responsibility is delegated to the electrical contractor engaged to carry out the installation work. Except where it is inappropriate for safety reasons, an ED is obliged to provide an earthing terminal for the consumer's use in the case of a new supply connection at LV (Regulation 24(4) of the ESQCR refers). An ED is not obliged to provide an earthing terminal in the case of an existing supply connection, but may be willing to do so. When an earthing terminal is offered, it is for the consumer (or their electrical contractor) to assess the nature of this facility and determine whether it is suitable for his purpose (e.g. a PME supply may not be acceptable in some cases). On existing installations, the means of earthing needs regular testing to confirm its continuing effectiveness and reliability. On rewires, where the supply is already installed, it would be foolish to assume that an earthing facility is available even where there is some evidence to suggest there has been a connection in the past. Under no circumstances should a consumer or their electrical contractor make a connection to an ED's supply cable. In all cases the ED should be requested, in writing at an early stage, to provide confirmation of whether or not a means of earthing can be provided.

12.2.3   Earthing: connection to Earth and system arrangements

Section 542 of BS 7671 deals with the requirements relating to the connection of the installation with Earth. Regulations 542.1.1–542.1.4 set out the requirements relating to the connection of the installation MET to the means of earthing depending on the system type. Figure 12.5 illustrates the connection arrangements for the TN-C-S, TN-S and TT system types. The link between the neutral and earthing terminal on TN-C-S systems is made by the ED and kept under seal, as is the bonding conductor terminal in certain cases.

On domestic supplies, a growing trend is for EDs to terminate their supplies in an isolator fed from their service cut-out via the meter and leave this for connection of the installation by the consumer (or their electrical contractor). In some cases, the isolator takes the form of a double-pole device with only the line conductor capable of being isolated. The other pole is used, where required, for a second tariff supply with the neutral solidly connected in a 'Henley' block. When such a device is used by the ED it should not be regarded as satisfying the requirements of Regulation 537.1.4, which demands a double-pole main switch for single-phase supplies where the main switch is intended for operation by ordinary persons (such as in domestic premises). When this method of supply termination is used, it is customary for the ED to attach instructive notices adjacent to the supply point to aid the consumer to make the necessary connection.

As demanded by Regulation 542.1.6, in every case the earthing arrangements must be such as to be considered permanent and reliable and be protected against thermal, thermo-mechanical and electro-mechanical stresses and all other external influences. Precautions need to be taken to avoid the risk of damage to other metallic parts (e.g. water and gas pipework) from electrolysis as required by Regulation 542.1.7, which may involve the use of sacrificial cathodic electrodes (see BS 7361, BS EN 13636 and BS EN 15112). Where a number of installations share protective conductors, the designer will need to consider the effects of fault current sharing and, as called for in Regulation 542.1.8, any common protective conductor must be capable of carrying the highest fault current available from any of the installations making use of the common protective conductor. Alternatively, the installations should be separated electrically by, for example, making use of the TT system arrangement.

12.2.4   Earthing conductors

As can be seen in Figure 12.6, and as defined in BS 7671, the earthing conductor is the conductor connecting the MET of the installation to the means of earthing. The means of earthing can be an earthing facility provided by the ED in the form of cable sheath or armouring earth (TN-S) or a neutral/earthing terminal on a PME of a TN-C-S system. It can also be a single or multiple earth electrode system (TT and IT systems), which might consist of driven rods and/or buried copper tape. On many TT systems, the installation earthing conductor is mechanically protected by conduit, and this allows for much smaller conductors to be used than would otherwise be dictated.

As Regulation 542.3.1 points out, earthing conductors are one form of protective conductor and must meet all the relevant requirements for those conductors. As required (amongst other things) by Regulation 542.3.1, the minimum cross-sectional area of the earthing conductor is to be not less than that required by Section 543 of BS 7671 and,

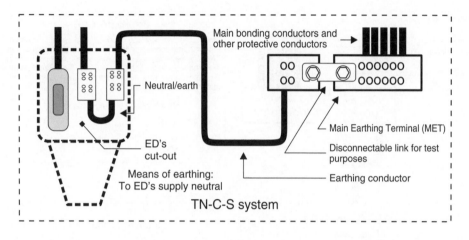

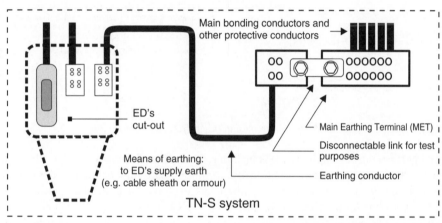

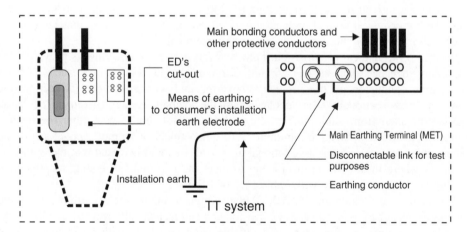

Figure 12.5   Connection arrangements depending on system type

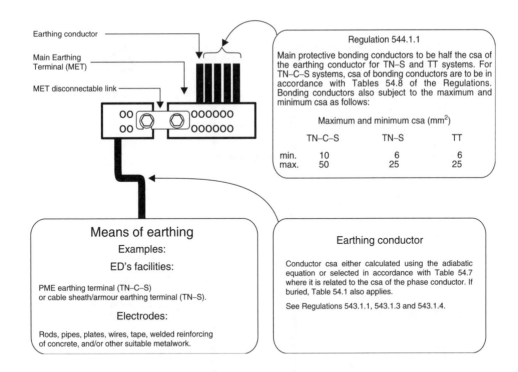

Figure 12.6 Earthing and other protective conductors

where PME conditions apply, not less than that required by Regulation 544.1.1 for a main protective bonding conductor. Additionally, for buried earthing conductors, the minimum cross-sectional areas must meet the requirements of Table 54.1. For tape and strip conductors, this regulation refers out to BS 7430, the CP for earthing. It also calls for the thickness of the conductor to be such as to withstand mechanical damage and corrosion. The sizes given for such conductors (in Table 54.1) depend on the conductor metal and whether or not additional protection is provided against mechanical damage (mainly impact and vibration) and corrosion (water and chemical).

Where no mechanical or corrosion protection (no protective covering) is provided, the minimum cross-sectional area is 25 mm^2 for copper and 50 mm^2 for steel. In cases of earthing conductors with a protective covering and with mechanical protection, the minimum cross-sectional area is calculated from the adiabatic equation given in Regulation 543.1.3 or taken from Table 54.7 of Regulation 543.1.4; but if no further mechanical protection is afforded, then the minimum is 16 mm^2 copper or 16 mm^2 *coated* steel. The values given are *minimum* cross-sectional area, and it should be noted that there may be instances where these will be insufficient to meet the requirement relating to protective conductors (e.g. compliance with Table 54.7 of BS 7671).

As called for in Regulation 542.3.2, all connections to the means of earthing, including electrodes, must be protected against corrosion and be mechanically and electrically sound in order to provide a satisfactory and reliable connection. As with all connections of protective conductors, a label reading '**Safety Electrical Connection – Do not remove**' must be provided in a visible position on or near the connection (see Figure 9.9).

12.2.5   Earthing electrodes

Regulations 542.2.1–542.2.5 deal with the particular requirements relating to earth electrodes and identify suitable and acceptable types, as follows:

- earth rods and pipes;
- earth tapes or earth wires;
- earth plates;
- underground structural metalwork embedded in the foundations;
- welded metal reinforcing of concrete embedded in the Earth (but not pre-stressed concrete);
- cable lead sheaths (or other metal coverings of cables, e.g. armouring) provided that the cable owner's consent has been obtained *and* the sheath is in effective contact with earth *and* arrangements have been made for the owner of the installation to be warned of any proposed changes to the cable which might affect the suitability of the sheath as an electrode; or
- any other *suitable* underground metalwork.

Whatever type of electrode is chosen, the design and construction must take account of all the external influences (e.g. corrosion, water table, mechanical impact and vibration) and, as called for in Regulation 542.2.3, allowances must be made for a possible increase in electrode resistance due to the effects of corrosion. Regulation 542.2.4 makes clear that the metalwork of pipes for gases or flammable liquids or the pipe of a utility supply (such as gas or water) must not, in any circumstances, be used as an earth electrode, but these must of course be connected to the MET to effect main protective bonding. The same regulation requires that other metallic water supply pipework is not used as an earth electrode unless precautions are taken to prevent its removal and it has been considered for such use.

Regulation 542.2.2 draws attention to the need to establish the effects on the electrode resistance of the soil drying or freezing. Where the designer has any options, they should choose wet, preferably permanently wet, locations for electrodes. Obvious areas to be avoided if possible are dry sand, gravel limestone and anywhere where rocks are near to the surface. For this reason, it may be necessary to locate an electrode at some distance from the building foundations, the trenches of which are often infilled with rubble.

There are many factors which affect the resistance of an earth electrode, including the size and shape of the electrode, the soil resistivity, the current density at the electrode surface and the step voltages in the ground around the electrode. The designer will find that reference to BS 7430 *Code of practice for earthing* will be invaluable in the design of effective and safe earthing systems, and this publication should be consulted in all but the simplest of earthing systems. The standard gives information, amongst other things, on the relationship between rod diameter and rod length to resistance. It will be seen that increasing the rod diameter has less effect on resistance than does increasing the number of rods or extending their length. Where multiple rods are required, information is provided relating to overlapping resistance areas and gives recommendations for spacing. Mechanical, chemical and electrolytic properties of materials are also addressed.

According to BS 7430, for the driven rod or similar electrode, the resistance to Earth is given by

$$R = \frac{1}{2\pi L}(\rho - \rho_c)\left(\log_e \frac{8L}{D} - 1\right) + \frac{\rho_c}{2\pi L}\left(\log_e \frac{8L}{d}\right) \tag{12.6}$$

where $\rho(\Omega\text{ m})$ is the resistivity of the soil, ρ_c (Ω m) is the resistivity of the infill material, d (m) is the diameter of the electrode, D (m) is the diameter of the infill and L (m) is the driven length of the electrode.

It can be readily seen from Equation (12.6) that, where the infill material is the same as the surrounding material, the first term goes to zero, leaving only the second term. This may, for example, apply to earth electrodes associated with TT and IT systems, where the electrode is driven directly in the ground. In this case, Equation (12.7) would apply:

$$R = \frac{\rho_c}{2\pi L}\left(\log_e \frac{8L}{D}\right) \tag{12.7}$$

As with all electrical equipment, it is essential that the earthing electrode, particularly when in the form of a single separate rod, is protected against mechanical, chemical and electrolytic damage. Figure 12.7 gives such an example of mechanical protection of a rod electrode, although a purpose-made concrete or brick housing with removable slab cover would also be acceptable, as would a suitable steel box-type enclosure. Where multiple

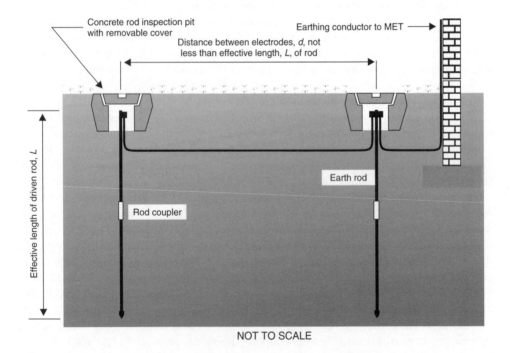

Figure 12.7   Electrodes with inspection pits

rods are required it is recommended that the distance between rods should not be less than the effective depth of the rod.

12.2.6 Main earthing terminals

As called for in Regulations 542.4.1 and 542.4.2, an MET is required for every installation. The MET must have provision for connecting separately the cpcs, the main bonding conductors, any functional earthing conductors and the main bonding conductor to the lightning protection system where used. One of the principal purposes of the MET is to provide for disconnection of the earthing conductor from cpcs and bonding conductors so that the external line–earth loop impedance Z_e can be measured. For this reason, the earthing conductor and other protective conductors should be clearly identifiable. The means of disconnection must involve the use of a tool (e.g. spanner or screwdriver) and may be provided in the form of a disconnectable link (see Figure 12.6). It goes without saying that METs must be mechanically robust and capable of providing long-term reliability and continuity.

Testing of Z_e must only be carried out by skilled and competent persons and the disconnectable link of the MET must never be removed whilst the installation is energized, since any earth fault current, however minor, could cause a voltage (up to U_o) between the disconnected parts of the MET. The designer may consider it prudent to affixed a notice adjacent to the MET warning of the dangers of disconnecting the link.

Where an installation is contained in more than one separate building, Regulation 411.3.1.2 calls for a MET (and equipotential bonding) for each building. Figure 12.8

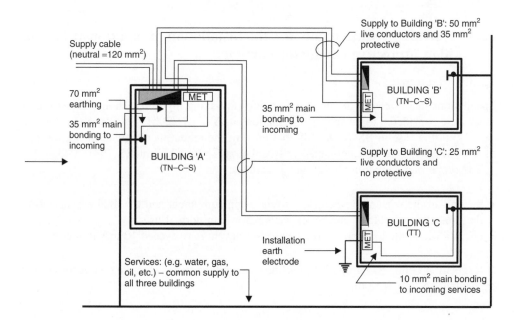

Figure 12.8 Typical arrangement for one installation spanning three separate buildings

shows a typical such arrangement for three separate buildings each with its own MET. Building 'A' is supplied with a PME service from the public network and houses the main distribution switchboard. Building 'B' makes use of the TN-C-S earthing facility, and the main bonding conductors here are of the same cross-sectional area as those of Building 'A'. The supply earthing is not utilized in Building 'C', which is treated as an installation which is part of a TT system and equipped with its own earth electrode.

12.2.7   Earthing: accessories and other equipment

As with all exposed-conductive-parts on installations where ADS is used, accessories require earthing via the cpc to the MET for compliance with Regulation 411.4.2 in TN systems or Regulation 411.5.1 in TT systems. This is also true for any metal boxes which house the accessory, including those installed flush with the wall and which are also deemed to be exposed-conductive-parts. Where the cpc(s) are separate, it is preferable to terminate them directly into the accessory and then earth the box by means of a 'fly-lead' from the earthing terminal of the accessory to the terminal in the box. Where the cpc is provided by steel conduit or steel trunking, all that is required is the 'fly-lead' (Regulation 543.2.7). Reliance of the contact between the accessory and a flush box for earthing purpose should not be regarded as providing a permanent and reliable connection owing to uneven surfaces and the possibility of corrosion over time.

The operating mechanism (if metal) of time switches needs earthing. It is not always recognized that, when such a mechanism is supplied with a conduit box enclosure, an earthing connection between the two is often not provided. Again, all that is required (provided the box is adequately earthed) is a 'fly-lead' between the box and the time-switch mechanism (Regulation 543.2.7).

12.3   Earthing requirements for the installation of equipment having high protective conductor currents

12.3.1   General

Installations incorporating or supplying equipment having high protective conductor currents were considered to be special installations in BS 7671:2001 (Section 607 referred). This is not the case in BS 7671:2008, but most of the requirements given in Section 607 of BS 7671:2001 are now contained in Regulation Group 543.7 of BS 7671:2008.

The scope of Regulation Group 543.7 includes:

- the part of the wiring between the current-using equipment and the final circuit, where the protective conductor current exceeds 3.5 mA in normal use;
- circuits where the total protective conductor current may exceed 10 mA in normal use.

Regulation Group 543.7 applies to *all* equipment having a protective conductor current exceeding 3.5 mA, including:

- information technology equipment complying with BS EN 60950;
- industrial and telecommunications equipment with radio-frequency interference suppression filtering; and
- heating elements.

The requirements of Regulation Group 543.7 relate to the particular circumstances where equipment has higher currents which flow to earth as a normal aspect of its functioning. All current-using equipment permits some protective conductor current, but this is not generally a problem unless this current is high enough to have physiological effects on a person coming into contact with charged metalwork and earth. Where such currents are higher than normal by design and are continuously in use, special precautions are required to minimize the risk of electric shock.

It will be appreciated that where such equipment is used in large numbers (e.g. as would be the case in commercial office buildings), there are additional risks associated with the accumulative earth-leakage current 'generated' by the equipment in use. These protective conductor currents pose additional risks for the users and possible problems with malfunctioning. Additionally, problems can occur with earth-leakage detection devices, such as RCDs, where the leakage may operate the protective devices, causing, for example, loss of data on computer-controlled systems.

Protective conductor currents are often the result of imbalance caused by low-order harmonic currents in, for example, oscillators and inverters (sometimes operating in the order of megahertz frequencies). Equipment containing sources of earth-leakage current will, generally, be the subject of BS EN 60950 (BS 7002) *Specification for safety of information technology including electrical business equipment*. However, industrial-type control equipment may also have such characteristics, as may domestic 'white' goods and other appliances, such as those containing heating elements. In consideration of all equipment, the designer would need to check with the relevant British or CENELEC standard to establish the maximum leakage current permitted by that standard.

Notwithstanding the user's requirements, the requirements of this Regulation Group 543.7 are paramount, because they relate to the safety in use of equipment and, in particular, to the earthing connection arrangements to prevent or minimize danger from electric shock. Of particular concern is the conduction of any remaining protective conductor currents to earth after the supply and the protective conductor have been disconnected, as might be the case where the plug to the equipment has been withdrawn.

12.3.2 Additional requirements for earthing of equipment in TN and TT systems

The requirements for system types other than IT systems are summarized in Table 12.9. The minimum cross-sectional areas of protective conductors of final and distribution circuits, where the total protective conductor current is likely to exceed 10 mA, are summarized in Table 12.10.

Table 12.9 Additional requirements for earthing of equipment in TN and TT systems

Ref	Regulation no.	Protective conductor current I_L	Additional protective conductor connection requirements (alternatives)
A	543.7.1.1	$3.5 \leq I_L \leq$ 10 mA	Permanently connected to the fixed wiring of the installation
B			Connected by means of an industrial-type plug and socket-outlet complying with BS EN 60309-2
C	543.7.1.2	$I_L < 10$ mA	Permanently connected to the fixed wiring of the installation, with a protective conductor complying with the particular cross-sectional area and other requirements. A flexible cable may be used for the permanent connection[a]
D			Flexible cable with an industrial-type plug and socket-outlet complying with BS EN 60309-2, providing the protective conductor meets specified cross-sectional area requirements
E			By a protective conductor complying with Section 543 (protective conductors) with an earth monitoring system to BS 4444
F	543.7.1.3		The wiring of every final circuit and *distribution circuit* where the protective conductor current is likely to exceed 10 mA must have a high-integrity protective connection
G	543.7.1.4		Where two protective conductors are used to provide a high-integrity protective connection, the ends of the conductors must be terminated independently at *all* points throughout the circuit, such as distribution boards, junction boxes and socket-outlets[b]
H	543.7.2.1		Three examples of high-integrity protective conductor arrangements for radial final circuits for socket-outlets: • the protective conductor may be connected as a ring; or • a separate protective conductor connection is provided at the final socket-outlet by connection to the metal conduit or ducting; or • where two or more similar radial circuits supply socket-outlet circuits in adjacent areas, then a second protective conductor may be provided at the final socket-outlet on one circuit by connection to the protective conductor of the adjacent circuit. However, conditions apply to the two socket-outlet circuits in this example. They must be fed from the same distribution board, have identical means of short-circuit and fault current protection, and have cpcs of the same cross-sectional area[c]
I	543.7.1.5	$I_L \geq 3.5$ mA	Visible information, identifying circuits having high protective conductor currents, to be given at the distribution board

[a]Item D is the preferred method.
[b]This requires accessories to be provided with two earthing terminals. Spurs are permitted in ring final circuits provided each spur includes a high-integrity protective connection.
[c]A socket-outlet circuit other than a ring final circuit or one of the example types of radial circuit can be used, provided that the protective conductor arrangements meet the general requirements of Regulation 543.7.1.

Table 12.10 Minimum cross-sectional area of protective conductors of final and distribution circuits, where the total protective conductor current is likely to exceed 10 mA[a,b]

Ref	Regulation no.	Protective conductor form	Min. csa (mm^2)	
			Protective conductors	Multicore cable
A	543.7.1.3	Single protective conductor meeting the requirements of Regulation Groups 543.2 and 543.3.	10	
B		Single copper protective conductor meeting the requirements of Regulation Groups 543.2 and 543.3 and enclosed to provide additional mechanical protection (e.g. within a flexible conduit).	4	
C		Two separate protective conductors meeting the requirements of Section 543 incorporated in a single multicore cable. One protective conductor may be sheath armour or wire-braid of cable meeting the requirements of Regulation 543.2.5.	–	10

[a]Where earth monitoring complying with BS 4444 is employed which provides for automatic disconnection in the event of a loss of protective conductor continuity, no additional requirements related to protective conductor cross-sectional areas are stipulated in the Section.
[b]Where the supply is derived from a double-wound transformer or equivalent source, the protective conductors must comply with A, B or C above.

12.3.3   Minimum cross-sectional area of protective conductors of final and distribution circuits

Regulation 543.7.1.3 gives a number of options for compliance in terms of cross-sectional areas of protective conductors, which are summarized in Table 12.10.

12.3.4   Residual current device compatibility

The requirement of Regulation 607-07-01 of BS 7671:2001 that the residual current which may be expected to occur in a circuit having a protective conductor current exceeding 3.5 mA, including switch-on surges, must not operate the RCD does not appear in Regulation Group 543.7 of BS 7671:2008.

Nevertheless, this is still necessary in order to comply with Regulation 531.2.4. Compliance necessitates careful consideration of all current-using equipment in terms of start-up inrush currents, as well as of their normal protective conductor currents. Where it is not possible to meet this requirement, the current-using equipment must be supplied through a double-wound transformer or equivalent device.

12.3.5   IT systems

Direct connection of equipment having high protective conductor currents to an IT system was precluded by Regulation 607-06-01 of BS 7671: 2001. However, no equivalent regulation appears in BS 7671:2008. Nevertheless, such equipment should not be directly connected to an IT system, as there would be no path for the protective conductor current to return to the distribution transformer or other source of supply, the live parts of which are not connected directly with Earth.

12.4   Protective bonding

Protective bonding will always be required where protection against electric shock is provided by ADS. There are essentially two forms of bonding (which should not be confused with earthing):

- main protective bonding (applicable to all installations where ADS is employed);
- supplementary bonding (applicable to some special locations and installations and other special-circumstance installations where ADS is employed).

Protective bonding should only be considered where conductive parts, including those of nonelectrical services, fall within the definition of extraneous-conductive-parts, i.e. *conductive parts liable to introduce a potential, generally Earth potential, and not forming part of the electrical installation*. Bonding of conductive parts which are not liable to introduce a potential, far from providing a measure of safety, can indeed, under certain circumstances, create a hazard that was not present hitherto.

12.4.1   Main protective bonding

The principal purpose of providing main protective bonding is to protect against dangers resulting from earth faults on the supply so that any fault voltage on the MET and all exposed-conductive-parts connected to it are substantially at the same potential (under these conditions) to extraneous-conductive-parts such as water pipes, etc.

Section 547 of BS 7671 deals with requirements for main protective bonding conductors, but it should not be forgotten that other relevant parts of BS 7671 are no less applicable to these conductors than to current-carrying conductors. In particular, precautionary measures need to be taken in relation to external influences (e.g. mechanical stress and the need for proper supports, and corrosion). If the conductor metal is other than copper, then the conductance of that material must not be less than that of copper of cross-sectional area appropriate to the particular installation.

The minimum cross-sectional area of main bonding conductors is related to the cross-sectional area of the earthing conductor (the conductor connecting the MET to the means of earthing), which itself is related to the line conductor cross-sectional area in all cases except where the supply is obtained from a network which uses PME. For installations supplied by a PME source (a TN-C-S system), the cross-sectional area of the main bonding conductors is related to the cross-sectional area of the neutral conductor of the supply. In all the cases, the cross-sectional area of the bonding conductor is subject to a minimum of 6 mm^2 (Regulation 544.1.1). For installations forming part of a TN-C-S system, Table 54.8 of BS 7671 gives data for selecting the cross-sectional area of these bonding conductors. Table 12.11 of this Guide summarizes the required cross-sectional area for main protective bonding conductors for both PME and non-PME supplies.

Where the protective measure of ADS is used, a main protective bonding conductor is required from the MET to each and every extraneous-conductive-part which enters the building. Regulation 411.3.1.2 cites a number of examples of such parts, but the list should not be regarded as exhaustive and there may be other conductive parts entering the premises which will require main bonding (e.g. oil pipe lines and compressed-air lines). As previously mentioned, bonding conductor connections at the MET should be made easily identifiable.

Figure 12.9 shows a typical installation with the main bonding connections. These connections should be made as near as practicable to the point (ideally within 600 mm) where the services enter the premises, as called for in Regulation 544.1.2. This regulation goes on to require that, where insulated parts are inserted into service pipes, etc., the bonding conductor termination should be made on the consumer's side of that insert. Where there is a significant length of pipework from the point of entry to an insulated insert it may be necessary to apply insulating material to that section of pipework.

It may not always be necessary or economical to take separate bonding conductors to the various extraneous-conductive-parts (incoming services) entering the premises. It may sometimes be appropriate for one conductor to connect a number of parts (possibly all),

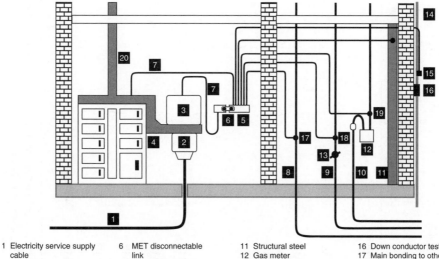

1 Electricity service supply cable	6 MET disconnectable link	11 Structural steel	16 Down conductor test point
2 ED's cut–out	7 Earthing conductor	12 Gas meter	17 Main bonding to other service pipework
3 Meter operator's metering equipment	8 Other service pipe (e.g. oil pipe)	13 Water stopcock	18 Main bonding to water service
4 Main switchboard	9 Water incoming service	14 Lightning protection down conductor	19 Main bonding to gas service
5 Main Earthing Terminal (MET)	10 Gas incoming service	15 Main bonding to down conductor	20 Distribution cable containment system

Figure 12.9 Typical main bonding connections

Table 12.11 Cross-sectional areas of main protective bonding conductors (nearest standard sizes)[a]

Ref	System supply	Csa (mm²) Line conductor	Neutral conductor	Earthing conductor	Non-PME supplies	PME supplies	Comments
A	Non-PME	6	–	6	6	–	Minimum csa applies. See Table 54.7 and Regulation 544.1.1
B	PME	–	6	6	–	10	See Table 54.8
C	Non-PME	10	–	10	6	–	See Table 54.7 and Regulation 544.1.1
D	PME	–	10	10	–	10	See Table 54.8
E	Non-PME	16	–	16	10	–	See Table 54.7 and Regulation 544.1.1
F	PME	–	16	16	–	10	See Table 54.8
G	Non-PME	25	–	16	10	–	See Table 54.7 and Regulation 544.1.1
H	PME	–	25	16	–	10	See Table 54.8
I	Non-PME	35	–	16	10	–	See Table 54.7 and Regulation 544.1.1
J	PME	–	35	16	–	10	See Table 54.8
K	Non-PME	50	–	25	16	–	See Table 54.7 and Regulation 544.1.1
L	PME	–	50	25	–	16	See Table 54.8
M	Non-PME	70	–	35	25	–	See Table 54.7 and Regulation 544.1.1
N	PME	–	70	35	–	25	See Table 54.8
O	Non-PME	95	–	50	25	–	See Table 54.7 and Regulation 544.1.1
P	PME	–	95	50	–	25	See Table 54.8.
Q	Non-PME	120	–	75	25	–	See Table 54.7 and Regulation 544.1.1
R	PME	–	120	75	–	35	See Table 54.8
S	Non-PME	150	–	75	25	–	See Table 54.7 and Regulation 544.1.1
T	PME	–	150	75	–	35	See Table 54.8
U	Non-PME	185	–	120	25	–	See Table 54.7 and Regulation 544.1.1
V	PME	–	185	120	–	50	See Table 54.8

[a]The cross-sectional area attributed to earthing conductors on installations supplied by a PME source assumes that the line conductor cross-sectional area equals that of the neutral conductor.

but it is advisable that the conductor is left continuous and is looped-terminated at every point but the last on the run when doing so.

As called for in Regulation 411.3.1.2, and again in Regulation 541.3, a main protective bonding conductor is required to connect the MET with the lightning protection system (see Figure 12.9) in accordance with BS EN 62305. The cross-sectional area of such a bonding conductor must be related to the earthing conductor and would normally be the same cross-sectional area (if of the same metal) as other main bonding conductors (see Table 12.11). BS 6651 (now superseded by BS EN 62305), which deals with lightning protection systems, has long called for such a conductor and requires the connection to the lightning protection system to be made above the test joint of the down conductors. Where, as in most cases, there are a number of down conductors, each will require effective bonding, but this would not preclude the lightning protection conductors also serving as bonding conductors provided they were of adequate conductance. Additional cross-bonding may be required where exposed and extraneous-conductive-parts are located within flash-over distance of the lightning-protection system conductors.

12.4.2 Supplementary bonding

Other than special locations or installations (embraced by Part 7 of BS 7671), there is generally little need for supplementary bonding except where circuit disconnection times cannot be met (Regulation 411.3.2.6). Supplementary bonding for special locations and installations is discussed in Chapter 16 under the relevant headings.

As called for in Regulation 544.2.1, supplementary bonding conductors between two or more exposed-conductive-parts must have equivalent conductance to that of the smaller or smallest cpc. Where the bonding conductor is not sheathed or otherwise protected against mechanical damage, it is subject to the minimum cross-sectional area of 4 mm². Bonding conductors between exposed-conductive-parts and extraneous-conductive-parts must, according to Regulation 544.2.2, be not less than half the cross-sectional area of the cpc of the circuit concerned; but again, the minimum cross-sectional area of 4 mm² applies if mechanical protection is not provided. Regulation 544.2.3 calls for the bonding connection between two or more extraneous-conductive-parts to be a minimum of 2.5 mm² if sheathed or otherwise mechanically protected, or 4 mm² if mechanical protection is not provided, except that Regulation 544.2.2 also applies to the conductor connecting the two extraneous-conductive-parts where one of the extraneous-conductive-parts is connected to an exposed-conductive-part. For most practical situations, a minimum cross-sectional area of 4 mm² would apply to all supplementary bonding conductors.

Regulation 544.2.4 acknowledges that the bonding conductors may take the form of separate conductors or conductive parts which provide a permanent and reliable conductive path or a combination of both. This would include metallic pipework (e.g. copper water pipes). Where it is necessary to effect a supplementary bonding connection to an item of Class I equipment (e.g. electric towel rail), Regulation 544.2.5 permits the cpc of the flexible cord making the final connection to the appliance to be used also as a bonding connection, thereby obviating the need for a separate conductor and separate termination. This dispensation is only applicable where the final connection is short (e.g. 150–200 mm).

As mentioned earlier, Regulations 415.2.1 and 415.2.2 reinforce the requirements contained in Chapter 54 relating to supplementary bonding and additionally call for the resistance R of the supplementary bonding conductor between conductive parts in a.c. systems to meet Equation (12.8) for the general situation:

$$R \; (\Omega) \leq \frac{50}{I_a} \tag{12.8}$$

where I_a (A) is the operating (tripping) current of the protective device, which if an RCD this is $I_{\Delta n}$ or if an overcurrent device this is the minimum current which causes disconnection within 5 s (in d.c. systems a value of 120 is used in Equation (12.8) instead of 50).

Transposing Equation (12.8) it will be seen that RI_a must not exceed 50 V Where there are a number of final circuits with exposed-conductive-parts requiring supplementary bonding, the operating current I_a to be used in Equation (12.8) will be the worst-case value. By way of example, if exposed-conductive-parts of three final circuits need supplementary bonding, then the designer would find the worst-case value of earth fault current from the time–current characteristics of the protective devices and insert it into the equation. Taking three circuits protected by Type B MCBs with nominal ratings of 6 A, 20 A and 32 A (for disconnection within 0.1 and 5 s), it can be determined from the time–current characteristics that the respective operating currents I_a are 30 A, 100 A and 160 A respectively. Clearly, the I_a of 160 A would be used to determine the supplementary bonding conductor resistance R, in this case limited to 0.31 Ω (50/160).

12.4.3   Bonding clamps

Protective bonding connections must be made to extraneous-conductive-parts (e.g. water pipes) using clamps manufactured to comply with BS 951 *Specifications for clamps for earthing and bonding purposes*. Clamps are available for securing to pipework, etc. of diameter of 6 mm or greater and for conductor sizes from 2.5 to 70 mm^2 (identified by the letters A, B, C, D, E, F, G, H and I for conductors of cross-sectional areas 2.5 mm^2, 4 mm^2, 6 mm^2, 10 mm^2, 16 mm^2, 25 mm^2, 35 mm^2, 50 mm^2 and 70 mm^2 respectively).

Wherever such clamps are used they must be accompanied by a suitable label to meet the requirements of Regulation 514.13.1 in the case of protection by ADS and to meet the requirements of Regulation 514.13.2 where one of the protective measures of earth-free equipotential bonding or electrical separation for the supply to more than one item of current-using equipment is employed (see Figures 9.9 and 9.10). Most commercially available clamps are acquired with a label attached (to meet Regulation 514.13.1) and are commonly of the form of a thin, twin-slotted or twin-hold aluminium embossed strip attached, for convenience of packing, to the strap of the clamp. Because of the risk of corrosion and electrolytic action between dissimilar metals (e.g. copper and aluminium), the label should be removed from the strap and attached either to the bonding conductor itself or below the locking nut of the clamp, taking care that the label cannot come into contact with the copper pipework.

Applying such clamps to lead-sheathed cables should not be done, because the inevitable expansion and contraction of the cable (under varying load conditions) is likely, in time, to make the clamp joint become loose and lose its effectiveness. Soldered joints should not be used under such circumstances. Regulation 544.1.2 demands that such bonding connections be made to the hard metal pipework of incoming services, and this would preclude a connection being made directly with a lead gas or water pipe and lead-sheathed cable.

Some manufacturers of BS 951 earthing clamps use a method of colour coding to indicate the suitability of the item for dry conditions and for wet or damp environments.

13

Specialized installations

13.1 General

It is not the intention here to provide a detailed specification for the specialized installations addressed, but rather to identify specific points which are worthy of note. An attempt is made, therefore, to distinguish some of the issues which, from experience, need consideration prior to carrying out the design. Where guidance on the subject is made in other publications, these are identified and their careful perusal is recommended.

13.2 Emergency lighting

The first point to make is that the designer of an emergency lighting installation should consult with other interested parties before commencing the detailed design. Interested parties would include, for example, the architect, the lighting engineer, the owner of the premises, the ED and the licensing and enforcing authority (e.g. Fire Authority, Local Authority, Health Authority). Some aspects which would be the subject of consultation are:

- number and position of fire alarm call points;
- location of fire-fighting equipment (e.g. hose reels);
- location of lifts, escalators, walkways, toilets, control room, accommodation, etc.;
- location of plant rooms and car parks and the siting of the generator, where required;
- designation of escape route(s), including an assessment of fire risks;
- assessment of potential risks along escape route;
- requirements for external escape lighting.

An emergency lighting installation is required to comply both with BS 7671 and BS 5266 *Emergency lighting* (see Regulation 110.1). Detailed guidance for emergency lighting installations is beyond the scope of this Guide. For those engaged in this type of work, BS 5266 Parts 1, 7 and 8 are considered to be essential reading and explain in detail all the necessary requirements. Further advice on luminaire specifications is obtainable from the Industry Committee for Emergency Lighting (ICEL) and the Lighting Industry

A Practical Guide to The Wiring Regulations: 17th Edition IEE Wiring Regulations (BS 7671:2008)
Fourth Edition Geoffrey Stokes and John Bradley
© Geoffrey Stokes and John Bradley. Published by John Wiley & Sons, Ltd

Federation. ICEL, in conjunction with the Assessment Department of BSI, have a certification scheme for emergency lighting luminaires and ICEL labels are affixed to certified products (Product Standard ICEL-1001).

BS 5266 calls for emergency lighting circuits to be segregated from all other cables, including those of fire detection and alarm systems. This requirement does not apply to the case of the circuit supplying self-contained luminaires with integral batteries, which is considered to be a normal LV circuit. Luminaires, like other equipment, need a means of interrupting the supply, and this can be done by a local 'key' switch either mounted adjacent to the luminaire or at a local gridswitch (suitably labelled). This switch can also provide the means for testing and discharging batteries. Devices for the control, isolation and operation of the system must be clearly identified and marked accordingly by, for example, labelling, as appropriate, '**Standby lighting**', '**Escape lighting**' or '**Emergency lighting**'.

Circuits which supply emergency lighting luminaires, including self-contained units fed from a general lighting circuit, need to be identified at their distribution board and distribution circuit feeder (if any) in the normal way.

On completion of such an installation, inspection, testing and certification must be completed and the necessary documentation handed to the person ordering the work. Testing for illuminance will involve the use of a suitable meter with an appropriate scale (lux). Records, including initial certification, must be kept in the form of a log book, detailing subsequent events and listing the checks that are made and the particulars of periodic inspection and testing.

13.3   Fire detection and alarm systems

A fire detection and alarm system installation is required to comply both with BS 7671 and BS 5839-1 *Code of practice for system design, installation, commissioning and maintenance* or, where applicable BS 5839-6 *Code of practice for the design, installation and maintenance of fire detection and alarm systems in dwellings* (see Regulation 110.1). Other parts of this standard relate to equipment specifications:

- Part 2 *Specification for manual call points*;
- Part 3 *Specification for automatic release mechanisms for certain fire protection equipment*;
- Part 4 *Specification for control and indicating equipment*;
- Part 5 *Specification for optical beam smoke detectors*;
- Part 8 *Code of practice for the design, installation and servicing of voice alarm systems*;
- Part 9 *Code of practice for the design, installation, commissioning and maintenance of emergency voice communication systems*.

Some other BSI Standards relevant to this subject are:

- BS 5306 *Fire extinguishing installations and equipment on premises*;
- BS EN 54 *Fire detection and fire alarm systems*;
- BS 5446 *Components of automatic fire alarm systems for residential premises*;

- BS 5588 *Fire precautions in the design, construction and use of buildings*;
- BS 6266 *Code of practice for fire protection of electronic data processing installations*.

Where these systems are installed to satisfy statutory requirements imposed, for example, by The Regulatory Reform (Fire Safety) Order 2005, the enforcing authority would normally require compliance with BS 5839. It is recommended that consultations with all interested parties, including the licensing and enforcing authority, take place prior to commencing detailed design work. Other interested parties who should be consulted include the building fire insurer, the architect, HSE, communications link provider, etc. Detailed consideration of such issues as the purpose of the system, occupants' escape time, fire brigade response time, the relationship with other building occupants and the servicing and maintenance procedures are required at an early stage. System operational requirements, planning of escape routes and sound levels of the alarm sounders, in the context of the ambient noise and occupancy patterns, will also need to be established at an early stage. Generally, a number of small sounders is usually more effective in terms of audibility levels than a lesser number of larger units. Verification of installations and their periodic testing will necessarily involve testing of the sounder(s) and a suitable meter (decibels) will be required. Particular attention needs to be paid to sound level attenuation caused by, for example, curtains, doors, carpeting and furniture.

There are two basic types of protection: the protection of life and the protection of property. The design approach for one system differs from the other, and Table 13.1 summarizes the principal features of the systems and their variations.

Clause 26.2 of BS 5839-1:2002 makes recommendations as to the acceptable types of wiring system. Table 13.2 summarizes this information.

Fire alarm systems come in many different types, and the installation planning and techniques will depend to a large extent on the particular type chosen. Open-circuit operation is used on many conventional systems, but can have the disadvantage that it cannot differentiate between a circuit fault condition and a genuine fire signal, although monitored open-circuit initiation circuits can make this distinction. Many addressable and time-division-multiplexing (TDM) systems can allow all devices to be wired on the same circuit, and others permit the trigger devices to be so wired with other power circuits to sounders wired through appropriate triggering relays. Generally, sounders are wired on radial circuits with more than one circuit per system. Where a single two-wire ring circuit is employed, it is normally necessary to fit each device with short-circuit 'isolators' (one each side of the device) so that a faulted unit will not prevent the rest of the system from operating.

It is generally necessary to split a building up into zones, if for no other reason than to be able to identify the location of a particular fire condition signal. In the consideration of these zones it would be necessary to decide on whether sounders should sound in all zones or groups of zones, bearing in mind the risk to occupants and their escape routes. In other words, zoning is vitally important from the fire protection planning viewpoint and it will be closely related to the 'fire compartments' or areas contained in fire-resisting material of, say, 30 minutes' resistance. As a rule of thumb, zones would not normally exceed $2000\,m^2$ and a person should not need to travel more than 30 m to identify the position of a fire visually. The use of remote indicating lamps repeating the signal of a detector device or group of devices can assist in this respect. Stairwells, lift shafts and the like

Table 13.1 Principal features of the systems for the protection of life and the protection of property

Protection required	Type	Description of type	Basic features	Possible additional features
Protection of property	P1	Covering all parts of the premises	A control and indicating panel, external and internal sounders, manual call points, automatic detectors	Fixed extinguishing system, additional sounders, repeat indicator panels, a manned-centre link
	P2	Covering only those parts of the premises having a high fire risk		
Protection of life	L1	Covering all parts of the premises	A control and indicating panel, alarm sounders, manual call points	Fixed extinguishing system, additional sounders, repeat indicator panels, a manned centre link, automatic detectors, automatic door release relay unit
	L2	Covering only those parts of the premises where there is a high risk to life if there is a fire anywhere in the building		
	L3	Covering only areas critical to free movement along escape routes		
	M	Provides only for manual initiation and relies on the presence of persons at all times		

should be treated as separate, vertical zones. Other than this, a zone would not normally cover more than one storey unless the whole building has a floor area of less than $300\,\mathrm{m}^2$. For multi-occupancy buildings, the zones should coincide with the occupancy boundaries, and whilst it is acceptable for a zone to extend over a number of fire compartments, it is not considered admissible to have zones overlapping.

Figure 13.1 shows a simple fire alarm system in which each zone has its own two-core circuit for sounders and triggering devices; it can be seen that this configuration has the advantage of limiting the effects of a faulted circuit to devices on that circuit. Figure 13.2 shows a typical fire alarm system supply schematic layout in which the fire alarm system supply is derived from the upstream side of the general installation main switch. The standard also permits the supply to be taken from the downstream side of the main switch – see Clause 25.2. In any event, warning notice labels are required as follows:

Table 13.2 Preferred wiring systems for fire alarm systems to BS 5839-1

Ref	BS no.	Cable type[a, b, c, d]	Additional protection against mechanical damage required?
A	BS EN 60702-1	Mineral-insulated copper-sheathed cables with an overall polymeric covering	Only in particularly arduous conditions
B	BS 7629	Multicore fire-resistant low smoke and fume cables	Yes. See BS 5839
C	BS 7846	Multi-pair fire-resistant low smoke and fume cables	Only in particularly arduous conditions if cables are armoured
D	As applicable	Cables rated at 300/500 V or greater that provide the same degree of safety to that afforded by compliance with BS 7629	Yes. See BS 5839

[a]BS 5839-1:2002 makes recommendations for two levels of fire resistance of fire-resisting cable systems, termed 'standard' and 'enhanced', according to the type of building and fire alarm system. Clause 26.2c of BS 5839-1:2002 refers.

[b]To provide standard fire resistance, in addition to meeting the British Standard in Table 12.3, the cable should meet the PH 30 classification when tested in accordance with BS EN 50200 and the 30 min survival time when tested in accordance with BS 8434-1.

[c]To provide enhanced fire resistance, in addition to meeting the British Standard in Table 12.3, the cable should meet the PH 120 classification when tested in accordance with BS EN 50200 and the 120 min survival time when tested in accordance with BS 8434-2.

[d]All fire alarm cables should be of a single, common colour that is not used for cables of general electrical services in the building. The colour red is preferred. (Clause 26.2o BS 5839-1:2002 refers.)

- FA system supply connected to the upstream side of the main switch: **Warning: this supply remains when the main switch is turned off**.
  Notice to be affixed to FA system switchfuse.
- FA system supply connected to the downstream side of the main switch: **Warning: this switch also controls the supply to the fire alarm system**.
  Notice to be affixed to general installation main switch.

Generally, fault protection by means of an RCD should be avoided if possible for fire alarm systems. Where RCD protection cannot be avoided (e.g. on TT systems), the device should generally not also protect general purpose circuits, although BS 5839-6 allows this in certain circumstances for systems in dwellings. Additionally, the use of the more sensitive RCDs ($I_{\Delta n} \leq 100$ mA) should be avoided wherever possible, although an RCD having $I_{\Delta n}$ not exceeding 30 mA must be used where required by BS 7671 for additional protection.

Clause 12.1 of BS 5839-1 calls for at least one sounder to continue to operate in the event of a fault condition occurring on others. In practice, for a simple system with radially wired sounder circuits, this means that at least two such circuits will be required. For the more complex systems (e.g. TDM) wired on a single ring circuit, each sounder device will need to be fitted with a short-circuit isolator on both 'legs' of the ring.

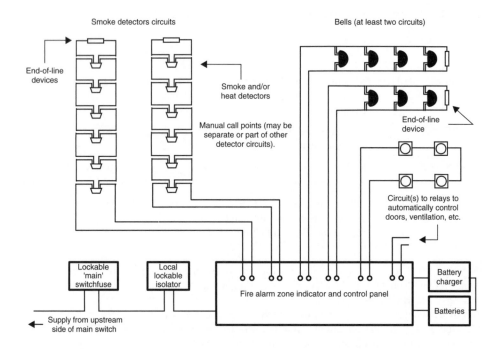

Figure 13.1   A schematic representation of a simple fire alarm system

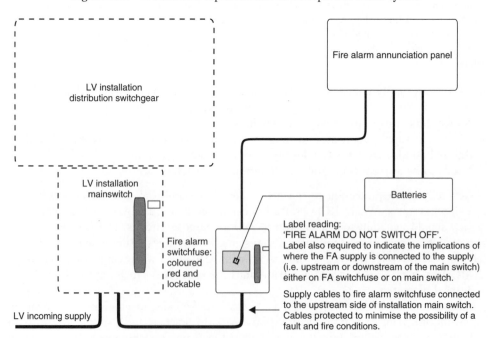

Figure 13.2   Fire alarm system supply schematic layout

Legislation, both on the UK mainland and in Northern Ireland, in the Building Regulations, requires provision of a smoke detection, heat detection (where applicable) and alarm on each floor of newly built houses and in dwellings converted from a building previously used for other purposes. This provision should be by means of a system complying with BS 5839-6, except for areas falling outside the scope of that standard, such as the communal parts of purpose-built sheltered housing and blocks of flats or maisonettes, where a system complying with BS 5839-1 should be provided. Smoke alarms and heat alarms must be mains powered with, in addition, a standby power supply in the form of a battery or capacitor. For the larger dwellings, the self-contained smoke alarms and heat alarms are not usually adequate and a full fire alarm and detection system should be used. In sheltered and warden-controlled housing, the warden's apartment should be equipped with a central monitoring annunciation panel to enable the warden to identify the location of the fire signal.

13.4   Petrol filling stations and liquid petroleum gas stations

Petrol filling stations are an obvious case of increased risk and require additional care in the design and installation of electrical equipment. There is no point in reiterating the excellent advice given in the publication *Guidance for the design, construction, modification and maintenance of petrol filling stations*, published jointly by the Association of Petroleum Explosives Administration (APEA) and the Institute of Petroleum (IP). A 'competent person' is defined as a person with enough practical experience and theoretical knowledge and actual experience to carry out a particular task safely and effectively. Persons who are not competent should not be engaged in this work. The publication identifies the hazardous zones, and determination of the extent of the zones should only be undertaken by a competent and responsible person, normally employed by the licensing and enforcing authority.

13.5   Installations in dusty environments

Regulation 522.4.2 requires, where there is a presence of dust in significant quantity (external influence AE4), that additional precautions are taken to prevent the accumulation of dust or other substances in quantities that could adversely affect the heat dissipation from the wiring system. If equipment can be placed outside dust-laden atmospheres, then this is obviously the best solution. However, where this is not possible, equipment should have at least the degree of protection IP5X. Where equipment installation in dusty atmospheres is contemplated, the equipment manufacturers should be consulted in order to clarify that the equipment satisfies the test requirements for the appropriate degree of ingress protection (IP5X) and that, by virtue of dust deposits forming on the enclosures, the temperature rise will not be such as to cause ignition of the collected dust. Further guidance is included in various British Standards and CENELEC, including:

- BS EN 50016 *Electrical apparatus for potentially explosive atmospheres*;
- BS EN 50281 *Electrical apparatus for use in the presence of combustible dust*;
- BS EN 61241 *Electrical apparatus for use in the presence of combustible dust*;
- BS 6713 *Explosion protection systems*;
- BS 7527 *Classification of environmental conditions*.

13.6   Installations in underground and multistorey car parks, etc

Underground car parks and multistorey car parks are not generally regarded as hazardous areas, though many should be so regarded. Although petrol is not dispensed in such locations, the accumulative quantity of petrol 'stored' in car petrol tanks warrants further consideration. One leaking tank together with an ignition source can spell disaster. Zone classification will depend to a great extent on the ventilation and other provisions, but it is not unusual for the whole of the volume occupied by parked cars to be classified as Zone 2 and the lower section of that volume, say from floor level up to a height of 225 mm, to be designated Zone 1. These matters need to be discussed with the Local Authority Fire Prevention Officer and the Petroleum Licensing Officer before carrying out detailed design work.

13.7   Installations in multi-occupancy blocks of flats

Generally speaking, multi-occupancy blocks of flats may be regarded as having a single earthing system, with main protective bonding conductors (as well as other protective conductors) connected to the MET at the main supply intake position in the normal way. However, where the ED provides distribution within the block of flats, main bonding may be required to individual flats, with each flat having its own MET. Bonding conductors are required to have not less than half the cross-sectional area required for the earthing conductor, subject to a maximum of $25\,\mathrm{mm}^2$, in the case of non-PME supplies and not less than that given in Table 54.8 for PME supplies (see Regulation 544.1.1), where the cross-sectional area is related to that of the supply neutral conductor.

Where the distribution of supply in the block of flats is by a lateral circuit provided with separate neutral and protective conductors, individual main bonding from every flat back to the main intake position is not generally required. Where the supply is PME and the main distribution within the block is carried out using a PEN conductor (combining the functions of neutral and protective conductors), main bonding in every flat is normally required. In this case, the cross-sectional area of all the main bonding conductors is usually related to the whole block incoming supply neutral. Figure 13.3 illustrates main bonding requirements where PEN distribution is employed. In all cases, the designer should consult, at an early stage, with the ED to ascertain the precise requirements.

Wiring systems for the flats, within the block, should be separate and, wherever possible, be contained within each flat in order to minimize the legal problems of entry when carrying out maintenance work in the years to come.

13.8   Installations in 'Section 20 buildings'

The phrase 'Section 20 buildings' is well known to designers working on projects within London, but may be less familiar to others who may from time to time be involved with such work. The reference is to Section 20 of the London Building Acts (Amendment) Act 1939 as amended by the Building (Inner London) Regulations 1985. The primary concern is with the danger from fire within certain classes of buildings owing to their use, height and/or cubic volume. Formerly administered by the Greater London Council, the implementation of the regulations is now the responsibility of the Inner London

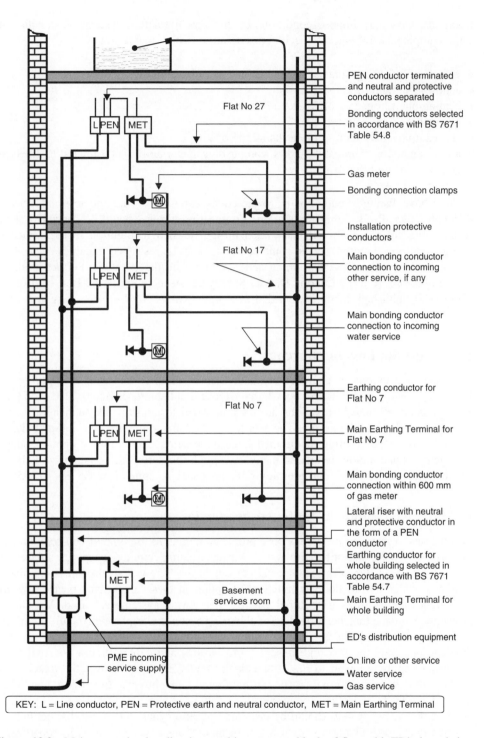

Figure 13.3 Main protective bonding in a multi-occupancy block of flats with ED's lateral riser incorporating a PEN conductor (PME supply)

Boroughs, who may impose conditions on fire and life-safety systems. Installation of some equipment will warrant special consideration, which may include:

- vehicle and parking areas at ground level and in basements;
- cellulose spraying rooms;
- heating plant in excess of 220 kW;
- generators in excess of 44 kW;
- HV equipment containing more than 250 l of oil;
- areas containing flammable and/or combustible materials in such quantities likely to constitute a fire hazard.

The above listing should not be regarded as exhaustive, and the relevant London Borough Council's Building Regulations Engineering Group should be consulted at the earliest stage of the electrical installation design. Particular attention to the choice of wiring system is essential, and in some cases this may be restricted to cables enclosed in screwed steel conduit, metal trunking and/or mineral insulated cables. Further guidance can be obtained from the London District Surveyors' Association's (PO Box 15, London SW6 3TU) published document entitled *Fire safety Guide No 1 – Fire safety in Section 20 buildings*.

13.9   Installations in churches

Electrical installations in churches are addressed in quite some detail in *Lighting and Wiring of Churches* published by Church House Publishing (ISBN 0 7151 7553 X). The designers of such installations are recommended to consult this publication before commencing the detailed design. The aspects covered are comprehensive and include, for example, power and lighting requirements, wiring systems, external flood lighting, supply arrangements and motor circuits. In many cases, lightning protection systems will also be desirable, particularly for high protruding features such as spires (see BS EN 62305).

13.10   Installations in thatched properties

Installations in thatched properties pose problems unique to the type of construction and, in particular, pose additional fire risks. Wiring systems running in spaces adjacent to the thatch (e.g. in the loft) need special consideration. Screwed steel conduit or mineral insulated cables are no doubt a much safer option in these areas, but special attention to scaling of enclosures is required where other wiring systems are used. Mechanical protection from the thatcher's fixings for all wiring systems except steel conduit is required (see Figure 13.4) in vulnerable areas. Joint boxes should be avoided in these areas (straw dust can penetrate the box and present a fire hazard) and cables should not, under any circumstances, be allowed to pass immediately under, through or over the thatch. This applies not only to the normal power and lighting circuits, but also to television and radio aerial down-leads and other communications cables. Where the thatch has wire-netting applied, the wiring system should be installed not closer than 300 mm to the netting.

A fire detection and alarm system would be an obvious advantage in such properties, as would an adequate number of fire extinguishers and fire blankets. For the larger buildings,

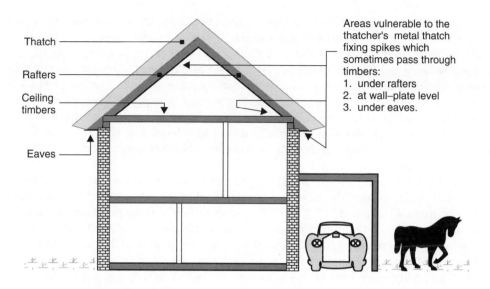

Figure 13.4 Mechanical protection in lofts of thatched properties

a sprinkler system may also be worth considering. Smoke detectors in high-risk areas, such as the loft, will provide some measure of reassurance.

Overhead wiring systems that pass over or near the thatch should be avoided, and consideration should also he given to providing a facility to switch off nonessential circuits at night and during periods when the building is unoccupied. Luminaires installed in roof spaces should be the totally enclosed type and mounted at sufficient distance from the thatch so as not to cause ignition of the thatch. Ideally, such luminaires should have some form of remote indication (e.g. indicator lamp near loft access) to indicate whether they are 'ON' or 'OFF'.

The designer will find it advantageous to consult with the Local Authority's Fire Prevention Officer and the building insurer prior to carrying out the initial installation design. The intervals between periodic inspection and testing may be significantly reduced for thatched properties.

The Dorset Model is a guide for builders and others on certain requirements and recommendations that should be met for thatched properties, including some concerning electrical installation matters. Information on the Dorset Model may be obtained from www.dorset-technical-committee.org.uk.

13.11 Extra-low voltage lighting

In addition to the information given in Chapter 11 of this Guide (see Table 11.12 in particular), there are one or two points worthy of mention concerning the installation of extra-low voltage lighting. It must be recognized that at these lower voltages ($U_n \leq 50\,V$ a.c. or $120\,V$ ripple-free d.c.) the currents are much higher for lamps of a similar power rating than at LV (e.g. $230\,V$). Protection against electric shock, overcurrent and thermal effects are just as important on these systems as on LV circuits.

In particular, proper consideration must be given to electric shock risks (basic protection and fault protection), fire hazards and thermal effects. Basically, there are two systems which can be used: SELV and PELV. Both of these systems use a safety source (e.g. a safety isolating transformer to BS EN 61558-2-6) as explained, amongst other electric shock requirements for these systems, in Section 5.6 of this Guide. The use of FELV is no longer recognized by BS 7671 for extra-low voltage lighting, as Regulation 559.11.1 precludes such use. (FELV relies on ADS, rather than extra-low voltage, for protection against electric shock, and all equipment used must have basic insulation corresponding to the voltage of the primary circuit of the transformer or other source; Regulations 411.7.2 and 411.7.3 refer.)

It is important, too, to consider the temperature of the luminaires with respect to the adjacent building fabric. The method of fixing should take account of the requirements of Regulations 421.2 and 559.5.1. In particular, provision must be made for the safe dissipation of heat with the luminaires mounted at a safe distance from adjacent material. Significantly, Regulation 421.7 requires that every termination of live conductors (see Figure 13.5) must be contained within an enclosure, irrespective of the nominal voltage; this requirement relates to protection against thermal effects and not necessarily to protection against electric shock.

Since much of this type of lighting equipment is used for decorative purposes and commonly installed in suspended ceilings, the proper mounting of transformers is important. Depending on the size, this type of transformer commonly weighs between 2 and 8 kg (for, say, a 50 VA and a 300 VA respectively) and secure fixings are always necessary. Many transformers, too, can be a major heat source (copper losses – I^2R) and the effects

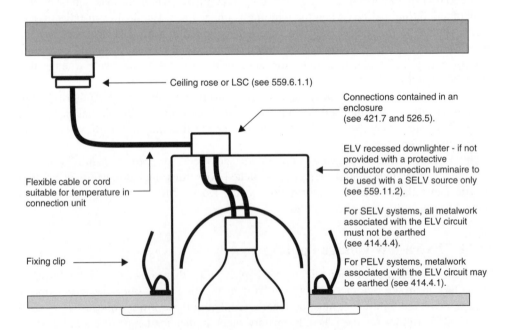

Figure 13.5 Typical final connection to extra-low voltage luminaire

on adjacent material must also be considered carefully. Overcurrent protection is required on both the primary and secondary circuits of the transformer.

This form of lighting is particularly sensitive to voltage variations, and lamp life is considerably shortened if lamps are energized outside their normal operating voltage tolerances. Voltages above and below the nominal voltage will have adverse effects on lamp life, and when designing such a system it is important to recognize the manufacturer's stated voltage variations which can be tolerated. The use of luminaires with their own integral transformer is probably the best way to avoid the effects of voltage regulation, but is quite an expensive solution. Alternatively, a number of luminaires may be supplied by a common transformer of adequate rating. In designing this system, account should be taken of the voltage drop with all lamps functioning and the voltage rise when one or more lamps fail.

Many extra-low voltage lighting systems operate on 12 V a.c. and many luminaires incorporate 50 W SBC halogen lamps. To illustrate some of the problems, Figure 13.6 shows a typical extra-low voltage lighting layout operating as an SELV system with six 50 W 12 V lamps supplied by a single 300 VA transformer. The secondary circuit full load current I_{sec} is given by

$$I_{\text{sec}} = \frac{\text{total power demand}}{\text{nominal voltage}} = \frac{6 \times 50}{12} = 25 \text{ A} \qquad (13.1)$$

If the wiring system used was PVC cables enclosed in nonmetallic conduit (reference method B), then a minimum cross-sectional area of 4 mm^2 would be required with a

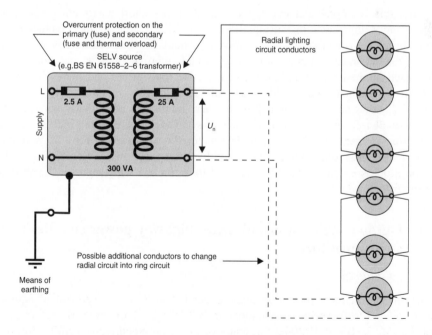

Figure 13.6 A problematic solution to extra-low voltage lighting

corresponding 11 mV/(A m) volt drop. This volt drop, together with the transformer voltage regulation, needs to be taken into account. For example, many 12 V halogen lamps normally have an acceptable voltage range of 11.7 to 12.3 V. This implies that the luminaire furthest from the transformer must have a voltage of at least 11.7 V and that the voltage at any luminaire at any time, including occasions when other lamps fail, is not more than 12.3 V. The transformer output voltage should not, therefore, be above 12.3 V, and this means a transformer voltage regulation not exceeding 2.5%, as given in Equation (13.2):

$$
\begin{aligned}
\text{Maximum voltage regulation} &= \frac{V_{\text{NL}} - V_{\text{FL}}}{V_{\text{NL}}} \\
&= \frac{12.3 - 12.0}{12.0} = 0.025 = 2.5\%
\end{aligned}
\tag{13.2}
$$

where V_{NL} is the secondary no-load output voltage and V_{FL} is the secondary full-load output voltage.

Similarly, the secondary circuit voltage drop should not exceed 0.3 V, and using the 4 mm² with a volt drop of 11 mV/(A m), this would limit the circuit length L to 1.09 m:

$$
\begin{aligned}
L &= \frac{\text{volt-drop limit}}{I_{\text{sec}} \times [\text{tabulated mV/(A m)}] \times 10^{-3}} \\
&= \frac{0.30}{25 \times 11 \times 10^{-3}} = 1.09 \text{ m}
\end{aligned}
\tag{13.3}
$$

Clearly, this is impractical due to the unrealistically short maximum length permitted by using 4 mm², and increasing the cross-sectional area of the conductor would present difficulties in effecting terminations, particularly into ceiling roses and LSCs. Forming the secondary circuit into a ring would improve matters, but not by much. Wiring each luminaire on a radial circuit from the transformer and increasing the conductor cross-sectional area would again present difficulties in terminating conductors. Figure 13.7 shows an alternative approach where luminaires are radially wired individually with smaller cross-sectional area conductors and each radial circuit individually fused near the transformer secondary output. The required design current for a 50 W luminaire is 4.2 A, and if the radial circuit is protected by a 5 A fuse, then the cross-sectional area could be little as 1 mm² for a 1.6 m run, 1.5 mm² for a 2.4 m length or 2.5 mm² for a 3.9 m circuit.

13.12 Outdoor lighting installations, highway power supplies and street furniture

13.12.1 General

Installations of highway power supplies, street furniture and street-located equipment were considered to be special installations in BS 7671:2001 (Section 611 referred). This is not the case in BS 7671:2008, but many of the requirements given in Section 607 of BS 7671:2001 are now contained in Regulation Group 559.10 of BS 7671:2008.

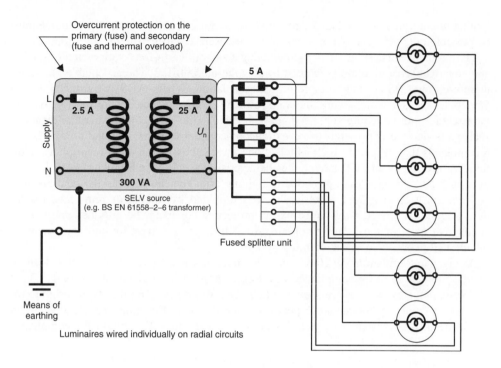

Overcurrent protection on the
primary (fuse) and secondary
(fuse and thermal overload)

5 A

L

2.5 A   25 A

U_n

Supply

N

300 VA

SELV source
(e.g. BS EN 61558–2–6 transformer)

Fused splitter unit

Means of
earthing

Luminaires wired individually on radial circuits

Figure 13.7   A practical solution to extra-LV lighting

The requirements of Regulation Group 559.10 apply not only to highway power supplies and street furniture, but also to outdoor lighting installations in general, such as those for parks, car parks, gardens and places open to the public, sporting areas and illumination of monuments and floodlighting (Regulation 559.3 refers). They do not apply to equipment of the owner or operator of a system for distribution of electricity to the public.

Designers involved in this type of installation will find reference to the Institution of Lighting Engineers' publication GPO3 *Code of Practice for Electrical Safety in Public Lighting Operations* essential.

13.12.2   Protection against electric shock

The protective measures of obstacles and placing out of reach are precluded for general application by Regulation 559.10.1. However, where maintenance work is restricted to skilled persons specially trained for the task and where LV overhead lines are placed beyond 1.5 m from street-located equipment, the protective measure of placing out of reach may be employed for those overhead lines.

The general requirement of Regulation 416.2 (requirements for barriers and enclosures) is not satisfied by the street furniture door. Regulation 559.10.3.1 goes on to call for an intermediate barrier (to IP2X or IPXXB) to be provided that is removable only by use of a tool. Additionally, doors located in such equipment at less than 2.5 m above ground level are to be locked with a key or secured by the use of a tool. Similarly, where a luminaire is located at less than 2.8 m above ground level, access to the light source must be possible only after removing a barrier or enclosure requiring the use of a tool.

The protective measures of non-conducting location, earth-free equipotential bonding and electrical separation are precluded by Regulation 559.10.2. The acceptable measures, therefore, are ADS or double or reinforced insulation. Where ADS is used, Regulation 559.10.3.1 states that metalwork, including structures, not connected to, or forming part of, the street furniture or equipment need not be bonded to the MET, though in other locations such parts may be considered to be extraneous-conductive-parts. Regulation 559.10.3.2 recommends that, in places such as telephone kiosks, bus shelters and town plans, additional protection by an RCD having the characteristics specified in Regulation 415.1.1 (see Section 5.7 of this Guide) is provided to lighting arrangements and similar equipment.

Where protection against electric shock is provided by double or reinforced insulation, Regulation 559.10.4 demands that no protective conductor be provided and that no conductive parts of the lighting column, street furniture or street-located equipment be intentionally connected to the earthing system.

As given in Regulation 559.10.3.3, the maximum disconnection time for circuits feeding fixed equipment is 5 s for compliance with Regulation 411.3.2.3 (TN systems) or 411.3.2.4 (TT systems). However, where socket-outlets are provided in or on the street furniture for use, for example, by maintenance personnel, the disconnection time for a 230 V circuit would need to be within 0.4 s in a TN system or 0.2 s in a TT system for compliance with Regulation 411.3.2.2.

13.12.3   Isolation and switching

Regulation 559.10.6.1 contains a provision for where isolation and switching in a TN system is intended to be carried out only by skilled or instructed persons and provisions are made to allow precautions to be taken to prevent equipment being inadvertently or unintentionally energized. Under these conditions, the regulation permits a suitably rated fuse carrier to be used as the means of switching the supply on load and as the means of isolation. Regulation 559.10.6.2 states that the ED's approval is first to be obtained where the ED's cut-out is used as a means of isolation for a highway power supply (as shown in Figure 13.8).

However, where more than one circuit emanates from a distribution board, a main linked switch or circuit-breaker should be installed for compliance with Regulation 537.1.4.

Some manufacturers market suitable cut-outs and double-pole switchfuse units for street furniture, and some units are available with built-in facilities for testing purposes and with lockable covers. One such unit has many features, including live parts made 'dead' on removal of cover, fuse insertion under load conditions restricted, separate sealing chamber for incoming supply, robust design using insulating material and facilities for locking and testing of Z_e without the necessity of removing covers.

13.12.4   Cable installation and identification

It is important that records are kept and maintained of the position and depth of underground cables supplying highway power supplies and other types of external lighting in order to satisfy the statutory requirements of, amongst other things, the Electricity at Work Regulations 1989.

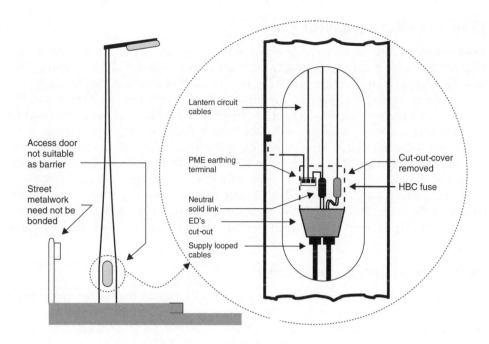

Figure 13.8   Street lighting column connections

Cable installations should be marked with ducting, cable tiles and/or marker tape and must meet the other requirements of Regulation 522.8.10. Where not installed in conduits or ducts giving an equivalent degree of mechanical protection cables must incorporate an earthed metallic armour and/or metallic sheath.

Cable identification must be distinct from the marking of other services, and advice on colour coding may be obtained from the National Joint Utilities Group and the Institution of Lighting Engineers' publication *Code of Practice for Electrical Safety in Public Lighting Operations*.

13.13   Security lighting

In recent years there has been a much increased use of security lighting, both in the industrial and commercial fields and in the domestic market. Generally, these installations involve the use of a passive infra-red sensor to detect movement by sensing heat and an integral or separate time-switch and/or photoelectric sensing device, and are often connected with a manual override switch. In addition to such devices, a means of isolation and a means for switching off for mechanical maintenance are required in order to, for example, change a lamp safely. These two functions, together with functional switching, may, if desired, be combined in one device, suitably labelled, located in some convenient place within the confines of the secure area. If necessary, the switch may be labelled to indicate that it should only be switched off for electrical and nonelectrical maintenance work to be carried out. A suitable form of words may be **Security lighting – Switch off only for maintenance**.

13.14   Welding equipment

Electric welding can often cause problems with the general electrical installation and, in particular, with protective conductors not associated with the circuit supplying the welding apparatus. Stray currents from the welder can find their way back to the supply transformer via a number of parallel paths, including those provided by other circuits. The problem can arise from poor equipment not properly maintained and from unsatisfactory working practices. Poor insulation on the welder return path conductor, poor connections, not placing the return lead as close as possible to the work being undertaken and incorrect earthing of the welding circuit are all major contributors to the problem. Cases have come to light where such stray currents have been sufficient to 'burn out' protective conductors of other circuits. Clearly, where welding apparatus is used on a day-to-day basis, regular cheeks on the continuity of protective conductors will be essential. The HSE guidance document HS(G)118 *Electrical safety in welding* addresses the use of welding apparatus.

13.15   Entertainers' equipment

Whilst strictly not specifically called for in BS 7671, socket-outlets for entertainer's equipment should be protected by an RCD with a residual operating current $I_{\Delta n}$ of not more than 30 mA and operating time not exceeding 40 ms at a residual current of $5I_{\Delta n}$. We have all seen examples of musical amplification equipment with wrongly connected plugs, loose terminals, frayed flexes, mismating of flex connectors and earthed protective conductors deliberately disconnected to prevent feedback on the audio system.

Even with RCD protection, it is essential that such installations are routinely maintained and regularly inspected and tested, and the licensing authority would normally require such procedures. Care must be taken in selecting RCDs where there are also electronic dimmers in circuits controlling lighting effects for example. Some RCDs are affected by such electronic devices, and in extreme cases can be rendered ineffective. Manual testing of RCDs is generally required quarterly; but, in such locations, consideration should be given to this test being carried out before every 'performance'. Those involved in design and maintenance of this type of installation should become familiar with HSE publications Guidance Note GS 50 *Electrical safety at places of entertainment* and booklet IND(G)102L *Electrical safety for entertainers*.

13.16   Generator sets

Section 551 of BS 7671 gives particular requirements for generating sets. The scope of Section 551 is given in Table 13.3. It does not apply to self-contained extra-low voltage source and load for which a specific product exists that includes electrical safety requirements (Regulation 551.1 refers).

Regulation 551.2.1 calls for the means of excitation and commutation to be appropriate for the intended use and such as not to impair the safe functioning of other sources. Regulation 551.2.2 demands that the prospective fault current be assessed for every independent source (or combination of sources), and also requires that the short-circuit capacity of protective devices be suitable for all the sources, or combination of a number of sources, likely to be used to supply the installation.

Table 13.3 Regulation Group 551.1: scope – LV and extra-LV generator sets

Aspect	Regulation	Application embraced in the scope
Scope – general	551.1	Generating sets supplying installations which are not supplied by the public supply Generating sets supplying installations as an alternative to being supplied by the public supply Generating sets supplying installations in parallel to the public supply Any appropriate combination of the above The note to this regulation states that the requirements of the ED must be ascertained before a generator set is installed in an installation connected to the public supply. The note also states that procedures for informing the ED that a generator has been connected in parallel with the public supply are given in the *Electricity Safety Quality and Continuity Regulations 2002* and that the ED's requirements for the connection of units rated at up to 16 A are given in BS EN 50438.
Scope – types of power source	551.1.1	Combustion engines Turbines Electric motors Photovoltaic cells Electrochemical accumulators Other suitable sources
Scope – electrical characteristics	551.1.2	Mains-excited and separately excited synchronous generators Mains-excited and self-excited asynchronous generators Mains-commuted and self-commuted static convertors with or without bypass facilities
Scope – utilization of generators	551.1.3	Supplies to permanent installations Supplies to temporary installations Supplies to portable equipment which is not connected to a permanent fixed installation Supplies to mobile units (Section 717 also applies)

Where a generating set supplies an installation, either as a sole source or as a standby, Regulation 551.2.3 requires that the capacity and operating characteristics must be such as not to be a danger or cause damage on connection or disconnection of the load. A means of load shedding must be provided where loading is beyond the generator capacity.

Extra-low voltage systems may be supplied from more than one source; where this is so, Regulation 551.3.1 requires that either all sources must be SELV or, alternatively, all sources must be PELV, and that the requirements of Regulation 414.3 (acceptable SELV and PELV sources) must be met for the sources. Where one of the sources does not comply with Regulation 414.3, then the requirements relating FELV must be applied (Regulation 411.7 – shock protection must be provided as specified in Regulations 411.7.1 to 411.7.3).

Regulation 551.3.2 stipulates that where it is necessary to maintain a supply to an extra-low voltage system when one or more parallel sources fail, it is essential that the remaining source is capable of supplying the load on its own. Additionally, provision must be made so that, on the loss of the LV supply to the extra-low voltage system, no danger or damage to other extra-low voltage systems can result.

With regard to fault protection, Regulation 551.4.1, stating the obvious, calls for this to be provided in respect of each source that may be run independently or combination of sources. This means, for example, that an installation fed by a standby generator must have suitable protective devices for automatic disconnection of supply bearing in mind the likely reduced fault level compared with that available on the public network supply. The regulation also requires steps to be taken so that, where fault protection is achieved in different ways according to the active sources of supply, there will be no influence or condition that could impair the effectiveness of the fault protection. A note to the regulation explains that this might require, for example, use of a transformer for electrical separation of parts of the installation using different earthing systems.

Regulation 551.4.2 identifies the need to make sure that fault protection, where it relies on the use of an RCD, remains effective for every intended combination of sources of supply. For example, one of the reasons why the winding of an a.c. microgenerator connected in parallel with the mains supply must not be earthed is that this could prevent the operation of an RCD in the circuit connecting the microgenerator to the consumer unit of the installation.

Regulation 551.4.3.1 demands that protection by ADS must be provided in accordance with Section 411, except as modified by Regulation Groups 551.4.3.2, 551.4.3.3 or 551.4.4.

Generators for standby systems must not rely on the public supply means of earthing, as indicated in Regulation 551.4.3.2.1. A suitable independent earth electrode must be provided.

Where installations incorporate static convertors, Regulation 551.4.3.3.1 stipulates that, in the case of fault protection which relies on automatic closure of a bypass switch and the protective devices on the supply side do not disconnect within the time required by Section 411, supplementary bonding must be carried out on the load side of the convertor (as per Regulation 514.2), and the following condition must be met:

$$R \leq \frac{50}{I_a}$$

where R is the resistance of the supplementary bonding conductor and I_a is the maximum fault current which can be supplied by the static convertor alone, for a period up to 5 s.

For parallel operation with the public supply, Regulations Group 551.7 also applies.

Regulation 551.4.3.3.2 calls for precautions to be taken so that the correct operation of protective devices is not impaired by d.c. generated by a static inverter, or by associated filters. Alternatively, selection of devices must take account of such d.c. For example, regarding the selection of RCDs, depending on the level and form of d.c. components, an RCD of Type B to IEC/TR 60755 Edition 2 may be required. This should not be necessary where a generator is, by construction, not able to feed d.c. into the electrical installation. The need or otherwise for a Type B RCD should be confirmed by reference to the installation instructions or by the supplier of the generator.

Additional requirements for protection by automatic disconnection apply where the installation and generating sets are not permanently fixed. Where a generating set which is not fixed is portable or is intended to be moved, Regulation 551.4.4.1 demands that protective conductors between separate items of equipment must be selected in accordance with Table 54.7 or incorporated in suitable cord or cable. Regulation 551.4.4.2 requires that, irrespective of the system type, an RCD (30 mA or less) must be employed in accordance with Regulation 415.1 to protect every circuit.

For protection against overcurrent, Regulation 551.5.1 simply calls for the means of detecting overcurrent in the generating set, as opposed to the conductors, to be located as near as practicable to the generator terminals.

As set out in Regulation 551.5.2, for a generating set operating in parallel, either with the public supply or with another generating set, the circulating harmonic currents must be limited to the thermal rating of the conductors. There are five options for limiting the circulating harmonic currents:

- generating sets with compensation windings;
- provision of suitable impedance in connection of the generator star points;
- provision of interlocking switches to interrupt the circulatory circuit, but which do not impair fault protection;
- provision of filters;
- other suitable means.

There are additional requirements for installations where the generating set provides a supply as a switched alternative to the public supply (standby systems). Regulation 551.6.1 requires precautions to be taken to prevent the generating set operating in parallel with the public supply by one or more of the options available, which include mechanical interlock with one key, electrical-controlled changeover device, three-position break-before-make changeover switch, or other means providing equivalent security.

For TN systems, where the neutral is not distributed, any RCD must be positioned so that malfunction due to any parallel neutral earth path is avoided, as indicated in Regulation 551.6.2.

Regulations 551.7.1 and 551.7.2 are concerned with making sure that protection against thermal effects and overcurrent in accordance with Chapters 42 and 43 (respectively) of BS 7671 remain effective where a generating set is used as an additional source of supply in parallel with the public supply. In particular, for the situation where a generator is installed on the load side of all the overcurrent protective devices for a final circuit, Regulation 551.7.2 requires the following condition to be met:

$$I_z \geq I_n + I_g$$

where I_z is the current-carrying capacity of the final circuit conductors, I_n is the rated current of the protective device of the final circuit and I_g is the rated output current of the generating set. The object of the above condition is to prevent the conductors of the final circuit being overloaded as a result of current from the generator, which is not 'seen' by the protective device of the circuit.

As called for in Regulation 551.7.3, where a generating set is to run in parallel with the public supply, care is necessary to avoid the adverse effects, to the public supply and other installations, in consideration of:

- power factor
- voltage changes
- harmonic distortion
- unbalance
- starting
- synchronizing
- voltage variation.

The use of automatic synchronizing systems which consider frequency, phase and voltage is preferred.

Regulation 551.7.4 requires means of automatic switching to be provided so that, in the event of loss of public supply or deviation in voltage or frequency of supply, the generating set is disconnected from the supply. For a generating set with output not exceeding 16 A, the regulation requires that the settings for the sensitivity and operating time of the automatic switching comply with BS EN 50438 – *Requirements for the connection of micro-generators in parallel with public low-voltage distribution networks* and that these settings, together with the type of protection, are agreed with the ED. For a static convertor, Regulation 551.7.4 also requires that the means of automatic switching to disconnect the convertor from the public supply is situated on the load side of the convertor.

Regulation 551.7.5 requires means to be provided to prevent connection of the generating set to the public supply if the generator voltage or frequency is on excursion outside of the values required by Regulation 551.7.4 (for the settings of the automatic switching).

Means of isolation, accessible to the ED at all times, must be provided for the generating set, as called for in Regulation 551.7.6. If the generating set has an output exceeding 16 A, then the accessibility of the means of isolation must comply with national rules and the ED's requirements. If the output of the generating set is not more than 16 A, then the accessibility of the means of isolation must comply with BS EN 50438.

Regulation 551.7.7 goes on to call for the requirements of Regulation Group 551.6 to apply also in the case where the generator set is also to serve as standby set.

Guidance on connecting a microgeneration system to a domestic or similar electrical installation (in parallel with the mains supply) is given in *Best Practice Guide No 3*, produced by the Electrical Safety Council in association with leading industry bodies for the benefit of electrical contractors and installers, and their customers, which may be downloaded free from www.esc.org.uk.

14

Safety services

14.1 Safety services: general

BS 7671 does not in itself call for safety services to be provided, but it does, where such services are installed, state additional requirements relating to the supplies, equipment and circuits for these services. The general requirements in Regulations 560.5.1–560.5.3 call for safety service supplies to be selected so that a supply of adequate duration is maintained. Where the service is required to operate under fire conditions, all equipment must be capable, by construction or by erection, of providing fire resistance for an adequate duration. In other words, the equipment must withstand the fire conditions for sufficient time for the safety service to perform its intended function. The last of these three regulations states a preference that the protective measure against electric shock does not automatically disconnect on first fault condition and where an IT system is used it must be provided with continuous insulation monitoring with provision for audible and visual indications of a first fault condition. As ever, it is important to arrange equipment so as to accommodate inspection, testing and maintenance, and Regulations 560.7.9–560.7.12 list diagrams, drawings and other information that must be provided to facilitate these operations and the use of the safety services.

Whilst BS 7671 does not stipulate where and for what purpose safety services may be required, there may be, in fact, a statutory requirement for a particular safety service to be provided. The Regulatory Reform (Fire Safety) Order 2005, for example, requires, in certain circumstances, a means of giving warning in the case of fire. Hospitals, schools, colleges, public places of entertainment, hotels, boarding houses, offices, factories and shops are all examples where a safety service may be needed to afford compliance with statutory requirements.

This chapter only addresses the additional regulatory requirements of BS 7671 relating to these services, and reference to the appropriate British Standard CP will be necessary to identify *all* the essential requirements relating to a particular safety service. Examples of safety services include:

- fire detection and alarm systems in buildings (BS 5839);
- emergency lighting (BS 5266);

A Practical Guide to The Wiring Regulations: 17th Edition IEE Wiring Regulations (BS 7671:2008)
Fourth Edition Geoffrey Stokes and John Bradley
© Geoffrey Stokes and John Bradley. Published by John Wiley & Sons, Ltd

- sprinkler systems (BS 5306);
- fire extinguishing systems (gas, foam, powders, etc.) (BS 5306);
- fire hydrant pumps;
- automatic door closing mechanisms (BS 5839);
- fire fighting lifts;
- fire services communication systems;
- smoke ventilation systems;
- essential medical systems;
- industrial safety systems.

The designer will find it advantageous, and in some cases obligatory, to discuss the design proposals with the local Fire Authority, the Health and Safety Executive and the building insurers. Any input to the design proposals by others should be welcomed but not be regarded as being the total design requirements.

14.2 Common sources

Regulations 560.6.1–560.6.12 lay down the requirements for safety service sources and give details of where and how the equipment is to be erected. The acceptable sources referred to are batteries, a generator capable of independent operation and separated independent feeders from a supply network, all of which must meet the specified requirements (Figure 14.1). Where separated independent feeders from a supply network are employed, it must be confirmed that both sources are unlikely to fail concurrently.

The source must always be fixed in position and installed in a suitable location accessible to skilled and instructed persons only. Installation of the source must be such that failure of the normal supply does not affect the safety source. A safety source must not be used to supply equipment other than that directly associated with the safety service,

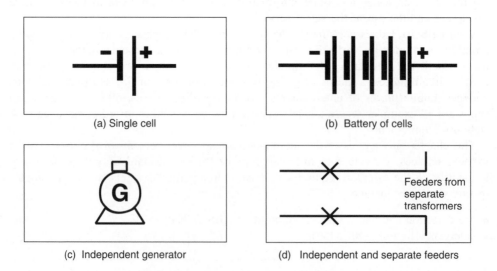

(a) Single cell

(b) Battery of cells

(c) Independent generator

(d) Independent and separate feeders

Feeders from separate transformers

Figure 14.1   Suitable safety sources

unless the availability of the safety services is not impaired as a result, and a fault in a circuit for purposes other than safety services does not cause interruption of a circuit for a safety service. This usually necessitates automatic off-loading of nonsafety service equipment and selectivity between protective devices. Where more than one source is available (e.g. public supply and generator) they may supply standby systems provided that, in the event of failure of one, the energy available from the other is sufficient to start and maintain operation of the safety service(s). Where necessary, nonessential load shedding will be required in order to reduce loading to a level within the capability of the remaining source. Figure 14.2 shows a typical arrangement for two sources and the load-shedding technique. It is essential that sources are adequately ventilated so as to prevent smoke, fumes and other exhaust gases, as required by Regulation 560.6.3.

Although there is no maximum time delay stated in BS 7671 for the standby supply to take over from the failed main feeder supply, this obviously needs to be as short as possible. Where, for example, emergency lighting is supplied, BS 5266 requires that the emergency lighting should reach the required level of illuminance within 5 s, or this time limit may be increased to 15 s with the agreement of the appropriate enforcing authority where the building is only likely to be occupied by persons familiar with the building layout. With modern equipment and techniques, very short changeover times can be achieved.

As mentioned earlier, BS 7671 expresses a preference for the safety service source to be such that automatic disconnection does not occur on first fault. One obvious solution would

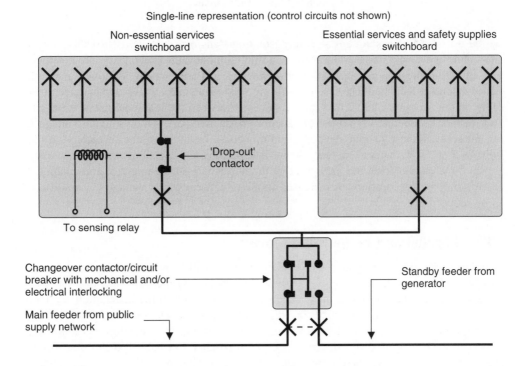

Figure 14.2 Typical standby arrangements

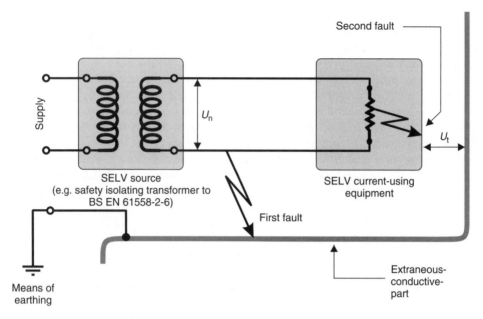

U_t is the voltage which a person would be subjected to in the event of a second fault.

Figure 14.3   SELV circuit

be the use of a SELV source (e.g. safety isolating transformer to BS EN 61558-2-6). As shown in Figure 14.3, the development of a first fault between a secondary live conductor and an extraneous-conductive-part, itself connected to Earth, would effectively make the SELV circuit a localized TN system with the faulted secondary live conductor earthed. Only on the occasion of a second fault condition, between the load equipment metal case (not earthed, deliberately) and the extraneous-conductive-part, would a touch voltage U_t appear between these conductive parts, and this could be, at worst, the secondary no-load voltage (e.g. 50 V). Other than for special installations and locations, this touch voltage (up to 50 V) could, from the viewpoint of the shock hazard, be tolerated indefinitely, at least in theory. For locations of increased shock risk, the nominal voltage U_n would need to be reduced accordingly.

14.3   Parallel and nonparallel sources

Regulation 560.6.7, by reference to Regulation 551.6, calls for safety service sources not capable of parallel operation to be effectively interlocked by mechanical and/or electrical means to ensure that not more than one source is connected to the installation served at any one time (see Regulation 551.6.1). Regulation 560.6.8 does, however, recognize that sources may be connected in parallel where that is the designer's intention. Particular care is needed in the selection of sources to operate in parallel, to establish their compatibility (e.g. voltage, frequency, rated current and impedance). The general requirements relating to protection against overcurrent and fault protection must be met in all cases, although where a source is used solely for a safety service, protection against overload may be omitted (as indicated in Regulation 560.7.3) provided the occurrence of an overload is indicated.

The general requirements for isolation and switching apply, and particular care is required to ensure that total isolation of the safety service from its sources can be achieved. Where more than one device is necessary to effect isolation, an appropriately worded notice is required (by Regulation 537.2.1.3) to be affixed on or near every isolating device, identifying devices and warning personnel of the correct isolation procedure.

In circumstances where there may be circulating currents (e.g. third harmonics and other triplens) in the connection between neutral points of sources, precautions need to be taken to obviate harmful effects.

14.4 Circuit and equipment requirements

A safety service circuit is required by BS 7671 to be independent of any other circuit (Regulation 560.7.1) and to be arranged so that a fault developing in any other circuit, including that of another safety service, should not be capable of adversely affecting the intended function of the circuit. Similarly, modifications or interventions to another circuit must not impair the correct functioning of the safety service circuit. Safety service circuit conductors must have adequate fire resistance appropriate for the locations through which they pass. Regulation 560.8.1 specifies the types of cable that may be used for safety services required to operate in fire conditions, although the requirements of the British Standard relevant to the particular safety service must, of course, also be complied with. In addition, Regulation 560.7.2 requires that circuits of safety services must not pass through any zones exposed to explosion risk (external influence BE3). With the exception of wiring for a fire rescue service lift, circuits of safety services must not be installed in a lift shaft or other flue-like opening (Regulation 560.7.8).

Overcurrent protective devices must be selected and installed so that an overcurrent in one circuit does not affect other circuits (Regulation 560.7.4). To this end, adequate circuits will need to be provided together with generous allowances for design current in the selection of protective devices and circuit conductors in order to avoid unwanted automatic disconnection. As called for in Regulation 560.7.5, switchgear and controlgear must be located in areas accessible only to skilled and instructed persons and must be clearly identified as to purpose. Similarly, alarms, controlling devices and indication equipment must also be clearly identified.

Where an item of equipment is supplied by two different circuits (e.g. normal supply and standby source), Regulation 560.7.6 requires that, should a fault occur in one circuit, the protection against electric shock must not be impaired nor the correct functioning of the other circuit affected in any way.

14.5 Protection against overcurrent and electric shock under fault conditions

As indicated in Regulation 560.6.7 and Regulation Group 560.6.8 (including through references to Section 551 in both cases), the general requirements relating to protection against electric shock under fault conditions and fault current apply to safety service circuits. Care is needed, particularly in the case of standby generator supplies, that the fault levels from all sources are adequate to operate protective devices. For example, the fault level of a standby generator may be much lower than that of a supply from the public network and, additionally, the generator prospective fault current may reduce rapidly. The designer would wish to consider these matters very carefully, and consultation with the generator manufacturer at an early stage will be advantageous. Similarly, a rigorous analysis of fault levels is essential where more than one source can simultaneously feed a fault.

15

The smaller installation

15.1 Scope

The guidance given in this chapter is limited in scope to the 100 A single-phase installation supplied from an ED's LV network and includes supplies forming part of TN-S, TN-C-S or TT systems. Consideration of special installations or locations is not addressed here, but Chapter 16 deals in some depth with these special requirements.

The general principle that every circuit and every item of equipment is required to be designed holds good for the smaller installation. Circuit designs for the smaller installation may utilize those given in the IEE *On-Site Guide* or the NICEIC *Domestic Electrical Installation Guide*, which, if implemented, will produce a circuit design that complies with BS 7671 but may not produce the most economic solution.

15.2 The IEE *On-Site* Guide and the NICEIC *Domestic Electrical Installation Guide*

The IEE *On-Site Guide* and the NICEIC *Domestic Electrical Installation Guide* are for competent electricians. Both publications are limited in scope to domestic installations (and small industrial and commercial installations in the case of the *On-Site Guide*) where circuit distribution is from distribution board(s) located near the ED's cut-out. Conventional circuits are covered, which are a legacy of the 15th edition of the IEE Wiring Regulations Appendix 5 'standard' circuits.

15.3 User's requirements

The first task for the designer is to ascertain the user's requirements and he/she is commonly called upon to assist in formulating and identifying what the customer actually wants. As a guide, Table 15.1 sets out some of the technical points which need to be clarified at an early stage. The table should not be regarded as exhaustive, as there may well be other points which need addressing depending on the particular circumstances.

A Practical Guide to The Wiring Regulations: 17th Edition IEE Wiring Regulations (BS 7671:2008)
Fourth Edition Geoffrey Stokes and John Bradley
© Geoffrey Stokes and John Bradley. Published by John Wiley & Sons, Ltd

Table 15.1   User requirements: points of clarification

Ref	Clarification required for:	Some common possible items to consider
A	Type of wiring system(s)	PVC-insulated and -sheathed cables PVC-insulated cables in steel conduit PVC-insulated cables in nonmetallic conduit PVC/PVC-insulated and -sheathed cables with nonmetallic conduit switch-drops PVC-insulated cables in steel trunking PVC-insulated cables in nonmetallic trunking Armoured cables Mineral-insulated cables Cable management system(s)
B	Installation methods	Flush wiring system(s) Surface run wiring system(s) Other environmental conditions
C	External influences	Identification of the required degrees of ingress protection (e.g. of water, dust) Identification of the ambient temperature, heat sources, and corrosive and polluting substances Identification of impact, vibration and other mechanical stresses Identification of the presence of flora and mould growth Identification of the exposure to solar radiation Identification of structure stability
D	Loadings	Load assessment of fixed current-using equipment together with all anticipated portable appliances Load assessment based on the rated supply
E	Luminaires	Identification of all lighting and switching requirements.
F	Other fixed electrical equipment	Identification of all fixed electrical equipment in terms of loadings and siting. Equipment may, for example, include: cooking equipment, immersion heater, boiler, waste-disposal unit, water heaters, space heating Information technology equipment, shower unit, shaver sockets, intruder alarms, fire alarms, emergency lighting, etc.
G	Socket-outlets	Number, siting and rating of socket-outlets Finished colour of socket-outlets Mounting height from finished floor level
H	Switches	Number, siting and rating of switches Finished colour of switches Mounting height from finished floor level
I	Other accessories	Number, siting and rating of accessories Finished colour of accessories Mounting height from finished floor level

Table 15.1   *(continued)*

Ref	Clarification required for:	Some common possible items to consider
J	Main distribution	Type and siting required Preferred type of protective devices (fuses, circuit-breakers and RCDs)
K	Special concerns	User's capabilities (e.g. disabled, infirm, elderly)

15.4   Wiring systems

The selection of the wiring system will depend to a large extent on the type of premises in which the installation is located and whether the building exists or is in the process of construction. For the small installation in domestic premises, thermoplastic (PVC) or thermosetting insulated and sheathed cables are likely to be the first choice because of the cost implications, whereas thermoplastic (PVC) or thermosetting insulated cables in conduit and/or trunking may be a more appropriate solution for small commercial premises. Often, a number of different types of wiring systems will be employed in a building, each with its own advantages and disadvantages. Chapter 10 of this Guide gives further information with regard to wiring systems.

15.5   Electricity distributor's requirements

As with all installations supplied from the public supply network, the ED will require the installation to be constructed to such a standard as to allow them to meet their statutory obligations under the *Electricity Safety, Quality and Continuity Regulations 2002*, as amended. The ED would be particularly interested in loading, earthing and protective bonding conductors and 'meter tails'.

15.6   Assessment of supply characteristics

As addressed in Chapter 4 of this Guide, an assessment of the general characteristics of the supply is necessary for all installation design. In addition to the nominal voltage, frequency and number of phases, the prospective fault current and external line–earth loop impedance are also required to be assessed together with the current rating of the service. This information is available from the ED, and the designer should obtain this at an early stage. An assessment of maximum demand will also be a necessary task, and it is essential that the supply is adequate for this demand, taking into account any allowable diversity. The most commonly used overcurrent protective device in the ED's cut-out is a BS 1361 Type II fuse. Cartridge fuses used may be 60, 80 or 100 A, but, if not identified, the highest rating must be assumed when considering overcurrent protection.

The prospective fault current (short-circuit and earth fault) for supplies up to 100 A may be taken as 16 kA generally, but the Electricity Association's Engineering Recommendation P25, available from the Energy Networks Association, provides a method for calculating the fault-level attenuation. Except for supplies provided in London, fault-level

attenuation can be correlated to the length of service cable from the public highway to
the intake position (i.e. the origin of the installation). The length of service cable from the
pavement (or roadway) to the ED's cut-out and the prospective fault current attenuation
are related as follows.

Length of service cable (m)	Up to 25 mm^2 Al and 16 mm^2 Cu (kA)	35 mm^2 Al and 25 mm^2 Cu (kA)
5	10.8	12.0
10	7.8	9.3
15	6.0	7.0
20	4.9	6.2
25	4.1	5.3
30	3.5	4.6
35	3.1	4.0
40	2.7	3.6
45	2.5	3.3
50	2.2	3.0

It is important to note that the lengths given above relate only to the final service cable.
No part of the roadway cable should be included, since the ED may, from time to time,
alter the network.

15.7 'Meter tails'

Usually, the only overcurrent protection afforded to the 'meter tails' is that provided by
the ED's cut-out fuse. In every case the designer should check with the ED for their
requirements for the cross-sectional area of the 'meter tails' and the maximum length of
'tails' which they will accept – commonly not more than 3 m. As a guide, the minimum
copper cross-sectional area of these conductors (to BS 6004) for Installation Method 20
(clipped direct), assuming no rating factors for ambient temperature, thermal insulation
or grouping, are given below:

Overcurrent protective device	Single-phase	Three-phase
60 A BS 1361 Type II rated	10 mm^2	16 mm^2
80 A BS 1361 Type II rated	16 mm^2	16 mm^2
100 A BS 1361 Type II rated	25 mm^2	25 mm^2

In all circumstances, the designer should check with the ED to confirm that the
above-quoted cross-sectional areas meet with their particular needs, which may exceed
the regulatory requirements.

In cases where the current rating of the cut-out fuse is unknown, the highest value should
be assumed, and it is generally considered good practice to install 'meter tails' for the

highest value in any case to allow for any fuse replacement errors. The cross-sectional areas given above may need to be increased where the 'meter tails' are run in close proximity to other load-carrying conductors and an appropriate grouping factor applied (see Table 4C1 of Appendix 4 of BS 7671). It should be noted in this respect that direct-reading electricity meters (i.e. those not employing current transformers) will not normally accept cables of cross-sectional area greater than 35 mm^2.

Mechanical protection of 'meter tails' is important, as is adequate support. Generally, thermoplastic (PVC) insulated and sheathed cables will be used, and care must be taken to terminate these into equipment so that the sheath is not removed excessively, leaving exposed PVC insulation only. It is sometimes necessary or desirable to enclose 'tails' in trunking using thermoplastic (PVC)-insulated cables only. Where this method is adopted, care must be taken to ensure that mechanical damage does not result from unsmooth edges of trunking and that the trunking system completely encloses the 'tails'. Additionally, where trunking is metal it must be earthed for fault protection, and this can only be effected by making use of the ED's cut-out fuse. The ED's express agreement must be obtained prior to adopting this method of protection of the 'tails' against mechanical damage. Metal trunking for the purpose of containing 'meter tails' is generally not an option for TT systems because the line–earth loop impedance is normally too high to effect fault protection (for the trunking).

It is important to note that no consumer equipment, such as a main switch, should be fixed to the ED's meter board or within their meter cabinet.

15.8   System earthing arrangements

System earthing arrangement is addressed in Chapter 12 of this Guide, but the following information will serve as a reminder for those contemplating the design of the smaller installation. Only TN-C-S, TN-S and TT systems are considered here, as TN-C is unlikely to be encountered on the smaller installation and IT systems are not available on the public supply network.

Figure 15.1 shows a typical arrangement for a TN-C-S system single-phase supply cut-out, 'meter tails', earthing and main protective bonding connections. Figures 15.2 and 15.3 indicate similar arrangements for TN-S and TT systems respectively. A supply forming part of a TN-C-S system (e.g. PME supply) will normally have a maximum external line–earth loop impedance of 0.35 Ω, compared with 0.80 Ω for a TN-S system (e.g. cable sheath means of earthing).

Irrespective of system type and earthing arrangements, the cross-sectional area of the earthing conductor must either be calculated using the adiabatic equation given in Regulation 543.1.3 or selected to meet the requirements of Table 54.7 of BS 7671. Since the length of such a conductor is generally relatively short, the favoured method of selection is by the use of the table. For a 100 A supply with line conductor(s) cross-sectional area (copper) of 25 mm^2 or 35 mm^2 an earthing conductor cross-sectional area of 16 mm^2 (copper) will be required. This cross-sectional area will also afford compliance with Table 54.1 for buried earthing conductors where protection against corrosion has been provided. In the case of a TT system, Regulation 411.5.3 requires that, where an RCD is used for fault protection (as is almost invariably necessary), Equation (15.1) is met (as well as the

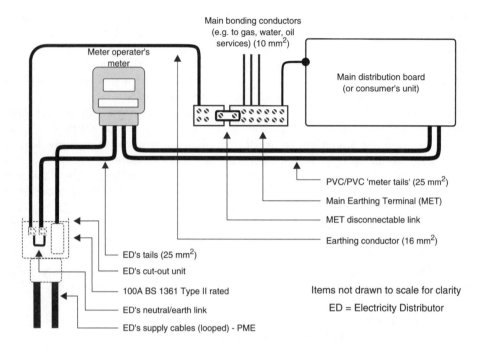

Figure 15.1   Typical arrangement for connection of a smaller installation on a TN-C-S system

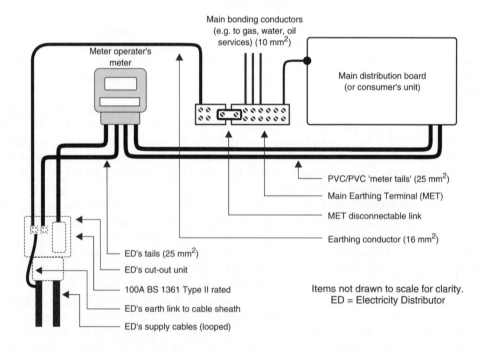

Figure 15.2   Typical arrangement for connection of a smaller installation on a TN-S system

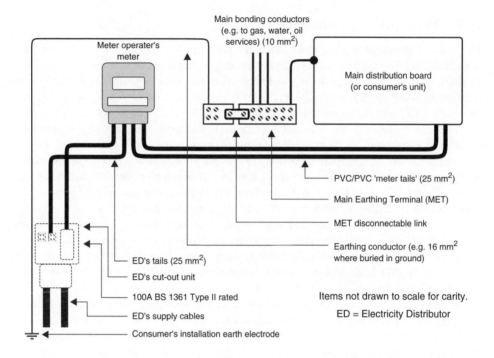

Figure 15.3 Typical arrangement for connection of a smaller installation on a TT system

required disconnection times having to be achieved in the installation):

$$R_A \times I_{\Delta n} \leq 50 \text{ V} \tag{15.1}$$

where R_A is the sum of the resistances of the installation earth electrode and the protective conductor(s) connecting the exposed-conductive-parts to the electrode and $I_{\Delta n}$ (A) is the rated residual operating current of the RCD.

In practice, Equation (15.1) is not difficult to meet. Generally, the resistance of the installation earth electrode is typically of the order of 30–150 Ω and constitutes the major contribution to R_A. To illustrate the point by example, take an earth electrode resistance of 149 Ω and a resistance of 1 Ω for the protective conductors. To meet Equation (15.1), the rated residual operating (tripping) current of the RCD would need not to exceed 0.33 A. Clearly, selecting a 300 mA RCD, at the front end, will more than satisfy this requirement.

It is important to recognize that an ED may refuse to provide an earthing facility where, for example, they cannot be satisfied that statutory requirements relating to PME (contained in the *Electricity Safety, Quality and Continuity Regulations 2002*, as amended) will be met. It is essential, therefore, that liaison with the ED is instigated at an early stage of the design to establish whether or not a means of earthing can be provided. When offered an earthing facility by the ED, it is the designer's responsibility to ensure that such a facility is suitable for the particular installation.

15.9   Main protective bonding

The general requirements for main protective bonding are discussed in Chapter 12 of this Guide, and the smaller installation has no dispensations in this respect. All incoming services, such as gas, water, oil, etc., will need to be main bonded to the MET. The bonding conductors, if not copper, must have the equivalent conductance to that of copper and must be selected to meet the requirements of Regulation 544.1.1. For TN-S and TT systems (i.e. non-PME supplies), the conductor cross-sectional area is subject to a minimum of $6\,mm^2$ and a maximum of $25\,mm^2$, and it must not be less than 50% of the cross-sectional area required for the installation earthing conductor. For a 100 A rated supply with an earthing conductor of $16\,mm^2$, the main bonding conductor cross-sectional area would need to be not less than $8\,mm^2$ and, therefore, a $10\,mm^2$ conductor will be required.

Where the installation forms part of a TN-C-S system (i.e. PME supply), the main bonding conductor cross-sectional area is related to the supply neutral as detailed in Table 54.8 of BS 7671, which stipulates a minimum of $10\,mm^2$ and a maximum of $50\,mm^2$. For a 100 A supply, a $10\,mm^2$ main bonding conductor will generally be adequate, but a check with the ED is always desirable to ensure that any particular requirements which it may have are met.

The connection of the bonding conductors to the incoming services must be made using BS 951 earthing/bonding clamps at a point as near as practicable to the entry of the service into the premises. For a water service, this point is generally immediately after the stopcock, and for the gas service it is within 600 mm of the meter if it is located within the building or at the point of entry if external. Other services should be treated in a similar way.

Where convenient and desirable, it is permissible to loop the main bonding conductors so that one conductor connects a number of incoming services. When this method is adopted, the conductor should be continuous and not cut at each connection. This is necessary to avoid any disconnection of any bonding connection (e.g. gas bond by gas fitter) affecting the connection of other services.

Further data for earthing and main bonding conductor cross-sectional areas for supplies other than 100 A are given in Section 15.10 and Table 15.2 of this chapter.

15.10   Minimum cross-sectional area of earthing and main protective bonding conductors

There are a number of regulations which place constraints on the cross-sectional area of earthing and main protective bonding conductors. Regulation 542.3.1 relates to earthing conductors buried in the ground and for most practical purposes only applies to TT and IT systems, since for TN-C-S and TN-S systems the means of earthing is provided by the ED. Regulations 543.1.1–543.1.4 give the requirements for protective conductors generally and Regulations 544.1.1 and 544.1.2 make additional demands for main protective bonding conductors. In the case of TT systems a further constraint is made on the earthing conductor, in that the product of the sum of the resistances of the earth electrode and connecting protective conductors and the minimum current causing automatic disconnection

Table 15.2 Earthing and main protective bonding conductors: minimum cross-sectional areas[a]

Line (or neutral) csa (mm²)	Earthing conductor Non-separate and separate conductors not buried or buried with corrosion and mechanical protection[b]	Main protective bonding conductor csa[c,d] Non-PME	Main protective bonding conductor csa[c,d] PME	Earthing conductor Separately buried with protection against corrosion but no additional mechanical protection[b]	Main protective bonding conductor csa[c,d] Non-PME	Main protective bonding conductor csa[c,d] PME	Earthing conductor Separately buried with no additional corrosion protection[b]	Main protective bonding conductor csa[c,d] Non-PME	Main protective bonding conductor csa[c,d] PME
4	*4 (10)[e]*	*6*	10	16	*10*	10	25	*16*	10
6	*6 (10)[e]*	*6*	10	16	*10*	10	25	*16*	10
10	*10*	*6*	10	16	*10*	10	25	*16*	10
16	*16*	*10*	10	16	*10*	10	25	*16*	10
25	*16*	*10*	10	16	*10*	10	25	*16*	10
35	*16*	*10*	10	16	*10*	10	25	*16*	10
50	25	16	16	25	16	16	25	16	16

[a] Data given in italics indicate that by use of the adiabatic equation given in Regulation 543.1.3 a smaller cross-sectional area may be achieved.

[b] Earthing conductor cross-sectional area taking account of Regulation 542.3.1 (Table 54.1) and Regulation 543.1.4 (Table 54.7).

[c] Main protective bonding conductor cross-sectional area taking account of Regulation 544.1.1 (and Table 54.8).

[d] Where TN-S and TT systems are employed, the data given for 'non-PME' should be used but where the system is TN-C-S, the larger value given in the two columns 'non-PME' and 'PME' should be used.

[e] Regulation 542.3.1 requires the earthing conductor cross-sectional area to be not less than the main bonding conductor cross-sectional area for installations with a PME supply.

of the protective device is limited to 50 V. This is not generally the overriding constraint, and, for most practical purposes, other constraints will take precedence. Table 15.2 summarizes the requirements and provides data for these conductors related to the supply line conductor for non-PME supplies and to the neutral conductors in the case of PME services.

It should be noted that, where the supply is derived from the public supply network, most EDs will stipulate the minimum line and neutral 'meter tails' that they require (generally not less than $6\,mm^2$), which will be related to the fault current protection afforded by their cut-out fuse. The smaller sizes of these conductors given in Table 15.2 are for use with connections to fire alarm systems and the like and would not generally be applicable to the installation main 'meter tails'.

15.11   Supplementary bonding

Supplementary bonding is addressed in Chapter 12 and, for special installations or locations, in Chapter 16. Generally, supplementary bonding will only be necessary in areas of increased risk of electric shock, such as bathrooms (although it may be omitted from a bathroom where certain conditions specified in Regulation 701.415.2 are met) and swimming pools. In such locations, both extraneous-conductive-parts and exposed-conductive-parts are required to be bonded together so that in the event of a fault no substantial voltage exists between these conductive parts. Generally, a conductor of $4\,mm^2$ is sufficient for most circumstances; but in any event, the bonding conductor cross-sectional area must not be less than that of the smaller cpc of any two circuits having exposed-conductive-parts requiring mutual connection. Where a bonding conductor is to connect an extraneous-conductive-part with an exposed-conductive-part, the cross-sectional area must not be less than half the cpc's cross-sectional area to the exposed-conductive-part.

Supplementary bonding conductors may be formed in part by a conductive part not forming a component of the electrical installation. For example, copper or iron pipework, provided that it is considered to be permanent and reliable, may be so used as a supplementary bonding conductor. Notwithstanding the minimum cross-sectional area of bonding conductors given above, a further constraint is placed on such conductors, in that their resistance R in ohms is limited so that $R \leq 50/I_a$ (where I_a is the operating current in amperes of the device used for fault protection – for an RCD, I_a is $I_{\Delta n}$).

15.12   Devices for protection against overcurrent and for fault protection

General guidance relating to protection against overcurrent is given in Chapter 7 of this Guide and for fault protection in Chapter 5. More often than not, a single device provides for both functions and where so used must satisfy all the relevant requirements.

Many smaller installations are served with a consumer unit to BS EN 60439-3 including Annex ZA. Subject to the following conditions, the outgoing circuit-breakers or fuses in a consumer unit complying with that British Standard have a conditional rated breaking capacity of 16 kA, which exceeds the likely prospective fault current at the origin of an installation in domestic premises. The conditions are that:

- the consumer unit is fed by a distributor's service cut-out having an HBC fuse to BS 1361 Type II, rated at not more than 100 A (confirmation of this should be obtained from the distributor or from their published notes of guidance);
- the supply is single-phase with a nominal voltage not exceeding 250 V.

Where the above two conditions are met, compliance with Regulation 434.5.1 is achieved for the rated breaking capacity of the circuit-breakers or fuses in a consumer unit, provided neither the prospective short-circuit current nor the earth fault current exceeds 16 kA.

In the smaller installation there will normally be a need to provide RCDs either for fault protection or for additional protection.

With regard to additional protection, RCDs with a rated residual operating current $I_{\Delta n}$ of not more than 30 mA and an operating time not exceeding 40 ms at a residual current of $5I_{\Delta n}$ will be required as specified in Regulation 415.1.1. Table 15.3 lists the principal applications where additional protection by an RCD is required in the smaller installation.

Generally, the whole of an installation forming part of a TT system will need to be protected by an RCD, but it is undesirable to protect the whole installation with one having $I\Delta n \le 30$ mA, since this is likely to be prone to unwanted tripping. Normally, it is necessary to provide a time-delayed RCD (say, $I_{\Delta n} = 200$ mA) at the 'front end' and 30 mA RCDs for circuits where additional protection is required (see Table 15.3), thus providing discrimination. Equipment prone to high earth-leakage (e.g. washing machines) and vulnerable items (e.g. freezers) should preferably be wired on their own dedicated circuit; for the latter, the more sensitive RCD protection should be avoided wherever possible.

Where the installation forms part of a TT system, the prospective earth fault current will be relatively low owing to the high line–earth fault loop impedance with a value of the order of 10–200 Ω not being uncommon; see Figure 15.4. The high impedance is due to a number of factors, including the fairly high resistance of the source earth electrode (e.g. 21 Ω) and that of the installation earth electrode (typically 30–150 Ω).

15.13   Devices for isolation and switching

Guidance in connection with isolation and switching is given in Chapter 8 of this Guide, and there are no dispensations for the smaller installation. Regulation 537.1.4 calls for a main linked switch or circuit-breaker to be provided for every installation. In the smaller installation, the consumer unit main switch will provide this function, as will a main circuit-breaker and, provided it is designed to meet the isolation requirements, an RCD.

15.14   Final circuit design

As mentioned earlier, every circuit has to be designed. It may be that the designer chooses to adopt the designs in the IEE *On-Site Guide* or the NICEIC *Domestic Electrical*

Table 15.3   The principal applications where RCDs will be needed to provide additional protection in the smaller installation[a]

Ref	Locations	Regulation	Requirement qualifications
A	Generally	411.3.3	Socket-outlets rated at 20 A or less for use by ordinary persons and intended for general use, except where for use under supervision of skilled or instructed persons or labelled/identified to supply a specific item of equipment
B		411.3.3	Mobile equipment with a rated current not exceeding 32 A
C		522.6.7 and 522.6.8	Cables concealed in a wall or partition at a depth of less than 50 mm from the surface, or, irrespective of depth, in a partition having an internal construction that includes metallic parts (except screws, etc.), in an installation not intended to be under supervision of a skilled or instructed person. This does not apply to cables having protection complying with Regulation 522.6.7 or 522.6.8, as applicable (such as earthed steel conduit)
D	Locations containing a bath or shower	701.411.3.3	All circuits of the location ($I_{\Delta n} \leq 30$ mA)
E	Swimming pools	702.55	Electric heating units embedded in the floor, except where provided by SELV meeting specified requirements
F	Rooms containing a sauna cabin with a sauna heater	703.411.3.3	All circuits of the sauna. Such protection need not be provided for the sauna heater unless recommended by the manufacturer
G	Floor and ceiling heating systems	753.415.1	Floor and ceiling heating systems

[a]PRCDs used to provide additional protection must have a rated residual operating current $I_{\Delta n}$ of not more than 30 mA and an operating time not exceeding 40 ms at a residual current of $5I_{\Delta n}$ (Regulation 415.1.1).

Installation Guide, which will provide for a safe design, if not the most economical. A number of conventional circuits, both radial and ring, are given in those guides.

Tables 15.4, 15.5, 15.6 and 15.7a–d of this Guide are based on tables given in the NICEIC *Domestic Electrical Installation Guide* and are reproduced by kind permission of The Electrical Safety Council. The tables give maximum circuit cable lengths for single-phase final circuits in domestic and similar installations, calculated to satisfy the requirements of BS 7671. The basis of the tables is as follows.

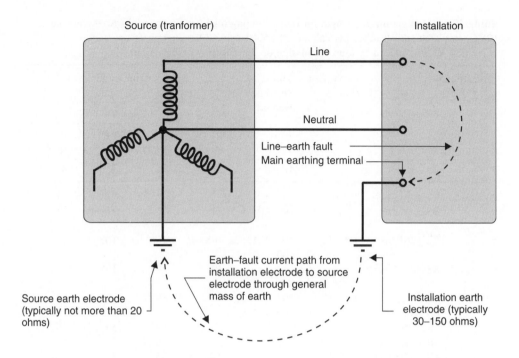

Figure 15.4 TT system earth fault path. Reproduced by kind permission of The Electrical Safety Council

Wiring system and reference method

Circuits are wired in 70 °C thermoplastic (PVC) insulated and sheathed flat twin-and-cpc cable to Table 8 of BS 6004.

Column 4 of each table gives the reference methods allowed for each combination of cable and overcurrent protective device, the reference methods being those in Table 4D5 of BS 7671 Appendix 4.

Ambient temperature

The ambient temperature is assumed not to exceed 30 °C.

Installation supply and consumer unit location

The nominal voltage U_0 of the supply is 230 V a.c. at a frequency of 50 Hz.

The installation is supplied from a TN-S system with a maximum external earth fault loop impedance Z_e of 0.8 Ω, or a TN-C-S system with a maximum Z_e of 0.35 Ω, or a TT system with an RCD for fault protection.

The consumer unit and main earthing terminal of the installation are located at the origin of the installation.

Table 15.4 Maximum cable length for radial lighting final circuits using 70 °C thermoplastic (PVC) insulated and sheathed flat cable to BS 6004 Table 8 having copper conductors. $U_o = 230$ V. Reproduced by kind permission of The Electrical Safety Council

Protective device		Cable size (mm^2)	Allowed Reference Method[a]	Maximum length[b] (m)			
				$Z_s \le 0.8$ Ω TN-S		$Z_s \le 0.35$ Ω TN-C-S	
Rating (A)	Type			RCD	No RCD	RCD	No RCD
1	2	3	4	5	6	7	8
5	BS 1361 fuse	1.0/1.0	103, 101, A,	71	71	71	71
	BS 3036 fuse		100, 102, C	71	71	71	71
5	BS 1361 fuse	1.5/1.0	103, 101, A, 100, 102, C	108	108	108	108
	BS 3036 fuse			108	108	108	108
6	BS 88-2.2,	1.0/1.0	103, 101, A,	59	59	59	59
	BS 88-6 fuse		100, 102, C				
	MCB Type B			59	59	59	59
	MCB Type C			59	59	59	59
	MCB Type D			25ss	25zs	36ss	36zs
	RCBO Type B			59	59	59	59
	RCBO Type C			59	59	59	59
	RCBO Type D			25ss	25ss	36ss	36ss
6	BS 88-2.2,	1.5/1.0	103, 101, A,	90	90	90	90
	BS 88-6 fuse		100, 102, C				
	MCB Type B			90	90	90	90
	MCB Type C			90	83zs	90	90
	MCB Type D			38ss	30zs	53ss	43zs
	RCBO Type B			90	90	90	90
	RCBO Type C			90	90	90	90
	RCBO Type D			38ss	38ss	53ss	53ss
10	BS 88-2.2,	1.0/1.0	101, A,	35	35	35	35
	BS 88-6 fuse		100.102, C				
	MCB Type B			35	35	35	35
	MCB Type C			35zs	35zs	35	35
	MCB Type D			8ss	8zs	18ss	18ss
	RCBO Type B			35	35	35	35
	RCBO Type C			34ss	34ss	35	35
	RCBO Type D			8ss	8ss	18ss	18ss
10	BS 88-2.2,	1.5/1.0	103, 101, A,	52	52	52	52
	BS88-6 fuse		100, 102, C				
	MCB Type B			52	52	52	52
	MCB Type C			51ss	41zs	52	52

Table 15.4 *(continued)*

	MCB Type D			12ss	9zs	27ss	22zs
	RCBO Type B			52	52	52	52
	RCBO Type C			51ss	51ss	52	52
	RCBO Type D			12ss	12ss	27ss	27ss
16	BS 88-2.2, BS88-6 fuse	1.5/1.0	100, 102, C	33	33	33	33
	MCB Type B			33	33	33	33
	MCB Type C			21ss	17zs	33	30zs
	MCB Type D			NPss	NPzs	12ss	10zs
	RCBO Type B			33	33	33	33
	RCBO Type C			21ss	21ss	33	33
	RCBO Type D			NPss	NPss	12ss	12ss
16	BS 88-2.2, BS88-6 fuse	2.5/1.5	101, A, 100, 102, C	53	53	53	53
	MCB Type B			53	53	53	53
	MCB Type C			35ss	27zs	53	46zs
	MCB Type D			NPss	NPzs	20ss	15zs
	RCBO Type B			53	53	53	53
	RCBO Type C			57	57	53	53
	RCBO Type D			NPss	NPss	20ss	20ss

[a]Any reference method giving a current-carrying capacity not less than that in column 4 may be used.
[b]NP: not permitted; zs: limited by earth fault loop impedance Z_s; ss: limited by line to neutral loop impedance (short-circuit).

Maximum circuit length

The tables give the maximum allowable circuit cable length measured from the consumer unit to the most distant end of the radial final circuit or all the way round the full length of the ring final circuit, as applicable. Cable lengths for ring final circuits do not include any spurs.

For lighting circuits, the circuit length includes the furthest switch drop.

Maximum circuit cable lengths given for a TN-S system may also be applied to a TT system with residual current device for fault protection.

Voltage drop

A maximum voltage drop of 3% has been allowed in circuits supplying lighting, and 5% in circuits supplying BS 1363 socket-outlets or other loads (Regulation 525.3 and Appendix 12 of BS 7671 refer).

For lighting circuits (Table 15.4), the circuit is assumed to have a load equal to the rated current I_n of the circuit protective device evenly distributed along the circuit. Where this is not the case, circuit lengths will need to be reduced where voltage drop is the limiting factor, and halved where the load is all at the extremity.

Table 15.5 Maximum cable length for radial final circuits supplying BS 1363 accessories using 70 °C thermoplastic (PVC) insulated and sheathed flat cable to BS 6004 Table 8 having copper conductors. $U_o = 230$ V. Reproduced by kind permission of The Electrical Safety Council

Protective device		Cable size (mm^2)	Allowed Reference Method[a]	Maximum length[b] (m)			
				$Z_s \leq 0.8\ \Omega$ TN-S		$Z_s \leq 0.35\ \Omega$ TN-C-S	
Rating (A)	Type			RCD	No RCD	RCD	No RCD
1	2	3	4	5	6	7	8
20	BS 88-2.2, BS 88-6 fuse	2.5/1.5	A, 100, 102, C	35	35	35	35
	BS 1361 fuse		A, 100, 102, C	35	35	35	35
	BS 3036 fuse		NP	NPol	NPol	NPol	NPol
	MCB Type B		A, 100, 102, C	35	35	35	35
	MCB Type C		A, 100, 102, C	19ss	14zs	35	34zs
	MCB Type D		A, 100, 102, C	NPss	NPzs	12ss	9zs
	RCBO Type B		A, 100, 102, C	35	35	35	35
	RCBO Type C		A, 100, 102, C	19ss	19ss	35	35
	RCBO Type D		A, 100, 102, C	NPss	NPss	12ss	12ss
20	BS 88-2.2, BS88-6 fuse	4.0/1.5	101, A, 100, 102, C	59	48zs	59	59
	BS 1361 fuse		101, A, 100, 102, C	59	44zs	59	59
	BS 3036 fuse		C	63	48zs	63	63
	MCB Type B		101, A, 100, 102, C	59	59	59	59
	MCB Type C		101, A, 100, 102, C	31ss	17zs	59	39zs
	MCB Type D		101, A, 100, 102, C	NPss	NPzs	20ss	11zs
	RCBO Type B		101, A, 100, 102, C	59	59	59	59
	RCBO Type C		101, A, 100, 102, C	31ss	31ss	59	59
	RCBO Type D		101, A, 100, 102, C	NPss	NPss	20ss	20ss
30	BS 1361 fuse	4.0/1.5	C	43	17zs	43	39zs
	BS 3036 fuse			NPol	NPol	NPol	NPol
32	BS 88-2.2, BS88-6 fuse	4.0/1.5	C	42	11zs	42	34zs
	MCB Type B			42	31zs	42	42
	MCB Type C			NPss	NPzs	33ss	18zs
	MCB Type D			NPss	NPzs	NPss	NPzs

Table 15.5 *(continued)*

RCBO Type B	42	42	42	42
RCBO Type C	NPss	NPss	33ss	33ss
RCBO Type D	NPss	NPss	NPss	NPss

[a]Any reference method giving a current-carrying capacity not less than that in column 4 may be used.
[b]NP: not permitted; ol: cable/device/load combination not allowed in any of the reference conditions; zs: limited by earth fault loop impedance Z_s; ss: limited by line to neutral loop impedance (short-circuit).

For radial circuits supplying BS 1363 accessories (Table 15.5), the circuit is assumed to have a total load equal to the rating of the protective device I_n. Of the load, 13 A is assumed to be at the farthest end of the circuit from the consumer unit, with the balance of the load at $0.75 \times L$ from the consumer unit, where L is the circuit length in metres from the consumer unit to the farthest end of the circuit.

For ring final circuits (Table 15.6), the circuit is assumed to have a load of 20 A at the farthest point and the rest of the load evenly distributed.

For radial circuits supplying a single load (Tables 15.7a–d), the circuit is assumed to have a load equal to the rated current I_n of the circuit protective device at the end of the circuit remote from the consumer unit.

Grouping of circuit cables

The circuit cable is assumed to be adequately spaced from other cables to give a current-carrying capacity I_z not less than the rated current I_n of the protective device (Regulation 433.1.1 refers). Where the protective device is a BS 3036 fuse and overload protection is to be provided, I_z must be not less than $I_n/0.725$ (Regulation 433.1.3 refers).

The grouping factors given in Table 4C1 of BS 7671 Appendix 4 allow all the cables in the group to carry their design load simultaneously and continuously. These circumstances are unlikely in household or similar circuits, except for circuits supplying water heaters of storage tanks exceeding 15 litres capacity and electric space heating systems, on and off peak. Cables of whole-house heating and water heating circuits are not to be grouped unless appropriately derated.

For cables of household or similar installations, except for heating and water heating, derating for grouping is not necessary provided:

- cables are not grouped where installed in or under thermally insulating material – that is, for Reference Methods 100, 101, 102, or 103 of Table 4A2 of BS 7671 Appendix 4;
- cables clipped direct (including in cement or plaster) are clipped side by side in one layer; preferably separated by one cable diameter;
- cables above ceilings are clipped to joists in accordance with Reference Methods 100 to 101.

Disconnection times

Where no RCD is presumed (columns 6 and 8 of the tables), for circuits with a device rating up to and including 32 A, circuit lengths are designed to limit earth fault loop

Table 15.6 Maximum cable length for ring final circuits supplying BS 1363 accessories using 70 °C thermoplastic (PVC) insulated and sheathed flat cable to BS 6004 Table 8 having copper conductors. $U_o = 230$ V. Reproduced by kind permission of The Electrical Safety Council

Protective device		Cable size (mm²)	Allowed Reference Method[a]	Maximum length[b] (m)			
Rating (A)	Type			$Z_s \le 0.8\ \Omega$ TN-S		$Z_s \le 0.35\ \Omega$ TN-C-S	
				RCD	No RCD	RCD	No RCD
1	2	3	4	5	6	7	8
30	BS 1361 fuse	2.5/1.5	A, 102, 100, C	111	59zs	111	111
	BS 3036 fuse			111	49zs	111	111
32	BS 88-2.2, BS 88-6 fuse	2.5/1.5	A, 102, 100, C	106	41zs	106	106
	MCB Type B			106	106	106	106
	MCB Type C			NPss	NPzs	82ss	63zs
	MCB Type D			NPss	NPzs	2ss	1zs
	RCBO Type B			106	106	106	106
	RCBO Type C			NPss	NPss	82ss	82ss
	RCBO Type D			NPss	NPss	2ss	2ss
30	BS 1361 fuse	4.0/1.5	101, A, 102, 100, C	183	69zs	183	159zs
	BS 3036 fuse			183	57zs	183	147zs
32	BS 88-2.2, BS 88-6 fuse	4.0/1.5	101, A, 102, 100, C	176	47zs	176	137zs
	MCB Type B			176	127zs	176	176
	MCB Type C			NPss	NPzs	133ss	73zs
	MCB Type D			NPss	NPzs	3ss	1zs
	RCBO Type B			176	176	176	176
	RCBO Type C			NPss	NPss	133ss	133ss
	RCBO Type D			NPss	NPss	3ss	3ss

[a]Any reference method giving a current-carrying capacity not less than that in column 4 may be used.

[b]NP: not permitted; zs: limited by earth fault loop impedance Z_s; ss: limited by line to neutral loop impedance (short-circuit).

impedances so that disconnection within 0.4 s is provided for fuses (Table 41.2 of BS 7671) and instantaneous operation for MCBs (Table 41.3 of BS 7671) in the event of a line to earth fault. For 40 and 45 A circuits, a disconnection time of 5 s is allowed for fuses (Regulation 411.3.2.3 refers).

Spurs on ring final circuits

Appendix 15 of BS 7671 provides guidance on conductor sizes, fuses and numbers of accessories for each spur connected to a ring final circuit, but only considers overload aspects. Requirements for protection against electric shock, thermal effects and fault current must also be considered. As a rule of thumb, the length of a spur cable that is connected to a given point on a ring should not exceed half the length of ring cable

Table 15.7a Maximum cable length for radial final circuits supplying a single load using 70 °C thermoplastic (PVC) insulated and sheathed flat cable to BS 6004 Table 8 having copper conductors. $U_o = 230$ V. Reproduced by kind permission of The Electrical Safety Council

Protective device		Cable size (mm²)	Allowed Reference Method[a]	Maximum length[b] (m)			
Rating (A)	Type			$Z_s \leq 0.8\ \Omega$ TN-S		$Z_s \leq 0.35\ \Omega$ TN-C-S	
				RCD	No RCD	RCD	No RCD
1	2	3	4	5	6	7	8
15	BS 1361 fuse	1.0/1.0	C	17	17	17	17
	BS 3036 fuse		NP	NPol	NPol	NPol	NPol
15	BS 1361 fuse	1.5/1.0	100, 102, C	26	26	26	26
	BS 3036 fuse		NP	NPol	NPol	NPol	NPol
15	BS 1361 fuse	2.5/1.5	101, A, 100, 102, C	43	43	43	43
	BS 3036 fuse		100, 102, C	45	45	45	45
15	BS 1361 fuse	4.0/1.5	103, 101, A, 100, 102, C	72	72	72	72
	BS 3036 fuse		101, A, 100, 102, C	75	75	75	75
16	BS 88-2.2, BS 88-6 fuse	1.0/1.0	C	16	16	16	16
	MCB Type B			16	16	16	16
	MCB Type C			14ss	14zs	16	16
	MCB Type D			NPss	NPzs	8ss	8zs
	RCBO Type B			16	16	16	16
	RCBO Type C			14ss	14ss	16	16
	RCBO Type D			NPss	NPss	8ss	8ss
16	BS 88-2.2, BS 88-6 fuse	1.5/1.0	100, 102, C	24	24	24	24
	MCB Type B			24	24	24	24
	MCB Type C			21ss	17zs	24	24
	MCB Type D			NPss	NPzs	12ss	10zs
	RCBO Type B			24	24	24	24
	RCBO Type C			21ss	21ss	24	24
	RCBO Type D			NPss	NPss	12ss	12ss
16	BS 88-2.2, BS 88-6 fuse	2.5/1.5	101, A, 100, 102, C	40	40	40	40
	MCB Type B			40	40	40	40
	MCB Type C			35ss	27zs	40	40
	MCB Type D			NPss	NPzs	20ss	15zs
	RCBO Type B			40	40	40	40
	RCBO Type C			35ss	35ss	40	40
	RCBO Type D			NPss	NPss	20ss	20ss

(*continued overleaf*)

Table 15.7a *(continued)*

Protective device		Cable size (mm^2)	Allowed Reference Method^a	Maximum length^b (m)			
Rating (A)	Type			$Z_s \leq 0.8\ \Omega$ TN-S		$Z_s \leq 0.35\ \Omega$ TN-C-S	
				RCD	No RCD	RCD	No RCD
1	2	3	4	5	6	7	8
16	BS 88-2.2, BS 88-6 fuse	4.0/1.5	103, 101, A, 100, 102, C	66	66	66	66
	MCB Type B			66	66	66	66
	MCB Type C			57ss	31zs	66	54zs
	MCB Type D			NPss	NPzs	33ss	18zs
	RCBO Type B			66	66	66	66
	RCBO Type C			57ss	57ss	66	66
	RCBO Type D			NPss	NPss	33ss	33ss

^{a}Any reference method giving a current-carrying capacity not less than that in column 4 may be used.
^{b}NP: not permitted; ol: cable/device/load combination not allowed in any of the reference conditions; zs: limited by earth fault loop impedance Z_s; ss: limited by line to neutral loop impedance (short-circuit).

between that point and the midpoint of the ring. This assumes that the ring is at the limit of its allowable length.

As with all circuitry, it is important to observe the requirements of Section 314 *Division of installation*. Every circuit must be electrically separate to facilitate safety in operation, inspection, testing and maintenance; implicitly, the line and neutral of each and every circuit must be separately connected in an easily identifiable sequence in the distribution board or consumer unit, and a neutral from one circuit must not be 'borrowed' for any other circuit. Regulation 514.1.2 requires that, as far as reasonably practicable, wiring is so arranged or marked that it can be identified for inspection, testing, repair and alteration of the installation. A way of complying with this requirement for the outgoing circuit conductors in the consumer unit or distribution board is to connect the neutral and protective conductors in the same sequence as the corresponding line conductors (that is, to the corresponding terminals of the neutral and earth bars).

It should not be overlooked that many, if not most, circuits in domestic and similar installations will require additional protection by RCDs to comply with the requirements of BS 7671. As mentioned earlier, Table 15.3 of this Guide lists the principal applications where additional protection by an RCD is required in the smaller installation.

To explain the use of Tables 15.4, 15.5, 15.6 and 15.7a–d, Figure 15.5 shows a typical distribution layout for a smaller installation with a consumer unit with 12 outgoing ways, one of which is a spare way. The design of some of the circuits is undertaken here to show the necessary procedures, including reference to the aforementioned tables. For this

Table 15.7b Maximum cable length for radial final circuits supplying a single load using 70 °C thermoplastic (PVC) insulated and sheathed flat cable to BS 6004 Table 8 having copper conductors. $U_o = 230$ V. Reproduced by kind permission of The Electrical Safety Council

Protective device		Cable size (mm²)	Allowed Reference Method[a]	Maximum length[b] (m)			
				$Z_s \leq 0.8\ \Omega$		$Z_s \leq 0.35\ \Omega$	
Rating (A)	Type			TN-S		TN-C-S	
				RCD	No RCD	RCD	No RCD
1	2	3	4	5	6	7	8
20	BS 88-2.2, BS 88-6 fuse	2.5/1.5	A, 100, 102, C	31	31	31	31
	BS 1361 fuse		A, 100, 102, C	31	31	31	31
	BS 3036 fuse		NP	NPol	NPol	NPol	NPol
	MCB Type B		A, 100, 102, C	31	31	31	31
	MCB Type C		A, 100, 102, C	19ss	14zs	31	31
	MCB Type D		A, 100, 102, C	NPss	NPzs	12ss	9zs
	RCBO Type B		A, 100, 102, C	31	31	31	31
	RCBO Type C		A, 100, 102, C	19ss	19ss	31	31
	RCBO Type D		A, 100, 102, C	NPss	NPss	12ss	12ss
20	BS 88-2.2, BS 88-6 fuse	4.0/1.5	101, A, 100, 102, C	53	48zs	53	53
	BS 1361 fuse		101, A, 100, 102, C	53	44zs	53	53
	BS 3036 fuse		C	57	48zs	57	57
	MCB Type B		101, A, 100, 102, C	53	53	53	53
	MCB Type C		101, A, 100, 102, C	31ss	17zs	53	39zs
	MCB Type D		101, A, 100, 102, C	NPss	NPzs	20ss	11zs
	RCBO Type B		101, A, 100, 102, C	53	53	53	53
	RCBO Type C		101, A, 100, 102, C	31ss	31ss	53	53
	RCBO Type D		101, A, 100, 102, C	NPss	NPss	20ss	20ss
20	BS 88-2.2, BS 88-6 fuse	6.0/2.5	103, 101, A, 100, 102, C	81	77zs	81	81
	BS 1361 fuse		103, 101, A, 100, 102, C	81	71zs	81	81
	BS 3036 fuse		A, 100, 102, C	85	77zs	85	85
	MCB Type B		103, 101, A, 100, 102, C	81	81	81	81
	MCB Type C		103, 101, A, 100, 102, C	47ss	27zs	81	63zs
	MCB Type D		103, 101, A, 100, 102, C	NPss	NPzs	30ss	17zs

Table 15.7b *(continued)*

Rating (A)	Type	Cable size (mm²)	Allowed Reference Method[a]	$Z_s \leq 0.8\ \Omega$ TN-S RCD	No RCD	$Z_s \leq 0.35\ \Omega$ TN-C-S RCD	No RCD
1	2	3	4	5	6	7	8
	RCBO Type B		103, 101, A, 100, 102, C	81	81	81	81
	RCBO Type C		103, 101, A, 100, 102, C	47ss	47ss	81	81
	RCBO Type D		103, 101, A, 100, 102, C	NPss	NPss	30ss	30ss
25	BS 88-2.2, BS 88-6 fuse	2.5/1.5	C	26	26	26	26
	MCB Type B			26	26	26	26
	MCB Type C			6ss	5zs	26	24zs
	MCB Type D			NPss	NPzs	6ss	4zs
	RCBO Type B			26	26	26	26
	RCBO Type C			6ss	6ss	26	26
	RCBO Type D			NPss	NPss	6ss	6ss

[a] Any reference method giving a current-carrying capacity not less than that in column 4 may be used.
[b] NP: not permitted; ol: cable/device/load combination not allowed in any of the reference conditions; zs: limited by earth fault loop impedance Z_s; ss: limited by line to neutral loop impedance (short-circuit).

purpose, it is assumed that the installation is in domestic premises and that the supply with the installation forms a TN-S system where the maximum prospective fault current is, say, 5.7 kA and the maximum external loop impedance Z_e is 0.8 Ω.

Circuit 1. radial circuit feeding a 230 V, 15.8 kW electric cooker (hob and oven) with cooker control that does not incorporate a 13 A socket-outlet. The cable is of the PVC-insulated and -sheathed flat twin-and-cpc type to BS 6004, enclosed in steel conduit in a thermally insulating wall for part of the run (Reference Method A) and clipped direct to a wooden joist above a plasterboard ceiling with thermal insulation more than 100 mm thick (Reference Method 101) for the remainder of the run, the total run being 46 m. Ambient temperature is 30 °C throughout and there is no grouping factor necessary.

The total rated (full-load) current is 68.7 A (15800 W/230 V) and applying diversity (see Appendix J of IEE Guidance Note No. 1) gives a design current I_b of 27.6 A:

$$
\begin{aligned}
I_b &= 10 + 0.3(\text{rated current} - 10) \\
&= 10 + 0.3(68.7 - 10) \\
&= 27.6\ \text{A}
\end{aligned}
\tag{15.2}
$$

Table 15.7c Maximum cable length for radial final circuits supplying a single load using 70 °C thermoplastic (PVC) insulated and sheathed flat cable to BS 6004 Table 8 having copper conductors. $U_o = 230$ V. Reproduced by kind permission of The Electrical Safety Council

Protective device		Cable size	Allowed	Maximum length[b] (m)			
Rating (A)	Type	(mm²)	Reference Method[a]	$Z_s \leq 0.8$ Ω TN-S		$Z_s \leq 0.35$ Ω TN-C-S	
				RCD	No RCD	RCD	No RCD
1	2	3	4	5	6	7	8
25	BS 88-2.2, BS 88-6 fuse	4.0/1.5	A, 100, 102, C	42	31zs	42	42
	MCB Type B			42	42	42	42
	MCB Type C			10ss	5zs	42	28zs
	MCB Type D			NPss	NPzs	9ss	5zs
	RCBO Type B			42	42	42	42
	RCBO Type C			10ss	10ss	42	42
	RCBO Type D			NPss	NPss	9ss	9ss
25	BS 88-2.2, BS 88-6 fuse	6.0/2.5	101, A, 100, 102, C	64	50zs	64	64
	MCB Type B			64	64	64	64
	MCB Type C			16ss	9zs	64	45zs
	MCB Type D			NPss	NPzs	14ss	8zs
	RCBO Type B			64	64	64	64
	RCBO Type C			16ss	16ss	64	64
	RCBO Type D			NPss	NPss	14ss	14ss
30	BS 1361 fuse	4.0/1.5	C	36	17zs	36	36
	BS 3036 fuse		NP	NPol	NPol	NPol	NPol
30	BS 1361 fuse	6.0/2.5	A, 100, 102, C	53	27zs	53	53
	BS 3036 fuse		C	57	23zs	57	57
30	BS 1361 fuse	10/4.0	101, A, 100, 102, C	88	45zs	88	88
	BS 3036 fuse		A, 100, 102, C	93	37zs	93	93
32	BS 88-2.2, BS 88-6 fuse	4.0/1.5	C	33	11zs	33	33
	MCB Type B			33	31zs	33	33
	MCB Type C			NPss	NPzs	33	18zs
	MCB Type D			NPss	NPzs	NPss	NPzs
	RCBO Type B			33	33	33	33
	RCBO Type C			NPss	NPss	33	33
	RCBO Type D			NPss	NPss	NPss	NPss
32	BS 88-2.2, BS 88-6 fuse	6.0/2.5	A, 100, 102, C	49	19zs	49	49
	MCB Type B			49	49	49	49
	MCB Type C			NPss	NPzs	49	29zs

(*continued overleaf*)

Table 15.7c *(continued)*

Protective device		Cable size (mm²)	Allowed Reference Method[a]	Maximum length[b] (m)			
				$Z_s \leq 0.8\ \Omega$		$Z_s \leq 0.35\ \Omega$	
Rating (A)	Type			TN-S		TN-C-S	
				RCD	No RCD	RCD	No RCD
1	2	3	4	5	6	7	8
	MCB Type D			NPss	NPzs	1ss	NPzs
	RCBO Type B			49	49	49	49
	RCBO Type C			NPss	NPss	49	49
	RCBO Type D			NPss	NPss	1ss	1ss

[a]Any reference method giving a current-carrying capacity not less than that in column 4 may be used.
[b]NP: not permitted; ol: cable/device/load combination not allowed in any of the reference conditions; zs: limited by earth fault loop impedance Z_s; ss: limited by line to neutral loop impedance (short-circuit).

It is necessary to decide whether additional protection by an RCD is required for the circuit. Such protection may have been required had the cable not been enclosed in steel conduit where it is concealed in the wall (Regulations 522.6.7 and 522.6.8 refer) or had the cooker control incorporated a socket-outlet (Regulation 411.3.3 refers). However, in this case, additional protection by an RCD is not required for the circuit.

The next step is to choose the protective device. The fault level of 5.7 kA is outside the capabilities of a semi-enclosed (rewirable) fuse had the circuit distribution been a distribution board. However, in the example shown, a consumer unit has been chosen and we could have used semi-enclosed fuses because of the conditional rating of such units. For the purposes of this exercise we shall choose an MCB with a rated current of 32 A. The MCB will need at least a rated breaking capacity of 6000 A if we are to avoid tiresome calculations of energy let-through of the upstream protective device. Cookers are not notorious for high inrush currents, so we could choose a Type B MCB, but for the purposes of this exercise we shall choose Type C. As the current rating of the circuit does not exceed 32 A and the installation forms part of a TN system, a disconnection time not exceeding 0.4 s is required (Table 41.1 of BS 7671 refers). From Table 41.3 of BS 7671 (for 0.4 s disconnection), we see that the limiting value for Z_s for a 32 A Type C MCB is 0.72 Ω (the values for 5 s and 0.4 s disconnection are the same for an MCB owing to its time/current characteristic). We can immediately see that this protective device is unsuitable, as the limiting Z_s (0.72 Ω) is less than Z_s (0.8 Ω). For this circuit it will obviously be necessary to select a 'more sensitive' MCB type. Let us consider a Type B MCB which, according to Table 41.3, has a limiting Z_s of 1.44 Ω.

The next step is to select the cable cross-sectional area. The minimum tabulated current rating $I_{t(min)}$ is given by

$$I_{t(min)} = \frac{I_n}{C_a \times C_g \times C_i} = \frac{32}{1 \times 1 \times 1} = 32\ \text{A} \qquad (15.3)$$

Table 15.7d Maximum cable length for radial final circuits supplying a single load using 70 °C thermoplastic (PVC) insulated and sheathed flat cable to BS 6004 Table 8 having copper conductors. $U_o = 230$ V.

Protective device		Cable size (mm²)	Allowed Reference Method[a]	Maximum length[b] (m)			
				$Z_s \leq 0.8\ \Omega$ TN-S		$Z_s \leq 0.35\ \Omega$ TN-C-S	
Rating (A)	Type			RCD	No RCD	RCD	No RCD
1	2	3	4	5	6	7	8
32	BS 88-2.2, BS 88-6 fuse	10/4.0	103, 101, A, 100, 102, C	81	31zs	81	81
	MCB Type B		103, 101, A, 100, 102, C 1	81	81	81	81
	MCB Type C		103, 101, A, 100, 102, C 1	NPss	NPzs	81	47zs
	MCB Type D		103, 101, A, 100, 102, C 1	NPss	NPzs	2ss	1zs
	RCBO Type B		103, 101, A, 100, 102, C 1	81	81	81	81
	RCBO Type C		103, 101, A, 100, 102, C 1	NPss	NPss	81	81
	RCBO Type D		103, 101, A, 100, 102, C 1	NPss	NPss	2ss	2ss
40	BS 88-2.2, BS 88-6 fuse	6.0/2.5	C	40	27zs	40	37
	MCB Type B			40	27zs	40	40
	MCB Type C			NPss	NPzs	30ss	17zs
	MCB Type D			NPss	NPzs	NPss	NPzs
	RCBO Type B			40	40	40	40
	RCBO Type C			NPss	NPss	30ss	30ss
	RCBO Type D			NPss	NPss	NPss	NPss
40	BS 88-2.2, BS 88-6 fuse	10/4.0	A, 100, 102, C	66	66	66	66
	MCB Type B			66	45zs	66	66
	MCB Type C			NPss	NPzs	51ss	29zs
	MCB Type D			NPss	NPzs	NPss	NPzs
	RCBO Type B			66	66	66	66
	RCBO Type C			NPss	NPss	51ss	51ss
	RCBO Type D			NPss	NPss	NPss	NPss
40	BS 88-2.2, BS 88-6 fuse	16/6.0	103, 101, A, 100, 102, C	104	104	104	104
	MCB Type B		103, 101, A, 100, 102, C 1	104	68zs	104	104
	MCB Type C		103, 101, A, 100, 102, C 1	NPss	NPzs	81ss	44zs
	MCB Type D		103, 101, A, 100, 102, C 1	NPss	NPzs	NPss	NPzs

Table 15.7d *(continued)*

Protective device		Cable size (mm²)	Allowed Reference Method[a]	Maximum length[b] (m)			
Rating (A)	Type			$Z_s \leq 0.8\ \Omega$ TN-S		$Z_s \leq 0.35\ \Omega$ TN-C-S	
				RCD	No RCD	RCD	No RCD
1	2	3	4	5	6	7	8
	RCBO Type B		103, 101, A, 100, 102, C 1	104	104	104	104
	RCBO Type C		103, 101, A, 100, 102, C 1	NPss	NPss	81ss	81ss
	RCBO Type D		103, 101, A, 100, 102, C 1	NPss	NPss	NPss	NPss
45	BS 1361 fuse	6/2.5	C	21	NPadia	35	22adia
	BS 3036 fuse		NP	NPol	NPol	NPol	NPol
45	BS 1361 fuse	10/4.0	100, 102, C	36ss	5adia	58	58
	BS 3036 fuse		C	62	62	62	62
45	BS 1361 fuse	16/6.0	101, A, 100, 102, C	57ss	31zs	91	91
	BS 3036 fuse		102,C	97	97	97	97

[a]Any reference method giving a current-carrying capacity not less than that in column 4 may be used.

[b]NP: not permitted; adia: limited by reduced cross-sectional area of protective conductor (adiabatic limit); ol: cable/device/load combination not allowed in any of the reference conditions; zs: limited by earth fault loop impedance Z_s; ss: limited by line to neutral loop impedance (short-circuit).

In this case, the rating factors C_a, C_g and C_i used in Equation (15.3) all have the value 1. C_a and $C_g = 1$ because, respectively, the ambient temperature is 30 °C throughout and there is no grouping factor necessary (other than $C_g = 1$). Less obviously, the rating factor for thermal insulation $C_i = 1$ because Table 4D5 of Appendix 4 of BS 7671 gives current-carrying capacities directly for the two reference methods (A and 101) used along the cable run, even though both methods involve contact with thermal insulation by the wiring system. Thus, it is not necessary to apply a rating factor C_i separately (except to give it the value 1 in Equation (15.3)).

From Table 4D5 of Appendix 4 of BS 7671 we see that the more onerous of the two reference methods (A and 101) used along the cable run is Reference Method 101. For this reference method, the smallest cable size with a tabulated current-carrying capacity of 32 A or more is 10 mm², I_t being 36 A.

We next have to look at the limiting Z_s for the Type B MCB, which, as previously mentioned, is 1.44 Ω. To find the limit on circuit length L, we need to divide $Z_s - Z_e$ by the $(R_1 + R_2)$ per metre (corrected for temperature rise under normal operating conditions), as given in Equation (15.4). The data for corrected $(R_1 + R_2)$ per metre values are given in Table 10.18 of this Guide: for 10/4 mm² with cpc as a core within

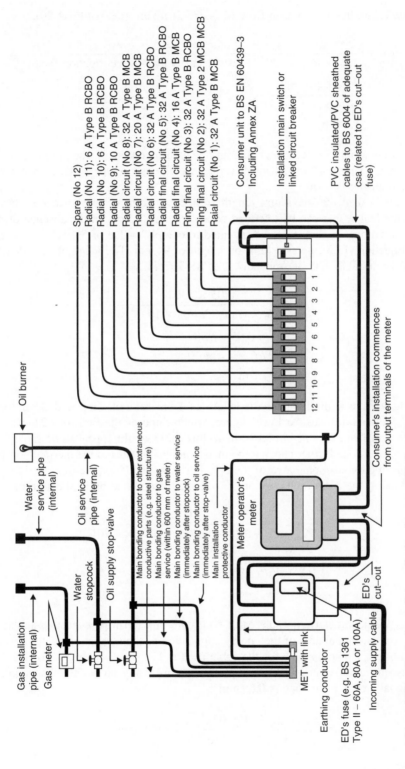

Figure 15.5 Typical small installation layout: TN-S system

a composite cable, this is given, in Row M, as $7.728\,m\Omega/m$ (i.e. 6.44×1.20).

$$L\ (\text{m}) = \frac{Z_{s(\text{Max})} - Z_e}{(R_1 + R_2) \times 10^{-3} \times \text{resistance/temperature correction factor}}$$

$$= \frac{1.44 - 0.8}{6.44 \times 1.20 \times 10^{-3}} \tag{15.4}$$

$$= 82.8\ \text{m}$$

From Equation (15.4) we can see that, in the consideration of the limiting Z_s, the maximum length permitted is 82.8 m; therefore, we can use a Type B MCB, because the length of circuit run is 46 m.

Having established the maximum length from a Z_s point of view, we now have to determine the length that can be tolerated considering volt-drop. In the absence of more precise information, we shall take the 'deemed to comply' value of 5% which, by application of Equation (15.5), gives a limit on length of 94.7 m. The value of 4.4 mV/(A m) is obtained from Table 4D5 of Appendix 4 of BS 7671.

$$L = \frac{0.05 \times 230}{I_b \times (\text{mV/(A m)}) \times 10^{-3}}$$

$$= \frac{0.05 \times 230}{27.6 \times 4.4 \times 10^{-3}} \tag{15.5}$$

$$= 94.7\ \text{m}$$

So we now have a limit on circuit length of 94 m for volt-drop and 82 m for Z_s. Obviously, we have to take the lower figure, which imposes the limit of 82 m on cable run. As the actual run for this circuit is 46 m, this cable would appear to be satisfactory. There is, however, one further check to be made, and that relates to the thermal constraint on the reduced cross-sectional area of the cpc. For this calculation we first need to know the magnitude of the earth fault current I_F, which is determined by application of

$$I_F = \frac{U_o}{Z_s} = \frac{U_o}{Z_e + L[(R_1 + R_2) \times 10^{-3} \times 1.20]}$$

$$= \frac{230}{0.8 + 46(6.44 \times 10^{-3} \times 1.20)} \tag{15.6}$$

$$= 199.05\ \text{A}$$

Now that we have determined the earth fault current we can proceed to establish the minimum cross-sectional area of the cpc by application of the adiabatic equation given in Regulation 543.1.3. The factor k for the cable used is 115 (from Table 54.3 of BS 7671) and the disconnection time is 100 ms (because we are using an MCB). The minimum cross-sectional area S of the cpc is $0.55\ \text{mm}^2$:

$$S = \frac{I_F \sqrt{t}}{k} = \frac{199.05 \sqrt{0.1}}{115} = 0.55\ \text{mm}^2 \tag{15.7}$$

In the case of MCBs, satisfying the thermal constraints relating to the cpc is not difficult owing to the rapid disconnection times. However, a check is always required where fuses are used to provide fault protection, particularly where disconnection times of 5 s are appropriate.

As an aid to the design process we could have used Table 15.7d of this Guide to obtain the maximum circuit length for this particular cable. We look to this table and find, in column 6, the limiting lengths for the various combinations of common protective devices, conductor sizes and reference methods corresponding to a circuit of a TN-S system where no RCD is used for additional protection. We can see at a glance from column 6 that, for cable with $10\,mm^2$ live conductors and a $4\,mm^2$ cpc, installed to Reference Methods A and 101, 32 A MCB Types C and D are unsuitable (as are 32 A RCBO Types C and D), and we can also see that a 32 A BS 88 fuse is limited to a 31 m circuit run. The limiting length for a 32 A Type B MCB is given as 81 m, and the absence of any letters (zs, ss, etc) after that length indicates that the volt-drop constraint is the limiting factor.

Circuit 2. a ring final circuit feeding BS 1363 13 A socket-outlets for general use over an area of $95\,m^2$ that does not include a kitchen or laundry room. The circuit length is estimated to be 64 m and the ambient temperature along the whole run is not expected to exceed 30 °C. No rating factors are applicable. Again, PVC-insulated and -sheathed flat twin-and-cpc cable to BS 6004 is to be used. The reference method in parts of the run is to be 101 (clipped direct to a wooden joist above a plasterboard ceiling with thermal insulation more than 100 mm thick) and in other parts of the run it is to be 102 (in a stud wall with thermal insulation, the cable touching the inner wall surface). See Figure 15.6 for typical ring final circuit.

Consulting Regulation 522.6.7, we find that the circuit cable requires additional protection by an RCD where it is concealed in the wall, as it is less than 50 mm from the

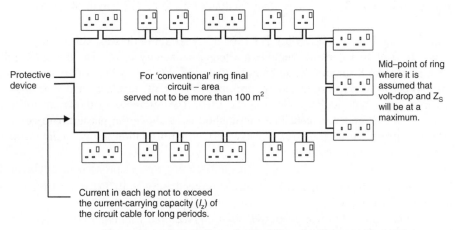

Figure 15.6 Typical ring final circuit

surface and does not have an earthed metallic covering, earthed metallic enclosure or suitable mechanical protection, and the installation (being in domestic premises) will not be under the supervision of a skilled or instructed person. Also, from Regulation 411.3.3, we find that the socket-outlets, too, require additional protection by an RCD, as they are rated at less than 20 A and are intended for general use, and, as has just been said, the installation will not be under the supervision of a skilled or instructed person. Thus, the circuit must be provided with additional protection by an RCD.

The first value to decide is the circuit design current. In the absence of better information, this is normally taken as the rated current I_n of the overcurrent protective device. It is important that a check is made that the chosen I_n is adequate for the anticipated load of the circuit. Because the area served is less than $100\,m^2$ and does not include areas of high load, such as a kitchen or laundry room, the number of socket-outlets for a 30 or 32 A ring final circuit is unlimited.

For the purposes of this design, let us select a 32 A RCBO. The overcurrent part of this device can be used to provide overcurrent protection for the circuit conductors. In addition, the RCD part of the device can be used to provide both fault protection and additional protection (the latter because the characteristics of RCD part of the device meet the requirements of Regulation 415.1.1 for additional protection).

As given in Regulation 433.1.5, the current-carrying capacity I_z of the circuit cable must be not less than 20 A. From Table 4D5 of BS 7671 we can see that for Reference Methods 101 and 102 we have $4\,mm^2$ with tabulated current-carrying capacities I_t of 22 A and 27 A respectively, which are more than adequate for this circuit, as the rating factors C_a, C_g and C_i all have the value 1 in this case, and so the current-carrying capacities I_z of this size of cable are also 22 A and 27 A respectively.

Looking now to Table 15.6 of this Guide, we can see in column 6 that, for a cable with $4\,mm^2$ live conductors and a $1.5\,mm^2$ cpc, we cannot use 32 A RCBOs of Types C or D because of considerations of thermal effects under short-circuit conditions. We could, however, use a 32 A Type B RCBO, which gives a maximum length of 176 m owing to volt-drop limitations. This is more than adequate for the circuit length of 64 m.

Circuit 3. a ring final circuit feeding BS 1363 13 A socket-outlets for general use over an area of $80\,m^2$ that does not include a kitchen or laundry room. The circuit length is estimated to be 86 m and the ambient temperature along the whole run is not expected to exceed $30\,°C$. No rating factors are applicable. PVC-insulated and -sheathed flat twin-and-cpc cable to BS 6004 is to be used. The reference methods are to be the same as for Circuit 2, except that the method used above the plasterboard ceiling is 100 (clipped direct to a wooden joist with thermal insulation *not exceeding* 100 mm thick), rather than method 101.

The circuit must be provided with additional protection by an RCD for the same reasons as explained for Circuit 2.

Similar consideration to that given to Circuit 2 is required with regard to design current. Selecting a 32 A RCBO, and bearing in mind that the reference methods for this circuit are 100 and 102 (as opposed to 101 and 102 for Circuit 2), we can see from Table 15.6, again column 6, that whilst we cannot use 32 A RCBOs of Types C or D because of considerations of thermal effects under short-circuit conditions, a 32 A Type B RCBO will allow us a circuit length of 106 m, which is more than adequate for the circuit length of 86 m.

Circuit 4. a radial final circuit feeding a 3 kW immersion heater that is not situated in a special location (such as a bathroom). The circuit length is estimated to be 50 m, the ambient temperature along the whole run is not expected to exceed 30 °C and there is no grouping factor necessary. PVC-insulated and -sheathed flat twin-and-cpc cable to BS 6004 is to be used. The installation method in parts of the run is to be 20 (clipped direct – Reference Method C), and in the other parts of the run the installation method is to be 8 (in trunking on a wall – Reference Method B).

There is no requirement for additional protection by an RCD for this circuit, as the cable is not concealed in a wall or partition and the circuit does not supply equipment in a special location. The design current for the circuit is 13.04 A (3000 W/230 V) and we may, therefore, select a 16 A MCB as the protective device. Furthermore, the MCB may be of Type B, as an immersion heater is not expected to have a high inrush current.

Table 15.7a of this Guide is the appropriate table of maximum cable length to use for this circuit. At first sight there seems to be a problem in using this table, in that column 4 does not appear to cover Reference Method B, which is one of the methods to be used in our circuit. However, table footnote *a* of the table tells us that any reference method giving a current-carrying capacity not less than that for the reference methods in column 4 may be used.

Consulting column 6 of Table 15.7a, we see that, with a 16 A Type B MCB, a 4 mm^2 cable will allow a circuit length of 66 m (which is more than adequate for our circuit length of 50 m), provided the current-carrying capacity of the cable is not less than given by one of the Reference Methods 103, 101, A, 100, 102 or C. Checking Table 4D5 of Appendix 4 of BS 7671 to find the tabulated current-carrying capacity of a 4 mm^2 cable installed to Reference Method B, we find that the table does include this method. Therefore, it is necessary to refer to Table 4D2A of the same appendix, which covers multicore 70 °C thermoplastic insulated and sheathed nonarmoured cables but is not limited to those of the flat twin-and-cpc type to BS 6004. From column 4 of this table we find that the tabulated current-carrying capacity of a 4 mm^2 two-core cable, with or without cpc, installed to Reference Method B is 30 A. Now, referring back to Table 4D5, we find that 30 A is indeed not less than the current-carrying capacity of a 4 mm^2 cable installed to Reference Methods 103 (17.5 A), 101 (22 A), A (26 A), 100 (27 A) and 102 (27 A). It follows, therefore, that the cable size of 4 mm^2 indicated by column 6 Table 15.7 is suitable for our circuit.

Circuit 5. a radial final circuit feeding BS 1363 13 A socket-outlets in a kitchen that are specifically labelled or otherwise suitably identified for the connection of a washing machine, a tumble dryer and a fridge-freezer. The circuit length is estimated to be 35 m, the ambient temperature along the whole run is not expected to exceed 30 °C and there is no grouping factor necessary. PVC-insulated and -sheathed flat twin-and-cpc cable to BS 6004 is to be used. The installation method in parts of the run is to be 20 (clipped direct – Reference Method C), and in the other parts of the run the reference method is to be 58 (buried direct in plaster on a brick wall – Reference Method C).

The socket-outlets do not require additional protection by an RCD, as they are specifically labelled or otherwise suitably identified for the connection of a particular item of equipment (Regulation 411.3.3 refers). However, the cable concealed in the wall requires such additional protection, for the same reasons as explained for Circuit 2.

It is not unreasonable to assume that a circuit rated at 32 A will be suitable for the diversified load of a washing machine, tumble dryer and fridge-freezer in domestic premises. Therefore, we may select a 32 A RCBO as the circuit protective device. From column 6 of Table 15.5 of this Guide, we see that, with a 32 A RCBO and a cable installed by Reference Method C, a cable size of 4 mm^2 will allow a circuit length of 42 m (which is more than adequate for our circuit length of 35 m), provided the RCBO is of Type B.

Circuits 6–8. all these circuits are radial and maximum length of runs can easily be determined from Tables 15.5 or 15.7a–d, as applicable.

Circuits 9, 10. these circuits are for lighting, and a similar approach should be made in the circuit design. It is generally considered necessary to provide at least two lighting circuits in order to avoid the entire premises being plunged into darkness as a result of a fault on a solitary lighting circuit.

Circuit 11. this circuit is dedicated to supplying the fire detection and alarm system, which in the case of new housing would probably consist of a number of mains-operated smoke alarms and heat alarms. Circuit design should be similar to that of any other circuit.

Where the supply and installation is a TN-C system, maximum circuit lengths should be taken from columns 7 or 8 (as applicable) or Tables 15.4–15.7 (as applicable). For TN-C-S systems, more limiting lengths are dictated by the volt-drop considerations than is the case for TN-S systems. For TT systems, with fault protection provided by an RCD, the limiting factor is nearly always the volt-drop.

15.15   Remote buildings

Where an installation supplied from a PME network also serves a remote building with an uninsulated floor (such as a garage), it may be impracticable to use the PME earth as the means of earthing for the circuits in that building. Indeed, the ED may not allow such use of the PME earth. Persons in such remote buildings have been known to experience a sensation of electric current when touching exposed-conductive-parts or extraneous-conductive-parts connected to the PME earth, due to the voltage that exists between 'true' earth and the PME earth when the network is supplying load. A solution to this problem is to make the remote building installation part of a TT system. The PME earthing would be discarded either at the main building or at the remote building. Where the former method is adopted, the outgoing distribution circuit (if it incorporates a cpc) must be protected by an RCD with its own installation earth electrode at the main building. Another possible solution, illustrated in Figure 15.7, is for the remote building supply cable to be terminated in an 'all-insulated' adaptable box at which point the PME earthing is 'lost'. The upstream overcurrent protective device will still provide fault protection up to the adaptable box, so that if, for example, an earth fault developed on the supply cable (say, SWA) between the buildings the main building device would operate. The remote

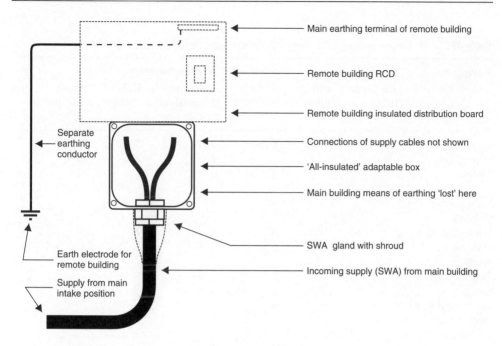

Main earthing terminal of remote building

Remote building RCD

Remote building insulated distribution board

Separate earthing conductor

Connections of supply cables not shown

'All-insulated' adaptable box

Main building means of earthing 'lost' here

SWA gland with shroud

Earth electrode for remote building

Incoming supply (SWA) from main building

Supply from main intake position

Figure 15.7   A method for earthing of installations in detached, remote buildings

building earthing would be by means of its own RCD and associated installation earth electrode, and circuits emanating from the remote building 'all-insulated' distribution board would be provided with fault protection by the local RCD. In the circumstances where an upstream series RCD was used, this would need to be of the selective (delay) type to provide for discrimination.

15.16   Minimum number of socket-outlets in domestic premises

Regulation 553.1.7 calls for adequate provision for the connection of portable equipment wherever such items are likely to be used, bearing in mind the length of flexible cord normally fitted to portable appliances and portable luminaires. Some portable appliances are supplied with flexible cords as short as 2 m. In practically all cases the designer will not know precisely what equipment is to be used when the installation is put into use, and still less about what the future may hold. Generally, the decision on the number of socket-outlets is dictated by budgetary considerations, but it should be appreciated that one can never have too many. Table 15.8 brings together recommendations by a number of bodies published since the early 1960s.

15.17   Modifications to existing installations

Additions and alterations to existing installations need careful consideration. Any alteration or addition must not impair the safe functioning of the existing installation. Additionally, if any parts of the existing installation are utilized to provide any protective

Table 15.8   Minimum number of socket-outlets in homes: recommendations[a]

Room type	Socket-outlet recommendations					
	This Guide[b] (1994)	NHBC[c] (1992)	EIILC[d] (1977)	Consumers' Association[e] (1974)	DoE[f] (1971)	Parker Morris[g] (1961)
Living room	4	4	6	6	3	3
Dining room	3	2	3	4	2	1
Kitchen	5	6	4	6	4	4
Utility	3	–	–	–	–	–
Main double bedroom	3	3	4	4	2	2
Double bedroom	2	2	4	4	2	2
Main single bedroom	2	3	3	≤ 4	2	2
Single bedroom	2	2	3	≥ 4	–	2
Study	2	–	3	≥ 4	–	2
Single bed/sitting room	4	–	4	–	–	–
Landing/stairs	1	1	–	–	–	–
Hall	1	1	1	1 or 2	1	1
Store/workroom	1	–	1	–	1	1
Garage	1	–	2	1 or 2	1	1

[a]All recommendations relate to socket-outlets for use with portable equipment. Fixed equipment will require additional supply points, such as switched and unswitched connection units.
[b]The recommended number of socket-outlets of this Guide relate to twin outlets in every case.
[c]NHBC is the National House-Building Council. Twin outlets count as two, but where used will result in a diminution of socket-outlet distribution.
[d]EIILC is the Electrical Installation Industry Liaison Committee.
[e]The recommendations published by the Consumers' Association are their most recent guidance
[f]DoE is the Department of the Environment, which published recommendations in the form of a leaflet. Each outlet recommended should be regarded as a twin outlet.
[g]The Parker Morris Report states that a twin socket-outlet should be regarded as two outlets with respect to recommendations made.

measures for the alteration or addition, then it is for the designer and the installer to check that those existing parts are both satisfactory and reliable. For example, where a new circuit is installed using ADS for protection against electric shock, it will be essential to determine that the earthing of the existing installation is adequate and that acceptable protective bonding is in place.

Generally, inspecting, testing, verification and certification are required for alterations and additions, however small. Direct replacement of equipment (e.g. lampholders, switches) involving no installation work does not warrant certification, but a certain amount of testing (e.g. polarity, Z_s) will inevitably be necessary and it may be in the contractor's interest to issue a minor electrical installation works certificate containing (amongst other things) the results of these tests.

15.18   Inspection, testing, verification and certification of the smaller installation

The procedures for inspection, testing, verification and certification for the smaller installation do not differ from those of any other installation and must be completed fully. Chapter 17 of this Guide gives guidance in this important aspect of electrical installation work.

16

Special installations and locations

16.1 General

Section 700 makes it clear that the particular requirements set out in the various sections of Part 7 supplement or modify the general requirements in the remainder of BS 7671. Though a design may meet all the requirements laid down in a particular section for a particular special installation or location, it is important to appreciate that it will not comply with BS 7671 unless all the other applicable requirements embodied elsewhere are also met.

The practical guidance given here assumes that the reader recognizes the full implications of the preceding paragraph, and the advice given for a particular location or installation is additional to guidance given elsewhere in this Guide.

16.2 Locations containing a bath or shower

16.2.1 General

A location containing a fixed bath or shower is considered to be a location where there is an increased risk of electric shock due to a reduction in body resistance caused either by bodily immersion or wet skin and likely contact between substantial areas of the body and Earth potential.

The requirements of Section 701 apply to a location containing a fixed bath or shower and the surrounding zones defined in that section. The requirements do not apply to emergency facilities, such as showers and similar decontamination facilities for emergencies in industrial areas and laboratories. Special requirements beyond those of BS 7671 may apply to a location containing a bath and shower for medical treatment or for disabled people. The requirements for such installations must be determined by the designer of the installation in consultation with the customer and/or the manufacturer of the equipment concerned.

A Practical Guide to The Wiring Regulations: 17th Edition IEE Wiring Regulations (BS 7671:2008)
Fourth Edition Geoffrey Stokes and John Bradley
© Geoffrey Stokes and John Bradley. Published by John Wiley & Sons, Ltd

Table 16.1 Particular requirements for bath and shower locations: zonal descriptions and demarcation dimensions

Zone	Zone description[a]	Dimensional information[b]
0	Interior of the bath tub or shower basin	For location containing a shower without a basin, zone 0 is the three-dimensional space limited by the floor, the horizontal plane 0.01 m above the floor and the vertical surface at a radius of 1.2 m horizontally from the centre point of the fixed water outlet on the wall or ceiling
1	Three-dimensional space above and below zone 0	Zone 1 is the three-dimensional space limited by the finished floor level, the horizontal plane corresponding to the highest fixed shower head or water outlet or the horizontal plane 2.25 m above the floor, whichever is higher, and the vertical surface circumscribing the bath tub or shower basin.
		For a shower without a basin, the limiting vertical surface of zone 1 is at a radius of 1.2 m horizontally from the centre point of the fixed water outlet on the wall or ceiling.
		Zone 1 does not include zone 0.
		Zone 1 includes the space below the bath tub or shower basin where that space is accessible without the use of a tool. Spaces under the bath accessible only with the use of a tool are outside the zones
2	Three-dimensional space adjacent to zones 0 and 1	Zone 2 is the three-dimensional space limited by the finished floor level, the horizontal plane corresponding to the highest fixed shower head or water outlet or the horizontal plane 2.25 m above the floor, whichever is higher, and the vertical surface forming the outer limit to zone 1 and a parallel vertical plane 0.6 m external to it.
		For a shower without a basin there is no zone 2 (but an increased zone 1 is provided by the horizontal dimension of 1.2 m mentioned above

[a]The zones are determined taking account of walls, doors, fixed partitions, ceilings and floors, where these effectively limit the extent of a zone.
[b]Where electrical equipment forms part of a wall or ceiling limiting a zone, the requirements for that zone apply to the equipment.

Whirlpool baths (e.g. Jacuzzis) should be treated as ordinary baths, and the general requirements and those contained in Section 701 of BS 7671 are applicable to installations incorporating this type of bath.

Table 16.1 sets out the descriptions and demarcation dimensions of the zonal system and Table 16.2 similarly sets out the degrees of ingress protection for the particular zones. Table 16.3 details switchgear and current-using equipment permitted in the various zones. Figure 16.1 illustrates a typical bathroom layout and the equipment often found in such a location.

Table 16.2 Particular requirements for bath and shower locations: minimum degree of ingress protection

Zone	Particular conditions, if any	Minimum degree of protection
0		IPX7
1 and 2	– [a]	IPX4
	Where water jets are likely to be used, such as for cleaning purposes	IPX5
Outside 1 and 2		General rules[b]

[a]Shaver supply units to BS EN 61558-2-5 may be installed in zone 2 provided they are located where direct spray from a shower is unlikely.

[b]Equipment to be suitable for the conditions likely to occur at the particular point of installation.

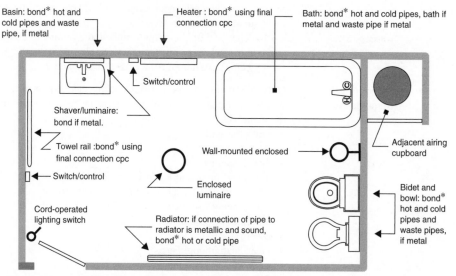

*Supplementary bonding is not required where all three conditions specified in Regulation 701.415.2 for its omission are met (see Section 16.2.6 of this Guide).

Figure 16.1 Typical bathroom layout and equipment

16.2.2   Zonal arrangements

The requirements for safety of a location containing a bath or shower are based on the application of a zonal concept similar to that used for swimming pools, the requirements for each zone and beyond being based on the perceived degree of risk of electric shock.

Whilst in BS 7671:2001 there were four zones (zones 0, 1, 2 and 3), in BS 7671:2008 there are only three (zones 0, 1 and 2). The demarcation dimensions for the zones are given in Table 16.1. The zones are limited by walls, doors, fixed partitions, ceilings and floors where these effectively limit the extent of a zone.

Table 16.3 Particular requirements for bath and shower locations: switchgear and current-using equipment permitted in zones.[a]

Zone	Equipment	Permitted equipment
0	Switchgear, controlgear and accessories	Switches and controls incorporated in fixed current-using equipment suitable for use in the zone
	Current-using equipment	Only fixed and permanently connected equipment that: • complies with the relevant standard; • is suitable for use on zone 0 according to the manufacturer's instructions for use and mounting; and • is protected by SELV at a nominal voltage not exceeding 12 V a.c. rms or 30 V ripple-free d.c., the safety source being installed outside of zones 0, 1 and 2
1	Switchgear, controlgear and accessories	Switches and controls incorporated in fixed current-using equipment suitable for use in the zone. Switches of SELV circuits supplied at a nominal voltage not exceeding 12 V a.c. rms or 30 V ripple-free d.c., the safety source being installed outside of zones 0, 1 and 2
	Current-using equipment	Only the fixed and permanently connected equipment listed below, provided it is suitable for use in zone 1 according to the manufacturer's instructions: • whirlpool units; • electric showers; • shower pumps; • equipment protected by SELV or PELV at a nominal voltage not exceeding 12 V a.c. rms or 30 V ripple-free d.c., the safety source being installed outside of zones 0, 1 and 2; • ventilation equipment; • towel rails; • water heating appliances; • luminaires
2	Switchgear, controlgear and accessories	Switches and controls incorporated in fixed current-using equipment suitable for use in the zone. Switches of SELV circuits, the safety source being installed outside of zones 0, 1 and 2. Shaver supply units to units to BS EN 61558-2-5, provided they are located where direct spray from a shower is unlikely
	Current-using equipment	The general requirements of BS 7671 apply
Outside 1 and 2	Switchgear, controlgear and accessories	Socket-outlets are not permitted within a distance of 3 m horizontally from the boundary of zone 1. However, this does not apply to SELV socket-outlets complying with Section 414 and shaver supply units to BS EN 61558-2-5
	Current-using equipment	The general requirements of BS 7671 apply

[a]The insulating pull cords of cord-operated switches are permitted in all zones and beyond.

16.2.3   Degrees of ingress protection

For locations containing a fixed bath or shower, the general requirements of Regulation Group 512.2 for degrees of protection against external influences are supplemented by Regulation 701.512.3, which are summarized in Table 16.2. Such influences are likely to include steam and condensation, falling drops of water and sprays and/or jets from shower nozzles.

16.2.4   Equipment permitted in and outside the various zones

The equipment sanctioned by Regulations 701.512.3 and 701.55 for installation in and outside the various zones is given in Table 16.3.

All the equipment listed in Table 16.3 should comply with the relevant requirements of the applicable standard (British or Harmonized) appropriate to the intended use of the equipment. Where equipment is not to a British or Harmonized standard, the designer or person responsible for specifying the installation is to verify that safety is not compromised.

Switches and controls, such as immersion heater switches, which fall within a particular zone would be required to meet the zonal requirements. However, the general requirements of BS 7671 are applicable to a location containing a bath or shower, and Regulation 512.2.1 calls for every item of equipment to be of a design appropriate to the situation in which it is to be used, or its mode of installation shall take account of the conditions likely to be encountered. Depending on the particular conditions of water and high humidity that are expected in the location, normal wall-mounted light switches and similar accessories may not have a degree of ingress protection appropriate for installation outside of zones 0, 1, and 2 and, in that case, would therefore not satisfy the requirements of BS 7671.

16.2.5   Electric shock

Protection against electric shock by obstacles and placing out of reach (Section 417) is not permitted by Regulation 701.410.3.5, even where such measures could be contemplated as potentially applicable. Similarly, protection by non-conducting location (Regulation 418.1) and protection by earth-free local equipotential bonding (Regulation 418.2) are precluded by Regulation 701.410.3.6.

Additional protection by RCDs

Regulation 701.411.3.3 requires additional protection to be provided to all circuits of the special location by one or more RCDs having the characteristics specified in Regulation 415.1.1 (a rated residual operating current $I_{\Delta n}$ not exceeding 30 mA and an operating time not exceeding 40 ms at a residual current of $5I_{\Delta n}$).

Although not explicitly stated in Regulation 701.411.3.3, for a circuit protected by SELV or PELV, additional protection by an RCD is required only for the circuit on the supply side of the source of SELV or PELV. Similarly, for a circuit protected by electrical separation, additional protection by an RCD is required only for the circuit on the supply side of the double-wound transformer or other source of the electrically separated supply.

Electrical separation

Regulation 701.413 stipulates that the protective measure of electrical separation may be used for only two purposes in the special location: a circuit supplying a single item of current-using equipment, or one single socket-outlet.

SELV and PELV

The only protective measure against electric shock permitted in zone 0 is SELV at a nominal voltage not exceeding 12 V a.c. rms or 30 V ripple-free d.c. with the safety source being located outside of zones 0, 1 and 2 (Regulation 701.55). The protective measure PELV may be used in a location containing a bath or shower outside zone 0. The preconditions for both SELV and PELV are summarized in Table 16.4.

Table 16.4 Particular requirements for bath and shower locations: preconditions for SELV or PELV circuits

Protective measure	Regulation	
SELV	701.414.4.5	Basic protection against electric shock must be provided by basic insulation complying with Regulation 416.1 or by barriers or enclosures complying with Regulation 416.2 giving at least a degree of protection of IPXXB or IP2X
	701.512.3, 701.55	The source of SELV must be located outside zones 0, 1 and 2
	701.55	The nominal voltage of an SELV circuit in zone 0 must not exceed 12 V a.c. or 30 V ripple-free d.c. The circuit must not have any switchgear or accessories in that zone
	701.55	The nominal voltage of an SELV circuit with switches installed in zone 1 (other than switches incorporated in fixed current-using equipment) must not exceed 12 V a.c. or 30 V ripple-free d.c. The nominal voltage of fixed current-using equipment in zone 1 protected by SELV must not exceed 25 V a.c. or 60 V ripple-free d.c.
PELV	701.414.4.5	Basic protection against electric shock must be provided by basic insulation complying with Regulation 416.1 or by barriers or enclosures complying with Regulation 416.2 giving at least a degree of protection of IPXXB or IP2X
	701.512.3, 701.55	The source of PELV must be located outside zones 0, 1 and 2
	701.55	The nominal voltage of fixed current-using equipment in zone 1 protected by PELV must not exceed 25 V a.c. or 60 V ripple-free d.c. The PELV circuit must not have any switchgear or accessories in that zone

16.2.6 Supplementary bonding

Although there is a general requirement in Regulation 701.415.2 for local supplementary bonding to be provided for a location containing a fixed bath or shower, the regulation allows this bonding to be omitted if all of the following three conditions are met:

1. All final circuits of the special location meet the requirements of Regulation 411.3.2 for automatic disconnection in the event of a fault (Section 5.3.3 of this Guide refers).
2. All final circuits of the special location have additional protection by means of an RCD in accordance with Regulation 701.411.3.3 (Section 16.2.5 of this Guide refers).
3. The special location is in a building where main protective bonding is provided in accordance with Regulation 411.3.1.2 (Section 5.3.3 of this Guide refers), and all extraneous-conductive-parts of the special location are connected to the MET of the installation by this main bonding.

As regards condition (3), a note to Regulation 701.415.2 points out that the effectiveness of the connection of extraneous-conductive-parts in the special location to the MET may be assessed by the application of Regulation 415.2.2. That is to say, in an a.c. system, this connection can be considered effective (at least in terms of its resistance) if Equation (16.1) is satisfied.

$$R \leq \frac{50}{I_a} \qquad (16.1)$$

where R is the resistance between extraneous-conductive-parts and extraneous-conductive-parts, between exposed-conductive-parts and exposed-conductive-parts, and between extraneous-conductive-parts and exposed-conductive-parts in the special location, and I_a (A) is the current that causes operation of the fault current protective device within 5 s (for an RCD this is the rated residual operating current $I_{\Delta n}$).

The value of I_a used in Equation (16.1) can be obtained from the time–current characteristics in Appendix 3 of BS 7671. It should be the highest of the I_a values for the protective devices at the consumer unit or distribution board that protect the circuits of the special location.

In a new installation, it is to be expected that conditions (1), (2) and (3) will all be met, by virtue of the whole installation having to satisfy the requirements of BS 7671 (although compliance with condition (3) should always be checked using Equation (16.1) where there is doubt). However, this may not be the case for an existing installation.

Suppose that a new circuit is being added that will supply equipment in the bathroom or shower room, or that an alteration or addition is being made to one or more of the existing circuits of that room. In such a case, conditions (1), (2) and (3) may not all be met. For example, although the new or altered circuit would have to be provided with additional protection by an RCD to comply with Regulation 701.411.3.3, one or more of the other (existing) circuits of the room may not have such protection, in which case condition (2) would not be met. In such a case, the designer of the alteration or addition would have to ensure that compliant supplementary bonding was provided for the special location, or that such bonding already existed.

Where supplementary bonding is to be provided, Regulation 701.415.2 requires the protective conductor of each circuit supplying Class I and Class II electrical equipment within the room containing the bath or shower, and accessible extraneous-conductive-parts within that room (for example, metallic pipework services, baths and shower basins), all to be connected together by local supplementary bonding conductors, which must comply with Regulation Group 544.2. It should be noted that the extent of the supplementary bonding required by BS 7671:2008 (that is, throughout the room containing the bath or shower) is greater than was required by BS 7671:2001 as amended. BS 7671:2001 only required the supplementary bonding to be within zones 1, 2 and 3 (zone 3 not being recognized in BS 7671: 2008), or in zones 1 and 2 in the case of a room other than a bathroom or shower room where a cabinet containing a shower and/or bath was installed.

Exposed-conductive-parts of SELV circuits are excluded from the supplementary bonding requirements and must not be connected to such bonding (Regulation 414.4.4).

Metallic door architraves, window frames and similar parts are not considered to be extraneous-conductive-parts unless they are electrically connected to metallic parts of the building and, therefore, are not required to be bonded. Normal accessible metallic structural parts of the building are required to be bonded (see Regulation 701.415.2).

The supplementary bonding may be carried out either inside or outside the room containing the bath or shower, but in either case it should preferably be done close to the points of entry of extraneous-conductive-parts into the room (Regulation 701.415.2 refers).

A supplementary bonding connection, as with all electrical connections or joints, is required by Regulation 526.3 to be accessible for inspection, testing and maintenance (apart from connections or joints made in one of the ways exempted by that regulation, such as joints made by welding, brazing or a compression tool).

Exposed-conductive-parts may include metallic accessory plates and stationary heating equipment, towel rails, etc., and floor heating element sheath.

Extraneous-conductive-parts may include metallic baths, metallic hot and cold pipework, metallic waste pipework, central heating radiators and associated pipework, floor heating grid, central heating tank and other pipework in the location, and any other conductive part entering the bathroom.

The cross-sectional area of supplementary bonding conductors is determined by a number of factors:

- Regulation 544.2.1 calls for the minimum conductance of a conductor connecting exposed-conductive-parts together to be not less than the conductance of the smaller cpc of the circuits concerned and, in any event, the cross-sectional area to be not less than $4 \, \text{mm}^2$ where mechanical protection is not provided;
- Regulation 544.2.2 calls for a conductor connecting exposed-conductive-parts together with extraneous-conductive-parts to be not less than half the conductance of the cpc of the exposed-conductive-part (the largest), and the cross-sectional area to be not less than $4 \, \text{mm}^2$ where mechanical protection is not provided; and
- Regulation 544.2.3 calls for a conductor connecting two extraneous-conductive-parts together to be not less than $2.5 \, \text{mm}^2$ if sheathed or otherwise mechanically protected, and to be not less than $4 \, \text{mm}^2$ where mechanical protection is not provided.

In most practical situations, the option of mechanical protection will not be feasible for the whole length, and the minimum cross-sectional area will therefore be $4 \, \text{mm}^2$. However, this may need to be increased because of the requirements of Regulations 544.2.1 and 544.2.2 where, for example, in rare cases a cpc of cross-sectional area greater than 4 mm^2 is employed.

Copper and other suitably conductive pipes and associated fittings, which represent permanent and reliable conductors, may be used as supplementary bonding conductors (or part conductors), as permitted by Regulations 544.2.4 and 543.2.6.

Regulation 544.2.5 permits supplementary bonding to fixed equipment, such as electric towel rails, to be effected by the cpc in a short length (say, 150 mm) of flexible cord used as the final connection to that appliance, so that ugly bonding clamps affixed to such equipment may be avoided. However, this 'relaxation' does not mean that the need to supplementary bond such equipment is obviated.

16.2.7 Shaver supply units and socket-outlets

In a location containing a bath or shower, shaver supply units complying with BS EN 61558-2-5 are permitted outside of zones 0 and 1 (see Regulation 701.512.3). However, such a shaver supply unit is not normally regarded as 'splash-proof'; therefore, it cannot simply be positioned anywhere outside those zones. It should be positioned well away from a bath or a shower, and normal lighting switch height will usually safeguard against splashing from a wash basin. Shaver supply units complying with BS EN 61558-2-5 incorporate a safety isolating transformer and afford protection against electric shock by electrical separation.

The long-established rule disallowing other types of LV socket-outlet (such as 13 A socket-outlets to BS 1363 and shaver socket-outlets complying with BS 4573 *Specification for 2-pin reversible plugs and shaver socket-outlets*) in a location containing a bath or shower has not been maintained in BS 7671:2008. However, Regulation 701.512.3 requires that any such socket-outlets are positioned at least 3 m outside the boundary of zone 1, and the circuits supplying them must be provided with additional protection by an RCD (see Section 16.2.5 of this Guide).

Regulation 701.512.3 also allows the installation of PELV socket-outlets, although, once again, they must be at least 3 m horizontally outside the boundary of zone 1.

Socket-outlets supplied by an SELV circuit are permitted anywhere outside zones 0 and 1 in the special location.

16.2.8 Mobile and nonfixed current-using equipment

Like all other circuits of the location containing the bath or shower, circuits to supply mobile or other nonfixed current-using equipment in the location must by provided with additional protection by RCDs (see Section 16.2.5 of this Guide).

As indicated in Section 16.2.7 of this Guide, socket-outlet provisions in the special location are limited to the following, whether for connection of mobile equipment or other nonfixed current-using equipment (Regulation 701.512.3 refers):

1. SELV socket-outlets installed outside zones 0 and 1, the safety source being outside zones 0, 1 and 2;
2. shaver supply units complying with BS EN 61558-2-5 installed outside zones 0 and 1;

3. Other socket-outlets, such as 13 A socket-outlets to BS 1363, installed at least 3 m horizontally outside the boundary of zone 1.

It is acceptable to install a means other than a socket-outlet, such as a 13 A fused connection unit to BS 1363, outside zones 0, 1 and 2 for the connection of nonfixed current-using equipment, such as a washing machine. Alternatively, such current-using equipment may be connected to an LV socket-outlet installed at least 3 m horizontally outside the boundary of zone 1 in accordance with (3) above.

Even where all the relevant electrical installation requirements of BS 7671 can be met, and in the absence of the equipment manufacturer's installation and/or user instructions for the appliance giving permission for such use, the express agreement of the manufacturer of a nonportable appliance should be sought as to the suitability of the appliance for use in a bathroom or shower room. In extreme cases, a manufacturer's agent may not be prepared to carry out a service on an appliance in a bathroom until it is moved to what might be considered a 'safer' location.

16.2.9   Electric heating embedded in the floor

An electric heating unit embedded in the floor below any zone for heating the floor of the location is permitted by Regulation 701.753, provided it is either covered by a fine mesh metallic grid or has a metallic sheath. The grid or sheath must be earthed (through the cpc), except where the protective measure SELV is used for the circuit.

The heating units may consist of either heating cables or thin-sheet flexible heating elements, and must comply with the relevant product standard.

As well as meeting the requirements of Section 701 of BS 7671 (and the general requirements of BS 7671), the floor heating system must meet the requirements of Section 753, which are explained later in this chapter of the Guide.

16.2.10   Other equipment

Other equipment, not readily associated with the fixed electrical installation, is sometimes considered for use in a bathroom, and care should be taken to avoid the use of potentially dangerous items in such a location. It has been known for telephones to be installed, and if this is contemplated then it should be borne in mind that this type of installation is covered by BS 7671 (see Regulation 110.1). Compliance with BS 6701 is also required for any installation subject to the Telecommunications Act 1984. Modern telephone systems typically operate at 50 V d.c. and the voltage is likely to exceed 80 V d.c. when the telephone rings, and where corded telephones are used it would obviously be desirable to prevent the telephone being accessible to the person using a bath even with the cord extended; in the typical UK bathroom, the installation of this type of telephone would be undesirable.

16.3   Swimming pools and other basins

16.3.1   General

Regulation 702.11 sets out the scope of Section 702 and states that the particular requirements relate to basins of swimming pools, basins of fountains, basins of paddling pools, and to their surrounding zones. Attention is drawn to the fact that this is a location in which there is an increased risk of electric shock by virtue of a reduction in body resistance (due mainly to the lower contact resistance associated with wet surfaces) and because of contact with Earth potential. This regulation also recognizes that swimming pools for medical treatment may need further special consideration. The regulation also explains that the requirements of Section 702 do not apply to natural waters, lakes in gravel pits, coastal areas and the like. The requirements for initial inspection, testing and certification of installations in the special location are no different from the general requirements.

Regulation 702.32 defines three zones (zones 0, 1 and 2) in terms of linear dimensions, and it is important to recognize that the zones are volume zones and the dimensions apply on the horizontal and vertical axes. Figure 16.2 illustrates the zones in section and plan views and gives a representation of the zones for a typical, small swimming pool. It should be noted that the increased height area of zone 1 depicted above and around the diving board in Section 'A–A' of Figure 16.2 applies wherever there are diving boards, springboards, starting blocks, chutes, accessible sculptures, viewing bays, decorative basins, etc. expected to be occupied by persons. In such cases, the increased upper limit of zone 1 is the horizontal plane 2.5 m above the highest surface expected to be occupied by persons, extending out to 1.5 m from the periphery of the diving board, etc. concerned.

Considerations of shock protection and selection and erection of equipment need to take account of the particular requirements peculiar to zones 0, 1 and 2, and one should not lose sight of the need to comply also with the general requirements of BS 7671.

16.3.2 Additional requirements relating to electric shock

Table 16.5 sets out the particular requirements in terms of protection against electric shock.

Supplementary bonding is required in zones 0, 1 and 2 if (and only if) there is any electrical equipment in those zones that has exposed-conductive-parts and for which protection against electric shock is provided by ADS. Where this is the case, Regulation 702.415.2 requires all extraneous-conductive-parts in the zones to be connected by supplementary bonding conductors to the protective conductors of the circuits supplying that equipment. The supplementary bonding must comply with Regulation 415.2, which requires (amongst other things) that the bonding connects to the main metallic reinforcement of constructional reinforced concrete, where practicable. The supplementary bonding must not connect to the exposed-conductive-parts of any equipment connected to a circuit protected by SELV or electrical separation.

Figure 16.3 gives an example of the connection of a supplementary bonding conductor to the metallic reinforcement of a concrete floor, though there are many other ways of achieving this link.

16.3.3 Additional requirements relating to selection and erection of equipment

Table 16.6 identifies the additional requirements relating to selection and erection of equipment. Table 16.7 details the degrees of protection required in terms of the three zones.

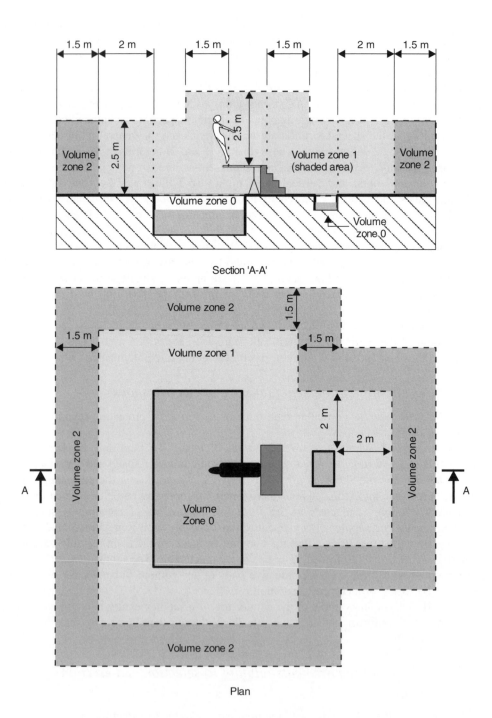

Figure 16.2 Swimming pool zones: section and plan. (See plate 1)

Table 16.5 Particular requirements for swimming pools – protection against electric shock

Aspect	Regulation	Particular requirements	Comments
General	702.11	Scope defined	
	702.3	Assessment of general characteristics	
Electric shock	702.410.3.4.1	In zone 0, except for fountains, only SELV at a nominal voltage not exceeding 12 V a.c. or 30 V d.c. is to be used	SELV safety source to be outside zones 0, 1 and 2
		In zone 1, except for fountains, only SELV at a nominal voltage not exceeding 25 V a.c. or 60 V d.c. is to be used	SELV safety source to be outside zones 0, 1 and 2
		In zones 0 and 1, except for fountains, circuits for supplying equipment for use in basins when people are not in zone 0 are to be supplied by one or more of:	Socket-outlet supplying the equipment, and the control device of the equipment, to have a notice not to use when persons are in the pool
		• SELV	SELV safety source to be outside zones 0, 1 and 2, but may be in zone 2 if supply circuit protected by RCD with characteristics specified in Regulation 415.1.1;
		• automatic disconnection of supply	using RCD with characteristics specified in Regulation 415.1.1;
		• electrical separation	Source of electrical separation to supply only one item of current-using equipment and to be outside zones 0, 1 and 2, but may be in zone 2 if supply circuit protected by RCD with characteristics specified in Regulation 415.1.1
	702.410.3.4.2	In zones 0 and 1 of fountains, one or more of the following is to be used:	
		• SELV	SELV safety source to be outside zones 0 and 1;
		• automatic disconnection of supply	using RCD with characteristics specified in Regulation 415.1.1;

(continued overleaf)

Table 16.5 *(continued)*

Aspect	Regulation	Particular requirements	Comments
	702.410.3.4.3	• electrical separation	Source of electrical separation to supply only one item of current-using equipment and to be outside zones 0 and 1
		In zone 2, one or more of the following is to be used (note: there is no zone 2 for fountains):	
		• SELV	SELV safety source to be outside zones 0, 1 and 2, but may be in zone 2 if supply circuit protected by RCD with characteristics specified in Regulation 415.1.1;
		• automatic disconnection of supply	using RCD with characteristics specified in Regulation 415.1.1; where PME conditions apply, it is recommended to install an earth mat of suitably low resistance (e.g. not more than 20 Ω) and connect to the protective equipotential bonding;
		• electrical separation	Source of electrical separation to supply only one item of current-using equipment and to be outside zones 0, 1 and 2, but may be in zone 2 if supply circuit protected by RCD with characteristics specified in Regulation 415.1.1
	702.410.3.5 and 702.410.3.6	These protective measures are precluded in zones 0, 1 and 2: • obstacles and placing out of reach; • non-conducting location and earth-free equipotential bonding	These protective measures are inappropriate and/or impracticable
	702.415.2	Local supplementary bonding required to extraneous-conductive-parts and cpcs of exposed-conductive-parts in zones 0, 1 and 2 (except equipment supplied by SELV or electrical separation	Only required where circuits protected by automatic disconnection of supply are used. The connection with the protective conductor may be made in close proximity to the location, such as at a local distribution board
	702.414.4.5	In zones 0, 1 and 2, basic insulation, or barriers and enclosure to at least IP2X or IPXXB, are required for SELV	Irrespective of nominal voltage

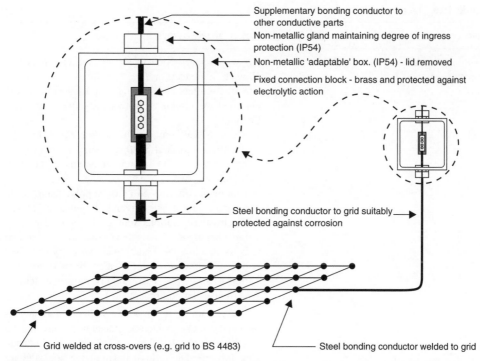

Supplementary bonding conductor to
other conductive parts

Non-metallic gland maintaining degree of ingress
protection (IP54)

Non-metallic 'adaptable' box. (IP54) - lid removed

Fixed connection block - brass and protected against
electrolytic action

Steel bonding conductor to grid suitably
protected against corrosion

Grid welded at cross-overs (e.g. grid to BS 4483)   Steel bonding conductor welded to grid

Figure 16.3 Typical example of the connection of a supplementary bonding conductor to the metallic reinforcement of a concrete floor

Table 16.6 Particular requirements for swimming pools: equipment

Aspect	Regulation	Particular requirements
Wiring systems (including where embedded ≤50 mm deep in walls, ceilings and floors)	702.522.21	Metallic sheaths and coverings of wirings systems in zones 0, 1 and 2 to be connected to supplementary bonding (note: cables preferably to be in nonmetallic conduit)
	702.522.22	Wiring systems in zones 0, 1 and 2 to be only those necessary for equipment in those zones
	702.522.23	Cables for equipment in zone 0 of a fountain to be installed as far as practicable outside basin rim, and run to the equipment by shortest practicable route Cables in zone 1 of a fountain to be suitably installed and protected for medium-severity stress (AG2) and submersion (AD8). See regulation for details
	702.522.24	No junction boxes are allowed in zone 0. Only SELV junction boxes are allowed in zone 1

(continued overleaf)

Table 16.6 *(continued)*

Aspect	Regulation	Particular requirements
Switchgear and controlgear	702.53	No switchgear, controlgear or socket-outlets are allowed in zones 0 or 1. Socket-outlets and switches are allowed in zone 2 provided the circuit is protected by one of the following: • SELV. SELV source to be outside zones 0 and 1. If source is located in zone 2, supply circuit to by protected by RCD having characteristics given in Regulation 415.1.1. • Automatic disconnection of supply, using RCD having characteristics given in Regulation 415.1.1. • Electrical separation. Source of electrically separated supply to be outside zones 0 and 1. If source is located in zone 2, supply circuit to by protected by RCD having characteristics given in Regulation 415.1.1. Source to supply only one item of current-using equipment. For swimming pools where switches and socket-outlets cannot be located outside zone 1, they may be installed in zone 1 if they are at least 1.25 m outside zone 0 border and at least 0.3 m above floor and are protected by one of the following. The accessories should preferably have non-conductive cover plates or covers. • SELV at nominal voltage not exceeding 25 V a.c or 60 V d.c. SELV source to be outside zones 0 and 1. • Automatic disconnection of supply, using RCD having characteristics given in Regulation 415.1.1. • Electrical separation. Source of electrically separated supply to be outside zones 0 and 1. If source is located in zone 2, supply circuit to by protected by RCD having characteristics given in Regulation 415.1.1. Source to supply only one item of current-using equipment
Current-using equipment of swimming pools	701.55.1	Fixed current-using equipment in zones 0 and 1 to be designed specifically for use in a swimming pool, in accordance with Regulations 702.55.2 and 702.55.4. Current-using equipment intended for operation only when no persons are in zone 0 may be used in all zones, but must be supplied from a circuit protected as required by Regulation 702.410.3.4.3 (see Table 16.5).

Table 16.6   (*continued*)

		Electric heating units embedded in the floor are allowed provided that they: • are protected by SELV, the SELV source being outside zones 0 and 1, and if located in zone 2, SELV source to be supplied by a circuit protected by RCD having characteristics given in Regulation 415.1.1; or • incorporate an earthed metallic sheath connected to the supplementary bonding required by Regulation 702.415.2 and are supplied by a circuit protected by RCD having characteristics given in Regulation 415.1.1; or • are covered by an embedded earthed metallic grid connected to the supplementary bonding required by Regulation 702.415.2 and are supplied by a circuit protected by RCD having characteristics given in Regulation 415.1.1
Underwater luminaires in swimming pools	702.55.2	Luminaires in contact with water to comply with BS EN 60598-2-18 and be fixed. Lighting behind watertight portholes and serviced from behind to comply with BS EN 60598 (applicable parts). Installation to be such that there will be no contact between exposed-conductive-parts of luminaires and conductive parts of portholes
Equipment in fountains	702.55.3	Equipment in zones 0 or 1 to be mechanically protected to medium severity (AG2). Luminaires in zones 0 and 1 to comply with BS EN 60598-2-18 and be fixed. Electric pumps to comply with BS EN 60335-2-41
Equipment in zone 1	702.55.4	Equipment such as filtration systems and jet stream pumps for swimming pools is allowed in zone 1 if all the following conditions are met: • The equipment is inside an insulating enclosure giving Class II or equivalent insulation and protection against medium severity impact (AG2). • Access to the equipment to be through a hatch or door, needing a key or tool for opening, which causes disconnection of all live conductors when opened. Installation of the supply cable and means of disconnection to such that Class II or equivalent insulation is provided.

(*continued overleaf*)

Table 16.6   (*continued*)

Aspect	Regulation	Particular requirements
		• The supply circuit to the equipment to be protected by: – SELV at nominal voltage not exceeding 25 V a.c or 60 V d.c. SELV source to be outside zones 0 and 1; or – ADS, using RCD having characteristics given in Regulation 415.1.1; or – Electrical separation. Source of electrically separated supply to be outside zones 0 and 1. If source is located in zone 2, supply circuit to by protected by RCD having characteristics given in Regulation 415.1.1. Source to supply only one item of current-using equipment. If the swimming pool has no zone 2, lighting equipment in zone 1 that is not protected by SELV is allowed if installed on a wall or ceiling, at least 2 m above the lower limit of zone 1, provided the circuit is protected by ADS, using RCD having characteristics given in Regulation 415.1.1. Every luminaire to have an enclosure that provides Class II or equivalent protection and protection against medium-severity impact.

Table 16.7   Particular requirements for swimming pools: IP degrees of protection (Regulation 702.512.2)

Circumstances	Zone 0	Zone 1	Zone 2
Swimming pool where water jets are likely to be used	IPX8	IPX5	IPX5
Swimming pool where water jets are not likely to be used – indoor pools	IPX8	IPX4	IPX2
Swimming pool where water jets are not likely to be used – outdoor pools	IPX8	IPX4	IPX4

Aide-mémoire (see BS EN 60 529 for full definitions of degrees of protection): IPX2 – protected against drops of water falling at up to 15° from the vertical; IPX4 – protected against projections of water from all directions; IPX5 – protected against jets of water from all directions.

16.4 Rooms and cabins containing sauna heaters

16.4.1 General

Regulation 703.1 sets out the scope of Section 703. The requirements of the section apply to a sauna cabin erected on site, such as in a room. They also apply to a room where a sauna heater or sauna heating appliances are installed, in which case the whole room is considered as the sauna. The requirements do not apply to prefabricated sauna cabins that comply with the relevant equipment standard. Where the sauna includes facilities such as a shower or bath, the requirements of Section 701 also apply.

As with swimming pools, the concept of zones is adopted for this location, but this time the zones relate to temperature. There are three such zones: 1, 2 and 3. The requirements for initial inspection, testing and certification of installations in this location are no different from the general requirements.

Figure 16.4 gives a sectional view and plan of a typical layout and Table 16.8 sets out the particular requirements in terms of protection against electric shock and selection and erection of equipment.

16.5 Construction-site installations

16.5.1 General

As with all special installations or locations, the general requirements of BS 7671 apply to construction- and demolition-site installations unless supplemented or modified by the particular requirements of Section 704 of BS 7671. Additionally, the regulations covered under 411.8 *Reduced low voltage systems* will be of particular interest to the designer of such installations, as will the recommendations contained in BS 7375 *Code of Practice for distribution of electricity on construction and building sites* for the installation. Site distribution equipment is addressed in BS EN 60439-4 *Low-voltage switchgear and controlgear assemblies. Particular requirements for assemblies for construction sites (ACS)*.

Regulation Group 704.1 outlines the scope of this section and states that Section 704 applies to temporary installations for construction and demolition sites during the period of the construction or demolition work. Examples of such sites that are listed are those where there is construction work of new buildings, where there is repair, alteration, extension or demolition of existing buildings or parts thereof, engineering works and earthworks.

Installations in administrative locations of construction sites are not required to meet the particular requirements of this Section 704, but would, of course, need to comply with the general requirements of BS 7671. Such fixed installations may include those of site offices, canteens, toilets, sleeping dormitories and meeting rooms, etc. where there is no requirement for temporary supplies. The requirements of Section 704 do not apply, either, to installations covered by IEC 60621 series 2, *Electrical installations for outdoor sites under heavy conditions (including open-cast mines and quarries)* where equipment similar to that used for surface mining applications is used.

Regulation 704.313.3 calls for equipment used on a particular supply to be identified with that supply and, of course, be compatible. Where more than one such supply unit exists on a site, marking of the equipment will be necessary and care must be taken that there are no interconnections between the supply units except for necessary control and signalling circuits, where required.

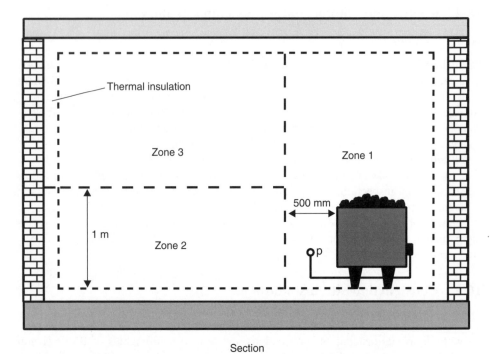

Section

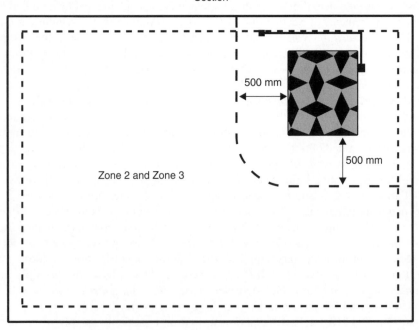

Plan

Figure 16.4 Hot air sauna: section and plan views

Table 16.8 Particular requirements for rooms and cabins containing sauna heaters

Aspect	Regulation	Particular requirements
Electric shock	703.414.4.5	Irrespective of nominal voltage (0–50 V), where SELV or PELV is used, basic protection must be provided by insulation (complying with Regulation 416.2) and/or barriers or enclosures to at least IP24 or IPX4B
	703.410.3.5	Protection not permitted by means of: • obstacles, and/or • placing out of reach
	703.410.3.6	Protection not permitted by means of: • non-conducting location, and/or • earth-free local equipotential bonding
	703.411.3.3	All circuits of the sauna must have additional protection by one or more RCDs having the characteristics specified in Regulation 415.1.1. Such protection need not be provided for the sauna heater unless recommended by the manufacturer
External influences	703.512.2	Equipment to have at least IPX4 degree of protection, or IPX5 where cleaning by use of water jets may be reasonably expected. In zone 1, only the sauna heater and associated equipment are allowed. In zone 2 there is no special requirement for heat resistance of equipment. In zone 3, equipment must withstand at least 125 °C and cables must withstand at least 170 °C
Wiring system	703.52	Wiring preferably to be installed outside the zones, on cold side of the thermal insulation. Any wiring in zones 1 or 3 must be heat resisting. Metallic cable sheath and containments not to be accessible in normal use
Isolation, switching	703.537.5	Switchgear and controlgear forming part of the sauna or other fixed equipment in zone 2 may be installed, to sauna manufacturer's instructions. Other switchgear and controlgear to be installed outside the special location. Socket-outlets not permitted in the special location
Sauna heating appliances	703.55	Must comply with BS EN 60335-2-53 and be installed to manufacturer's instructions

The requirements for initial inspection, testing and certification of installations in this location are no different from the general requirements.

16.5.2 Protection against electric shock

Table 16.9 sets out the particular requirements of Section 704 in terms of protection against electric shock. These requirements are far less extensive, and far less complex, than those of the corresponding section of BS 7671:2001, Section 604. For example, the reduced disconnection times for TN systems, and the reduced touch voltage requirements, specified

Table 16.9 Particular requirements for construction and demolition sites: protection against electric shock

Aspect	Regulation	Particular requirements	Comments
General	704.410.3.5	Protection by obstacles or placing out of reach (Section 417) is not permitted	These protective measures are inappropriate and/or impracticable. Although not mentioned in the regulation, the measures of non-conducting location and earth-free local equipotential bonding are not suitable either
	704.410.3.10	Protection by one of the following measures is required for a circuit supplying either socket-outlets rated at 32 A or less or hand-held equipment rated at 32 A or less: • Reduced low volatge (Regulation 411.8).	Strongly preferred for supplies to portable hand lamps for general use, portable hand tools and local lighting up to 2 kW
		• Automatic disconnection of supply with additional protection by an RCD having the characteristics given in Regulation 415.1.1.	
		• Electrical separation (Section 413). Each socket-outlet and item of hand-held equipment to be supplied by a separate transformer or separate winding thereof.	It is essential that flexible cables and cords are visible where they are liable to mechanical damage
		• SELV or PELV (Section 414)	Strongly preferred for supplies to hand lamps in confined and damp locations
Automatic disconnection of supply	704.411.3.1	A TN-C-S system is not to be used.	This restriction refers to a TN-C-S system where PME is applied (i.e. supplied from the public LV network), and is due to the impracticability of main bonding all extraneous-conductive-parts to comply with Regulation 411.3.1.2.

Table 16.9   (*continued*)

			Such a TN-C-S system is prohibited by legislation (ESQCR) for supplies to caravans or similar units
	704.411.3.2.1	For a circuit supplying one or more socket-outlets rated at more than 32 A: • it is not permitted to apply Regulations 411.3.2.5 and 411.3.2.6, allowing the disconnection times required by Chapter 41 to be exceeded in certain circumstances; and	
		• an RCD having rated residual operating current not exceeding 500 mA is to be provided for fault protection	The disconnection time is not required to be any shorter than required by the general requirements of Regulations 411.3.2.2–411.3.2.4.
SELV and PELV systems	704.414.4.5	Basic protection by means of basic insulation or barriers and enclosures is always required.	Irrespective of the nominal voltage

in Section 604 of BS 7671:2001 where earthed equipotential bonding and automatic disconnection (EEBAD) was used (the equivalent of ADS in BS 7671: 2008), are not required by Section 704 of BS 7671: 2008.

Many, if not most, site 'short-term' installations will take their supply from the public network through an ED. Most EDs will not be willing to provide earthing facilities, particularly as the use of a TN-C-S system with PME is prohibited by Regulation 704.411.3.1. This will usually necessitate the use of a TT system, which will inevitably lead to the need to use RCDs to meet the requirements of Regulation Group 411.5 to provide protection against electric shock under fault conditions. Where a TT system is employed, a connection to Earth will be required for the exposed-conductive-parts of the installation. It is important to establish that this connection is made with a permanent and reliable earth electrode which is suitably protected against any mechanical damage to which it may be subjected, bearing in mind the more onerous conditions prevailing on such sites and the robust work methods of site-workers.

16.5.3   Selection of equipment

There are additional requirements for construction-site installations in the selection and erection of equipment. These requirements, supplementing or modifying the general requirements, are summarized in Table 16.10.

Table 16.10 Construction site: additional requirements in terms of selection and erection of equipment

Aspect	Regulation	Particular requirements
Switchgear and controlgear	704.511.1	All assemblies for supply and distribution of electricity to comply with BS EN 60439-4. All socket-outlets rated at more than 16 A to comply with BS EN 60309-2
	704.537.2.2	Each assembly for construction sites to incorporate suitable devices for switching and isolating the incoming supply. Devices for isolation and switching to be suitable for securing in the off position, such as by a padlock or location of the device in a lockable enclosure. All current-using equipment to be supplied from assemblies for construction sites that comprise: • overcurrent protective devices; • devices affording fault protection (electric shock); and • socket-outlets, where required. Supplies for safety services and standby supplies to be connected by devices arranged to prevent interconnection of different supplies
Wiring systems	704.522.8.10	Cables not to be installed across roads or walkways unless adequately protected against mechanical damage
	704.522.8.11	Cables for reduced low voltage systems and flexible cables for applications exceeding reduced low voltage are to comply with the British Standards specified in this regulation

Regulation 704.522.8.10 stresses the need to protect cables from mechanical damage, particularly where they cross walkways and roadways where they may suffer major damage by safety boots, spades, forks, etc., not to mention compression from heavy vehicles.

16.5.4 Isolation and switching

The general requirements of Section 537 are applicable to construction sites, but Regulation 704.537.2.2 makes further demands for this location. An assembly housing the main switchgear and controlgear must be provided at the origin (see Figures 16.5 and 16.6). This assembly must include the means to switch and isolate the incoming supply and incorporate facilities to secure (lock off) all isolating devices in the OFF position. It is a requirement for all construction-site temporary circuits to be fed from a distribution assembly enclosing the necessary overcurrent protective devices, devices for fault protection, isolating devices and, where required, socket-outlets to BS EN 60309-2.

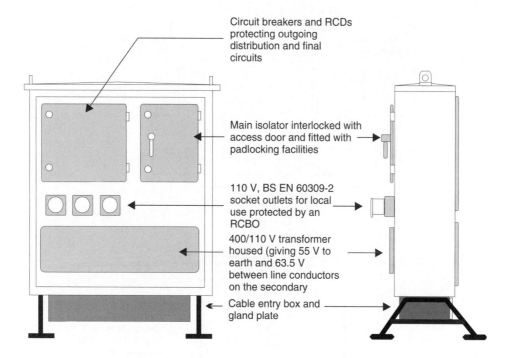

Circuit breakers and RCDs protecting outgoing distribution and final circuits

Main isolator interlocked with access door and fitted with padlocking facilities

110 V, BS EN 60309-2 socket outlets for local use protected by an RCBO

400/110 V transformer housed (giving 55 V to earth and 63.5 V between line conductors on the secondary

Cable entry box and gland plate

Figure 16.5 A 200 A construction-site distribution unit

16.6 Agricultural and horticultural premises

16.6.1 General

As with all special installations or locations, the general requirements apply unless supplemented or modified by the particular requirements. Regulation 705-1 defines in the scope that fixed installations, indoors and outdoors, in premises for agricultural and horticultural use are subject to the particular requirements of Section 705. Fixed installations in buildings intended solely for use by humans (farmhouses, farm offices, canteens, etc.) are not expected to meet these particular requirements though they do, of course, need to meet the general requirements.

This location is one in which there is an increased risk of electric shock by virtue of a reduction in body resistance (due mainly to the lower contact resistance associated with wet surfaces) and because of contact with Earth potential.

16.6.2 Protection against electric shock

Table 16.11 sets out the particular requirements of Section 705 in terms of protection against electric shock. These requirements are far less extensive, and far less complex,

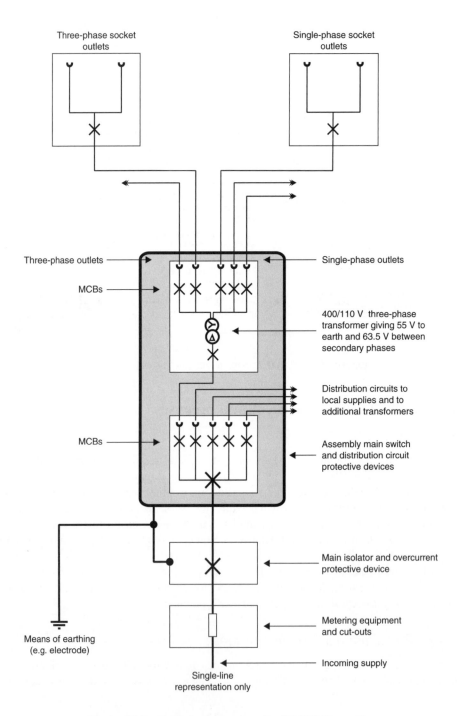

Figure 16.6   Typical construction-site 400/110 V supplies

Table 16.11 Particular requirements for agricultural and horticultural premises – protection against electric shock

Aspect	Regulation	Particular requirements	Comments
General	705.410.3.5	Protection by obstacles or placing out of reach (Section 417) is not permitted	These protective measures are inappropriate and/or impracticable
	705.410.3.6	Protection by of non-conducting location or earth-free local equipotential bonding (Regulation 418.1) is not permitted	
Automatic disconnection of supply.	705.411.4	A TN-C system is not to be used. This applies both to the installation of the agricultural premises and in the installation of residences or other locations that have a conductive connection to the agricultural or horticultural premises by protective conductors or extraneous-conductive-parts	In any event, the use of a TN-C system is prohibited by legislation (ESQCR) for an installation connected to the public electricity network
	705.411.1	RCDs are to be used as the disconnecting devices in circuits as follows ($I_{\Delta n}$ denotes rated residual operating current): • in final circuits supplying socket-outlets rated at 32 A or less, RCD with $I_{\Delta n} \leq 30$ mA; • in final circuits supplying socket-outlets rated at more than 32 A or less, RCD with $I_{\Delta n} \leq 100$ mA; • in all other circuits, RCDs with $I_{\Delta n} \leq 300$ mA	The disconnection time is not required to be any shorter than required by the general requirements of Regulations 411.3.2.2 to 411.3.2.4
SELV and PELV systems	705.414.4.5	Basic protection by means of basic insulation or barriers and enclosures is always required.	Irrespective of the nominal voltage
Supplementary bonding	705.415.2.1	Supplementary bonding to be provided in locations intended for livestock	Refer to Section 16.6.2 of this Guide

than those of the corresponding section of BS 7671:2001, Section 605. For example, the reduced disconnection times for TN systems, and the reduced touch voltage requirements, specified in Section 604 of BS 7671:2001 where EEBAD was used (the equivalent of ADS in BS 7671: 2008), are not required by Section 705 of BS 7671:2008.

Many, if not most, installations of premises for agricultural and horticultural use will take their supply from the public network through an ED. Some EDs may not be willing to provide PME earthing facilities for premises where livestock are kept, particularly where floors intended to be occupied by the livestock do not incorporate a metal grid connected to the supplementary bonding. This will usually necessitate the use of a TT system, which will inevitably lead to the need to use RCDs to meet the requirements of Regulation Group 411.5 to provide protection against electric shock under fault conditions. Where a TT system is employed, a connection to Earth will be required for the exposed-conductive-parts of the installation. It is important to establish that this connection is made with a permanent and reliable earth electrode which is suitably protected against any mechanical damage to which it may be subjected, bearing in mind the more onerous conditions prevailing on such sites.

Supplementary bonding is called for by Regulation 705.415.2.1 in locations intended for livestock. This must connect together all exposed-conductive-parts and extraneous-conductive-parts that can be touched by livestock. Extraneous-conductive-parts that the regulation particularly refers to in the context of supplementary bonding are those in or on the floor, such as concrete reinforcement in general and reinforcement of cellars for liquid manure, as well as reinforcement prefabricated concrete elements of spaced floors. A note to the regulation points out that the use of a TN-C-S system is not recommended where there is not a metal grid in the floor. The particular need for supplementary bonding conductors to be protected against mechanical damage and corrosion in agricultural and horticultural locations is underlined by Regulation 705.544.2, which also requires the conductors to be selected to avoid electrolytic effects. Examples of suitable bonding conductors that are given in the regulation are hot-dip galvanized steel strip with dimensions of at least 30 mm × 3mm, hot-dip galvanized round steel of at least 8 mm diameter and copper conductors of at least 4 mm^2 cross-sectional area.

16.6.3   Fire and harmful thermal effects

Regulation 705.422.7 calls for all equipment to be protected with an RCD of rated operating current not exceeding 300 mA to protect against fire. Where continuity of service is required, the RCDs should be of the delay type (S) where they are not protecting socket-outlets. As indicated in the note to Regulation 705.422.7, RCDs provided for fault protection of final circuits in accordance with Regulation 411.1 may also be used to meet the requirement of this regulation for protection against fire.

Careful siting of heating appliances is called for in Regulation 705.422.6, and such equipment needs to be located at a safe distance from livestock and combustible materials bearing in mind the likely temperatures that such equipment may attain. In the case of radiant heaters, the distance must be at least 500 mm or some greater distance specified by the manufacturer. Electric heating appliances for breeding and rearing livestock are required to comply with BS EN 60335-2-71 *Household and similar electrical appliances.*

Safety. Particular requirements for electrical heating appliances for breeding and rearing animals.

For extra-low voltage circuits in locations where there is a risk of fire, Regulation 705.422.8 calls for live conductors to be provided with barriers and enclosures affording at least the degree of protection IPXXD or IP4X, or be enclosed in insulating material (such as insulating conduit) in addition to their basic insulation. A note to the regulation points out that cables of the type H07RN-F to BS 7919 for outdoor use will meet this requirement.

The general requirements relating to protection against thermal effects of Chapter 42 of BS 7671 and to the spread of fire contained in Section 527 also apply.

16.6.4   Selection of equipment

There are additional requirements for installations in agricultural and horticultural premises in the selection and erection of equipment. These requirements, supplementing or modifying the general requirements, are summarized in Table 16.12. Generally speaking, the use of Class II equipment is recommended, but careful selection of all equipment is necessary to establish, as far as is practicable, that the installation remains safe during its lifetime in what must be considered to be difficult environmental conditions. The choice of equipment with adequate IP rating is essential, and particular care in installation is needed to guarantee that this rating is not impaired during installation. Wiring system entries to other equipment should be kept to a minimum and sealed as appropriate to maintain the necessary protection. Unused holes should similarly be blanked off and sealed.

16.6.5   Automatic life support for high-density livestock rearing

For the first time in the IEE Wiring Regulations, Regulation 705.559 requires particular account to be taken of systems for the life support of livestock where there is high-density livestock rearing.

The regulation requires that a standby generator or other secure source of supply, of adequate capacity, is provided to ensure the supply of food, water, air and lighting to livestock where this is not ensured in the event of a power supply failure. The regulation also calls for separate, dedicated final circuits to be provided for the supplies to ventilation equipment and lighting.

As an alternative to a standby electrical source in the case of ventilation equipment, the regulation permits the use of devices that monitor the temperature and the supply voltage. The devices must give a visual or audible warning that can easily be observed by the persons responsible for the welfare of the livestock in the event of the temperature or supply voltage falling outside of the acceptable limits. The operation of the warning must be independent of the normal source of supply.

To avoid the loss of supplies to ventilation equipment in the event of a fault on another circuit, the regulation requires discrimination to be ensured between the main circuits supplying ventilation equipment and those used for other purposes. Section 11.7 of this Guide gives guidance on achieving discrimination, as does Section 16.6.6 in the case of RCDs in series.

Table 16.12 Agricultural and horticultural premises: additional requirements in terms of selection and erection of equipment

Aspect	Regulation	Particular requirements	Comments
External influences	705.512.2	The degree of protection afforded to electrical equipment in normal use must be not less than IP44. Where equipment affording this protection is not available, the equipment must be installed in an enclosure affording not less than IP44.	Any additional enclosure must be selected so that it does not cause equipment to overheat
		Socket-outlets (and other equipment) must be selected and erected so as to withstand influences such as water jets, water sprays or impact to which they may be subjected, and they must be installed where likely to come into contact with combustible material.	Combustible material could include straw, for example
		Attention is drawn to the need to select and erect equipment to withstand corrosive substances where expected to be present, such as in dairies and cattle sheds	Animal waste is corrosive
Accessibility to livestock	705.513.2	Equipment should be installed where it is inaccessible to livestock where practicable. Where this is unavoidable, such as for equipment at feeding basins for watering, the equipment must be suitably selected and erected to avoid damage to it and injury to livestock	
Diagrams	705.514.9.3	Documentation to be provided, including a plan showing the location of all electrical equipment, the routing of all concealed cables, a single-line distribution diagram, and the locations of bonding connections	This is in addition to the diagram or chart, etc. required by Regulation 514.9.1
Wiring systems (in relation to external influences)	705.522	Wiring systems in locations accessible to or enclosing livestock must be installed where they are inaccessible to livestock or be suitably protected against mechanical damage.	

	Overhead lines must be insulated. Cables buried in the ground must be at a depth or at least 600 mm, or at least 1 m in arable or cultivated land. Self-supported suspension cable must be at a height of at least 6 m	
Protection against vermin	705.522.10 Special attention must be given to the presence of fauna	Where appropriate, wiring systems must be suitably selected and erected to prevent harmful effects by vermin, such as gnawing of cables by rats, mice and squirrels etc., or entry of vermin into enclosures to nest. (Nesting materials can be a fire risk)
Conduit, trunking and ducting systems	705.522.16 External influences in locations where livestock is kept are to be classified as AF4. Conduits to have at least Class 2 or Class 4 protection to BS EN 61386-21 for indoor and outdoor locations respectively. External influences to be classified as AG3 in locations where wiring systems may be exposed to impact or mechanical shock from vehicles or farm machines, etc. Impact protection provided to be 5 J to BS EN 61386-21 by conduits and 5 J to BS EN 50085-2-1 by trunking and ducting systems.	AF4 relates to continuous presence of corrosive and polluting substances AG3 relates to high severity mechanical stress
Isolation and switching	705.537.2 The installation in each building or part of a building to be isolated by a single device. For circuits used occasionally (such as during harvest time), means of isolation of all live conductors, including neutral, to be provided.	

(continued overleaf)

Table 16.12 (*continued*)

Aspect	Regulation	Particular requirements	Comments
		All isolation devices to be clearly marked to show the part of the installation they control. All isolation and switching devices, including those for emergency stopping, to be sited so they are inaccessible to livestock and access to them will not be impeded by livestock	
Supplementary bonding conductors	705.544.2	Bonding conductors to be protected from mechanical damage and corrosion, and of a type resistant to electrolytic effects. The regulation gives examples of suitable types of conductor	See Section 16.6.2 of this Guide
Socket-outlets	705.553.1	Socket-outlets to comply with: • BS 1363, BS 546 or BS 196 where rated at 20 A or less, or • BS EN 60309-1, or • BS EN 60309-2 where interchangeability is required	
Luminaires and lighting installations	705.559	Luminaires to comply with BS EN 60598 series and be selected regarding their: • degree of protection against ingress of solid objects and dust (such as IP54); • suitability for mounting on a normally flammable surface; • limited surface temperature. Limited surface temperature luminaires to have IP54 ingress protection	Denoted by ▽F marked on the luminaire Denoted by ▽D marked on the luminaire

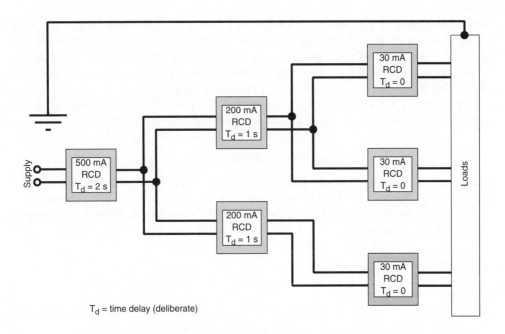

T_d = time delay (deliberate)

Figure 16.7   Typical arrangement providing discrimination of RCDs

16.6.6   RCDs in series

It is inevitable that installations in this location will involve RCDs in series, and the designer also needs to consider discrimination of these devices. For compliance with Regulation 705.411.1, every circuit will be protected by an RCD and, realistically, an RCD will protect a number of circuits. The total number of circuits protected by a single RCD will depend on the likely equipment to be used. Bearing in mind that some equipment may be less than adequately maintained and subject to poor environmental conditions, the maximum number of circuits protected by an RCD, by 'rule of thumb', should ideally not exceed three. RCDs in series should be avoided wherever possible, and it may be beneficial to employ combined MCB–RCDs (RCBOs) to protect individual circuits. Where RCDs have to be connected in series, careful consideration is required so as to provide effective discrimination. Figure 16.7 shows a typical arrangement providing discrimination of RCDs by the use of devices with deliberate time-delayed operation.

16.7   Conducting locations with restricted movement

16.7.1   General

As with other special locations, the particular requirements of Section 706 of BS 7671 supplement or modify the general requirements. As indicated in Regulation 706.1, this section applies to fixed equipment in conducting locations where movement of persons is restricted by the location, and to supplies for mobile equipment for use in such locations. The location is defined as one comprised mainly of metallic or other conductive surrounding parts, within which it is likely that a person will come into contact through a substantial portion of the body with the metallic or other conductive surrounding parts and where the possibility of interrupting this contact is limited. Examples of such locations would include large boilers, storage tanks, large pipework or any conductive locations which restrict the physical movement of a person working in it.

The requirements of Section 706 do not apply to locations that allow freedom of bodily movement to a person to work, enter and leave the location without physical constraint.

16.7.2 *Protection against electric shock*

Not surprisingly, Regulation 706.410.3.5 precludes the use of the protective measures of obstacles and placing out of reach, as clearly they are unsuitable for such locations.

There are a number of options, given in Regulation Group 706.41, for protection against electric shock. These are:

- SELV;
- PELV;
- ADS;
- electrical separation supplying one item of equipment (each socket-outlet and each item of equipment connected to a separate winding of an isolating transformer);
- double or reinforced insulation (additionally protected by an RCD with $I_{\Delta n} \leq 30$ mA and operating time ≤ 40 ms at $5I_{\Delta n}$).

Supplementary bonding conductors are required in cases where a functional earth is required, to connect together the exposed-conductive-parts of the installation and the conductive parts of the location (Regulation 706.411.1.2 refers).

As indicated in Regulation 706.410.3.10 and Regulation Groups 706.413 and 706.414, further demands are made on supplies to equipment within the location as summarized in Table 16.13.

16.8 Electrical installations in caravan/camping parks and similar locations

16.8.1 *General*

As with all special locations or installations, the general requirements apply and are supplemented or modified by the particular requirements embodied in Section 708. As stated in Regulation 708.1, the additional requirements relate to electrical installations in caravan/camping parks and similar locations providing facilities for supplying leisure accommodation vehicles (including caravans) and tents. The nominal voltage of the installation for providing the supplies must be 230 V single-phase a.c. or 400 V three-phase and neutral a.c., as indicated in Regulation 708.313.1.2.

Initial inspection and testing and certification for caravan parks are essentially no different from general installations (see Chapter 17 of this Guide). As far as periodic inspection and testing are concerned, the interval between inspections is largely a matter of judgement and would depend, to a great extent, on the degree of maintenance of the park

Table 16.13 Summary of requirements relating to supplies to equipment within the restrictive conductive location

Regulation	Equipment	Requirement
706.413.1.2	An unearthed source of supply for a circuit where protection against electric shock is provided by electrical separation	The source must provide at least simple separation (e.g. a double-wound transformer). It must be situated outside the special location, unless it is part of the fixed installation within that location, for the supply to fixed equipment
706.414.3(ii)	A source of SELV or PELV	The source must be situated outside the special location, unless it is part of the fixed installation within that location, for the supply to fixed equipment
706.414.4.5	Basic protection for SELV and PELV circuits	Irrespective of the nominal voltage, basic protection must be provided by basic insulation complying with Regulation 416.1 or barriers or enclosures complying with Regulation 416.2
706.410.3.10	The supply to a hand-held tool or item of mobile equipment	The measure for protection against electric shock must be one of the following: • electrical separation complying with Section 413, with only one item of equipment being connected to a secondary winding of a transformer (note: the transformer may have more than one secondary winding); • SELV complying with Section 414
	The supply to handlamps	The measure for protection against electric shock must be SELV complying with Section 414. The SELV circuit may supply incorporating a built-in double-wound step-up transformer
	The supply to fixed equipment	The measure for protection against electric shock must be one of the following: • Automatic disconnection of supply complying with Section 411, with supplementary bonding complying with Regulation 415.2 connecting exposed-conductive-parts of fixed equipment and conductive parts of the special location. • Double or reinforced insulation complying with Section 412. All circuits must also have additional protection by RCDs having the characteristics specified in Regulation 415.1.1. • Electrical separation complying with Section 413, with only on item of equipment being connected to a secondary winding of a transformer (note: the transformer may have more than one secondary winding).

(*continued overleaf*)

Table 16.13   (*continued*)

Regulation	Equipment	Requirement
		• SELV complying with Section 414.
		• PELV complying with Section 414. Supplementary bonding complying with Regulation 415.2 must also be provided between all exposed-conductive-parts and extraneous-conductive-parts in the special location, and the connection of the PELV system with Earth

installation. The generally accepted maximum interval would be 12 months and, where the park is subjected to licensing (see Table 17.21), the maximum period would be a mandatory condition.

16.8.2   Requirements for safety

The protective measures of obstacles and placing out of reach are, not surprisingly, precluded by Regulation 708.410.3.5; similarly, the measures of non-conducting location and earth-free local equipotential bonding are precluded by Regulation 708.410.3.6.

16.8.3   Equipment: selection and erection

Regulation Group 708.5 makes additional demands relating to the selection and erection of equipment; Table 16.14 summarizes these additional requirements.

16.8.4   Typical caravan park distribution layout

Figure 16.8 shows a typical park layout (one of the many possible) in which the supply is derived from a PME source. From a central switchroom, distribution cables run underground, following routes which do not encroach on the pitches, to a number of pitch distribution pillars. Each pillar, serving four pitches, is constructed predominately of non-metallic material (with degree of protection IP54) and embodies a main isolating switch, four 16 A BS EN 60309-2 socket-outlets and six combined MCB–RCDs (RCBOs). Each socket-outlet is individually circuited and protected against overcurrent and the electric shock hazard under single fault conditions by an RCBO.

  BS 7671 requires (Regulation 708.553.1.14) that, where a PME service is provided, the protective conductors for the socket-outlets must not be connected to the PME earthing terminal (which is connected to the PEN conductor of the supply). This reflects the statutory requirement of Regulation 9(4) of the *Electricity Safety, Quality and Continuity Regulations 2002* that prohibits EDs from offering connections to earthing terminals from PME networks for consumers' installations in caravans or boats. Figure 16.9 illustrates one example of how armoured cables may be terminated into the gland plate of the pillar and how continuity may be maintained for the underground cables' protective conductors (e.g. SWA). Provided there are no metallic parts within the pillar which need earthing,

Table 16.14 Additional requirements relating to equipment selection and erection in caravan/camping parks

Equipment	Regulation	Additional requirement	Comments
Equipment installed outside	708.512.2	The equipment must provide at least the following degrees of protection: • protection against ingress of water: IPX4 to BS EN 60529; • protection against ingress of solid foreign bodies – IP3X; • protection against mechanical stress – IK08 to BS EN 62262	IK08 corresponds to an impact energy of 5 J applied under specified conditions
Wiring systems	708.521.1, 708.521.1.1 and	The following wiring systems are suitable for distribution circuits serving the supply equipment for caravan and tent pitches:	
	708.521.1.2	• *Underground.* The underground cables must be buried at least 600 mm deep unless additional mechanical protection is provided, and routed outside caravan pitches and areas where tent pegs or ground anchors may be driven.	The buried depth of cables could be less than 600 mm where contained in a pipe duct crossing beneath a road of concrete construction
		• *Overhead.* The height of conductors above ground must be at least 6 m in area were vehicles move and 3.5 m elsewhere. Poles and other supports must be so located and/or protected that damage by vehicles is unlikely	The conductors of an overhead wiring system must be insulated
Caravan pitch supply equipment	708.530.3	The equipment to be located not more than 20 m from the supply inlet connection device of the leisure accommodation vehicle or tent	No more than four socket-outlets should be grouped in one location
	708.553.1.8	Pitch socket-outlets and their enclosures to comply with BS EN 60309-2 and have degree of protection at least IP44	

(continued overleaf)

Table 16.14 (*continued*)

Equipment	Regulation	Additional requirement	Comments
	708.553.1.9	Lowest part of socket-outlet height to between 0.5 and 1.5 m above ground, but may be higher than 1.5 m where there is risk of flooding or heavy snowfall	
	708.553.1.10	The rated current of socket-outlets to be not less than 16 A (higher than 16 A where greater demands are expected)	
	708.553.1.11 708.553.1.12, 708.553.1.12	At least one socket-outlet to be provided for each pitch Each socket-outlet to be provided with individual overcurrent protection and individual RCD protection. RCD to have the characteristics specified in Regulation 415.1.1 and disconnect all live conductors (including the neutral)	A suitably rated two-pole RCBO to BS EN 61009-1 could be used to provide both overcurrent protection and RCD protection for each socket-outlets
	708.553.1.14	Socket-outlet protective conductors must not be connected to a PME facility (the PEN conductor of a public electricity supply). Where PME conditions apply to the supply, the protective conductor of each socket-outlet must be connected to an installation earth electrode and the requirements for a TT system must be met (Regulation 411.5)	

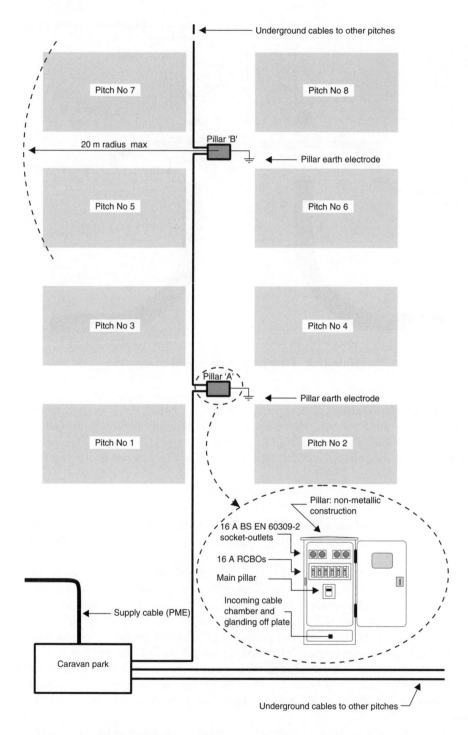

Figure 16.8   A typical caravan park installation layout

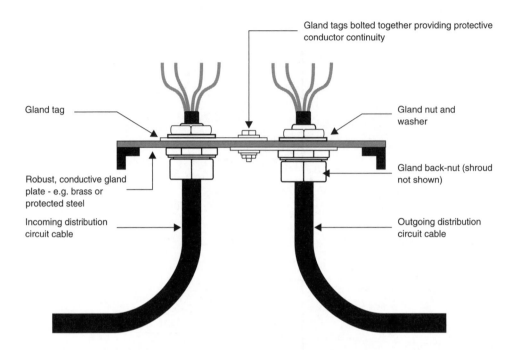

Figure 16.9   Pillar gland plate: typical arrangement

the PME earthing need not be brought further into the pillar and a local earth electrode will be used to protect all the socket-outlet circuits. The installation in the pillar would then form part of a TT system, where Regulation Group 411.5 applies. This includes the requirement for the resistance of the earth electrode to meet the following formula:

$$R_A I_a \leq 50\text{V} \tag{16.2}$$

where R_A is the sum of resistances of the earth electrode and protective conductor between the socket-outlet and the earth electrode and I_a is the rated operating current of the RCBO ($I_{\Delta n}$).

Had the supply been a TN-S system, the underground cable protective conductor could have been used to connect the protective conductors of the socket-outlet circuits to Earth and the local electrodes would have been unnecessary. However, the designer will need to consider the potential risk associated with the loss of this underground protective conductor on the provisions for fault protection; the loss of protective conductors in the underground site distribution cables would render all RCBOs inoperative and an earth fault on any connected caravan would raise the potential on every caravan to Earth up to line conductor voltage to Earth U_o. For this reason, the designer should consider the advantages of employing an earth-monitoring system at least for the site distribution circuits.

Park distribution may be single-phase or three-phase, depending on the availability of supply and park size in terms of loading. Where three-phase is used, the load should be balanced as far as is possible and the necessary through connections provided in the pillars.

Regulation 708.553.1.14 requires that each socket-outlet is individually protected by an RCD having the characteristics specified in Regulation 415.1.1 (rated residual operating current $I_{\Delta n} \leq 30$ mA and an operating time ≤ 40 ms at a residual current of $5I_{\Delta n}$). The pillar shown in Figure 16.8 shows six RCBOs, one each for the four socket-outlets, one perhaps for path and roadway lighting and one spare.

16.9 Marinas and similar locations

16.9.1 General

Section 709 was introduced into BS 7671 with the publication of BS 7671:2008. Like all sections of Part 7, its contents supplement and modify the general requirements of BS 7671. As stated in Regulation 709.1, the requirements of this section apply only to circuits intended to supply pleasure craft or houseboats. They do not apply to other parts of the installation at the marina or similar location, to which the general requirements of BS 7671 apply together with the relevant requirements of other applicable sections of Part 7. The requirements of Section 709 do not apply, either, to the supply to houseboats directly supplied from the public network or to the internal electrical installations of pleasure craft or houseboats.

Initial inspection and testing and certification for installations in marinas and similar locations is essentially no different from general installations (see Chapter 17 of this Guide).

Throughout the Section 709, and in this part of the Guide, the term 'marina' refers to a marina or similar location.

The nominal voltage of the installation for providing supplies to pleasure craft and houseboats must be 230 V single-phase a.c. or 400 V three-phase and neutral a.c. to comply with Regulation 709.313.1.2.

16.9.2 Protection against electric shock

The risk of electric shock at a marina is increased by the presence of water, the reduction in body resistance due to wetness from water and/or perspiration and contact by the body with Earth potential.

Not surprisingly, the use of the protective measures of obstacles and placing out of reach is prohibited, as is the use of non-conducting location and earth-free local equipotential bonding, as indicated in Regulations 709.410.3.5 and 709.410.3.6 respectively.

Regulation 709.411.4 requires that the final circuits for the supply to pleasure craft and houseboats do not include a PEN conductor. This reflects the statutory requirement of Regulation 8(4) of ESQCR), that the consumer shall not combine the neutral and protective functions in a single conductor in their installation.

It should also be noted that Regulation 9(4) of ESQCR prohibits EDs from offering connections to earthing terminals from PME networks for consumers' installations in caravans or boats. This means that the protective conductors of socket-outlets intended to provide supplies for pleasure craft or houseboats, and the protective conductors of circuits intended for the fixed connection of supplies to houseboats, must not be connected to a

PME earthing facility (the PEN conductor of a public electricity supply). Each such protective conductor must instead be connected to an installation earth electrode and the requirements for a TT system must be met (Regulation 411.5).

16.9.3 External influences

The external influences that the electrical installation of a marina may have to withstand can be harsh and deserve special mention. These include: water splashes, jets and waves; the likelihood of corrosive elements, such as exposure to salt-laden atmosphere and salt water at coastal locations; movement of jetties, piers and pontoons; the possibility of mechanical damage, including that from the movement of vehicles and boats; and the presence of flammable fuel.

16.9.4 Equipment: selection and erection

Regulation Group 709.5 makes additional demands relating to the selection and erection of equipment, and Table 16.15 summarizes these additional requirements.

16.10 Exhibition shows and stands

16.10.1 General

Section 711 is another new section added into BS 7671 with the publication of BS 7671:2008. It covers temporary electrical installations in exhibitions, shows and stands, including mobile and portable displays and equipment, to protect users. As with the other sections in Part 7, its requirements supplement and modify the general requirements of BS 7671.

Section 711 does not apply to exhibits unless indicated otherwise in the text. Neither does it apply to the fixed electrical installations of buildings where the exhibitions or shows may be held. Finally, Section 711 does not cover the electrical systems defined in BS 7909 *Code of practice for design and installation of temporary distribution systems delivering a.c. electrical supplies for lighting, technical services and other entertainment related purposes*, which relates to television, film, theatre, pop concerts and similar events, either indoor or outdoor.

Inspection and testing and certification for temporary installations in exhibitions, shows and stands are essentially no different from general installations (see Chapter 17 of this Guide) except that, as stated in Regulation 711.6, these installations must be inspected and tested after each assembly on site.

The nominal voltage of the temporary installations in an exhibition, show or stand must not exceed 230/400~V a.c. or 500 V d.c. to comply with Regulation 711.313.

16.10.2 Definitions

In order to apply the requirements of Section 711 properly it is important understand the definitions of the particular terms used in that section, which are given in Part 2 of BS 7671, some of the main ones of which are:

Table 16.15 Additional requirements relating to equipment selection and erection in marinas and similar locations

Equipment	Regulation	Additional requirement	Comments

All equipment – suitability for external influences	709.512.2.1.1 and 709.512.2.1.2	The equipment must provide at least the following degrees of protection to BS EN 60529.
		• Protection against ingress of water:
		  – IPX4 where splashes are expected;
		  – IPX6 where water jets are expected to be used;
		  – IPX6 where exposed to waves.
		• Protection against ingress of solid foreign bodies for equipment on or above a jetty, wharf, pier or pontoon: IP3X	

Equipment on or above a jetty, wharf, pier or pontoon – suitability for external influences	709.512.2.1.3	The equipment to be suitable for use where corrosive or polluting substances may be present in the atmosphere. Where hydrocarbons may be present, equipment to be suitable for occasional or accidental exposure (AF3)	Corrosive or polluting substances may include salt laden air in coastal locations Hydrocarbons may be present in fuel
	709.512.2.1.4	The equipment to be protected against medium-severity mechanical damage by one or more of the following: • being positioned to avoid damage; • provision of local or general protection against damage; • use of equipment providing at least the degree of protection IK08 to BS EN 62262	IK08 corresponds to an impact energy of 5 J applied under specified conditions

(continued overleaf)

Table 16.15 *(continued)*

Equipment	Regulation	Additional requirement	Comments
Wiring systems	709.521.1.4	Wiring systems suitable for distribution circuits of marinas include: • underground cables;	Buried at sufficient depth to avoid damage by heavy vehicle movement (e.g. at least 500 mm deep) unless additionally mechanically protected. (Regulation 709.521.7 refers)
		• overhead cables or overhead insulated conductors;	At a height of at least 6 m where subject to vehicle movement and 3.5 m elsewhere. Support to be located or protected to avoid damage by vehicles (Regulation 709.521.8 refers)
		• cables with copper conductors and thermoplastic or elastomeric insulation or sheath installed in a cable management system suitable for external influences such as movement, impact, corrosion and ambient temperature; • mineral-insulated cables with PVC covering; • armoured cables with thermoplastic or elastomeric serving; • wiring systems providing equivalent protection	
	709.521.1.6	Cables to be so selected and installed so as to prevent damage due to tidal and other movement of structures.	
	709.521.1.5	None of the following wiring systems are to be used on or above a jetty, wharf, pier or pontoon: • cables in free air suspended from or incorporating a support wire;	

		• nonsheathed cables in conduit or trunking, etc.; • cables with aluminium conductors; • mineral-insulated cables The cables must be sheathed, for protection against chafing against the cable management system due to movement
Fault protection devices	709.531.2	Each socket-outlet and each final circuit intended for fixed connection for the supply to a houseboats to be individually protected by an RCD having the characteristics specified in Regulation 415.1.1 and disconnect all live conductors, including neutral A suitably rated two-pole RCBO to BS EN 61009-1 could be used to provide both overcurrent protection and RCD protection
Overcurrent protection devices	709.533	Each socket-outlet and each final circuit intended for fixed connection for the supply to a houseboats to be individually protected by an overcurrent protective device
Isolation	709.537.2.1	At least one means of isolation to be provided for each distribution cabinet. Socket-outlets to be controlled in groups of not more than four by isolating switching devices
Plugs and socket-outlets	709.553.1.8	Socket-outlets to comply with BS EN 60309-1 above 63 A and BS EN 60309-2 up to 63 A. Degree of protection to be not less than IP44, or IPX5 or IPX6 where exposed to water jets or waves respectively In general, single-phase socket-outlets rated at 16 A 200–250 V to be provided, but rated current to be higher where greater demands are envisaged (Regulation 709.553.1.12 refers)

(continued overleaf)

Table 16.15 (*continued*)

Equipment	Regulation	Additional requirement	Comments
	709.553.1.9	Socket-outlets to be located as close as practicable to the berths they supply	An instruction notice giving at least the information included in Figure 709.3 of BS 7671 should be placed adjacent to each group of socket-outlets
	709.553.1.10	A maximum of four socket-outlets to be grouped in a single enclosure, to avoid long connection cables	
	709.553.1.11	A socket-outlet must supply only one pleasure craft or houseboat	
	709.553.1.13	Height of socket-outlets to be not less than 1 m above highest water level, except for socket-outlets on floating pontoons or walkways, where the height may be reduced to 300 mm above highest water level	Additional measures to be taken against the effects of splashing where the height is reduced from 1 m

- *Exhibition.* Event intended for the purpose of displaying and/or selling products, etc., which can take place in any suitable location, either a room, a building or a temporary structure.
- *Show.* Display or presentation in any suitable location, either a room, a building or a temporary structure.
- *Stand.* Area or temporary structure used for display, marketing or sales.
- *Temporary electrical installation.* Electrical installation erected for a particular purpose and dismantled when no longer required for that purpose.
- *Temporary structure.* A unit or part of a unit, including mobile or portable units, situated indoors or outdoors, designed and intended to be assembled or dismantled.

16.10.3   Protection against electric shock and protection against fire

The additional requirements of Section 711 for protection against electric shock and fire are summarized in Table 16.16.

16.10.4   Equipment: selection and erection

Regulation Group 711.5 makes additional demands relating to the selection and erection of equipment, and Table 16.17 summarizes these additional requirements.

16.11   Solar photovoltaic power supply systems

16.11.1   General

Section 712 is a further new section added into BS 7671 with the publication of BS 7671:2008. Its requirements apply to the electrical installations of photovoltaic (PV) power supply systems, including systems with a.c. modules.

As with all sections of Part 7, the requirements of Section 712 supplement and modify the general requirements of BS 7671. In particular, it should not be forgotten that, as an assembly of PV arrays is considered to be a generator, the electrical installation of a PV power supply system is subject to the requirements of Section 551 (generators), the content of which is covered in Section 13.16 of this Guide.

Initial inspection, testing and certification for the electrical installations of PV power supply systems are, at least as far as BS 7671 is concerned, essentially no different from general installations (see Chapter 17 of this Guide). However, it is likely that care will have to be taken to avoid damage to any part of the PV system as a result of testing (such as insulation resistance testing), and in this respect the instructions of the manufacturer of the PV system should be followed carefully. It is also likely that a number of additional tests, over and above those required by BS 7671, will need to be carried out, including functional tests and tests for commissioning purposes. Once again, strict adherence to the manufacturer's instructions will be necessary for these additional test procedures.

Where carrying out any work on a PV system or its associated installation, including inspection and testing, it should be remembered that PV equipment on the d.c. side must be considered to be energized even when the system is disconnected from the a.c. side (Regulation 712.410.3 refers). A safe system of work must, therefore, be followed to avoid the risk of electric shock or burns.

Table 16.16 Additional requirements relating to protection against electric shock and fire: exhibitions, shows and stands

Aspect	Regulation	Additional requirement	Comments
RCD protection of supply cable	711.410.3.4	Protection by RCD of rated residual operating current $I_{\Delta n} \leq 300$ mA to be provided to a cable intended to supply a temporary structure. To discriminate with RCDs protecting final circuits, the RCD to have a delay in accordance with BS EN 60947-2 or be of S-type to BS EN 61008-1 or BS EN 61009-1	The RCD protection is required due to the increased risk of damage to cables in temporary installations
Protective measures not to be used	711.410.3.5 and 711.410.3.6	Use of the protective measures of obstacles and placing out of reach, and non-conducting location and earth-free local equipotential bonding, is not permitted	Use of these electric shock protection measures is inappropriate and/or impracticable
ADS	711.411.3.1.2	Main protective bonding conductors to connect structural metallic parts accessible from within a stand, vehicle, wagon, caravan or container to the main earthing terminal of the temporary installation in that unit	This requirement envisages that the structural metallic parts are extraneous-conductive-parts (e.g. liable to introduce earth potential)
	711.411.3.1.3	Additional protection by RCD having the characteristics specified in Regulation 415.1.1 to be provided to each socket-outlet circuit of rating 32 A or less and all final circuits except for emergency lighting	Regulation 415.1.1 requires rated residual operating current $I_{\Delta n} \leq 30$ mA and operating time ≤ 40 ms at residual current of $5I_{\Delta n}$
	711.411.4	Where the type of earthing system is TN, the installation must be TN-S Note to regulation points out that ESQCR prohibits use of a TN-C-S system for supply to a caravan or similar construction	As required by Regulation 8(4) of ESQCR, a PEN conductor must not be used in the installation Protective conductor of installation in a caravan or similar not to be connected to PME terminal of supply, but be connected instead to an earth electrode, and the requirements for TT system must be met (Section 16.8.4 of this Guide refers)

Table 16.16 (*continued*)

Aspect	Regulation	Additional requirement	Comments
SELV and PELV	711.414.4.5	Basic protection to be provided by basic insulation or barriers and enclosures to at least IP4X or IPXXB, irrespective of nominal voltage	
Protection against fire	711.422.4.2	Lamps and other equipment/appliances with high-temperature surfaces to be suitably guarded, and installed/located according to relevant standard. Showcases and signs to be of material with adequate heat resistance, mechanical strength, electrical insulation and ventilation. Account to be taken of combustibility of exhibits and heat generation. Unless adequate ventilation is provided, no stand installations to contain concentration of heat-producing equipment.	

Table 16.17 Additional requirements relating to equipment selection and erection: exhibitions, shows and stands

Equipment	Regulation	Additional requirement
Location of switchgear and controlgear	711.51	Switchgear and controlgear to be placed in closed cabinets openable only with a tool or key, except for parts intended and designed for operation by ordinary persons
Wiring systems	711.52	Cables to be armoured or otherwise suitably protected where necessary to protect from mechanical damage. Wiring cables to have copper conductors not less than 1.5 mm^2 and be of thermoplastic or thermosetting insulated type. Flexible cord not to be laid in publicly accessible areas unless protected from mechanical damage
	711.521	In buildings where no fire alarm system is installed, cables to be: • flame-retardant type tested to BS EN 60332-2-1 or relevant part of BS EN 50266 series; or • unarmoured single-core or multicore cables in nonmetallic conduit or trunking to BS EN 61386 series or BS EN 50085 series respectively with degree of protection at least IP4X
Electrical connections	711.526.1	No joints permitted in cables except for connection into circuit. Any joints to be in enclosure providing at least IP4X or IPXXD protection or use connectors to relevant standards. (Note: all connections must be enclosed to comply with Regulation 526.5)
Isolation	711.537.2.3	A dedicated, readily accessible means of isolation to be provided for: • each temporary structure (e.g. vehicle, stand and unit) intended to be occupied by a specific user; and • each distribution circuit supplying outdoor installations
Other equipment	711.55.1.5	All equipment to be so fixed and protected as to prevent focusing or concentration of heat that may cause ignition of material
Electric motors	711.55.4.1	A motor likely to cause a hazard must be provided with means of isolation of all live conductors, adjacent to the motor
Extra-LV transformers and convertors	711.55.6	Secondary circuit of each transformer and convertor to have manually reset protective device. Electronic convertors to comply with BS EN 61347-1 (*Lamp controlgear. General and safety requirements*). Transformers not to be mounted within arm's reach of the public, but must be accessible to competent person for testing and must have adequate ventilation

Table 16.17   (*continued*)

Equipment	Regulation	Additional requirement
Socket-outlets and plugs	711.55.7	The number of socket-outlets must be adequate to meet user requirements safely. Floor-mounted socket-outlets to be protected from water ingress and suitable to withstand traffic load expected
Luminaires and lighting installations	711.559.4.3	Extra-LV systems for filament lamps to comply with BS EN 60598-2-23
	711.559.4.3	Insulation-piercing lampholders not to be used except with compatible cables and subject to lampholders not being removed
	711.559.4.4 to 711.559.4.7	Luminous tubes, lamps or signs with normal supply voltage exceeding 230/400 V on stands or exhibits to be installed only onto nonignitable background, out of arm's reach and protected to reduce risk of injury to persons. Supply circuit not to supply other equipment, and be controlled by an easily visible and accessible emergency switch, clearly marked
	711.559.5	Luminaires accessible to accidental contact, such as below 2.5 m above floor, to be firmly fixed and located or guarded to prevent risk of injury to persons or ignition of materials.

16.11.2   Definitions

To apply the requirements of Section 712 properly it is important understand the definitions of the particular terms used in that section, which are given in Part 2 of BS 7671, the main ones of which are:

- *PV*. Solar PV.
- *PV a.c. module*. Integrated module invertor assembly where the electrical interface terminals are a.c. only. No access is provided to the d.c. side.
- *PV array*. Mechanically and electrically integrated assembly of PV modules, and other necessary components, to form a d.c. power supply unit.
- *PV array cable*. Output cable of a PV array.
- *PV array junction box*. Enclosure where PV strings of any PV array are electrically connected and where devices can be located.
- *PV cell*. Basic PV device which can generate electricity when exposed to light, such as solar radiation.
- *PV d.c. main cable*. Cable connecting the PV generator junction box to the d.c. terminals of the PV invertor.
- *PV generator*. Assembly of PV arrays.
- *PV generator junction box*. Enclosure where PV arrays are electrically connected and where devices can be located.
- *PV installation*. Erected equipment of a PV power supply system.
- *PV invertor*. Device which converts d.c. voltage and current into a.c. voltage and current.

- *PV module.* Smallest completely environmentally protected assembly of interconnected PV cells.
- *PV string.* Circuit in which PV modules are connected in series, in order to form a PV array to generate the required output voltage.
- *PV string cable.* Cable connecting PV modules to form a PV string.
- *PV supply cable.* Cable connecting the a.c. terminals of the PV invertor to a distribution circuit of the electrical installation.

16.11.3   Earthing a live conductor on the d.c. side

Where a generator is to run in parallel with the public electricity supply, it is usually necessary that none of the live conductors of the generator is earthed, if for no other reason than this could prevent the operation of an RCD in the circuit connecting the generator to the distribution board or switchboard. However, as given in Regulation 712.312.2, it is permitted to earth one of the live conductors of a PV supply system on the d.c. side, provided there is at least simple separation between the a.c. side and the d.c. side (equivalent to the separation provided by a double-wound transformer). A note to the regulation points out that any connection with Earth on the d.c. side should be electrically connected in order to avoid corrosion.

16.11.4   Protection against electric shock, overcurrent and electromagnetic interference

The additional and supplementary requirements of Section 712 for protection against electric shock, overcurrent and electromagnetic interference are summarized in Table 16.18.

16.11.5   Equipment: selection and erection

Regulation Group 712.5 makes additional demands relating to the selection and erection of equipment, and Table 16.19 summarizes these additional requirements.

16.12   Mobile or transportable units

16.12.1   General

Section 717 was added into BS 7671 with the publication of BS 7671:2008. The scope of the section is given in Regulation 717.1, where it is explained that the requirements apply to the electrical installations of:

- mobile units, such as vehicles (whether self-propelled or towed) and
- transportable units, such as containers and cabins

in which all or part of the electrical installation is contained, this being provided with a temporary supply of electricity by means such as a plug and socket-outlet.

As indicated in the regulation, mobile and transportable units within the scope of the Section 717 may be used for purposes such as accommodating technical facilities for the

Table 16.18 Additional requirements relating to protection for safety: PV power supply systems

Aspect		Regulation	Additional requirement	Comments
Electric shock	Protective measures not to be used	712.410.3.6	Use of the protective measures of non-conducting location and earth-free local equipotential bonding, is not permitted on the d.c. side	Use of these measures is inappropriate and/or impracticable.
	ADS	712.411.3.2.1.1	The PV supply cable must be connected to the supply side of the protective devices for automatic disconnection of circuits supplying current-using equipment	In practice, this means that the option given in Regulation 551.7.2 to connect to a final circuit (subject to certain conditions being met) is not available for a PV power supply system.
		712.411.3.2.1.2	An RCD used for fault protection must be of Type B to IEC/TR 60755 Edition 2 if the PV power supply system does not have at least simple separation between the a.c. and d.c. sides	The need or otherwise for a Type B RCD should be confirmed by reference to the installation instructions or to the supplier of the generator
	Double or reinforced insulation	712.412	Class II or equivalent insulation is the preferred method of shock protection on the d.c. side	
	SELV and PELV	712.414.1.1	For a SELV or PELV circuit on the d.c. side of the PV power supply system, U_{0STC} is to be taken as the voltage to Earth, and this must not exceed 120 V	STC stands for standard test conditions

(continued overleaf)

Table 16.18 (*continued*)

Aspect	Regulation	Additional requirement	Comments
Overload protection to cables (Note. But also see manufacturer's instructions for protection of PV modules)	712.433.1	Overload protection need not be provided to PV string and PV array cables if their $I_z \geq 1.25 I_{scSTC}$ at any location	I_z is the current-carrying capacity of the cables.
	712.433.1	Overload protection need not be provided to PV main cable if its $I_z \geq 1.25 I_{scSTC}$ of the PV generator	I_{scSTC} is the short-circuit current under standard test conditions
Fault current protection	712.434.1	Fault current protection to the PV supply cable must be provided by the overcurrent device where the cable connects to the a.c. mains	
Protection against electromagnetic interference	712.444.4.4	The area of all wiring loops to be kept as small as possible, to minimize voltages induced by lightning	

Table 16.19 Additional requirements relating to equipment selection and erection: PV power supply systems

Equipment	Regulation	Additional requirement
PV power supply system equipment	712.511.1	PV modules to comply with relevant standard, e.g. crystalline PV modules to BS EN 61215. Where U_{ocSTC} of string exceeds 120 V d.c., PV modules of Class II construction (or with equivalent insulation) recommended. PV array junction box, PV generator box and switchgear to comply with BS EN 60439-1
	712.512.1.1	Connection of PV modules in series must be such that the maximum operating voltage does not exceed that of the PV modules or that of the PV convertor, whichever is the lower. Equipment specifications to be obtained from the manufacturer. Reverse voltage rating of any blocking diodes to be $\geq 2U_{ocSTC}$, the diodes being connected in series with PV strings
	712.512.2.1	PV modules to be erected in a way specified by manufacturer that permits adequate heat dissipation even when solar radiation is at the maximum expected for the site
Accessibility	712.513	All equipment to be selected and erected to facilitate safe maintenance, without adversely affecting provisions made by manufacturer for maintenance/servicing work to be safely done.
Wiring systems (suitability for external influences)	712.522.8.1	Selection and erection of PV sting cables, PV array cables and PV d.c. main cables to such that risk of earth fault and short-circuits is minimized. One method of achieving this is by use of single-core sheathed cables
	712.522.8.3	Wiring systems to be selected and erected to withstand expected conditions, such as wind, ice formation, temperature and solar radiation
Isolation	712.537.2.1.1	Means to be provided for isolating to for the PV convertor from both the d.c. side and the a.c. side, for maintenance of PV convertor
	712.537.2.2.1	The public supply is to be considered as the source, and the PV installation considered as the load, when selecting and erecting isolation and switching devices installed between the PV installation and the public supply
	712.537.2.2.5	A switch disconnector to be provided on d.c. side of PV convertor
	712.537.2.2.5.1	Warning label to be provided on all junction boxes (of PV generator and PV array), indicating parts inside the boxes may still be live after isolation from PV convertor
Protective bonding conductors	712.54	Any protective bonding conductors installed must be parallel to, and in close contact with, d.c. cables, a.c. cables and accessories

entertainment industry, medical services, advertising or fire fighting; or to accommodate workshops, offices, or catering units.

The requirements of Section 717 do not apply to generating sets, marinas or pleasure craft, mobile machinery to BS EN 60204-1 (relating to the electrical equipment of machines), caravans or motor caravans (as covered in Section 721), traction equipment of electric vehicles, or electrical equipment to allow a vehicle to be driven safely or used on the highway (such as that covered by The Road Vehicles Lighting Regulations 1989).

The requirements of other sections of Part 7 may also apply to the electrical installation of a mobile or transportable unit, such as where the unit contains a shower (in which case Section 701 would apply).

16.12.2   Source of supply

The electricity supply to the installation of a mobile or transportable unit may be from either an LV generating set located inside or outside the unit, or an LV distribution network, such as the public network.

16.12.3   Protection against electric shock

Not surprisingly, it is not permitted to use the protective measures of obstacles and placing out of reach, or the protective measures of non-conducting location or earth-free local equipotential bonding.

Where the protective measure of ADS is to be used, the type of earthing arrangement for the installation may be TN, TT or IT. However, the use of a TN-C-S system is not permitted by Regulation 717.411.4 unless the installation is continuously under the supervision of a skilled or instructed person and the suitability and effectiveness of the means of earthing has been confirmed before the installation is connected to the supply. In most TN-C-S systems the supply is from a PME network. Where this is the case, the ED may refuse to offer a connection from the network to the earthing terminal of the installation in a mobile or transportable unit for safety reasons. This is because of the possibility of the supply neutral conductor becoming disconnected from Earth, possibly causing the metalwork in the mobile or transportable unit to rise to line potential (assuming that the unit does not benefit from an independent connection with Earth). Persons entering or exiting the unit would then be at risk of electric shock.

Where an IT system is to be used, then, in order to comply with Regulation 411.6.1, the neutral or star point of the source must be either insulated from Earth or connected to earth through a suitably high impedance. Regulation 717.411.6.2 gives two choices for the type of source for the IT system. The first is an LV generating set with an insulation monitoring device installed. The second is a transformer that provides at least simple separation, such as a transformer conforming to BS EN 61558-1. In the second case, the associated requirements of Regulation 717.411.6.2 must be met relating to the provision of either an insulation monitoring device with or without an earth electrode or RCD protection with an earth electrode meeting certain requirements specified in the regulation.

Whatever type of earthing arrangement is employed where ADS is used as the protective measure, main protective bonding must be installed from the main earthing terminal of the installation of the mobile or transportable unit to accessible conductive parts of the unit, such as the chassis. As stated in Regulation 717.411.3.1.2, the bonding conductors must be of the finely stranded type, such as cables of types H05V-K and H07V-K to BS 6004.

It should be noted that RCDs, rather than overcurrent protective devices, must be used as the devices for automatic disconnection in the event of a fault where ADS is used (Regulation 717.411.1 refers). Furthermore, for all socket-outlets intended to supply current-using equipment outside the mobile or transportable unit, additional protection must be provided by an RCD having the characteristics specified in Regulation 415.1.1 (Regulation 717.415 refers). The only socket-outlets that are excused from the requirement for additional protection are those supplied from extra-low voltage circuits protected by SELV or PELV.

16.12.4   Equipment: selection and erection

Regulation Group 717.5 gives additional requirements for the selection and erection of equipment, and Table 16.20 summarizes these additional requirements.

16.13   Electrical installations in caravans and motor caravans

16.13.1   General

As with all special locations or installations, the general requirements apply and are supplemented or modified by the particular requirements embodied in Section 721. As indicated in Regulation 721.1, the requirements relate to installations, on single-phase or three-phase supplies of nominal voltage not exceeding 230/400 V, in caravans and motor caravans.

Installations in mobile homes, residential park homes and transportable units are subject to the general requirements, but not to the particular requirements of this section, as is also true for installations in caravans intended for habitation purposes.

The requirements of Section 721 do not apply to vehicle electrical systems employed to meet the requirements of The Road Vehicle Lighting Regulations 1989, except in the sense of the necessary segregation of the LV system from the vehicle road lighting systems as required by Section 528 (proximity to other services) of BS 7671 and Regulation 721.528.1. Similarly not covered in the scope of Section 721 are installations covered by BS EN 1648-1 *12 V direct current extra-low voltage electrical installations. Caravans* and BS EN 1648-2 *12 V direct current extra-low voltage electrical installations. Motor caravans*, except, again, to the extent of segregation. Where such accommodation includes a bath or a shower, the particular requirements of Section 701 also apply.

Initial inspection and testing and certification for caravans and motor caravans is essentially no different from general installations (see Chapter 17 of this Guide). As far as periodic inspection and testing are concerned, the intervals between inspections are largely a matter of judgement and would depend to a great extent on the extent of usage and the degree of maintenance. The generally accepted maximum interval would be 3 years (see Table 17.21).

Table 16.20 Additional requirements relating to equipment selection and erection: mobile or transportable units

Equipment	Regulation	Additional requirement	Comments
Notice adjacent to inlet connector	717.514	Permanent, durable notice to be fixed adjacent to the supply inlet connector or prominent position on the unit, giving details specified in the regulation, type of supply, voltage, number of phases, earthing arrangement and maximum power requirement	
Wiring systems	717.52.1	Flexible cables for temporary supply to unit to have conductor size not less than 2.5 mm^2 and be of type H07RN-F (BS 7917) or equivalent. Cable to enter unit by means of insulated insert, arrange to minimize risk of insulation damage or fault	
	717.52.2	Wiring system for internal wiring of the unit to be either nonsheathed cables in conduit, or, subject to precautions being taken to prevent damage by sharp edges or abrasion, insulated and sheathed cables. Regulation gives more detail of cable and conduit types.	Flexible or multi-strand conductors (at least seven strands) likely to be necessary, to withstand vibration in service
Equipment in gas cylinder storage compartment	717.528.3.5	No electrical equipment permitted in compartment except extra-low voltage equipment for gas supply control Any cable running through compartment to be at least 500 mm above height of base of gas cylinders, in continuous gas-tight conduit or duct, able to withstand high-severity impact (AG3) without visible damage	
Plugs, connectors and socket-outlets	717.55.1	Plugs and connectors for temporary supply to the unit to comply with BS EN 60309-2. Plugs to have enclosure of insulating material. Degree of protection for appliance inlets and for plugs and socket-outlets used outdoors to be IP44 or better	Appliance inlet to be mounted on the unit. Socket-outlet to be mounted outside the unit, in location near where unit is to be situated
	717.55.2	Socket-outlets located outside unit to be in an enclosure providing degree of protection of IP44 or better	

16.13.2   Requirements for safety

The protective measures of obstacles and placing out of reach are, not surprisingly, precluded by Regulation 721.410.3.5 and, similarly, the protective measures of non-conducting location, earth-free local equipotential bonding are precluded by Regulation 721.410.3.6. The use of the protective measure of electrical separation is not permitted except for shaver socket-outlets, as given in Regulation 721.410.3.3.2.

Regulation 721.411.1 calls for an RCD (see Figure 16.10) where protection by ADS is used (the normal method) with a protective conductor connecting the exposed-conductive-parts of the installation, including socket-outlet protective contacts, with the earthing contact of the caravan inlet plug. The RCD must comply with BS EN 61008-1 or BS EN 61009-1, interrupt all live (line and neutral) conductors and have the characteristics specified in Regulation 415.1.1 ($I_{\Delta n} \leq 30$ mA and operating time ≤ 40 ms at a residual current of $5I_{\Delta n}$).

All extraneous-conductive-parts accessible from within the caravan (or motor caravan) must be bonded with a main protective bonding conductor to the main earthing terminal of the installation in the caravan, as called for in Regulation 721.411.3.1.2. All that will be required, in many cases, will be a 'main' bond on to the vehicle chassis.

Regulation 721.43.1 calls for each final circuit to be protected by an overcurrent protective device which disconnects all live (line and neutral) conductors (see Figure 16.10). This will of necessity involve the use of single-pole-and-neutral (SP&N) or double-pole

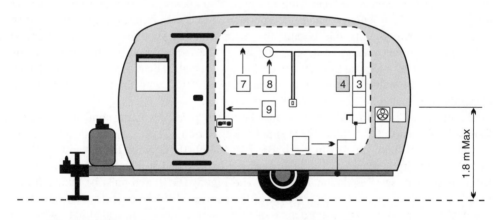

1	DP main isolating switch	2	RCD (30 mA)
3	Overcurrent protective devices	4	Periodic inspection and test notice
5	Power inlet notice	6	Power inlet (16 A 2 pole and E)
7	Horizontal cable run to socket-outlet	8	Luminaire fixed to caravan wall
9	Vertical cable run to socket-outlet	10	Main bonding conductor to chassis

Figure 16.10   Installation in a typical caravan

(DP) circuit-breakers. DP devices will detect overcurrent in both line and neutral and interrupt both on operation, whereas SP&N devices detect only overcurrent in one pole (line) but will disconnect both poles on operation.

Where more than one electrically independent installation is present, each will require its own independent inlet plug, as indicated in Regulation 721.510.3.

Any part of the caravan installation operating at extra-low voltage must comply with the requirements of Section 414 (SELV and PELV) and for d.c. be at one of the standard voltages (12, 24 or 48 V). For a.c., permissible standard voltages are 12, 24, 42 and 48 V.

16.13.3   Wiring systems

A number of the regulations in Section 721 supplement and modify the general requirements relating to wiring systems in caravans and motor caravans, and these are summarized in Table 16.21.

16.13.4   Main isolating switch, caravan inlets and connection leads

Regulation 721.537.2.1.1 calls for the installation to be equipped with a main isolating switch that disconnects all live (line and neutral) conductors, which must be placed within the caravan and positioned for ready operation (see Figure 16.10). Isolators located in linen cupboards or under bed-box covers or blanket chests, and which need some effort to locate, are unlikely to meet the requirement for ready operation. Additionally, such devices may be the subject of adverse effects from the lack of proper ventilation and consequential temperature rise, particularly where incorporated with overcurrent protective devices. In cases where there is only one circuit, the main isolating switch may be incorporated with the overcurrent protective device, and where the latter also meets the requirements for isolation (e.g. adequate contact separation and visual indication of contacts) it may also serve as the main isolating switch. Similarly, an RCD with the necessary isolation characteristics may also serve the combined function.

The electrical inlet to the caravan must be a two pole plus earth appliance inlet (or four pole plus earth for a three-phase supply; Regulation 721.55.1.1), of adequate rating for the caravan load, complying with BS EN 60309-01, or with BS EN 60309-02 if interchangeability is required. The inlet must not be installed at a height exceeding 1.8 m from the ground (see Regulation 721.55.1.2 and Figure 16.10) and must be mounted in a readily accessible position on the exterior of the caravan and housed in an enclosure with a suitable cover, so that at least a degree of protection of IP44 is provided both with and without the connector engaged. A legible and durable notice as required by Regulation 721.537.2.1.1.1 must be affixed on or near the inlet (see Figure 9.23 for example of label details) giving details of nominal voltage and frequency together with the installation rated current.

It is a requirement of Regulation 721.55.2.6 for the caravan installation to be equipped with a flexible cord or cable to connect the caravan inlet to the park-pitch socket-outlet. The flexible link, to be complete with plug and connector both of two pole and earth with key position '6h', must not be longer than 27 m nor shorter than 23 m. Rubber-insulated cables to HO7RN-F or equivalent are acceptable and should be sized according to the rating of the caravan installation:

Table 16.21 Additional requirements relating to wiring systems in caravans and motor caravans

Regulation	Aspect	Requirements
721.521.2	Type of wiring system	Must be one of the following three types; • single-core cables with flexible class 5 conductors in nonmetallic conduit; • single-core cables with stranded class 2 conductors (at least seven strands) in nonmetallic conduit; • sheathed flexible cables. Cables to comply at least with BS EN 60332-1-2, relating to test for vertical flame propagation. (Types of cable meeting the relevant requirements include those to BS 6004 (thermoplastic), BS 7211 and BS 5467 (thermosetting) and BS 6500 (flexible cords).) Conduits and cable management systems to comply with BS EN 61386-21 and BS 61386 respectively
721.522.7.1	Protection of wiring from mechanical damage	All wiring to be protected, either by location or by enhanced protection, against mechanical damage, particularly where cables pass through metalwork
721.522.8.1.3	Cable supports	Where cables are not contained within rigid conduit, to be supported at spacings not exceeding 400 mm for vertical runs and 250 mm for horizontal runs. (Note: designers may consider these spacings for safety are too generous to provide for an aesthetically pleasing surface installation.) See Figure 16.10
721.524.1	Minimum csa of live conductors	1.5 mm^2
721.528.1	Separation of low voltage and extra-low voltage wiring systems	LV and extra-low voltage wiring systems to be run separately from each other so there is no physical contact between them
721.528.3.5	Fuel storage	Except for extra-low voltage equipment for gas supply control, no electrical equipment (including wiring systems) to be installed in any compartment intended for gas cylinder storage. Cables running through compartment to be at least 500 mm above base of gas cylinders and enclosed in gas-tight conduit or duct able to withstand sever impact
721.543.2.1	Protective conductors	All protective conductors to be incorporated with associated live conductors in multicore cable or conduit

- 16 A – 2.5 mm^2
- 25 A – 4 mm^2
- 32 A – 6 mm^2
- 63 A – 16 mm^2
- 100 A – 35 mm^2.

Most, if not all, small touring caravans will not need a greater rating than 16 A.

16.13.5   Luminaires and accessories

There are additional requirements relating to accessories and luminaires in terms of selection, and these are summarized in Table 16.22.

Figure 16.10 identifies in graphical representation some of the requirements relating to caravans and motor caravans.

16.13.6   Annex A of Section 721: guidance on 12 V d.c. installations

As already mentioned in Section 16.13.1, Section 721 does not apply to installations covered by BS EN 1648-1 and BS EN 1648-2, relating to 12 V d.c. installations for habitation aspects of caravans and motor caravans respectively. However, Section 721 does include an informative annex – Annex A – with content very similar, but not identical, to that of BS EN 1648-1:2004 and BS EN 1648-2:2005. Like BS EN 1648-1:2005, Annex A also covers the design and integration of a caravan electrical system with that of a towing vehicle.

Table 16.22   Additional requirements for luminaires and accessories in caravans and motor caravans

Regulation	Equipment	Additional requirement
721.55.2.1	LV socket-outlets	Unless supplied by an individual winding of an isolating transformer, to incorporate a protective contact
721.55.2.2		All extra-low voltage socket-outlets to be clearly marked with nominal voltage
721.55.2.3	Accessories – generally	Where accessories are exposed to moisture, at least IP44 degree of protection is required either for the accessory itself or by a suitable enclosure
721.55.2.4	Luminaires – generally	Caravan luminaire preferably to be affixed directly to the structure or lining (see also Figure 16.10). Where pendant luminaires are installed, provision for securing them in transit must be made. Accessories for the suspension of luminaires must be suitable for the suspended mass
721.55.2.5		Where intended for dual-voltage operation, must comply with appropriate product standard

16.14 Temporary electrical installations for structures, amusement devices and booths at fairgrounds, amusement parks and circuses

16.14.1 General

Section 740 is another of the sections added into BS 7671 with the publication of BS 7671:2008. It applies to the electrical installations of temporarily erected or transportable electrical machines and structures that incorporate electrical equipment, intended to be installed repeatedly at fairgrounds, amusement parks and the like. The requirements do not apply to the permanent electrical installation or to the internal electrical wiring of machines, which is covered by BS EN 60204-1 *Safety of machinery. Electrical equipment of machines. General requirements*.

As with all other sections of Part 7, the requirements of Section 740 supplement or modify the general requirements of BS 7671. The requirements for inspection and testing are essentially the same as for any other installation, except that, as stated in Regulation 740.6, the installation must be inspected and tested after each assembly on site, between its origin and the connected electrical equipment.

16.14.2 Electrical supplies and protection against electric shock

The nominal voltage of the temporary electrical installations must not exceed 230/400 V a.c. or 400 V d.c. (Regulation 740.313.1.1). Supplies may be derived either from LV generating sets or from the public network, but where they are taken from the public network the live conductors from different sources must not be interconnected within the installation (Regulation 740.313.3).

It is not permitted to use the protective measures of non-conductive location or earth-free local equipotential bonding, as stated in Regulation 740.410.3.6. Likewise, the use of the protective measure of obstacles is precluded by Regulation 740.410.3, which also confines the use of the protective measure of placing out of reach to dodgems.

Where the protective measure against electric shock is ADS, as will generally be the case, one or more RCDs with rated residual operating current $I_{\Delta n}$ not exceeding 300 mA must be provided at the origin of the temporary installation as devices for automatic disconnection in the event of a fault (Regulation 740.410.3). As would be expected, these RCDs are required to incorporate a time delay in accordance with BS EN 60947-2 or be of the selective type (S) in accordance with BS EN 61008-1 or BS EN 61009-1, in order to discriminate with RCDs protecting final circuits. Any RCD used on the supply to a motor should be of the time-delayed type or S type where this is necessary in order to prevent unwanted tripping.

Regulation 740.411.4 precludes the use of a PEN conductor in the temporary electrical installation, but this is precluded in any event by Regulation 8(4) of ESQCR, which states that a consumer shall not combine the neutral and protective functions in a single conductor in the consumer's installation. The use of an IT system is precluded by Regulation 740.411.6 if an alternative system is available, although the regulation does permit an IT system to be used for d.c. applications where continuity of service is needed.

All final circuits for lighting, socket-outlets rated at up to 32 A and for mobile equipment connected through a flexible cable or flexible cord of current-carrying capacity up to 32 A are required by Regulation 740.415.1 to be provided with additional protection by RCDs having the characteristics specified in Regulation 415.1.1 ($I_{\Delta n} \leq 30$ mA and operating time ≤ 40 ms at $5I_{\Delta n}$). The supply circuit to battery-operated emergency lighting must be connected to the same RCD as the general lighting circuit in the area served by the emergency lighting, so that operation of the RCD will also bring the emergency lighting into operation. This requirement does not apply to a circuit protected by SELV, PELV or electrical separation, or to a circuit that is outside the limit of arm's reach (as defined in Part 2 of BS 7671) supplied other than through a socket-outlet for household or similar purposes (such as a 13 A socket-outlet to BS 1363) or a socket-outlet to BS EN 60309-1.

In any locations intended for livestock, supplementary bonding is required, as specified in Regulation 740.415.2.1, in the manner as described in Section 16.6.2 of this Guide for parts of agricultural and horticultural premises that are intended for livestock.

16.14.3   Motors: protection against excess temperature

Because of the likelihood of their being situated near combustible materials, motors that are automatically or remotely controlled are required by Regulation 740.422.3.7 to be fitted with an excess temperature protective device that has to be manually reset in the event of its operation. The requirement for such a device to be provided does not apply where the motor is continuously supervised.

16.14.4   Equipment: selection and erection

There are additional requirements for the selection and erection of equipment, and Table 16.23 summarizes these.

16.15   Floor and ceiling heating systems

16.15.1   General

Section 753 was added into BS 7671 with the publication of BS 7671:2008. The section applies to the installation of indoor electric floor and ceiling heating systems erected either as thermal storage (e.g. utilizing off-peak tariff electricity and concrete or similar material to store heat) or direct heating systems. The section does not apply to wall heating systems. However, a sloping ceiling located under the roof of a building down to a height of 1.5 m above floor level is considered to be a ceiling, rather than a wall, for the purposes of Section 753.

There are two main reasons why electrical installations of floor and ceiling heating systems are considered to be installations of increased electric shock. One is the possibility of a heating unit being penetrated by a nail or screw or the like, leading to the risk of electric shock to a person due to direct contact with a live part of the heating unit. The other reason is the possibility of an earth fault occurring part way along the length of a heating unit. In such a case, the fault current may to be too small to cause the circuit overcurrent protective device to disconnect the supply within the maximum time permitted by Chapter 41 of BS 7671 for fault protection, due to the resistance of the heating unit between the point of fault and the line conductor terminal of the heating unit.

Table 16.23 Additional requirements relating to equipment selection and erection: temporary installations for structures, amusement devices and booths at fairgrounds, amusement parks and circuses

Equipment	Regulation	Additional requirement
Equipment – generally	740.512.2	To have at least a degree of protection of IP44
Wiring systems	740.521.1	Cables to meet the requirements of BS EN 60332-1-2, relating to test for vertical flame propagation. (Types of cable meeting the relevant requirements include those to BS 6004 (thermoplastic), BS 7211 and BS 5467 (thermosetting), and BS 6500 (flexible cords).) Rated voltage of cables to be at least 450/750 V, or 300/500 V within amusement devices. Cables buried in the ground to be protected from mechanical damage and their routes marked at suitable intervals (on the surface of the ground). Cables subject to risk of mechanical damage to be armoured or otherwise suitably protected. Wiring systems likely to be subjected to movement are to be of flexible construction, such as cables of type H07RNF or H07BN4-F in flexible conduit
Joints in wiring	740.526	Not permitted except as a connection into a circuit. Joints to have cable anchorages for strain relief where needed and be made in enclosures protecting to at least IP4X or IPXXD or using connectors to relevant British Standard
Switchgear and controlgear	740.51	Except where intended to be operated by ordinary persons, to be within cabinets openable only by a tool or key.
	740.537.1	Readily accessible means of isolation, switching an overcurrent protection to be provided for each booth, stand and amusement device. Readily accessible, properly identified means of isolation to be provided for each temporary installation for amusement devices and each distribution circuit for outdoor installations
	740.537.2.2	An isolation device must disconnect all live conductors (line(s) and neutral)
Lighting installations	740.55.1.1	Luminaires and decorative lighting chains to have suitable IP rating and be installed in a way that does not impair ingress protection and properly supported. Where less than 2.5 m above floor level or otherwise liable to accidental contact, luminaires and decorative lighting chains to be fixed and guarded to prevent injury to persons or ignition of materials. Fixed light source to be accessible only by removing a barrier or enclosure requiring use of a tool. Lighting chains to use H05RN-F cable or equivalent

(*continued overleaf*)

Table 16.23   (*continued*)

Equipment	Regulation	Additional requirement
	740.55.1.2	Insulation-piercing lampholders only to be used with compatible cable type and not to be removed
	740.55.1.3	Lamps to be protected against accidental damage from projectiles, such as in shooting galleries
	740.55.1.4	Luminaires of transportable floodlights to be inaccessible and supplied by flexible cables protected from accidental damage
	740.55.1.5	Luminaires, including floodlights, to be arranged to prevent ignition of material due to focusing or concentration of heat
	740.55.3.1 and 740.55.3.2	Luminous tubes, signs or lamps operating at a voltage >230/400 V on booths, stands or amusement devices to be installed out of arm's reach and supplied by a dedicated circuit controlled by an emergency switch that is easily visible and accessible and marked to local authority requirements

Safety isolating transformers and electronic convertors	740.55.5	Secondary circuit of each transformer or electronic convertor to be protected by manually reset device.
		    Transformers to be located out of arm's reach or where accessible only to skilled or instructed persons. Must be accessible for testing and for maintenance of protective device.
		    To be adequately ventilated and vents not obstructed when in use.
		    Transformer to conform to BS EN 61558-2-6 and convertors to BS EN 61347-2-2

Plugs and socket-outlets	740.55.7	Adequate numbers of socket-outlets to be installed to safely meet user's requirements.
		    Socket-outlets to be marked or encoded to show their purpose if dedicated to lighting circuits that do not have additional protection by an RCD (as is permitted by Regulation 740.415.1(iii) subject to certain conditions).
		    Plugs and socket-outlets used outdoors to comply with BS EN 60309-1, or BS EN 60309-2 where interchangeability is required, or, if rated at ≤16 A and provide equivalent mechanical protection, may comply instead with relevant national standard

Connection point on amusement devices	740.55.8	Each amusement device to have an electricity supply connection point marked with its rated current, voltage and frequency
Supplies to dodgems	740.55.9	Voltage to be ≤50 V a.c. or 120 V d.c. Circuit to be separated from mains supply by or an isolating transformer to BS EN 61558-2-4 or a motor-generator set

Table 16.23 (*continued*)

Equipment	Regulation	Additional requirement
LV generating sets	740.551.8	Measures to be taken to prevent danger or injury to persons from hot surfaces or dangerous part, by suitable location or protection of generator. Electrical equipment associated with generator to be securely mounted and on anti-vibration mountings if necessary. Earthing arrangements and, where applicable, earth electrodes to comply with relevant requirements of BS 7671, given in Regulations 542.1 and 542.2 respectively (even though generator installation is temporary). Except in an IT system, neutral (or star) point of generator to be connected to exposed-conductive-parts of generator (e.g. frame and casing).

As with all the sections in Part 7, the requirements of Section 753 supplement or modify the general requirements of BS 7671. General requirements relating to the selection and erection of heating conductors and cables are given in Regulation Group 554.4, referred to in Section 11.8.4 of this Guide.

For a floor or ceiling heating installation in a location containing a bath or shower, or in a location containing a swimming pool or other basin, the relevant requirements of Section 701 or Section 702 of BS 7671 respectively, covered in sections 16.2 and 16.3 respectively of the Guide, must be met in addition to those of Section 753.

16.15.2   Heating units

Two types of heating unit are considered in Section 753: heating cables and flexible-sheet heating elements. Both types have rigidly fixed cold tails or terminal fittings connected to the terminal of the electrical installation. Heating cables may be single-core or two-core, with or without a metallic sheath or braid. Regulation 753.511 states that they shall comply with the BS 6351 series *Electric surface heating*. However, heating cables can also be manufactured to IEC 60800 *Heating cables with a rated voltage of 300/500 V for comfort heating and prevention of ice formation* or IEC 61423-1 *Heating cables for industrial applications* or BS EN 60335-1 *Household and similar electrical appliances*.

Flexible-sheet heating elements are heating elements consisting of sheets of electrical insulation laminated with electrical resistance material, on a base material on which electrically insulated heating wires are fixed. Regulation 753.511 requires that they shall comply with BS EN 60335-2-96 *Household and similar electrical appliances. Safety. Particular requirements for flexible sheet heating elements for room heating*.

Whichever of the two types of heating unit is used, Regulation 753.512.2.5 requires that they have a degree of protection not less than IPX1, or IPX7 where installed in a floor of concrete or similar material, and have adequate mechanical properties where installed in such a floor.

Regulation 753.515.4 points out that heating units must not be installed across constructional expansion joints, as this could result in damage to the units where there is differential movement at such a joint.

16.15.3   Protection against electric shock

Not surprisingly, it is not permitted to use the protective measures of obstacles or placing out of reach, or the protective measures of non-conducting location or earth-free equipotential bonding. Such use is prohibited by Regulations 753.410.3.5 and 753.410.3.6 respectively. It is also not permitted to use the protective measure of electrical separation, as stated in Regulation 753.413.1.2.

Where the protective measure of ADS is used, Regulation 753.411.3.2 requires that the disconnecting devices shall be RCDs with rated residual operating current $I_{\Delta n}$ not exceeding 30 mA. A note to the regulation points out that unwanted tripping of the RCDs due to leakage capacitance may be avoided by limiting the rated heating power per RCD to 7.5 kW at a nominal voltage of 230 V or 13 kW at a nominal voltage of 400 V.

Regulation 753.411.3.2 also requires that where the heating units are delivered from the manufacturer without exposed-conductive-parts (such as an earthed metallic sheath), a suitable conductive covering, such as a grid with a spacing of 30 mm or less, is to be installed above floor heating units or below ceiling heating units and connected to the cpc. However, it should be appreciated that this requirement does not apply to heating units of Class II construction or equivalent insulation, as electric shock protection for such units does not rely on ADS. (Different requirements apply in locations covered by Sections 701 and 702 of BS 7671. See Section 16.2.9 and Table 16.6 respectively of this Guide).

Where heating units of Class II construction or equivalent insulation are used, the circuit is required by Regulation 753.415.1 to be provided with additional protection by an RCD having the characteristics specified in Regulation 415.1.1 ($I_{\Delta n} \leq 30$ mA and operating time ≤ 40 ms at a residual current of $5I_{\Delta n}$). It would be permissible to rely on the RCD required by Regulation 753.411.3.2 (for ADS) for this purpose, provided it was not of the nondelay type.

16.15.4   Protection against harmful thermal effects

Floor heating systems must be arranged so that the surface temperature is limited to a suitable value if contact with skin or footwear is possible, as would normally be the case. Regulation 753.423, which requires this temperature limitation, suggests a limiting temperature of 35 °C. In any event, the temperature of both floor heating units and ceiling heating units must be limited so that it cannot rise above 80 °C. The acceptable measures to prevent overheating beyond this temperature are given in Regulation 753.424.1.1. They include appropriate design of the heating system, appropriate installation of the system according to the manufacturer's instructions and use of protective devices. Connection of the heating system to the electrical installation must be either by cold tails or suitable terminals. The connection of the heating units to the cold tails must be by crimped connections or some other inseparable means.

Regulation 753.424.1.2 is concerned with the danger that could arise under fault conditions where heating units are installed close to easily ignitable building materials, such as some kinds of timber or plastics. The regulation calls for measures to be taken to avoid harmful thermal effects in such circumstances, such as by placing the heating units on a metal plate or, where applicable, in metallic conduit, or at a distance of not less than 10 mm in air from the ignitable material.

16.15.5   Heating-free areas

Regulation 753.520.4 requires there to heating-free areas (areas of floor or ceiling where there are no heating units) to allow for the attachment of room fittings, so that the heat emission of heating units is not prevented by these. Regulation 753.522.4.3 requires heating-free areas to be provided where necessary to facilitate drilling and fixing, by screws and the like, such as for built-in furniture. The installer is required to inform other contractors that such drilling and fixing is not to take place in the areas where floor and ceiling heating units are installed.

16.15.6   Wiring systems

Where installed in the zone of the heated surfaces, cold tails (the circuit wiring connecting heating units to the fixed installation) and control leads must be suitable to withstand the temperature of the medium in which they will be surrounded (Regulation 753.522.1.3 refers).

16.15.7   Identification and notices

A plan for each heating system must be provided adjacent to the heating system distribution board, giving the information referred to in Regulation 753.514 relating to the heating units and the associated electrical installation. The same regulation requires information to be provided for the owner or user of the heating system, as detailed in Figure 753 of Section 753. This information includes a description of the system and instructions for its use, including at least the particulars listed in Figure 753.

17

Inspection, testing, certification and reporting

17.1 Inspection, testing, certification and reporting: general

Part 6 of BS 7671 sets out the requirements relating to the inspection and testing of an electrical installation and identifies specific aspects as follows:

- initial verification;
- periodic inspection and testing;
- certification and reporting.

Figure 17.1 identifies the various chapters, sections and regulations of Part 6.

17.2 Test instruments

17.2.1 General

To comply with Regulation 612.1, test instruments must be chosen according to the relevant parts of BS EN 61557 or, if not, they must provide at least the same degree of performance and safety. Beyond this, however, it will be the purchaser who will have to evaluate the relative merits of the instruments offered and make a decision on what is best for their particular circumstances.

BS 7671 does not demand regular calibration of test instruments. However, the installation designer may wish to stipulate that the test engineer has in place an effective system to confirm the continuing accuracy and consistency of all test instruments used for certification and reporting purposes. Such a system may consist of maintaining records of the formal calibration/recalibration of test instruments as recommended by the instrument manufacturers, supported by calibration certificates issued by recognized organizations with measurements traceable to national standards.

A Practical Guide to The Wiring Regulations: 17th Edition IEE Wiring Regulations (BS 7671:2008)
Fourth Edition Geoffrey Stokes and John Bradley
© Geoffrey Stokes and John Bradley. Published by John Wiley & Sons, Ltd

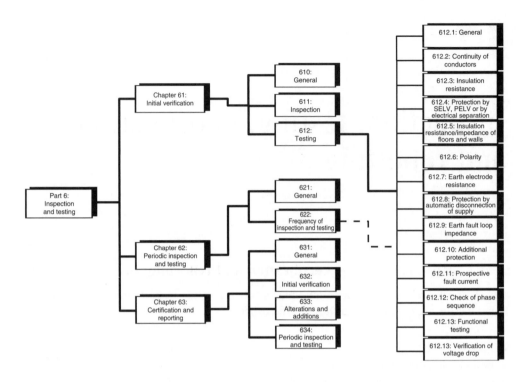

Figure 17.1   Part 6: inspection and testing

Otherwise, a system recognized by an assessment body could be used that includes maintaining records over time of comparative cross-checks with other test instruments used by the business, and of measurements of the characteristics of designated reference circuits or devices. However, such alternative systems can only provide a measure of confidence in the consistency of test measurements over time. The accuracy of each test instrument will need to be confirmed before any reliance can be placed in such systems.

It should be noted that new or repaired test instruments may not be supplied with calibration certificates unless specifically requested.

It is essential for the test engineer to be fully versed in the capabilities and limitations of instruments and to be aware when a test is likely to create a dangerous situation. As called for in Regulation 610.1, a test must not lead to danger to persons or damage to property or equipment even when the circuit being tested is faulty.

It is considered good practice that a record of test instruments is maintained showing the serial numbers and the dates of calibration and checks on accuracy and consistency. It can also be advantageous to record the particular instrument(s) used for making tests for certification purposes in order that, should a recorded reading be challenged or give cause for concern later, the particular instrument is traceable. Using such a procedure would also enable other tests (on other installations) undertaken using the same instrument(s) to be checked should there be any doubt about the test results.

17.2.2   Insulation test instruments

An instrument for testing insulation must have a d.c. test voltage and be capable of supplying that test voltage (1 kV, 500 V or 250 V, as appropriate to the circuit) when loaded at 1 mA, as required by BS EN 61557-2.

Resolution for these instruments should present little problem, with a 100 kΩ resolution being adequate for most applications.

17.2.3   Continuity test instruments

For continuity testing, Regulation 612.2.1 recommends a supply having a no-load voltage between 4 and 24 V d.c. or a.c. and a short-circuit current of not less than 200 mA. As far as resolution is concerned, 10 mΩ will normally be adequate. A test instrument complying with BS EN 61557-4 will meet with these requirements.

17.2.4   Earth loop impedance test instruments

Generally speaking, a loop impedance test instrument with a resolution of 10 mΩ should be sufficient. Most commercially available instruments make comparative measurements between the no-load voltage and a test circuit loaded with 10Ω and, consequently, operate with a test current of 20–25 A. The test is normally limited to a duration of approximately four half cycles (40 ms on 50 Hz). From a safety viewpoint, the potential on the cpc could rise to dangerous levels (particularly where a defective circuit loop is involved, where the potential could rise to line conductor voltage with respect to earthy parts) and for this reason the instrument would not normally require a test duration exceeding 40 ms.

17.2.5   Applied voltage test instruments

There are no specific regulatory requirements in terms of performance of applied voltage test instruments, but these instruments should be capable of steadily increasing the applied voltage and should have an accuracy of about 5%. Unless there is a need for higher test currents, the output current should be limited to 5 mA. The duration of the test should be at least 60 s and the maximum applied test voltage should be 4 kV to accommodate all tests stipulated in BS 7671. As with all testing, the methods and procedures of testing should be such that the risk of electric shock for test personnel and bystanders (and livestock) is kept to a minimum. This is particularly important when test currents exceeding 5 mA are used.

17.2.6   Earth electrode test instruments

BS 7671 itself makes few demands with regard to electrode test instruments. Generally speaking, a proprietary make of instrument with a three- or four-terminal arrangement should be used in order that the resistance of the test leads and test electrode may be evaluated. Instruments may be hand-cranked or powered by batteries, including rechargeable ones. Compliance with BS EN 61557-5 is likely to provide all the necessary instrument facilities and safety requirements. Resistance ranges from 10 mΩ to 2 kΩ should meet most of the practical test circumstances. Reference to BS 7430 *Code of practice for earthing* will provide additional information relating to earthing electrodes and associated tests.

17.2.7 Residual current device test instruments

Regulations 612.8.1 and 612.10 require the use of an RCD test instrument to BS EN 61557-6 to test the operation RCDs used for fault protection and additional protection respectively. The RCDs will need to be tested at the rated residual operating current $I_{\Delta n}$ and at $5I_{\Delta n}$, where applicable. Additionally, a test at $0.5I_{\Delta n}$ will be desirable. The in-service accuracy of displayed disconnection times should be within $\pm 10\%$ and will depend to some extent on supply voltage variations. Some contemporary test instruments have a facility for testing the general (nondelay)-type and the selective (delay)-type RCDs and also have a capability for test initiation at the start of either the positive-going or negative-going cycles of the a.c. supply.

17.2.8 Voltage indication

Every test engineer will need some form of voltage indication for, if nothing else, proving a circuit dead. Indicators can be of the form of a voltmeter or other no less effective and reliable device (see also Section 17.3).

17.2.9 Phase sequence test instruments

Phase sequence test instruments should comply with BS EN 61557-7 or provide at least the same degree of performance and safety.

17.3 Safety in electrical testing

Perhaps the most important, and often overlooked, aspect of safety in electrical testing is that of the skill of the operator undertaking the testing. Testing is not a task for the unskilled person, and persons engaged in such work must be skilled and have received proper and adequate training relating to procedures and test equipment utilized for the particular testing activity envisaged.

It is essential that test instruments and test leads are maintained in a safe condition and that the test methods do not create danger either for the operator or for other persons or livestock. An unsafe or unsuitable instrument is akin to handling a bomb, and it is paramount that instruments are obtained from reputable manufacturers with a good instrument safety record. Test leads are of crucial relevance to safety and, when making voltage measurements on potentially live circuits, should always be fused or fitted with current-limiting devices. Test lead fuses should be capable of breaking the prospective fault current and be inserted in both leads.

Where tests are required to circuits that have not to be energized, it is always necessary to prove that the circuit is dead before applying the test. Regulation 14 of the *Electricity at Work Regulations 1989* requires that no person shall be engaged in work activities on or so near any live conductor that danger may arise, unless (a) it is unreasonable in all the circumstances for it to be dead; and (b) it is reasonable in all the circumstances for him to be at work on or near it while it is live; and (c) suitable precautions (including

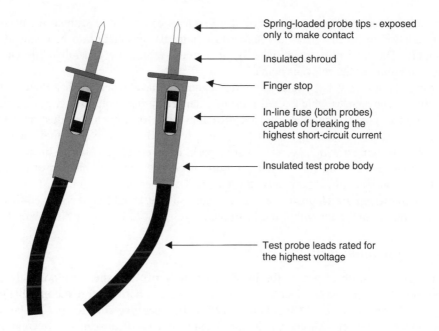

Spring-loaded probe tips - exposed only to make contact

Insulated shroud

Finger stop

In-line fuse (both probes) capable of breaking the highest short-circuit current

Insulated test probe body

Test probe leads rated for the highest voltage

Figure 17.2   Safety features for fused test probes

where necessary the provision of suitable protective equipment) are taken to prevent injury. Tests involving live conductors (such as earth loop impedance testing) should therefore be restricted to those occasions where these three conditions are simultaneously met. Testing generally should be the subject of a well thought out procedure and system of work in order to provide safety for the test engineer and others.

Some environments need special consideration relating to safety of testing. For example, in potentially explosive gas atmospheres, most generally available instruments will produce sufficient current to ignite potentially explosive gases and vapours. Where testing in such environments is envisaged, the test equipment and procedures are critically important, and the test engineer should consult with the instrument manufacturer and the person responsible for safety at the site and, indeed, where necessary, the Health and Safety Executive. With regard to petrol filling stations, *Guidance for the design, construction, modification and maintenance of petrol filling stations*, published jointly by the Association for Petroleum Explosives Administration (APEA) and the Institute of Petroleum, is essential reading for the designer, installer and testing engineer involved in such work. Part 14 of that publication deals with the electrical installation aspects, including inspection and testing.

17.4   Test methods

17.4.1   General

It is essential that a certain sequence is followed in the testing procedures to minimize the risk of a defective circuit component creating a dangerous condition when a test is made and to ensure that the test results are valid. Most tests can be made without the need for the installation to be energized, but some do require the installation to be connected to the supply. The sequences for initial and for periodic testing are generally different and are identified under the relevant items.

Testing should only be carried out after enquiries have revealed that there are no particular circumstances which could present a danger during such testing. For example, many organic materials, plastics and metal oxides may form explosive dust, and some very serious cases of dust explosions have been reported. It is accepted that some dust clouds may be ignited with an electrical arc (spark) of an energy level of as little as 5 mJ. A typical energy level of, for example, a loop impedance test instrument is of the order of 5 J – a thousand times more than the energy level needed for ignition. Instruments are available which are claimed to be intrinsically safe; it should be recognized that this generally relates to arcing within the instrument itself, not to its use or test method.

17.4.2 Insulation tests

The principal purpose of testing the insulation is to verify that there are no inadvertent connections between live conductors and between live and Earth, before the installation is energized. Tests are required (Regulation 612.3.1) between live conductors (e.g. between line conductors and between line conductor(s) and neutral) and between all live conductors and the protective conductor connected to the earthing arrangement. In TN-C systems, the PEN conductor is considered to be part of the earth. Insulation testing does not require the installation, or the part of it under test, to be connected to the electricity supply, and the test engineer must ensure that the circuit(s) under test and associated equipment are de-energized and proved to be 'dead' before carrying out any such testing.

As called for in Regulation 537.2.1.7, provision must be made for disconnecting the neutral conductor from the supply. This is to facilitate the necessary insulation resistance testing between neutral and Earth, which is not possible without such disconnection. As regards testing between conductors, it will be necessary to remove any loads (e.g. lamps, capacitors, inductors) connected between conductors under test in order to avoid obtaining a distorted test value for insulation. As required by Regulation 612.3.2, surge protective devices and any other equipment likely to be damaged by an insulation test or influence the test results (such as electronic equipment and other solid-state devices, dimmer switches, electronic starters and the like) should be disconnected before the test is carried out. Where disconnection of such equipment is not reasonably practicable, Regulation 612.3.3 requires that only a measurement between live conductors connected together and the earthing arrangement is made; furthermore, Regulation 612.3.2 allows the test voltage for the circuit concerned to be reduced to 250 V d.c., provided the insulation resistance is at least 1 MΩ. It should be noted, however, that such measures may not prevent damage to electronic devices, which can only be completely avoided by disconnecting them.

The minimum insulation resistances that are acceptable are given in Regulation 612.3.2 and Table 61 of BS 7671 and are summarized, albeit in a somewhat different format, in Table 17.1. The normal insulation tests for LV systems apply equally to FELV systems.

Where feasible, installations should be tested for insulation resistance as a whole, but where this is not possible, or desirable, the installation should be divided into convenient sections. In any event, tests should be carried out with the upstream supply isolation

Table 17.1 Minimum insulation resistance values as they relate to circuit nominal voltages

Ref	Type of circuit (or parameter)	Nominal voltage U_n (V)	Test instrument output voltage (V)	Minimum resistance (MΩ)
A	LV circuit	$500 < U_n \leq 1000$	1000 d.c.	1.00
B	LV circuit (except extra-low voltage)	$0 < U_n \leq 500$	500 d.c.	1.00
C	SELV and PELV circuits	$0 < U_n \leq 50$	250 d.c.	0.50
D	FELV	$0 < U_n \leq 50$	500 d.c.	1.00

device in the open position (OFF) with all downstream protective devices (e.g. fuses, MCBs) in the closed position (ON). It should be remembered that, notwithstanding the values given in BS 7671 for minimum insulation resistances, the designer may consider a much higher value is desirable; indeed, such low values may suggest some impending latent defect. Generally speaking, most designers would consider a value of less than 100 MΩ on a final circuit and 10 MΩ on a distribution circuit to be unacceptably low and requiring further investigation.

Referring to Figure 17.3, the sequence of testing the insulation resistance of a three-phase circuit (or group of circuits) may, for example, be:

1. test between line 1 and line 2;
2. test between line 1 and line 3;
3. test between line 1 and neutral;
4. test between line 1 and protective conductor;
5. test between line 2 and line 3;
6. test between line 2 and neutral;
7. test between line 2 and protective conductor;
8. test between line 3 and neutral;
9. test between line 3 and protective conductor;
10. test between neutral and protective conductor.

Alternatively, line conductors may be grouped together for testing insulation to neutral and to protective conductor. For single-phase circuits the tests are much simplified, and only tests 3, 4 and 10 apply. Where the circuit incorporates two-way or multiple-position switching, separate tests are required for each and every switch position (i.e. one test for each setting of a switching device, with a different switch operated between tests).

17.4.3 Barriers and enclosures

For barriers and enclosures to be effective in providing basic protection they must provide at least the degree of protection IP2X or IPXXB (see Regulation 416.2.1). Additionally, the top horizontal surface of such enclosures and barriers must afford at least IP4X or IPXXD (see Regulation 416.2.2). It goes without saying that all these barriers and enclosures must be properly affixed and secured in place (see Regulation 416.2.3). Regulation 612.4.5

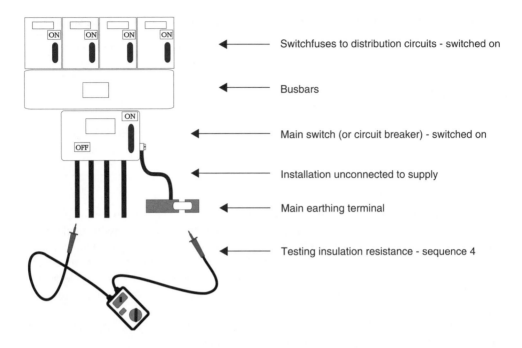

Switchfuses to distribution circuits - switched on

Busbars

Main switch (or circuit breaker) - switched on

Installation unconnected to supply

Main earthing terminal

Testing insulation resistance - sequence 4

Figure 17.3   Testing insulation

requires a test to be carried out to verify that the necessary degree of protection is provided. This test is applicable only where barriers or enclosures are provided during erection.

Where equipment is the subject of an appropriate British or Harmonized Standard and has been subjected to type-testing or partial type-testing, compliance with the relevant requirements for barriers and enclosures will be assured providing the equipment has been properly installed in accordance with the manufacturer's recommendations. However, where enclosures and barriers have been purposely fabricated and fitted on site, a test is required to determine that these degrees of protection have been met. The test involves the use of the standard finger (see BS EN 60 529 and Figure 17.7). The IP2X or IPXXB degrees of protection relate approximately to prevention of insertion of a finger or similar object not exceeding 80 mm long. For IP4X or IPXXD, this relates to wires and strips of thickness more than 1 mm and solid objects greater than 1 mm diameter.

In establishing that a barrier or enclosure affords the IP2X or IPXXB protection, it is necessary to insert the metallic test finger through the aperture and, in doing so, the finger may be bent through 90° to the axis of the finger with both of the two joints bent in the same direction. This test should use only normal force (not exceeding 10 N), and the test finger, when it enters the aperture, should be placed in every conceivable position and orientation. An extra-low voltage supply of between 40 and 50 V is connected between the live parts of the equipment and the test finger. Before carrying out such a test, it is essential that the supply to the equipment under test is isolated and that the live parts are proved to be dead. Where live conductors have had varnishes, paint or other covering applied to their surfaces, a metallic foil should be applied to such parts before carrying out this test. The protection can only be considered acceptable if the test lamp does not light up during testing in all the positions; and on completion, the lamp itself must be tested to ensure that it is still in working order.

The test for IP4X or IPXXD is made with a straight, rigid steel wire, free of burrs, of between 1.05 and 1.00 mm diameter applied with a force of 1 N ± 10%. The equipment affords the degree of protection if the wire is prevented from entering.

17.4.4 Non-conducting location tests

Protection by non-conducting location is a little-used protective measure which may provide fault protection under conditions where constant trained supervision can ensure that earthed metalwork is not inadvertently introduced into the location. When this protective measure is so used, Regulation 612.5.1 calls for the resistance of the insulating floors and insulated walls to be tested. The test involves measurements to be taken at not less than three points on each surface, one of which is within 1.0–1.2 m from any extraneous-conductive-parts (e.g. pipes) in the location. Figure 17.4 illustrates the test point arrangements.

As stipulated in Regulation 418.1.5, the acceptable lower limit for resistances of walls and floors is 50 kΩ for installations operating on voltages not exceeding 500 V and is 100 kΩ for operating voltages exceeding 500 V. As called for in Regulation 612.5.2, a resistance value of 1 MΩ for extraneous-conductive-parts is the minimum that is acceptable when tested at 500 V d.c. Where these resistance values are not met, the parts tested (e.g. walls and floors) must be considered to be conductive and this particular protective measure would be rendered invalid. Where any part which would otherwise be an extraneous-conductive-part is insulated, an applied voltage test of 2 kV rms a.c. (output current limited to 1 mA) is required. If no flashover is experienced and a further insulation test at 500 V produces a 1 MΩ (minimum) reading, then the insulation may be considered satisfactory. HV test procedures are similar to those given for site-applied insulation testing.

17.4.5 Polarity tests

Regulation 612.6 calls for a polarity test to be carried out to establish that every fuse and every single-pole device (e.g. MCB, switch) is located in the line conductor only. Additionally, tests are required to confirm that the polarity of socket-outlets, other accessories and points of utilization (e.g. luminaires) is correct. Also, a check is required to ensure that the centre contacts of Edison screw (ES) lampholders are connected to the line conductor, though E14 and E27 ES lampholders to BS EN 60238 are exempted from this test as there is no safety issue involved for these specific lampholders. Polarity testing must be carried with the supply isolated from the installation or the part of it under test.

There are a number of ways that polarity can be tested, one of which can be the use of a continuity test instrument (e.g. a low-resistance ohm-meter) connected to the outgoing way (at the distribution board) of the circuit under test with a long extension test lead connected to the test instrument and the other test lead probe used to test the various points of the circuit (e.g. switches, luminaires). Another method would involve the bridging out of the outgoing way in the distribution board and the cpc marshalling terminal block and

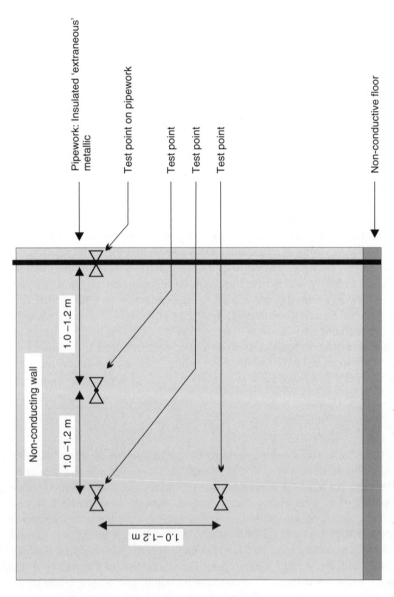

Figure 17.4 Nonconducting wall: example of test point positions

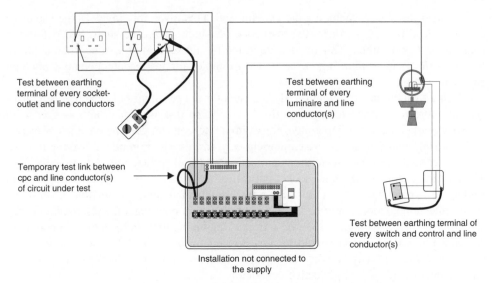

Test between earthing terminal of every socket-outlet and line conductors

Test between earthing terminal of every luminaire and line conductor(s)

Temporary test link between cpc and line conductor(s) of circuit under test

Test between earthing terminal of every switch and control and line conductor(s)

Installation not connected to the supply

Circuit protective conductors and line conductors only shown, for clarity.

Figure 17.5   Testing protective conductor continuity

then making tests between the line conductor and the earthing terminal of all the points of the circuit (see Figure 17.5 showing the link). There may be instances where testing of polarity can be incorporated in tests relating to other parameters.

17.4.6   Continuity tests

The continuity of every protective conductor must be tested, as called for in Regulation 612.2.1, to verify that it is electrically sound and correctly connected. The inductive reactance for protective conductors of $35\,\text{mm}^2$ or less can be ignored; therefore, it is acceptable to test with an instrument with a d.c. output. The inductive reactance becomes significant (i.e. the $X : R$ ratio is much greater) for protective conductors of greater than $35\,\text{mm}^2$; for these conductors, a test instrument with an a.c. output (frequency as that of the installation supply being preferable) is desirable. All such tests are required to be made with supplementary bonding (if any) disconnected from the protective conductor under test.

There are a number of ways in which tests may be carried out, one of which involves the strapping of the earthing terminal block in the distribution board (or consumer unit) with the outgoing line conductor of the circuit under test, as shown in Figure 17.5. The instrument's test leads should then be applied at points along the circuit, including switches, socket-outlets and points of utilization. The measurement at the extreme end of the circuit will provide the value for $(R_1 + R_2)$, and this should be recorded to verify compliance with BS 7671.

Another equally acceptable method of testing protective conductor continuity is by connecting one test lead to the distribution board's earth terminal and the other to various points along the circuit, including the extreme end. This invariably involves the use of an extension lead on one test lead, the resistance of which should be subtracted before the measurement is recorded. This method does not, of course, permit a value to be obtained for $(R_1 + R_2)$ and is more appropriate, for example, for testing equipotential bonding conductors.

For ring final circuits, Regulation 612.2.2 calls for the continuity of each live conductor and the protective conductor to be verified. This is particularly important, as even with properly marked cables it is possible, for example, to connect the line conductors of two different ring final circuits into one outgoing way of a distribution board, leaving the two circuits without adequate overcurrent protection and fault protection. Additionally, it is not unheard of for one or more socket-outlets to be 'lost' (e.g. during plastering), again leading to a potentially dangerous situation through inadequate protection. To test the circuit, line conductors should be split at the distribution board and the test applied to the ends of the ring. This should then be repeated for the neutral and protective conductor. Any open-circuit condition should be investigated further. The $(R_1 + R_2)$ value to the midpoint of the ring may be taken as a quarter of the sum of the values measured for the line conductor and for the protective conductor.

It must be borne in mind that testing of ring final circuits as described above will not reveal the situation where a ring has been 'bridged' by the connection of a cable between outlets other than from one outlet to the next. The $(R_1 + R_2)$ value obtained by measurement of such a bridged ring circuit would be grossly inaccurate and overoptimistic.

Where a protective conductor is formed by metal conduit, metal trunking, MICC, SWA or other conductive enclosure, continuity tests become extremely difficult because of the fortuitous parallel paths. For this reason, the test described above becomes impracticable. This would also apply to the situation where exposed-conductive-parts of equipment (e.g. luminaires) are affixed to parts of the building which themselves provide conductive paths back to the earthing terminal (e.g. steel girders, ceiling tile frames). Where luminaires are connected by LSC, disconnection for test purposes is fairly easy. Where this is not possible, tests are required to be undertaken before final connection of the luminaires. To test conductive enclosures (e.g. steel conduit) it is necessary first to apply a standard continuity test followed by a careful examination of the earth fault current path back to the earthing terminal of the distribution board. When satisfied that such a path is going to provide a permanent and reliable path, the test engineer may then use a standard line–earth loop impedance test instrument to the protective conductor with one test lead connected to the line conductor of the circuit for which the protective conductor is provided and the other test lead applied to the protective conductor. In cases where there is any doubt, a further test will be necessary using an a.c. instrument with output voltage not exceeding 50 V and a test current of 1.5 times the design current I_b or 25 A, whichever is the less. It is worth remembering that no test has yet been developed to predict the performance of a fault current path under actual fault conditions. A single strand of flexible cord making a connection between otherwise separated parts of a fault current path would undoubtedly produce good results under a continuity test (low-current test) but would obviously not withstand fault currents, which may be of the order of many kilo-amps.

An assessment of the circuit lengths of radial circuits may be made from the measured values of R_1 or $(R_1 + R_2)$ using one of the following methods and utilizing data given in Table 10.17 of this Guide. First, taking the measured resistance value for a line R_1 and dividing by the value given for the particular csa given in Table 10.17, the length in metres may be obtained. Similarly, the same procedure will give the length when the measured value is that of $(R_1 + R_2)$ when the appropriate tabulated value is used. Take a simple example; the $(R_1 + R_2)$ value of a 10/4 mm^2 cable has been measured as 0.4Ω. From Table 10.17 (Row M) it will be seen that this cable has a $(R_1 + R_2)$ per metre of 6.44 mΩ/m. The circuit length is therefore 62 m (0.4/0.00644). These assessments are only valid where the conductor temperature is about 20 °C.

Every protective conductor (including protective bonding conductors, both main and supplementary) needs testing (Regulation 612.2.1) for continuity. It is important to recognize that the impedance of a conductor is made up of two constituent parts, namely resistance and inductive reactance. For conductors of csa of 35 mm^2 or less the inductance is small compared with the resistive component (i.e. the ratio $X : R$ is small) and in most cases can be ignored. Conversely, the inductance cannot be ignored for conductors of csa greater than 35 mm^2, as it represents a significant contribution to the impedance. It must also be remembered that resistance, unlike inductive reactance, is a function of temperature.

Where the inductance of a conductor can be neglected, continuity testing of protective conductors may be carried out using a d.c. test instrument. Where the inductance cannot be ignored, an a.c. instrument must be used.

17.4.7 Earth loop impedance and prospective fault current tests

Where protective measures are used that require knowledge of the earth fault loop impedance, as is the case where ADS is used, Regulation 612.9 requires the relevant impedances to be measured or determined by an alternative method. By implication, this makes a demand for testing line–earth fault loop impedances.

Tests of the line–earth fault loop impedance Z_s must be carried out at the most remote end of the circuit (i.e. the longest cable length from the circuit overcurrent protective device). The line–earth fault loop path is as shown in Figure 17.6. For testing of socket-outlet circuits it is more convenient to use a test lead with an appropriate plug top fitted, and it is considered good practice, in addition to the prerequisite tests for continuity of conductors, for every socket-outlet to be tested.

It can be seen from Figure 17.6 that measuring the external line–earth loop impedance Z_e involves the partial loop comprising the transformer impedance Z_o, the impedance of the supply line conductor Z_L and that of the supply earthing conductor Z_E in the TN-S system shown. For a TN-C-S the supply earthing conductor impedance Z_E is replaced by that of the neutral. The total loop impedance Z_s of a TN-S system is represented by Equation 17.1 and the resultant earth fault current I_F is given in Equation (17.2), where U_o is the nominal circuit voltage to Earth:

$$Z_s(\Omega) = Z_o + Z_L + Z_1 + Z_2 + Z_E \tag{17.1}$$

$$I_F(A) = \frac{U_o}{Z_o + Z_L + Z_1 + Z_2 + Z_E} \tag{17.2}$$

It should be recognized that the measured value of total earth loop impedance Z_s is likely to be less than that which would pertain under fault conditions. The conductors of

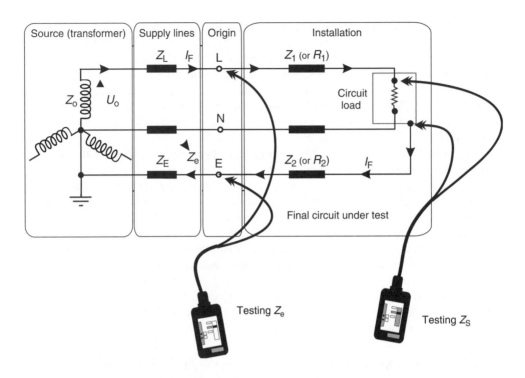

Figure 17.6 Testing earth loop impedance on a TN–S system

the loop when tested are going to be at some temperature not exceeding their maximum operating temperature, but are more likely to be at ambient temperature. In any event, the temperature at testing is going to be much less than the maximum final temperature of the conductors under fault conditions. Because the resistance component of impedance is a function of temperature, some correction needs to be made to the values obtained from testing so that a realistic evaluation of Z_s for comparison with Tables 41.2, 41.3 and 41.4 of BS 7671 (limiting values of Z_s) may be made. The correcting factor will depend to some extent on the relative contributions to the total loop impedance that each part of the whole circuit makes. For example, the fault current will have more effect on the final circuit conductors than on the supply conductors and the distribution circuit (sub-main), if any, though the final circuit conductor impedances may make a minor contribution to the loop impedance. However, the impedance of the final circuit will, in most cases, be a significant contributor to the total loop impedance.

To obviate this potential problem, it often makes sense to measure the loop (or more accurately, the partial loop) at the distribution board end of the final circuit, where, for most practical situations, this part may be ignored from the aspect of temperature-related increase in resistance. This measured value can then be subtracted from the value measured for Z_s with the resultant value representing $(Z_1 + Z_2)$, which can then be corrected accordingly. For conductors of csa not exceeding $35 \, \text{mm}^2$ the inductive reactance may be ignored and $(Z_1 + Z_2)$ becomes $(R_1 + R_2)$ and can be corrected by applying a correction factor relating to the temperature at which tests are carried out and a correction factor relating to the conductor insulation (see Table 10.17). The temperature correction factor is based on the resistance–temperature coefficient of copper and aluminium conductors taken as 0.004/°C (approximately) taking 20 °C as the datum. So, for tests carried out at 5 °C, the correction factor is 1.06 derived as $1/\{1 - [(20 - 5) \times 0.004/°C]\}$. Similarly, for testing temperatures of 10 °C, 15 °C, 25 °C and 30 °C, the correction factors are 1.04, 1.02, 0.98 and 0.96 respectively.

To illustrate by example, take a final circuit of PVC-insulated cables of live conductor csa of 4 mm^2 and cpc 2.5 mm^2, where the cpc does not form part of a composite cable. At the circuit overcurrent protective device position, the line–earth loop impedance has been measured as 0.3 Ω and Z_s (at the load end of the circuit) has been tested as 0.7 Ω. This indicates that the uncorrected value of $(R_1 + R_2)$ is 0.4 Ω (0.7 − 0.3). Had the test been taken at an ambient temperature of 10 °C a correcting factor of 1.04 must be applied to the 0.4 Ω previously derived, giving $(R_1 + R_2)$ as 0.416 Ω (0.4 × 1.04). Additionally, the factor relating to the particular conductor insulating material also needs applying. In this case the factor is 1.1. This is given by dividing 13.238 by 12.020, where 13.238 and 12.020 are values of $(R_1 + R_2)$ per metre taken from Row H of Table 10.17; 13.238 is the value corresponding to the particular conductor insulating material where the cpc does not form part of a composite cable and 12.020 is the value at the reference temperature of 20 °C. The real value of $(R_1 + R_2)$ for the circuit under earth fault conditions now becomes 0.46 Ω (0.416 × 1.1), giving a corrected value for Z_s under fault conditions as 0.76 Ω, representing an increase of 9% on the originally tested value.

In many cases, where more detailed investigation is unwarranted, a rough guide can be used and this may be taken as the measured value of Z_s which represents 80% of the true value under fault conditions, as indicated in Appendix 14 of BS 7671. In other words, the measured value should be multiplied by 1.25 and then compared with values given in Tables 41.2, 41.3 and 41.4 of BS 7671 for the appropriate overcurrent protective device. As a guide only, Table 17.2 lists the values of Z_s for overcurrent protective devices adjusted to 80% of those given in BS 7671 for comparison with test results. This table should be used with much caution and it should be borne in mind that it is intended as a rough guide only.

It is not unheard of when testing lighting circuits and the like for the earth loop impedance for the test probes to be applied across the line conductor of a convenient socket-outlet and exposed-conductive-parts of a lighting circuit (e.g. metalwork of a Class I luminaire). Such a test does not measure the line–earth loop impedance, since the loop under test comprises the line conductor of the socket-outlet circuit and the protective conductor of the lighting circuit. Such a test serves no useful purpose and should not be undertaken because of the obvious risks to safety and the inherent inaccuracies.

Where an installation or a circuit incorporates an RCD it is still necessary to evaluate the line–earth loop impedance Z_s at the remote end of the circuit in order to confirm that the requirements of Regulation 411.4.5 (TN systems) or Regulation 411.5.3 (TT systems), as applicable, have been met. Tables 5.9 and 5.10 of this Guide list the maximum values of Z_s for RCDs used for fault protection, at a nominal voltage U_o of 230 V, in TN

Table 17.2 Anticipated values of earth loop impedance Z_s (Ω) for common circuit protective devices, for fault protection operating on 230 V under test conditions (80% of limiting values).[a,b]

Ref	Nominal rating (A)	BS 88 'gG' Parts 2 and 6 (0.4 s)	BS 88 'gG' Parts 2 and 6 (5 s)	BS 1361 (0.4 s)	BS 1361 (5 s)	BS 3036 (0.4 s)	BS 3036 (5 s)	BS 1362 (0.4 s)	BS 1362 (5 s)	MCB Type 1 (0.1 and 5 s)	MCB Type 2	MCB Type B	MCB Type 3 and C	MCB Type D	MCB Type 4
A	3	–	–	–	–	–	–	13.12	18.56	–	–	12.26	–	–	–
B	5	–	–	8.36	13.12	7.66	14.16	–	–	9.20	5.25	7.36	3.68	1.84	0.73
C	6	6.81	10.80	–	–	–	–	–	–	7.66	4.37	6.13	3.06	1.52	0.60
D	10	4.08	5.93	–	–	–	–	–	–	4.60	2.62	3.68	1.84	0.92	0.36
E	13	–	–	–	–	–	–	1.93	3.06	–	–	–	–	–	–
F	15	–	–	2.62	4.00	2.04	4.28	–	–	3.06	1.75	2.44	1.22	0.60	0.24
G	16	2.16	3.34	–	–	–	–	–	–	2.87	1.64	2.29	1.15	0.56	0.22
H	20	1.41	2.32	1.36	2.24	1.41	3.06	–	–	2.29	1.31	1.84	0.92	0.45	0.18
I	25	1.15	1.84	–	–	–	–	–	–	1.84	1.04	1.47	0.73	0.36	0.14
J	30	–	–	0.92	1.47	0.87	2.11	–	–	1.52	0.87	–	–	–	–
K	32	0.83	1.47	–	–	–	–	–	–	1.43	0.81	1.15	0.57	0.28	0.11
L	40	–	1.08	–	–	–	–	–	–	1.14	0.65	0.92	0.45	0.22	0.08
M	45	–	–	–	0.76	–	1.27	–	–	1.01	0.58	–	–	–	0.08
N	50	–	0.83	–	–	–	–	–	–	0.92	0.52	0.73	0.36	0.18	0.07

Table 17.2 (continued)

Fuses columns are grouped under **Fuses**; the last six columns are grouped under **MCBs to BS 3871 or BS EN 60898[2]**.

Ref	Nominal rating (A)	BS 88 'gG' Parts 2 and 6		BS 1361		BS 3036		BS 1362		Type 1	Type 2	Type B	Type 3 and C	Type D	Type 4
		0.4 s	5 s	0.4 s	5 s	0.4 s	5 s	0.4 s	5 s	0.1 and 5 s					
O	60	–	–	–	0.56	–	0.89	–	–	–	–	–	–	–	–
P	63	–	0.65	–	–	–	–	–	–	0.72	0.41	0.58	0.28	0.14	–
Q	80	–	0.45	–	0.40	–	–	–	–	0.56	0.32	0.45	0.23	0.11	–
R	100	–	0.33	–	0.28	–	0.42	–	–	0.45	0.25	0.36	0.18	0.08	–
S	125	–	0.26	–	–	–	–	–	–	–	–	0.29	0.14	0.07	–
T	160	–	0.20	–	–	–	–	–	–	–	–	–	–	–	–
U	200	–	0.15	–	–	–	–	–	–	–	–	–	–	–	–
V	250	–	0.11	–	–	–	–	–	–	–	–	–	–	–	–
W	315	–	0.08	–	–	–	–	–	–	–	–	–	–	–	–
X	400	–	0.06	–	–	–	–	–	–	–	–	–	–	–	–
Y	500	–	0.04	–	–	–	–	–	–	–	–	–	–	–	–
Z	630	–	0.03	–	–	–	–	–	–	–	–	–	–	–	–
A1	800	–	0.02	–	–	–	–	–	–	–	–	–	–	–	–

aThe entries denoted by a dash represent either that the device is not commonly available or that by virtue of its characteristics it is not appropriate for fault protection.

bThe impedance values are based on the 'worst-case' limits allowed by the relevant standard and in certain cases where the manufacturer can claim closer limits than the standard permits the values may be modified accordingly.

and TT systems respectively. Although these are design values, they may be used for comparison with test results. This is because, for TN systems, the measured values of Z_s should be much lower than those in Table 5.9, as the earth loop path in a TN system is wholly metallic; and for TT systems, values of Z_s that exceed more than a few tens of ohms (as all those in Table 5.10 do) generally consist mainly of the resistances of the installation earth electrode and source earth electrode and, therefore, are affected very little, in percentage terms, by the temperature of the circuit conductors.

Many loop impedance test instruments will operate (trip) the RCD on line–earth loop testing before a measurement of Z_s can be taken. However, one can measure the line–earth loop impedance upstream of the RCD, such as at the origin of the installation or at the distribution board if this is not at the origin, and add the $(Z_1 + Z_2)$ values to this to give a resultant value for Z_s. Where the downstream conductors are of 35 mm^2 or less, $(Z_1 + Z_2)$ becomes $(R_1 + R_2)$.

This same procedure of adding the $(R_1 + R_2)$ values (or $(Z_1 + Z_2)$ values where applicable) to the measured value of line–earth loop impedance at the origin or distribution board can be used irrespective of whether the circuit incorporates an RCD. The procedure is particularly useful as an alternative to carrying out earth loop impedance tests with an earth loop impedance test instrument at points in a circuit, such as lighting points and lighting switches. Such live testing often cannot be justified on grounds of safety, due to the restrictions that Regulation 14 of the Electricity at Work Regulations 1989 places on working on or near live conductors (see Section 17.3 of this Guide).

17.4.8   Applied voltage tests

Where double or reinforced insulation is used as the protective measure against electric shock, live parts that have basic insulation only are required by Regulation 412.2.1.2 to have supplementary insulation applied to them on site in the process of erecting the installation. Similarly, live parts that are uninsulated are required by Regulation 412.2.1.3 to have reinforced insulation applied to them on site. In both cases the insulation is required, amongst other things, to provide a degree of safety equivalent to that provided by Class II electrical equipment. Except where the insulation has already been type-tested, as may be the case for a proprietary insulating enclosure or for insulation in the form of prefabricated parts provided by the electrical equipment manufacturer, such as applies to sectionalized switchgear assemblies, it must be established by testing that the insulation provides this degree of safety. It should be verified that the insulation provides a degree of protection not less than IPXXB or IP2X, by using a British Standard test finger as that shown in Figure 17.7. The insulation should also be subjected to applied voltage testing to demonstrate that there is no flashover and no breakdown of the insulation. Insulation that has been applied on site to provide fault protection should also be subjected to applied voltage testing.

Where there is no applicable British Standard or there are no test procedures laid down, the applied voltage test must be at 3.75 kV, at the supply frequency, and the test duration should be 60 s. Figure 17.8 shows some typical applied voltage testing. It should be noted that the test voltage is applied between the live conductors temporarily connected together and a metallic foil closely wrapped around the insulation.

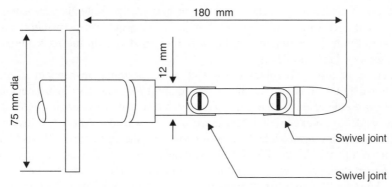

Not to scale - refer to BS EN 61032 for full dimensions

Figure 17.7 British Standard test finger

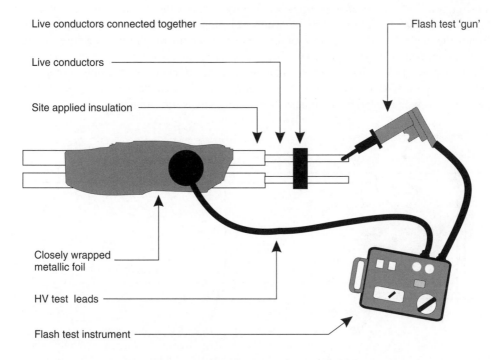

Figure 17.8 Applied voltage testing

17.4.9 Earth electrode tests

Testing of earth electrodes is required by Regulation 612.7, where the protective measures employed to protect against electric shock (e.g. in TT and IT systems) dictate a knowledge of that resistance. In practice, this generally applies to installations forming part of a TT system or parts of an installation which employ separate electrodes. The electrode, in such cases, is a critically important constituent part of the protective measure, and the designer would need to be satisfied that it is going to provide a permanent and reliable connection to Earth and maintain a sufficiently low resistance during its required lifetime (see Chapter 12 of this Guide).

One method of testing involves the use of a four-terminal instrument together with two additional test spikes and is shown in Figure 17.9. The test instrument terminals C1 and P1 are connected to the electrode under test and these connections should be made independently to exclude the resistance of the test leads. Where test lead resistance is insignificant, it is permissible to connect together the terminals (C1 and P1) at the instrument. Terminals C2 and P2 of the instrument are connected to the *potential* spike and the *current* spike respectively. A three-terminal instrument can also be used for this purpose. The output voltage for such instruments is commonly 50 V on open circuit and 25 V applied voltage to the measured loop and at a frequency of 128 Hz. Some instruments operate on reversed d.c., which assists in overcoming problems with electrolytic effects, and some instruments incorporate phase-sensitive detectors, which minimize the effects of stray currents. Most instruments employ a facility to check the resistance of test spikes. In the interest of safety and reliability of results, it is important when using test equipment to follow the methods and procedures recommended by the instrument manufacturer. Further information is contained in BS 7430 *Code of practice on earthing*.

It is important to position the test spikes so that the effects of overlapping of resistance areas are minimized. Normally, this would require the distance between the test spike and the electrode being tested to be not less than 10 times the major dimension of the electrode.

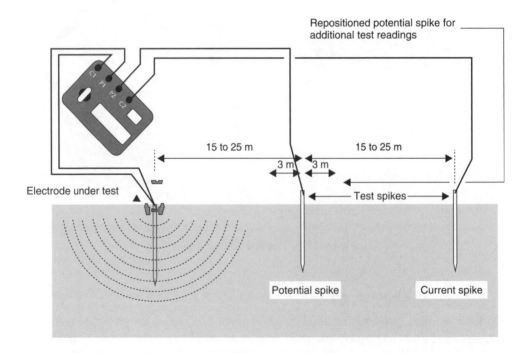

Figure 17.9 Earth electrode testing

For example, for a 15 mm diameter 1.2 m deep electrode, the separating distance would be not less than 12 m. However, as a general rule, the distances shown in Figure 17.9 should be used where possible in order for the test engineer to have confidence that the test results are reliable.

It is necessary for the three following readings to be taken.

- A measurement is taken with the *potential* spike midway (i.e. 50% of distance) between the electrode under test and the *current* spike giving electrode resistance reading R_1.
- A measurement is taken with the *potential* spike 40% of the distance between the electrode under test and the *current* spike giving electrode resistance reading R_2.
- A measurement is taken with the *potential* spike 60% of the distance between the electrode under test and the *current* spike giving electrode resistance reading R_3.

Provided reliable readings are taken, the resistance of the electrode R_A may be taken as the average as follows:

$$R_A \approx R_{A(av)} = \frac{R_1 + R_2 + R_3}{3} \tag{17.3}$$

where $R_{A(av)}$ is the average of the three measured resistance values R_1, R_2 and R_3.

Before finally accepting the resistance value obtained by using Equation (17.3), a check needs to be made on the percentage deviation of all the three readings by applying

$$\text{Percentage deviation} = \frac{\text{electrode reading} - R_{A(av)}}{R_{A(av)}} \times 100 \tag{17.4}$$

If the deviation in any of the three readings from the average is less than 5% then the readings may be regarded as reliable. If not, then the tests should be repeated with a greater separation between the electrode under test and the current spike until a more reliable measurement is made.

As an alternative method, where an installation forms part of a TT system and employs RCDs for protection for the whole of the installation (which is normally the case), it is permissible to use an earth loop impedance test instrument to measure the total line–earth loop impedance Z_s. As the resistance of the earth electrode R_A is usually much greater than that of other constituent parts of the loop, the measured Z_s may be taken as being equal to R_A. It is necessary for the supply to be energized for this test, which involves a measurement using the test probes of the loop impedance test instrument between the supply side of the main switch and the MET which is connected to the electrode by the earthing conductor. Before a test is made the test engineer must ensure that the supply to the installation is OFF and the main protective bonding conductors are disconnected from the MET. The measured value should not exceed that required to comply with Regulation 411.5.3 (i.e. $Z_s I_a \leq U_o$ and $R_A I_{\Delta n} \leq 50$ V; see Table 5.10 for installations where U_o is 230 V). On completion of testing, the protective bonding conductors must be reconnected to the MET *before* the installation is energized.

In practice, in most locations in the UK, the necessary earth electrode resistance will be met with ease. However, where values of 200 Ω or more are found, further efforts should be made to reduce this value by the use of additional rods and/or longer rods or the possibility of using plates should be considered. Electrodes with resistance exceeding this value should, in the absence of better information, be considered unreliable and prone to instability.

17.4.10   Residual current device tests

Where RCDs form part of the installation, whether they are for fault protection or for additional protection or both, it is a requirement of Regulations 612.8.1 and 612.10 respectively that such devices must be tested. This test is in addition to the test carried out by the use of the manual test button incorporated within the RCD, which, as previously mentioned, only tests the mechanism, which does not include external parts such as protective conductors and earth electrode. Before RCD testing is contemplated, other prerequisite testing must have been carried out and any connected loads either disconnected or switched off.

Using an RCD test instrument (as previously referred to), the test is made using the instrument's probes connected to the line conductor on the load side of the RCD and the associated protective conductor again on the load side, as shown in Figure 17.10, or by using the test instrument fitted with a test lead plug and inserted into a convenient

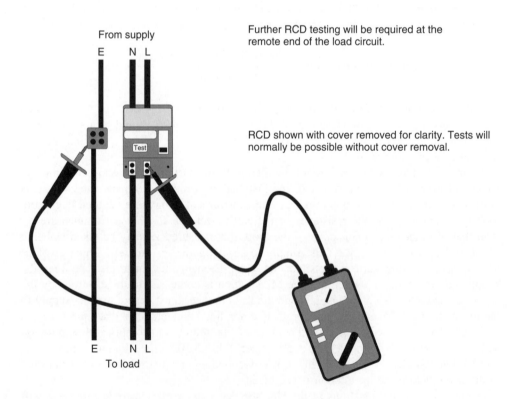

Figure 17.10   RCD testing

socket-outlet. The test results should meet those laid down in BS 4293 and other standards which are summarized in Table 17.3.

17.4.11   Protection by SELV, PELV or electrical separation

Where the protective measures of SELV, PELV or electrical separation are used, Regulation 612.4 requires the separation of circuits in accordance with Regulations 612.4.1, 612.4.2 and 612.4.3 respectively to be verified by insulation resistance measurements. The resistance values obtained must be not less than that specified for the circuit with the highest voltage present, in accordance with the requirements of Table 61 of BS 7671, which are summarized in Table 17.1 of this Guide.

For SELV circuits, Regulation 612.4.1 requires the live parts to be separated from other circuits and from earth. For PELV circuits, Regulation 612.4.2 requires the live parts to be separated from other circuits. For protection by electrical separation, Regulation 612.4.3 requires the live parts to be separated from other circuits and from Earth.

A further requirement applies where the protective measure of electrical separation is used to supply more than one item of current-using equipment. Regulation 612.4.3 requires it to be verified that if two coincidental faults occur from different line conductors to the nonearthed protective bonding conductor of the system or to the exposed-conductive-parts connected to it, at least one of the faulty circuits will be automatically disconnected. The disconnection time must not exceed that required for where the protective measure of ADS is used in a TN system. For example, if the nominal voltage between the line conductors of the system is 230 V, then the disconnection time must not exceed 0.4 s for a final circuit rated at 32 A or less, or 5 s for final circuit of higher rated current and for distribution circuits (see Regulations 411.3.2.2 and 411.3.2.3). Compliance with the disconnection time requirement may be verified by measuring the loop impedance between line conductors with an earth loop impedance test instrument of the appropriate voltage rating. The test result should not exceed $0.8 \times (U/I_a)$, where U is the nominal voltage in volts between line conductors and I_a is the current in amperes causing automatic operation of the protective device within the maximum permitted disconnection time.

17.4.12   Verification of phase sequence

In a three-phase or other multiphase installation, the phase sequence should be checked at the origin and in each multiphase distribution circuit and multiphase final circuit, to the extent necessary to verify that the phase sequence has been maintained.

This is particularly important where there is equipment that will function incorrectly if connected to a supply having other than the intended phase sequence. For example, the direction of rotation of a three-phase motor will be reversed if the polarity of any two line conductors of the circuit is reversed, and in some cases this can result in danger or in damage to the mechanically driven load.

17.4.13   Functional testing

The button marked 'T' or 'Test' of each RCD should be pressed while the device is energized to establish that the device trips and, therefore, that the test facility incorporated in it is effective.

Table 17.3 Test results for RCDs

Ref	Residual operating current of device $I_{\Delta n}$ (mA)	No-trip test $0.5I_{\Delta n}$ (mA) Trip Time[b] (ms)			Trip test $I_{\Delta n}$ (mA) Trip Time[c] (ms)				Trip test $5I_{\Delta n}$[a] (mA) Trip Time[c] (ms)			
					BS 4293 and BS 7288			BS EN 61008-1 and BS EN	BS 4293 and BS 7288			BS EN 61008-1 and BS EN 61009-1
1	2	3	4	5	6	7	8	9	10	11	12	13
RCDs without time-delayed operation												
A	10	5	No	2000	10	Yes	200	300	50	Yes	40[a,d]	40[a,e]
B	30	15	No	2000	30	Yes	200	300	150	Yes	40[a,e]	40[a,e]
C	100	50	No	2000	100	Yes	200	300	500	Yes	40[d]	40[e]
D	300	150	No	2000	300	Yes	200	300	1500	Yes	40[d]	40[e]
E	500	250	No	2000	500	Yes	200	300	2500	Yes	40[d]	40[e]
RCDs with time-delayed T_d operation, and RCDs of the delay type (S) to BS EN 61008-1 and BS EN 61009-1												
F	100	50	No	2000	100	Yes	$(1/2T_d + 200)$ to $(T_d + 200)$	500	500	Yes	$1/2T_d$ to $(T_d + 40)$[d]	40–150[e]
G	300	150	No	2000	300	Yes	$(1/2T_d + 200)$ to $(T_d + 200)$	500	1500	Yes	$1/2T_d$ to $(T_d + 40)$[d]	40–150[e]
H	500	250	No	2000	500	Yes	$(1/2T_d + 200)$ to $(T_d + 200)$	500	2500	Yes	$1/2T_d$ to $(T_d + 40)$[d]	40–150[e]

[a]Test necessary where RCD provides additional protection (see Regulation 415.1.1).
[b]Maximum test duration.
[c]Maximum time permitted for operation of RDC under type-test conditions.
[d]Test may be necessary to prove effective automatic disconnection where RCD provides fault protection for a final circuit rated at 32 A or less and a maximum disconnection time less than the time in column 8 of this table is required by Table 41.1 BS 7671 (see Table 5.5 of this Guide).
[e]Test may be necessary to prove effective automatic disconnection where RCD provides fault protection for a final circuit rated at 32 A or less and a maximum disconnection time less than the time in column 9 of this table is required by Table 41.1 BS 7671 (see Table 5.5 of this Guide).

All switchgear and controlgear assemblies, drives, controls, interlocks and the like that form part of the installation should be functionally tested to verify that they are properly mounted, adjusted and installed in accordance with the relevant requirements of BS 7671 and the electrical installation designer's intentions.

17.4.14   Verification of voltage drop

As pointed out in the note to Regulation 612.14, verification of voltage drop is not normally required during the initial verification of an electrical installation. This is not surprising, as the electrical installation designer is expected to have sized the circuit conductors so that the voltage drop requirements of Section 525 of BS 7671 will have been met.

However, the situation may be different in the case of the periodic inspection and testing of an existing installation. Here, the inspector may consider it necessary to check the voltage drop for a particular circuit because he or she suspects that, due to excessive voltage drop, the voltage at the terminals of fixed current-using equipment may be low enough to impair the safe or effective functioning of the equipment.

Regulation 612.14 gives two examples of methods that may be used to verify voltage drop, although other methods no less safe and effective may also be used. One of the methods mentioned in the regulation is to evaluate the voltage drop by measuring the circuit impedance. For a single-phase circuit, this would essentially entail measuring the impedance of the line and neutral conductors of the circuit (or, for conductors of $35\,\text{mm}^2$ or less, their resistance only - neglecting inductive reactance) and, after allowing for the increase in the resistance of the conductor due to the increase in temperature caused by normal load current, multiply the resistance by the current that the conductors are expected to carry in service.

The other method of verifying voltage drop mentioned in Regulation 612.14 is to evaluate the voltage drop by using calculations. The regulation mentions the use of diagrams or graphs intended for voltage drop verification purposes; but, in the absence of such diagrams or graphs, the voltage drop could be calculated by using the tabulated values of voltage drop given in Appendix 4 of BS 7671, as explained in Section 10.4.3 of the Guide, based on the installed cable sizes and estimated circuit lengths and load currents.

17.5   Initial verification

17.5.1   General

As required by Regulation 610.1, every installation must be inspected and tested during erection and on completion before it is put into service and verified as being compliant with all the relevant requirements of BS 7671. In order that the person may carry out such verification, it is necessary for that person to have available all the initial design data relating to the assessment of general characteristics, which is summarized in Table 17.4.

Table 17.4   Information required by the person carrying out the initial verification

Ref	Regulations		Information required
A	610.2	311.1	Maximum demand of the installation
B	610.2	312.2.1	The number and types of live conductor
C	610.2	312.3.1	Type of earthing arrangement
D	610.2	313.1	Nominal voltage(s)
E	610.2	313.1	The nature of current and frequency
F	610.2	313.1	The prospective fault current at the origin (short-circuit and earth fault currents)
G	610.2	313.1	The external line–earth loop impedance Z_e
H	610.2	313.1	The suitability of the supply for the particular installation
I	610.2	313.1	The type and nominal rating of the overcurrent protective device at the origin
J	610.2	514.9.1	A diagram, chart or table providing adequate information relating to the type and composition of each and every circuit including points of utilization, number and size of circuit conductors and type(s) of wiring system
K	610.2	514.9.1	Information relating to the identification of every device and their location, providing protection and isolation and switching
L	610.2	514.9.1	Information and identification of any circuit and/or any equipment liable to damage by a typical test
M	610.2	514.9.1	Identification of the method used for, and means of, compliance with Regulation 410.3.2 (protection against electric shock)

17.5.2   Inspection

It is essential that a detailed inspection is carried out prior to testing with the installation (or any part of it being inspected) isolated from the supply to meet the requirements of Regulation 611.1. Tables 17.5–17.19 summarize some of the many facets of inspections which are demanded by Regulations 611.2 and 611.3. Whilst it is not possible to list all aspects for inspection on every installation, the tables may serve as a useful *aide mémoire* for the inspector, who should appreciate that the listings are not exhaustive and should be supplemented with additional items where appropriate. Inspections should, of course, be made on completion; but, under close supervision, these can be done, in part at least, as the work progresses, provided detailed records are maintained. Table 17.5 sets out the items mentioned in Regulation 611.3 which need to be included for inspection. Not all the items will apply to every installation, and in some cases additional items may need to be included.

Tables 17.6–17.19 relate to specific aspects of the typical installation, where the measure of protection against electric shock is ADS, as follows:

- Table 17.6 – general checklist for equipment;
- Table 17.7 – checklist for switchgear;
- Table 17.8 – general checklist for cables and conductors;
- Table 17.9 – checklist for flexible cables and cords;
- Table 17.10 – checklist for protective devices;

Table 17.5 Examples of installation items requiring inspection (see Regulation 611.3)

Ref	Item	Section or chapter[a]	Comment
A	Connection of conductors	526	Applicable to all installations
B	Identification of conductors	514	Applicable to all installations
C	Cables routed through 'safe' zones or protection against mechanical damage	522	Applicable to all installations
D	Conductors suitable for current-carrying capacity	433	Applicable to all installations
E	Conductors suitable for voltage drop	525	Applicable to all installations
F	Connection of single-pole switching devices (including fuses) in line conductor only (unless linked)	530	Applicable to all installations
G	Connection of equipment (including polarity)	General	Applicable to all installations
H	Connection of ES lampholders (including polarity)	559	Applicable to all installations
I	Connection of track lighting (including polarity)	559	Applicable to all installations
J	Presence of fire barriers and suitable seals	527	Applicable to all installations
K	Protection against thermal effects	42	Applicable to all installations
L	Basic protection by insulation of live parts	416	Applicable to all installations
M	Basic protection by barriers and enclosures	416	Applicable to all installations
N	Basic protection by obstacles, where applicable	417	Applicable only to installations where this measure has been employed
O	Basic protection by placing out of reach, where applicable	417	Applicable only to installations where this measure has been employed
P	Fault protection – presence of earthing conductor	542	Applicable only to installations where ADS has been employed
Q	Fault protection – presence of protective conductors	543	Applicable only to installations where ADS has been employed
R	Fault protection – presence of main protective bonding conductors	544	Generally applicable to installations where ADS has been employed, but also see electrical separation (Regulation 418.3)
S	Fault protection – presence of supplementary bonding conductors	544	Generally applicable only to installations where ADS has been employed, but see also earth-free equipotential bonding (Regulation 418.2)
T	Fault protection – presence of suitable combined functional and protective conductors	543	Applicable only to installations where ADS has been employed

(continued overleaf)

Table 17.5 (*Continued*)

Ref	Item	Section or chapter[a]	Comment
U	Basic and fault protection – proper use of double or reinforced insulation	412	Applicable only to installations where this measure has been employed.
V	Basic and fault protection – proper use of SELV and PELV	414	Applicable only to installations where these measures have been employed
W	Fault protection – proper use of non-conducting location	418	Applicable only to installations where this measure has been employed
X	Fault protection – proper use of earth- free equipotential bonding	418	Applicable only to installations where this measure has been employed
Y	Fault protection – proper use of electrical separation	413 and 418	Applicable only to installations where this measure has been employed
Z	Prevention of mutual detrimental influences	515	Applicable to all installations
A1	Suitable devices for isolation and switching	537	Applicable to all installations
B1	Suitable devices for undervoltage protection	535	Applicable to all installations
C1	Suitable rated (or setting) of overcurrent protective and monitoring devices	533	Applicable to all installations
D1	Suitable rated (or setting) of protective devices for ADS	531	Applicable to all installations
E1	Adequate access to equipment and switchgear	513	Applicable to all installations
F1	Equipment suitable for external influences	522	Applicable to all installations
G1	Protective measures suitable for external influences	522	Applicable to all installations
H1	Suitable equipment erection methods	510	Applicable to all installations
I1	Proper identification of switchgear	514	Applicable to all installations
J1	Proper identification of circuits, fuses, switches and terminals	514	Applicable to all installations
K1	Presence of voltage warning notices	514	Applicable to all installations
L1	Presence of isolation and danger notices	514	Applicable to all installations
M1	Presence of diagrams and instructions	514	Applicable to all installations

[a]Reference to BS 7671 sections and chapters includes only the most appropriate; reference to the particular regulation will also be necessary.

Table 17.6 General checklist for items of equipment that require visual inspection prior to carrying out testing

Ref	Regulation references	Facet to inspect
A	133.1	Equipment correctly selected and affords compliance with the relevant British Standard or harmonized European Standard
B	132.5	Equipment correctly selected for atmospheric and temperature conditions
C	132.5	Equipment correctly selected for environmental conditions (including external influences). See BS EN 60529 for IP requirements
D	134.1.2	Equipment not been damaged or is not otherwise defective
E	134.1.1	Equipment installed in a good workmanlike manner
F	Chapter 41	All necessary provisions for basic protection against electric shock have been effected
G	513.1	Equipment accessible for operation, inspection and maintenance
H	527.2.1–527.2.7	Equipment provided with fire barriers, as necessary
I	Chapter 42	Equipment protected from thermal effects
J	Section 515	Equipment protected against mutual detrimental influences
K	133.2.2	Equipment correctly selected in terms of rated current and is suitable for the ambient temperature
L	411.3.1.1	Equipment properly earthed (see also Chapter 54 of BS 7671)
M	134.1.1	Equipment properly fixed and is secure
N	537.2	Adequate means of isolation
O	537.3	Adequate means of switching off for mechanical maintenance
P	537.4	Adequate means of emergency switching
Q	510.1	Equipment voltages properly identified
R	514.1.1	Equipment properly identified as to purpose
S	512.1.1	Equipment suitable for voltages present
T	514.11.1	Equipment properly identified for isolation purposes where isolation is required by more than one device
U	514.12.1	Periodic test notice affixed
V	514.12.2	RCD test notice affixed, where appropriate
W	521.5.2	Due consideration paid to electromagnetic effects
X	521.5.1	Due consideration paid to electromechanical stresses
Y	515.1 515.2	Circuits segregated as required to prevent mutual detrimental influences
Z	528.1 528.2	Circuits segregated from other electrical services, including communications and cables, as required
Z1	528.3.1–528.3.5	Circuits segregated from nonelectrical services as required
Z2	515.1 515.2	There are no mutual detrimental influences between electrical and non-electrical services (see also Section 528 of BS 7671)

Table 17.7   Checklist for items of switchgear that require visual inspection prior to carrying out testing

Ref	Regulation references	Facet to inspect
A	Chapter 53	Switchgear correctly selected and installed
B	133.1	Switchgear meets the appropriate standard(s)
C	133.2.2	Switchgear correctly selected in terms of rated current and is suitable for the ambient temperature
D	Chapter 53	Isolation, switching for mechanical maintenance and emergency switching requirements met as necessary
E	411.3.1.1	Switchgear properly earthed (see also Chapter 54 of BS 7671)
F	134.1.1	Switchgear properly fixed and is secure
G	Chapter 41	Adequate basic protection against electric shock (such as insulation of live parts)
H	417.1	Where basic protection is provided by obstacles, restriction to skilled and/or instructed persons is evident
I	417.1	Where basic protection is provided by placing out of reach, restriction to skilled and/or instructed persons is evident
J	530.3.2	Single–pole devices in line conductor only
K	522.8.1	No sharp edges to damage cable insulation, etc.
L	Chapter 53	Protective devices correctly selected and installed
M	513.1	Switchgear accessible for operation, maintenance and testing
N	514.1.1	Switchgear properly identified as to purpose
O	514.10.1	Switchgear properly selected as to voltages present
P	701.512.3	Switchgear suitable for zone in locations containing a bath or shower
Q	416.2.1 416.2.5 Section 522	Degree of protection (IP) is appropriate
R	537.2	Adequate means of isolation
S	537.3	Adequate means of switching off for mechanical maintenance
T	537.4	Adequate means of emergency switching
U	537.6	Firefighter's switch installed where required
V	537.4.2.4	Emergency switches suitably colour identified
W	514.11.1	Switchgear properly identified for isolation purposes where isolation is required by more than one device
X	514.8.1 514.9.1	Adequate diagrams, instructions for circuit and protective device identification
Y	514.9.1	Vulnerable equipment and circuits identified
Z	514.12.1	Periodic inspection and testing notice has been affixed on or near main switchgear
Z1	134.1.1	Where applicable, check cable glands and gland plates have been correctly selected and installed (see BS 6121 and BS EN 50262)
Z2	543.2.5	Where switchgear, trunking, conduit and cable sheaths or armouring have been used as a protective conductor, the requirements of Regulation 543.2.4 have been satisfied

Table 17.8 General checklist for cables and conductors that require visual inspection prior to carrying out testing

Ref	Regulation references	Facet to inspect
A	Section 511 Section 521	Cables have been correctly selected for the particular circumstances and that they are compliant with the relevant standard
B	Section 522	Suitable for all external influences, e.g. temperature, water, corrosive substances, mechanical stresses, flora and fauna, solar radiation and building structural design
C	Section 523	Correctly current rated
D	Section 525	Voltage drop not excessive
E	311.1	Diversity correctly applied
F	526.5	All connection enclosed
G	527.2	All necessary sealing of wiring system penetrations has been effected
H	522.6	All cables either run in 'safe' zones or other means of protection provided
I	523.7	Where run in thermal insulation, suitably derated
J	523.1	Conductor operating temperature not exceeding that given in Table 52.1 of the Wiring Regulations
K	524.1	Conductor cross-sectional area is not less than that given in Table 52.3 of the Wiring Regulations
L	522.6 522.8	Protected against mechanical damage including abrasion and suitable for location
M	522.8	Cables adequately supported
N	134.1.1	Cable glands correctly selected and fitted
O	522.1.1	Suitable for the highest and lowest temperatures
P	521.10.1	Nonsheathed cables enclosed in conduit, trunking, etc.
Q	514.4.2	Core colour identification of protective conductors (green–yellow)
R	514.4.5	The single colour green has not been used
S	514.1.1	Core colour identification of live cables and conductors (see also Table 51 of BS 7671)
T		Deleted
U	130.1.1	Where applicable, cable glands and gland plates have been correctly selected and installed (see BS 6121)
V	Section 511	Correct selection and installation of cables (see Appendix 4 of BS 7671)
W	543.2.5	Where required, MICC earth-tail pots have been correctly fitted
X	314.4	Check that circuits are separate, e.g. no 'borrowed' neutrals
Y	522.11.1	Where exposed to direct sunlight, of suitable type
Z	522.3.1	Where exposed to water, of suitable type
A1	528.3.5	Unless forming part of the lift installation complying with BS 5655, cables not run in lift shaft
B1	Deleted	Deleted
C1	543.4.1	Earthed concentric wiring used only as permitted

(*continued overleaf*)

Table 17.8 (*Continued*)

Ref	Regulation references	Facet to inspect
D1	514.4.4 and Table 51	Fixed-wiring line conductors identified brown, neutral conductors blue switched line conductors (switchwires) brown and protective conductors green–yellow
E1	543.3.2	Bare protective conductors of composite cables covered with green–yellow sleeving to BS EN 60684 series, at terminations
F1	543.2.7	Where protective conductor formed by conduit, trunking, etc., a 'fly lead' linking the wiring system protective conductor to the earthing terminal of the accessory is provided
G1	526.9	Unsheathed cores of cables contained within enclosure
H1	522.8.5	Where applicable, cord grips properly utilized
I1	526.1	All conductor terminations tight
J1	526.2	All strands of conductor properly terminated
K1	Appendix 4	Cables correctly installed to appropriate installation method and having sufficient current-carrying capacity and within voltage-drop constraints

Table 17.9 Checklist for flexible cables and cords that require visual inspection prior to carrying out testing

Ref	Regulation references	Facet to inspect
A	514.4	Core identification of flexible cords and flexible cables with Table 51
B		Deleted
C	514.4.5	The single colour green has not been used
D	134.1.1	Where applicable, cable glands and gland plates have been correctly selected and installed (see BS 6121 and BS EN 50262)
E	514.4 Table 52	Flexible cables and cords line conductors identified brown, neutral conductors blue and protective conductors green–yellow
F	526.9	Unsheathed cores of cables and cords contained within enclosure
G	522.8.5	Where applicable, cord grips properly utilized
H	526.1	All conductor terminations tight
I	526.2	All strands of conductor properly terminated
J	Appendix 4 Table 4F3A	When used to suspend luminaires, suitable for the suspended mass

- Table 17.11 – checklist for conduit systems;
- Table 17.12 – checklist for trunking systems;
- Table 17.13 – checklist for lighting points of utilization;
- Table 17.14 – general checklist for accessories;
- Table 17.15 – checklist for lighting switches;
- Table 17.16 – checklist for socket-outlets;
- Table 17.17 – checklist for joint boxes;

Table 17.10 Checklist for protective devices that require visual inspection prior to carrying out testing

Ref	Regulation references	Facet to inspect
A	Section 511 432.1	Overcurrent protective devices correctly selected and are compliant with the relevant standard
B	434.5	Protective devices have adequate short-circuit capacity
C	433.1.1	Protective devices correctly selected with regard to load current and current-carrying capacity of connected conductors
D	411.4.7	Where MCBs, the correct type
E	432.2	Overload protection is afforded, where applicable
F	432.3	Fault current protection is afforded, where applicable
G	411.3.2	Protection against electric shock (fault protection) is afforded, where applicable
H	411.3.2	Disconnection times are likely to be met
I	536.2	There is discrimination between overcurrent devices in series
J	536.3	There is discrimination between RCDs in series
K	415.1.1	RCDs have been installed where they are necessary – additional protection
L	411.4.5	RCDs have been installed where they are necessary – TN systems
M	411.5.3	RCDs have been installed where they are necessary – TT systems
N	411.6.4	RCDs have been installed where they are necessary – IT systems
O	411.3.3	RCDs have been installed where they are necessary – socket-outlets and supplies for portable equipment outdoors
P	522.6	RCDs have been installed where they are necessary – cables concealed in walls and partitions
Q	Part 7	RCDs have been installed where they are necessary – special installations or locations
R	514.1.2	Neutral conductors and cpcs are in an easily identifiable sequence related to the associated line conductor(s)
S	514.9.1	Identification of protective devices is adequate
T	526.1 526.2	Conductor connections are tight and connect all conductor strands

- Table 17.18 – checklist for cooker control units and fused connection units;
- Table 17.19 – checklist for protective conductors.

The detailed inspection envisaged by Regulation 611.2 is threefold in purpose. First, the inspection must determine that the equipment complies with Section 511 of BS 7671, in that it conforms to the relevant standard confirmed by suitable marking or certificate provided by the equipment manufacturer or installer. Second, the inspection must confirm that the equipment has been correctly selected and installed in such a manner as to provide conformity with the requirements of BS 7671. Finally, the inspection is intended to identify equipment which has been damaged or is otherwise defective so as to compromise safety.

Table 17.11 Checklist for conduit systems that require visual inspection prior to carrying out testing

Ref	Regulation references	Facet to inspect
Conduits generally		
A	511.1 521.6	Complies with the relevant standard
B	514.2.1	Where required to be distinguishable from other services, identified by the colour orange
C	522.8.2	Adequate means of access (drawing-in boxes) and, if sunken, complete before circuit cables are installed
D	522.8.5	Fixings made securely and spacings between fixings not excessive (see Chapter 10 of this Guide)
E	522.6.1	Protected against impact
F	522.7.1	Protected against vibration
G	522.8	Protected against other mechanical stresses
H	522.5.1	Corrosive resistant in damp locations
I	522.8.3	Adequate radii of bends
J	522.1	Protected against other external influences including ambient and working temperatures
K	522.6.1 522.8.1	Conduit ends properly reamed and cleaned out and bushed
L	522.8.3 522.8.6	Solid elbows and 'tees' used only where permitted
M	522.8.6 513.1	Conduit boxes and other inspection points to be accessible
N	522.3.2	Conduit system provided with adequate drainage points, where necessary
O	528.3	Conduit system does not contain nonelectrical services (pipes, etc.)
P	Deleted	Deleted
Q	Deleted	Deleted
R	521.10.1	Conduit ends blanked-off where unused
S	527.2.4	Fitted with fire barriers to prevent the spread of fire
T	522.8.1	Conduits not filled excessively with cables
Rigid metallic conduits		
U	543.2.4	Electrically continuous
V	411.3.1.1	Earthed
W	612.2.1	Continuity satisfactory
X	543.1.1	Adequacy for use as a cpc where so used
Y	Deleted	Deleted
Z	522.5	Adequate protection for conduits in damp and/or corrosive environment
A1	522.8.4	Where used as overhead link, span not excessive (see Chapter 10 of this Guide)
B1	521.5.2	Line(s) and neutral conductors of a circuit to be enclosed by the same conduit
C1		Deleted

Table 17.11   *(Continued)*

Ref	Regulation references	Facet to inspect
Flexible metallic conduits		
D1	543.2.1	Flexible metal conduit not used as a protective conductor
E1	522.8.4	Flexible conduits adequately supported
F1	134.1.1	Appropriate conduit gland-adaptors are used
Rigid non-metallic conduits		
G1	522.1.1	Suitable for highest and lowest temperatures
H1	522.8.1	Adequate provision for expansion and contraction
I1	559.6.1.5	Conduit boxes suitable for the mass of suspended luminaires where so used

Certain specialized installations are not only subject to the requirements of BS 7671, but also those contained within the relevant British Standard or CP, which are listed in Table 17.20.

17.5.3   Testing

Initial testing of an electrical installation is essential, not least to enable the verifying engineer to sign the Electrical Installation Certificate with confidence that the installation meets all the relevant requirements of BS 7671. On any except a very small installation, it will be necessary for some testing to be carried out during the course of construction, but a record of such test results needs to be kept so that duplication does not occur. Even so, some tests will need to be repeated on final completion to ensure that essential features have not changed.

As called for in Regulation 612.1, tests must be carried out in a logical sequence and the results compared with the relevant criteria, such as loop impedance limits and those associated with insulation resistance. Where a particular test indicates that the item tested does not comply, that failure is required to be remedied and the test, and those preceding it, need repeating until a satisfactory condition is obtained. Figure 17.11 summarizes that sequence.

17.5.4   Initial certification: general

Regulation 631.1 calls for an Electrical Installation Certificate to be completed for every installation and handed to the person ordering the work, who may be an agent of the consumer (e.g. architect or consulting engineer). This regulation also embodies a require-ment to record details of the inspections undertaken and the test results on a schedule. Regulation 631.1 also requires the schedule of test results to identify every circuit and its protective device(s). The certificate must be in the form as set out in Appendix 6 of BS 7671 and must be signed by a competent person each for the design, the construction and the inspection, testing and verification of the installation. The three required signato-ries may in fact be one and the same person where, for example, an installation has been designed, constructed, inspected, tested and verified by an electrical contractor who has

Table 17.12 Checklist for trunking wiring systems that require visual inspection prior to carrying out testing

Ref	Regulation references	Facet to inspect
A	511.1 521.6	Complies with BS EN 50085 series
B	522.8.5	Fixings made securely and spacings between fixings not excessive (see Chapter 10 of this Guide)
C	522.6.1	Protected against impact
D	522.7.1	Protected against vibration
E	522.8	Protected against other mechanical stresses
F	522.5.1	Corrosive resistant in damp locations
G	522.8.3	Adequate radii of bends
H	522.1	Protected against other external influences including ambient and working temperatures
I	543.2.4	Where metallic, electrically continuous
J	411.3.1.1	Where metallic, earthed
K	612.2.1	Where metallic, continuity satisfactory
L	543.2.4	Where metallic, joints electrically and mechanically sound
M	543.1.1	Adequacy for use as a cpc where so used
N	Deleted	Deleted
O	522.5	Adequate protection for trunking in damp and/or corrosive environment
P	522.6.1 522.8.1	Trunking ends properly deburred and cleaned out and bushed
Q	521.5.2	Line(s) and neutral conductors of a circuit to be enclosed by the same trunking, if metallic
R		Deleted
S	522.3.2	Trunking system provided with adequate drainage points, where necessary
T	528.3	Trunking system does not contain nonelectrical services (pipes, etc.)
U	Deleted	Deleted
V	528.1	Band I and Band II circuits segregated or appropriately insulated
W	Deleted	Deleted
X	522.8.1	Trunking not filled excessively with cables
Y	527.2.4	Fitted with fire barriers to prevent the spread of fire
Z	528.1	Trunking accessories, boxes and common outlets for Band I and Band II circuits provided with barriers or partitions

been responsible for all three aspects. Conversely, the three elements may have been the responsibility of three or more persons where, for instance, the design has been undertaken jointly by a consulting engineer and an electrical contractor, the construction by an electrical contractor, and the inspection, testing and verification by a third party engaged only for that purpose. There are, of course, many other possibilities and combinations of various personnel who might be involved in an installation, but in all cases a minimum of three signatures are required from the competent person(s).

Table 17.13 Checklist for lighting points of utilization that require visual inspection prior to carrying out testing

Ref	Regulation references	Facet to inspect
A	559.6.1.9	Luminaires controlled by a lighting switch to BS 3676/BS EN 60669-1 and/or BS EN 60669-2-1 or automatic control system
B	559.6.1.1	Connection to luminaires by one of the means specified in Regulation 559.6.1.1
C	559.6.1.9	Ceiling roses comply with BS 67
D	559.6.1.2	Ceiling roses not installed in any circuit operating at normal voltages exceeding 250 V
E	559.6.1.3	Ceiling roses not connecting more than one flexible cord unless specifically so designed
F	522.8.1	Ceiling rose's flex support utilized
G	559.6.1.5	Fixing means for pendant luminaire to be suitable to support 5 kg, or more if necessary
H	527.2.1	Wiring system penetration through building fabric and ceiling rose made good to minimize the spread of fire
I	559.6.1.4	Luminaire supporting couplers not used for connection of equipment other than luminaires
J		Deleted
K	559.6.1.1	Luminaire supporting couplers comply with BS 6972 or BS 7001
L	559.6.1.1	Batten lampholders comply with BS EN 61184 or BS EN 60238
M	559.6.1.7	BC lampholders comply with temperature rating 'T2'
N	559.6.1.2	Lampholders for filament lamps not used in circuits operating at a voltage exceeding 250 V
O	559.6.1.7	Bayonet cap lampholders (B15 and B22) comply with BS EN 61184
P		Deleted
Q	559.6.1.8	ES lampholders (and track-mounted systems), except E14 and E27 to BS EN 60238, with outer contact connected to neutral in TN and TT systems
R	559.4.1	Luminaires compliant with the relevant standard and selected to manufacturer's instructions
S	559.4.4	Lighting track compliant with BS EN 60570
T		Deleted
U	559.6.1.5	Luminaires adequately supported and no strain on suspended ceilings, etc. or cables/cords
V	559.6.2	Only luminaires designed for the purpose to be through wired, and through-wiring cables to be suitably selected for temperature within luminaires
W	559.5.1	Thermal effects of radiant and convected energy taken into account in selecting and erecting luminaires
X	559.7	Lamp controlgear not used external to a luminaire unless marked as suitable for independent use
Y	559.6.1.6	Type B15, B22, E14, E27 and E40 lampholders have overcurrent protective device not exceeding 16 A
Z	559.11	Particular requirements for extra-LV lighting installations are met

Table 17.14 General checklist for accessories that require visual inspection prior to carrying out testing

Ref	Regulation references	Facet to inspect
A	512.2.1–512.2.4	Suitable for all external influences, e.g. temperature, water, corrosive substances, mechanical stresses, flora and fauna, solar radiation and building structural design
B	512.1.2	Suitable current ratings. Every item suitable for design and fault currents
C	134.1.1	Polarity visually correct
D	513.1	Readily accessible
E	511.1	Accessories compliant with relevant standard
F	134.1.1	Box and/or enclosures securely fixed
G	411.3.1.1	Metallic enclosures, metallic front plates and grids properly earthed
H	527.2.1 526.3	Sunken boxes flush with finished building fabric (e.g. plaster)
I	522.8.1	No evidence of sharp edges within enclosure, including those caused by the use of countersunk screws instead of round heads
J	514.4 Table 51	Fixed-wiring line conductors identified brown, neutral conductors blue and switched line conductors (switchwires) brown and protective conductors green–yellow
K	514.4 Table 52	Flexible cables and cords line conductors identified brown, neutral conductors blue and protective conductors green–yellow
L	543.3.2	Bare protective conductors of composite cables covered with green–yellow sleeving to BS EN 60684 series, at terminations
M	543.2.7	Where protective conductor is formed by conduit, trunking, etc., a 'fly lead' linking the wiring system protective conductor to the earthing terminal of the accessory
N	526.9	Unsheathed cores of cables contained within enclosure
O	522.8.3	Where applicable, cord grips properly utilized
P	526.1	All conductor terminations tight
Q	526.2	All strands of conductor properly terminated
R	528.1	Accessories boxes and common outlets for Band I and II provided with barriers or partitions

An electrical contractor would wish to establish at the outset who is going to carry out the design before contracts are signed and sealed. They should also establish precisely the designer(s) responsible and obtain that person's signature certifying the design before installation work commences. Certification without the three signatures is considered invalid.

Competency of persons is difficult, if not impossible, to define except in general terms. A competent person will have suitable technical and professional qualifications in addition to a sound knowledge, and relevant experience, of the particular aspect of the installation in which they are involved (including the technical standards demanded by BS 7671). They will, for the verification, be fully versed in inspection and testing and be equipped with adequate test equipment. Regulations 632.1 and 632.4 confirm the obvious by stating

Table 17.15 Checklist for lighting switches that require visual inspection prior to carrying out testing

Ref	Regulation references	Facet to inspect
A	559.6.1.9 511.1	Lighting switches compliant with BS 3676/BS EN 60669-1 and/or BS EN 60669-2-1
B	559.6.1.9	Lighting switches controlling discharge lighting circuits and other inductive loads to be suitably rated
C	411.3.1.1	Switch box, grid and plate adequately earthed
D	530.3.2	Single-pole switches in line conductor only
E	701.512.3	Switches suitable for zone bathroom or shower room
F	411.3.1.1 412.2.3.2	Where insulated enclosures and boxes are used, cpc connected to an earth terminal
G	514.4 Table 51	Fixed-wiring line conductors identified brown, neutral conductors blue and switched line conductors (switchwires) brown and protective conductors green−yellow
H	514.10.1	Where switch unit is connected to more than one phase, label indicating presence of voltages above 250 V gives warning to skilled personnel
I	514.1.1	Where not obvious, a designation label indicating the function of the switch

Table 17.16 Checklist for socket-outlets that require visual inspection prior to carrying out testing

Ref	Regulation references	Facet to inspect
A	553.1.3	Socket-outlets comply with BS 1363, BS 546, BS 196 or BS EN 60 309-2 (BS 4343) (see Table 55.1 of BS 7671)
B	553.1.4	Socket-outlets for household or similar use are of the shuttered type
C	411.3.1.1	Box and plate adequately earthed
D	134.1.1	Polarity visually correct
E	Deleted	
F	543.2.9	Where cpc is formed by other than conduit or trunking, etc. it is of a ring if the circuit is also in the form of a ring
G	543.2.7	Where metal conduit serves as cpc, a 'fly lead' connection to box from socket, otherwise cpc terminated directly into socket-outlet earthing terminal
H	553.1.6	Mounted at sufficient height above floor or working surface
I	554.3.3	Not used to supply water heater with immersed elements
J	701.512.3	Not located within 3 m of zone 1 in room containing a bath (except SELV or shaver supply unit)
K	Deleted	
L	553.2.1	Excepting SELV and Class II, nonreversible couplers to be to BS 196, BS EN 60 309-2, BS 6991, BS EN 61535 or BS EN 60320 with provision for protective conductor connection
M	553.2.2	The coupler of cable coupler fitted to the end remote from the supply

Table 17.17 Checklist for joint boxes that require visual inspection prior to carrying out testing

Ref	Regulation references	Facet to inspect
A	512.2.1–512.2.4	Suitable for all external influences, e.g. temperature, water, corrosive substances, mechanical stresses, flora and fauna, solar radiation and building structural design
B	133.2.2 512.1.2	Suitable current ratings
C	513.1	Readily accessible
D	526.7	Joints mechanically protected
E	526.9	Unsheathed cores of cables contained within enclosure
F	543.1.1	If protective conductor is located outside joint box enclosure, to be not less than 2.5 mm^2
G	526.1	All conductor terminations tight
H	526.2	All strands of conductor properly terminated
I	702.522.24	No joint boxes in zones 0 or 1 of a swimming pool (except SELV in zone 1)

Table 17.18 Checklist for cooker control units and fused connection units that require visual inspection prior to carrying out testing

Ref	Regulation references	Facet to inspect
A	512.2.1–512.2.4	Suitable for all external influences, e.g. temperature, water, corrosive substances, mechanical stresses, flora and fauna, solar radiation and building structural design
B	133.2.2 512.1.2	Suitable current ratings
C	433.1.1 434.2	Suitably fused, where appropriate
D	Deleted	
E	537.3.2.4 537.4.2.5	Readily accessible
F	526.9	Unsheathed cores of cables contained within enclosure
G	526.1	All conductor terminations tight
H	526.2	All strands of conductor properly terminated
I	530.3.2	Single-pole fused connection unit switches in line conductor only
J	514.1.1	Where not obvious, a designation label indicating the function of the unit is affixed

that the inspection must be in accordance with Chapter 61 and any defects made good prior to the issuing of a certificate, and that the certificate signatories must be competent persons.

Here in the United Kingdom we have traditionally accepted self-certification (first-party certification), and by and large this has worked well and, compared with third-party

Table 17.19 Checklist for protective conductors that require visual inspection prior to carrying out testing

Ref	Regulation references	Facet to inspect
A	542.1	Earthing arrangements properly assessed
B	542.4.1	Main earthing terminal provided
C	542.4.2	Main earthing terminal facility for disconnection of the earthing conductor for test purposes provided
D	Section 511 Section 521	Conductors have been correctly selected for the particular circumstances and that they are compliant with the relevant standard
E	512.2.1–512.2.4	Suitable for all external influences, e.g. temperature, water, corrosive substances, mechanical stresses, flora and fauna, solar radiation and building structural design
F	543.2.1 543.2.2	Protective conductor of suitable type
G	543.2.3	Protective conductors of $10 \, mm^2$ or less of copper
H	543.3.1	Protective conductors protected against mechanical, chemical and electro-dynamic effects
I	543.2.4	Where protective conductor formed by metal enclosure, electrical continuity assured
J	542.3	Earthing conductors of suitable material, size and adequately protected
K	Deleted	
L	542.2	Earth electrodes correctly selected and installed
M	542.3.2	Connection to earth electrodes correctly protected against corrosion and mechanical damage
N	543.2.5	Cable sheaths and/or armouring are thermally protected
O	543.1.3 543.1.4	Protective conductors of adequate cross-sectional area (see Tables 54.1, 54.2, 54.3, 54.4, 54.5, 54.6, 54.7 and 54.8 of BS 7671)
P	543.3.2	Copper conductors of $6 \, mm^2$ or less, other than strip, covered by insulation/covering
Q	Deleted	
R	543.2.1	Gas and oil pipe lines not used as a protective conductor
S	524.1	Conductor cross-sectional area not less than that given in Table 52.3 of BS 7671
T	413.3.1.2	Main bonding correctly carried out and connects all extraneous-conductive-parts
U	544.1	Main protective bonding conductors correctly sized
V	544.2	Supplementary bonding conductors correctly sized
W	Deleted	
X	701.415.2	Supplementary bonding carried out in bathrooms, where required
Y	702.415.2	Supplementary bonding carried out in swimming pools
Z	Part 7	Supplementary bonding carried out in other special installations or locations where livestock is kept, where required
A1	543.3	Electrical continuity preserved throughout
B1	543.4	Combined protective and neutral (PEN) correctly used and sized
C1	522.6.1 522.8.1	Protected against mechanical damage including abrasion and suitable for location

(continued overleaf)

Table 17.19   (Continued)

Ref	Regulation references	Facet to inspect
D1	525.8.5	Cables adequately supported
E1	522.1.1	Suitable for the highest and lowest temperatures
F1	514.13.1 514.13.2	Protective conductor connection appropriately labelled
G1	514.4.2	Colour identification of protective conductors (green–yellow)
H1	514.4.5	The single colour green has not been used
I1	543.2.1	Flexible conduit not used as a sole cpc

Table 17.20   Specialized installations subject also to the requirements of a British Standard and CP

Ref	BS/CP reference	Type of installation
A	BS EN 60079 and BS EN 50014	Installations in potentially explosive atmospheres
B	BS 5266	Emergency lighting
C	BS 5839	Fire detection and alarm systems in buildings
D	BS 559, BS EN 50107	Electric signs and HV discharge tube installations
E	BS 6701	Telecommunications installations
F	BS 6351	Electric surface heating systems
G	BS EN 61241, BS EN 50281	Installations for use in the presence of combustible dust

certification, has been relatively inexpensive. It is important for this reason, if for no other, that installations are properly and thoroughly inspected, tested, verified and certified, so that the pressure for third-party certification is kept at bay, which from many points of view would be highly desirable but would in all probability lead to more expense for the consumer.

Figure 17.12 is a reproduction of a suitable Electrical Installation Certificate, by kind permission of The Electrical Safety Council, of the form based on the model given in Appendix 6 of BS 7671. The Electrical Safety Council also publishes 'red' versions of many of these forms of certification for use by electrical contractors enrolled as Approved Contractors with its subsidiary, NICEIC. A Domestic Electrical Installation Certificate is also produced by the Electrical Safety Council, including a 'green' version for general use, and 'red' and 'purple' versions for use by NICEIC Approved Contractors and NICEIC Domestic Installers respectively.

For competent engineers, compiling the Electrical Installation Certificate will present few difficulties. However, for those less familiar with the necessary form compilation, the following guidance will hopefully assist.

Details of the client (Figure 17.12a). This section is self-explanatory and needs no further illustration, requiring as it does the name and address of the client for whom the work has been carried out.

Details of the installation (Figure 17.12a). The address of the installation together with all relevant information relating to the extent of the installation should fully identify the

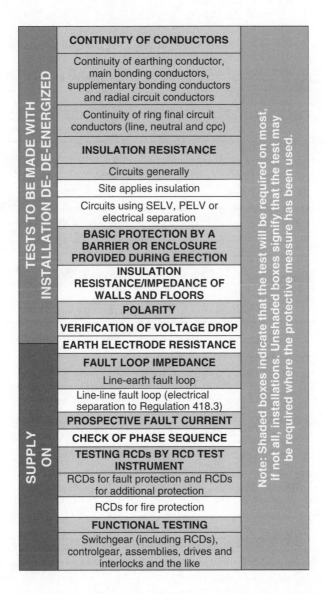

Figure 17.11   Sequence of tests for initial testing. (See plate 2)

scope and nature of the electrical installation work which is the subject of the certificate. The installation should also be declared *new, an addition* and/or *an alteration*, as appropriate.

Design, construction, inspection and testing boxes (Figure 17.2a). Each of the three elements of the installation work, namely *Design, Construction* and *Inspection and testing*, must be certified in the boxes provided for this purpose. In cases where the design element has been shared between two bodies, both bodies responsible must certify that the design

ICM4

Original (To the person ordering the work)

This safety certificate is an important and valuable document which should be retained for future reference

ELECTRICAL INSTALLATION CERTIFICATE

Issued in accordance with *British Standard BS 7671- Requirements for Electrical Installations*

SPECIMEN

DETAILS OF THE CLIENT

Client / Address:

DETAILS OF THE INSTALLATION

The installation is:

Address: | New

Extent of the installation covered by this certificate: | An addition | An alteration

DESIGN

I/We, being the person(s) responsible for the design of the electrical installation (as indicated by my/our signatures below), particulars of which are described above, having exercised reasonable skill and care when carrying out the design, hereby CERTIFY that the design work for which I/we have been responsible is to the best of my/our knowledge and belief in accordance with BS 7671: amended to (date) except for the departures, if any, detailed as follows:

Details of departures from BS 7671, as amended (Regulations 120.3,120.4):

The extent of liability of the signatory/signatories is limited to the work described above as the subject of this certificate.

For the **DESIGN** of the installation: ***(Where there is divided responsibility for the design)*

Signature | Date | Name (CAPITALS) | Designer 1

Signature | Date | Name (CAPITALS) | ** Designer 2

CONSTRUCTION

I/We, being the person(s) responsible for the construction of the electrical installation (as indicated by my/our signature below), particulars of which are described above, having exercised reasonable skill and care when carrying out the construction, hereby CERTIFY that the construction work for which I/we have been responsible is to the best of my/our knowledge and belief in accordance with BS 7671: amended to (date) except for the the departures, if any, detailed as follows:

Details of departures from BS 7671, as amended (Regulations 120.3,120.4):

The extent of liability of the signatory is limited to the work described above as the subject of this certificate.

For the **CONSTRUCTION** of the installation:

Signature | Date | Name (CAPITALS) | Constructor

INSPECTION AND TESTING

I/We, being the person(s) responsible for the inspection and testing of the electrical installation (as indicated by my/our signatures below), particulars of which are described above, having exercised reasonable skill and care when carrying out the inspection and testing, hereby CERTIFY that the work for which I/we have been responsible is to the best of my/our knowledge and belief in accordance with BS 7671: amended to (date) except for the departures, if any, detailed as follows:

Details of departures from BS 7671, as amended (Regulations 120.3,120.4):

The extent of liability of the signatory/signatories is limited to the work described above as the subject of this certificate.

For the **INSPECTION AND TESTING** of the installation: | Reviewed by †

Signature | Date | Signature | Date

Name (CAPITALS) | Inspector | Name (CAPITALS)

DESIGN, CONSTRUCTION, INSPECTION AND TESTING *

* This box to be completed only where the design, construction, inspection and testing have been the responsibility of one person.

I, being the person responsible for the design, construction, inspection and testing of the electrical installation (as indicated by my signature below), particulars of which are described above, having exercised reasonable skill and care when carrying out the design, construction, inspection and testing, hereby CERTIFY that the said work for which I have been responsible is to the best of my knowledge and belief in accordance with BS 7671, amended to (date) except for the departures, if any, detailed as follows:

Details of departures from BS 7671, as amended (Regulations 120.3, 120.4):

The extent of liability of the signatory is limited to the work described above as the subject of this certificate.

For the **DESIGN**, the **CONSTRUCTION** and the **INSPECTION AND TESTING** of the installation. | Reviewed by †

Signature | Date | Signature | Date

Name (CAPITALS) | Name (CAPITALS)

† *The completed schedules of inspection and testing should preferably be reviewed by another competent person to confirm that the recorded results are consistent with electrical installation work conforming to the requirements of BS 7671*

Page 1 of

This form is based on the model shown in Appendix 6 of BS 7671: 2008.

Please see the 'Notes for Recipients' on the reverse of this page.

ICM4/1

Figure 17.12a Electrical Installation Certificate, page 1. Reproduced by kind permission of The Electrical Safety Council. (See plate 3)

meets the requirements of BS 7671. In addition to the necessary signatures, the competent person's name and date of signature are also required. Where the design, construction and inspection and testing have been the responsibility of one person, the composite box towards the bottom of the page may be used instead of the three separate boxes.

Where the use of new materials or inventions leads to departures from BS 7671, details of these departures are to be recorded in the data-entry box provided in the *Design* section of the certificate (Regulations 120.3 and 120.4), and repeated in the data-entry boxes provided in the *Construction* and the *Inspection and testing* sections to confirm the designer(s)' intent. Where no departures have been sanctioned by the designer(s), the data-entry boxes should be completed by entering 'None'.

In each of the four boxes, provision has been made to insert the date of the latest amendment to BS 7671 to which the certificate relates. It is important that this data is inserted in order to obviate disputes over amended regulatory requirements.

Particulars of the organisation(s) responsible for the electrical installation (Figure 17.12b). These details will identify to the recipient the organization(s) responsible for the work certified by their representatives on page 1. The organizations, addresses and postcodes must be given. Where a contractor has been responsible for elements of the work in addition to construction, repetition of the *Particulars of the organisation(s) responsible for the electrical installation* is unnecessary and, in such cases, the words 'Details as given for construction' may be inserted, where appropriate.

Supply characteristics and earthing arrangements (Figure 17.12b). The *System type* must be identified in terms of TN-S, TN-C-S, TN-C, TT or IT. There may be circumstances where more than one system type is involved, such as TN-C-S and TT, in which case a positive indication should be made in each of the relevant data-entry boxes.

The *Number and type of live conductors* is self-explanatory, but it is also necessary to identify the supply current in terms of a.c. or d.c. The number of phase(s) will usually be one (two-wire) or three (four-wire) or, on rare occasions, a supply with two phases (three-wire) may be encountered. However, other combinations of a.c. supplies are possible, and provision is made on the form for most a.c. and d.c. system supplies.

The details of the *Nature of the supply parameters* will have been determined already by the designer(s) in consultation with the electricity distributor, and all such data should be readily available. The nominal voltage U_0 will be the voltage between line and Earth and, if the supply is derived from the public supply network, this will normally be 230 V in England, Wales and Scotland and in Northern Ireland. Similarly, the nominal voltage U is that between unearthed live conductors, and for many supplies this will simply be $\sqrt{3}U_0$ (400 V). Both these nominal voltages can only be ascertained from enquiry to the electricity distributor – measured values are not nominal values. The frequency in the United Kingdom will normally be 50 Hz for a.c. supplies.

Prospective fault-current I_{pf} relates to the maximum current that would flow if a fault of negligible impedance (both short-circuit and earth-fault prospective currents) occurred at the origin of the installation. These values are needed to ensure that protective devices have adequate short-circuit capacity to interrupt the fault current without sustaining damage or causing a hazard to its surroundings or to persons.

The *External line–earth loop impedance* Z_e at the origin is the impedance of the part of the loop external to the installation (i.e. the supply source and the supply lines coming into the installation upstream of the origin).

ICM4

Original (To the person ordering the work)

SPECIMEN

PARTICULARS OF THE ORGANISATION(S) RESPONSIBLE FOR THE ELECTRICAL INSTALLATION

DESIGN (1) Organisation †

Address:

Postcode

DESIGN (2) Organisation †

Address:

Postcode

† **CONSTRUCTION** Organisation

Address:

Postcode

INSPECTION AND TESTING Organisation †

Address:

Postcode

SUPPLY CHARACTERISTICS AND EARTHING ARRANGEMENTS

Tick boxes and enter details, as appropriate

✣ System Type(s): TN-S / TN-C-S / TN-C / TT / IT

✣ Number and Type of Live Conductors

a.c. / d.c.

1-phase (2 wire) / 1-phase (3 wire) / 2 pole

2-phase (3 wire) / 3-pole

3-phase (3 wire) / 3-phase (4 wire) / other

Other (Please state)

Nature of Supply Parameters

Nominal voltage(s): U [1] V; U_0 [1] V

Nominal frequency, f [1] Hz

Prospective fault current, I_{pf} [2][3] kA

External earth fault loop impedance, Z_e [2][3] Ω

Number of supplies

Notes:
(1) by enquiry
(2) by enquiry or by measurement
(3) where more than one supply, record the higher or highest values

✣ Characteristics of Primary Supply Overcurrent Protective Device(s)

BS(EN)

Type

Rated current A

Short-circuit capacity kA

PARTICULARS OF INSTALLATION AT THE ORIGIN

Tick boxes and enter details, as appropriate

✣ Means of Earthing

Distributor's facility:

Installation earth electrode:

Details of Installation Earth Electrode (where applicable)

Type: (e.g. rod(s), tape etc)

Location:

Electrode resistance, R_A: Ω

Method of measurement:

✣ Main Switch or Circuit-Breaker

* *(applicable only where an RCD is suitable and is used as a main circuit-breaker)*

Type: BS(EN)

Voltage rating V

No of poles

Rated current, I_n A

Supply conductors: material

RCD operating current, $I_{\Delta n}$* mA

Supply conductors: csa mm²

RCD operating time (at $I_{\Delta n}$)* ms

Maximum Demand (Load): kVA / Amps *Delete as appropriate

Protective measure(s) against electric shock:

Earthing and Protective Bonding Conductors

Earthing conductor: Conductor material; Conductor csa mm²; Continuity check (✓)

Main protective bonding conductors: Conductor material; Conductor csa mm²; Continuity check (✓)

Bonding of extraneous-conductive-parts (✓): Water service; Gas service; Oil service; Structural steel; Lightning protection; Other incoming service(s)

COMMENTS ON EXISTING INSTALLATION

In the case of an alteration or additions see Section 633

Note: Enter 'NONE' or, where appropriate, the page number(s) of additional page(s) of comments on the existing installation.

NEXT INSPECTION

§ *Enter interval in terms of years, months or weeks, as appropriate*

I/We, the designer(s), RECOMMEND that this installation is further inspected and tested after an interval of not more than §

† *Where the electrical contractor responsible for the construction of the electrical installation has also been responsible for the design **and** the inspection and testing of that installation, the 'Particulars of the Organisation Responsible for the Electrical Installation' may be recorded only in the section entitled 'CONSTRUCTION'.*

✣ *Where a number of sources are available to supply the installation, and where the data given for the primary source may differ from other sources, a separate sheet must be provided which identifies the relevant information relating to each additional source.*

Page 2 of

This form is based on the model shown in Appendix 6 of BS 7671: 2008.

Please see the 'Notes for Recipients' on the reverse of this page.

ICM4/3

Figure 17.12b Electrical Installation Certificate, page 2. Reproduced by kind permission of The Electrical Safety Council. (See plate 4)

Both these parameters (I_{pf} and Z_e) have to be determined by measurement or enquiry, although for design purposes the often higher values quoted by the ED should have been used for the design. Where supplies are derived from the public supply network, both parameters are obtainable upon request from the distributor, and this will often be the preferred method. Whilst calculation is given as an option in BS 7671, such a procedure is often impractical because all the necessary data relating to upstream impedances are not known. Measurements produce a value only for the particular time of measurement, and, as we all know, the network must be considered liable to change (within the normal range) without notice being given to the consumer. The technique of calculation should only be used where the designer has at their disposal all the relevant characteristics of the supply and is confident that these will remain unchanged. However, confirmation by measurement will always be necessary to ensure that the means of earthing is adequate and reliable.

The *Number of supplies* must be identified in the data-entry box even where there is only one. Where the installation can be supplied by more than one source, such as the public supply and a standby generator, the higher or highest values of prospective fault current I_{pf} and external earth fault loop impedance Z_e must be recorded in the data-entry boxes provided for this purpose. Where a number of sources are available to supply the installation, and where the data given for the primary source may differ from other sources, an additional page must be included in the certificate which gives the relevant information relating to each additional source.

The *Characteristics of the primary supply overcurrent protective device(s)* are required to be identified in terms of BS or BS EN number (e.g. BS 1361), type (e.g. Type II), rated current (e.g. 100 A) and short-circuit capacity (e.g. 16.5 kA) to be recorded. It is important to be assured that the rated current is not less than the maximum demand and that the short-circuit capacity is not less than the highest prospective fault current I_{pf} at the origin.

Particulars of the installation at the origin (Figure 17.12b). The *Means of earthing* will have already been determined by the designer in consultation with the ED. There are two options for providing a means of earthing, namely an installation earth electrode (for TT and IT systems – no earthing facility provided by the distributor) or an earthing facility made available by the distributor. The most common means of earthing which might be provided by the distributor will be those associated with TN-C-S (e.g. PME supply) or TN-S (separate supply earth provided, usually in the form of cable sheath or separate protective conductor).

Details of the earth electrode are normally only applicable to TT and IT systems. Where so used, the details of the type of electrode (e.g. rod(s), tape and/or plate(s)) and its location(s) are required to be recorded, as is the resistance of the electrode R_A and the method of measurement of that parameter (see Section 17.4.9 of this Guide).

A main linked switch or main linked circuit-breaker, as demanded by Regulation 537.1.4, is required for every installation, and its details, in terms of its BS or BS EN number, voltage rating, the number of poles (e.g. two-, four-pole) and current rating must be recorded on the certificate. Additionally, provision has been made on the form to identify the supply conductors to the main switch (often the meter tails) in terms of conductor material (e.g. copper) and cross-sectional area. Where an RCD (meeting all the necessary requirements for isolation) has been employed as a main switch, the rated residual operating current $I_{\Delta n}$ must also be recorded together with the operating time (in milliseconds) when subjected to a test current equal to the rated residual operating current $I_{\Delta n}$.

Maximum demand, in terms of amperes per line conductor or kVA, will be that which the designer has assessed from their design calculations, taking into account allowable diversity, and will depend to a large extent on the size of the installation and any load management measures adopted to limit peak loads.

The *Protective measure against electric shock* is that measure which has been used for protection for the majority of the installation, although other measures may have been used on certain parts of the installation (e.g. Class II luminaire). By far the most common measure of protection against electric shock is ADS. However, there are alternative measures for providing this protection (e.g. double or reinforced insulation), but these are much less extensively employed.

Provision has been made under *Earthing and protective conductors* to record the particulars of the earthing conductor in terms of conductor material (e.g. copper) and cross-sectional area, together with a facility to record that a continuity check has been carried out on this conductor. Similar provision has to be made for the *Main protective bonding conductors*. Additionally, the various services to which main protective bonding conductors are connected are to be identified in this section.

Comments on existing installations (Figure 17.12b). Such comments are applicable only where an existing installation has been altered or has had an addition made to it. As mentioned in Regulation 610.4, it is imperative that an alteration and/or addition to an installation complies in full with BS 7671. Where, for example, a new circuit makes use of any protective measures within the existing installation (e.g. an existing fuse for protection against overcurrent and for fault protection), the person carrying out such an addition would need to be assured that all the requirements relating to that protection are adequate (e.g. main protective bonding). Additionally, any alteration and addition must not adversely affect the safety of the existing installation. This part of the certificate gives the compiler the opportunity to draw the consumer's attention, as required by Regulation 633.2, to any aspects of unrelated parts of the existing installation which require attention. The entry here should be 'None' or the additional page number(s) on which the comments are set out.

Next inspection (Figure 17.12b). The recommendation for the interval to the next inspection is for the designer(s) to complete. The entry should be an appropriate interval, in terms of weeks, months or years, bearing in mind all the relevant circumstances relating to the installation.

Schedule of items inspected (Figure 17.12c). This schedule identifies most of the items that need inspection prior to testing, and data-entry boxes are provided so that necessary actions (that of inspections) may be recorded. All data-entry boxes are to be completed, as appropriate for the particular installation, by inserting a '$\checkmark$' indicating that the inspections have been undertaken on all applicable parts of the installation and that the result of inspection was satisfactory. It is unlikely that all items will apply, and the range of applicable inspections will depend on the particular installation covered by the certificate. If an inspection is not applicable, then 'N/A' (meaning 'Not applicable') should be recorded in the box. No boxes should remain blank.

Schedule of items tested (Figure 17.12c). Again, this schedule identifies most of the items that need testing, and data-entry boxes are provided so that necessary actions (that of testing) may be recorded. All data-entry boxes are to be completed, as appropriate for the particular installation, by inserting a '√' indicating that the tests have been undertaken on all applicable parts of the installation and the results are satisfactory. It is unlikely that all items will apply, and the range of applicable tests will depend on the particular installation covered by the certificate. If a test is not applicable, then 'N/A' should be recorded in the box. No boxes should remain blank.

Schedule of additional records (Figure 17.12c). Where additional records, such as as-fitted drawings, are to be submitted as part of the certificate, these are required to be identified in the *Schedule of additional records* section near the bottom of the page, by entering the additional page (or sheet) numbers of these records.

Schedule of circuit details for the installation (Figure 17.12d). This schedule begins with *Location of distribution board* and *Distribution board designation*. Information required here is self-explanatory and needs to be completed for every installation, irrespective of where the distribution board is located. Where the distribution board is located at a position other than at the origin, the requested information should be recorded, which includes:

- 'Supply to distribution board is from:'
- 'No of phases:'
- 'Nominal voltage:'
- 'Overcurrent protective device for the distribution circuit: Type BS (EN):'
- 'Overcurrent protective device for the distribution circuit: Rating:'
- 'Associated RCD (if any): BS (EN):'
- 'RCD No of poles:'
- '$I_{\Delta n}$:'.

A separate *Schedule of circuit details for the installation* will be required for each and every distribution board, which should be identified using the same designation identification as used on the distribution board itself.

Moving on to the *Circuit details* section, the entry for *Circuit number and phase(s)* needs no explanation nor does the *Circuit designation*, except to say that identification should be succinct and unambiguous in order to avoid confusion later (e.g. at subsequent periodic inspections).

The *Type of wiring* should be identified in accordance with the code at the bottom of the form. For ease of reference the code is repeated here:

A: PVC/PVC cables
B: PVC cables in metallic conduit
C: PVC cables in non-metallic conduit
D: PVC cables in metallic trunking
E: PVC cables in non-metallic trunking
F: PVC/SWA cables
G: XLPE/SWA cables
H: Mineral-insulated cables
O: Other.

ICM4

Original (To the person ordering the work)

SCHEDULE OF ITEMS INSPECTED   † *See note below*

PROTECTIVE MEASURES AGAINST ELECTRIC SHOCK

Basic and fault protection

Extra low voltage

☐ SELV   ☐ PELV

Double or reinforced insulation

☐ Double or Reinforced Insulation

Basic protection

☐ Insulation of live parts   ☐ Barriers or enclosures

☐ Obstacles **   ☐ Placing out of reach **

Fault protection

Automatic disconnection of supply

☐ Presence of earthing conductor

☐ Presence of circuit protective conductors

☐ Presence of main protective bonding conductors

☐ Presence of earthing arrangements for combined protective and functional purposes

☐ Presence of adequate arrangements for alternative source(s), where applicable

☐ FELV

☐ Choice and setting of protective and monitoring devices (for fault protection and/or overcurrent protection)

**Non-conducting location **

☐ Absence of protective conductors

**Earth-free equipotential bonding **

☐ Presence of earth-free equipotential bonding

Electrical separation

☐ For one item of current-using equipment

☐ For more than one item of current-using equipment **

Additional protection

☐ Presence of residual current device(s)

☐ Presence of supplementary bonding conductors

** *For use in controlled supervised conditions only*

Prevention of mutual detrimental influence

☐ Proximity of non-electrical services and other influences

☐ Segregation of Band I and Band II circuits or Band II insulation used

☐ Segregation of Safety Circuits

Identification

☐ Presence of diagrams, instructions, circuit charts and similar information

☐ Presence of danger notices and other warning notices

☐ Labelling of protective devices, switches and terminals

☐ Identification of conductors

Cables and Conductors

☐ Selection of conductors for current carrying capacity and voltage drop

☐ Erection methods

☐ Routing of cables in prescribed zones

☐ Cables incorporating earthed armour or sheath or run in an earthed wiring system, or otherwise protected against nails, screws and the like

☐ Additional protection by 30mA RCD for cables concealed in walls (where required in premises not under the supervision of skilled or instructed persons)

☐ Connection of conductors

☐ Presence of fire barriers, suitable seals and protection against thermal effects

General

☐ Presence and correct location of appropriate devices for isolation and switching

☐ Adequacy of access to switchgear and other equipment

☐ Particular protective measures for special installations and locations

☐ Connection of single-pole devices for protection or switching in line conductors only

☐ Correct connection of accessories and equipment

☐ Presence of undervoltage protective devices

☐ Selection of equipment and protective measures appropriate to external influences

☐ Selection of appropriate functional switching devices

SCHEDULE OF ITEMS TESTED   † *See note below*

☐ External earth fault loop impedance, Z_e

☐ Installation earth electrode resistance, R_A

☐ Continuity of protective conductors

☐ Continuity of ring final circuit conductors

☐ Insulation resistance between live conductors

☐ Insulation resistance between live conductors and Earth

☐ Protection by SELV, PELV or by electrical separation

☐ Basic protection by barrier or enclosure provided during erection

☐ Insulation of non-conducting floors or walls

☐ Polarity

☐ Earth fault loop impedance, Z_s

☐ Verification of phase sequence

☐ Operation of residual current devices

☐ Functional testing of assemblies

☐ Verification of voltage drop

SCHEDULE OF ADDITIONAL RECORDS* (See attached schedule)

Page No(s) _____

Note: Additional pages(s) must be identified by the Electrical Installation Certificate serial number and page number(s).

† *All boxes must be completed. '✓' indicates that an inspection or a test was carried out and that the result was **satisfactory**. 'N/A' indicates that an inspection or test was **not applicable** to the particular installation.*

Page 3 of _____

* *Where the electrical work to which this certificate relates includes the installation of a fire alarm system and/or an emergency lighting system (or a part of such systems), this electrical safety certificate should be accompanied by the particular certificate(s) for the system(s).*

This form is based on the model shown in Appendix 6 of BS 7671: 2008.
© Copyright The Electrical Safety Council (Jan 2008).

IC M4/5

Figure 17.12c   Electrical Installation Certificate, page 3. Reproduced by kind permission of The Electrical Safety Council. (See plate 5)

Figure 17.12d Electrical Installation Certificate, page 4. Reproduced by kind permission of The Electrical Safety Council. (See plate 6)

The *Reference method* is that which is appropriate to the particular installation method, both of which are given in Table 4A2 of Appendix 4 of BS 7671.

The *Number of points* served is self-explanatory and refers to the number of points such as socket-outlets and/or items of current-using equipment (e.g. luminaires).

Provision is made to record the cross-sectional area of *Circuit conductors* both for live conductors (line(s) and neutral) and cpcs.

A record of the *Max. disconnection time permitted by BS 7671* is required for every circuit. Such disconnection times will, of course, depend on the type of circuit and the location of equipment served.

Details of the *Overcurrent protective devices* are required in terms of their BS (EN) specification reference (e.g. BS EN 60 898), type (e.g. Type C), rating (e.g. 32 A), and short-circuit capacity (e.g. 6 kA).

Where an RCD is employed, the residual operating current $I_{\Delta n}$ must be recorded for each circuit. Likewise, provision is made for recording the *Maximum Z_s permitted by BS 7671*, which completes the *Schedule of circuit details for the installation*.

Schedule of test results for the installation (Figure 17.12e). This schedule begins with provision for recording parameters at the distribution board where it is located at other than the origin. Information required here is self-explanatory and needs to be completed for every installation where the distribution board is located at a position remote from the origin. The requested information that should be recorded includes:

- Earth fault loop impedance Z_s at the distribution board.
- Prospective fault current I_{pf} at the distribution board.
- Operating time of RCD when subjected to a test current equal to $I_{\Delta n}$.
- Operating time of RCD when subjected to a test current $5I_{\Delta n}$, if applicable.

Where the installation can be supplied by more than one source, such as a primary source (e.g. public supply) and a secondary source (e.g. standby generator), the higher or highest value earth fault loop impedance Z_s and prospective fault current I_{pf} at the distribution board must be entered.

Provision has been made to make the essential record of the test instruments used for such testing. This is an important aspect, since inaccuracies may be revealed later that necessitate the tracing of the test instruments used.

A separate schedule will be required for each and every distribution board, which should be identified using the same designation identification as used on the distribution board itself. Generally, where entry to a column is not relevant the data-entry box should be marked 'N/A'.

Moving on to the *Test results* section, the entry for *Circuit number and phase(s)* needs no explanation and will be identical to that given on the *Schedule of circuit details for the installation*.

Five columns are provided for recording *Circuit impedances*, which for most final circuits will only warrant recording the resistance component of impedance. The three columns headed 'r_1', 'r_n' and 'r_2' relate to the end-to-end resistances of the line conductor, the neutral conductor and the cpc of a ring final circuit respectively. Where the circuit is not a ring final circuit, the columns should be marked 'N/A'. The two following columns headed '$R_1 + R_2$' and 'R_2' apply to all circuits, but it is essential only to complete one or the other.

ICM4

**SCHEDULE OF TEST RESULTS
FOR THE INSTALLATION**

Original (To the person ordering the work)

TO BE COMPLETED ONLY IF THE DISTRIBUTION BOARD IS NOT CONNECTED DIRECTLY TO THE ORIGIN OF THE INSTALLATION
Characteristics at this distribution board

Confirmation of supply polarity

★ See note below

Z_s ★ Ω Operating times of associated   At $I_{\Delta n}$ ms

I_{pf} ★   kA   RCD (if any)   At 5$I_{\Delta n}$ (if applicable)   ms

Test instruments (serial numbers) used:

Earth fault loop impedance		RCD	
Insulation resistance		Other	
Continuity		Other	

TEST RESULTS

Circuit number and phase	Circuit impedances (Ω)					Insulation resistance † Record lower or lowest value				Polarity	Maximum measured earth fault loop impedance, Z_s ★ See note below	RCD times	
	Ring final circuits only (measured end to end)			All circuits (At least one column to be completed)		Line/Line †	Line/Neutral †	Line/Earth †	Neutral/Earth			at $I_{\Delta n}$	at 5$I_{\Delta n}$ (if applicable)
	r_1 (Line)	r_n (Neutral)	r_2 (cpc)	$R_1 + R_2$	R_2	(MΩ)	(MΩ)	(MΩ)	(MΩ)	(✓)	(Ω)	(ms)	(ms)

★ Note: Where the installation can be supplied by more than one source, such as a primary source (eg public supply) and a secondary source (eg standby generator), the higher or highest values must be recorded.

TESTED BY

| Signature: | | Position: | | Page 5 of |
| Name: (CAPITALS) | | Date of testing: | | |

This form is based on the model shown in Appendix 6 of BS 7671: 2008.
© Copyright The Electrical Safety Council (Jan 2008).

See previous page for Circuit Details

ICM4/5

Figure 17.12e Electrical Installation Certificate, page 5. Reproduced by kind permission of The Electrical Safety Council. (See plate 7)

Four columns are provided for *Insulation resistance* and are all self-explanatory, except to note that the lower or lowest values must be recorded where multiphase systems are employed. A record that *Polarity* has been checked must be acknowledged by ticking the data-entry box.

The *Maximum measured earth fault loop impedance, Z_s*, must be recorded. Where the installation can be supplied by more than one source, such as a primary source (e.g. public supply) and a secondary source (e.g. standby generator), the higher or highest value must be entered. It should be acknowledged that the measured value of earth fault loop impedance should normally be less than that allowed by BS 7671 in recognition of the likely lower temperature of the circuit conductors under test conditions than when operating under full load. It is generally accepted that, where the earth fault loop impedance values do not exceed 80% of the limits given for the overcurrent protective device in BS 7671, the earth fault loop impedance of the circuits will be acceptable under all conditions, including when fully loaded. Circuits where values exceed 80% will need closer examination to be confident that the circuit complies in this respect, as explained in Appendix 14 of BS 7671.

To complete this schedule, provision is made to record the operating times of RCDs, when subjected to a test current of $I_{\Delta n}$ and, if applicable, at $5I_{\Delta n}$.

17.5.5   Initial certification: caravans

The Electrical Safety Council, in consultation with the National Caravan Council, has devised a certificate solely for the purposes of certification of new leisure accommodation vehicles, including caravans and motor caravans. This certificate is reproduced, by kind permission of The Electrical Safety Council, in Figure 17.13 and is self-explanatory.

17.5.6   Initial certification: minor works

The Minor Electrical Installation Works Certificate, shown in Figure 17.14, is for the very small alteration work and/or small additions to an electrical installation. The scope of the certificate is limited to small alterations and additions that do not extend to the introduction of a new circuit. Its use, therefore, should be restricted to the very small jobs, and full certification will be required for all other installation work. This form of certification for alterations and additions which do not extend to a new circuit is recognized in Regulations 631.3 and 633.2.

17.6   Periodic inspection and testing

17.6.1   General

As set out in Regulations 621.1–622.2, the periodic inspection and testing of every installation must be carried out in accordance with Chapter 62 of BS 7671 and involve careful scrutiny without dismantling, or with partial dismantling, and accompanying inspection and testing to determine, as far as is reasonably practicable, whether the installation is in a satisfactory condition for continued service. Additionally, safety risks to persons and livestock from electric shock and burns, and risks relating to damage to property from fire and heat, must be identified, as must any damage or deterioration of installation equipment

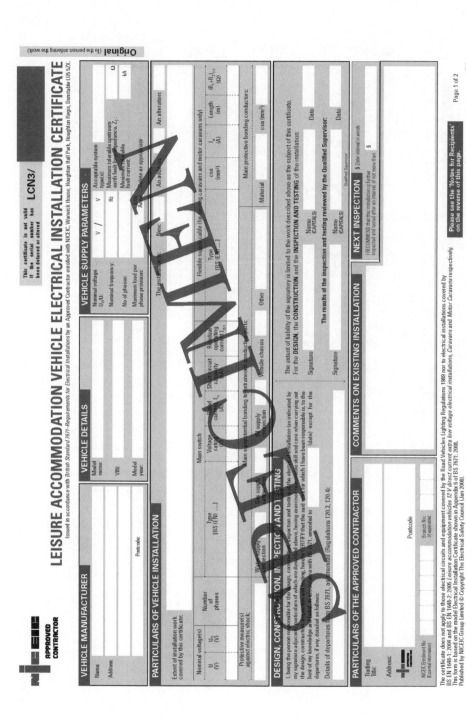

Figure 17.13a Leisure Accommodation Vehicle Electrical Installation Certificate, page 1. Reproduced by kind permission of The Electrical Safety Council. (See plate 8)

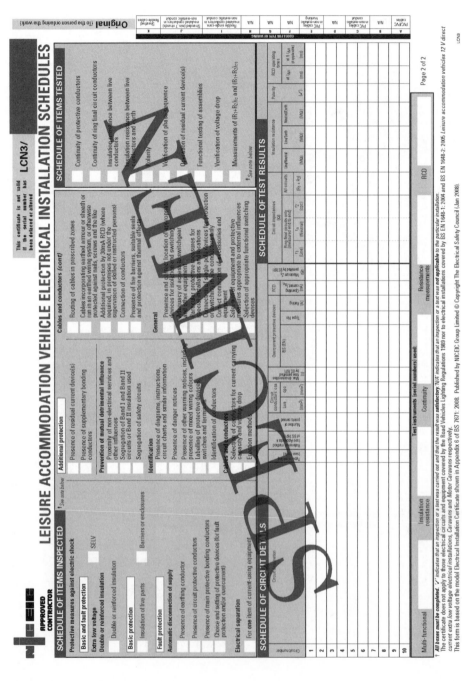

Figure 17.13b Leisure Accommodation Vehicle Electrical Installation Certificate, page 2. Reproduced by kind permission of The Electrical Safety Council. (See plate 9)

IMM4

Original (To the person ordering the work)

This safety certificate is an important and valuable document which should be retained for future reference

MINOR ELECTRICAL INSTALLATION WORKS CERTIFICATE

Issued in accordance with *British Standard BS 7671 - Requirements for Electrical Installations*

To be used only for minor electrical work which does not include the provision of a new circuit

PART 1: DETAILS OF THE MINOR WORKS

Client:

Date minor works completed:

Contract reference, if any:

Details of departures, if any, from BS 7671 (as amended):

Description of the minor works:

Location/address of the minor works:

Postcode

PART 2: DETAILS OF THE MODIFIED CIRCUIT

System type and earthing arrangements: TN-C-S TN-S TT TN-C IT

Protective measures against electric shock:

Overcurrent protective device for the modified circuit: BS(EN) Type Rating A

Residual current device (if applicable): BS(EN) Type $I_{\Delta n}$ mA

Details of wiring system used to modify the circuit: Type Reference method csa of live conductors mm² csa of cpc mm²

Where the protective measure against electric shock is ADS, insert maximum disconnection time permitted by BS 7671: s Maximum Z_S permitted by BS 7671 Ω

Comments, if any, on existing installation:

PART 3: INSPECTION AND TESTING OF THE MODIFIED CIRCUIT AND RELATED PARTS

† Essential inspections and tests

† Confirmation that necessary inspections have been undertaken	(✓)	† Confirmation of the adequacy of earthing	(✓)
† Circuit resistance: $R_1 + R_2$ Ω or R_2	Ω	† Confirmation of the adequacy of protective bonding	(✓)
Insulation resistance: (* In a multi-phase circuit, record the lower or lowest value, as appropriate) Line/Line*	MΩ	† Confirmation of correct polarity	(✓)
Line/Neutral*	MΩ	† Maximum measured earth fault loop impedance, Z_S	Ω
Instrument Serial No(s): † Line/Earth*	MΩ	† RCD operating time at $I_{\Delta n}$ (if RCD fitted)	ms
† Neutral/Earth	MΩ	RCD operating time at $5 I_{\Delta n}$, if applicable	ms

Agreed limitations, if any, on the inspection and testing:

PART 4: DECLARATION

I/We CERTIFY that the said works do not impair the safety of the existing installation, that the said works have been designed, constructed, inspected and tested in accordance with BS 7671 (IEE Wiring Regulations), amended to and that the said works, to the best of my/our knowledge and belief, at the time of my/our inspection complied with BS 7671 except as detailed in Part 1.

Name (CAPITALS)

Signature

Position

Date

For and on behalf of (*Trading title of electrical contractor*)

Address and Postcode

This form is based on the model shown in Appendix 6 of BS 7671

Please see the 'Notes for Recipients' on the reverse of this page.

IMM4/1

Figure 17.14 Minor Electrical Installation Works Certificate. Reproduced by kind permission of The Electrical Safety Council. (See plate 10)

which may impair safety. Finally, any defect or noncompliance with BS 7671 which may give rise to danger must be identified.

It is the responsibility of the person certifying the initial verification of an installation to make a recommendation for the subsequent inspection and testing, and this would be based on the whole circumstances relating to the particular installation. Regulation 634.1 calls for a report to be issued to the person ordering such an inspection. The report must draw the attention of the consumer (or their agent) to any dangerous conditions which are observed and must also state the scope and limitations of such an inspection (Regulation 643.2). In the absence of any stated limitations it would be fair for the consumer to assume that every part of the installation had been inspected and tested, including, for example, cables concealed within the building interfloor spaces. A suitable report form is reproduced by kind permission of The Electrical Safety Council in Figure 17.15.

Any periodic inspection report of an installation which has been carried out to an earlier Edition of the Wiring Regulations should only refer to departures (if any) from the current edition (BS 7671) and not to the edition to which it was constructed. Should an inspection reveal deficiencies in the installation which present an immediate danger, the consumer (or their agent) should be advised immediately (verbally initially and preferably in writing subsequently) of the safety risks and detail any corrective action required.

It is self-evident that all installations should be regularly inspected and tested periodically. However, BS 7671 cannot impose such a requirement on the consumer, who may, if so disposed, ignore completely the recommendations of the initial verifier, and this may be particularly true of domestic consumers, some of whom are unable to see any benefit of such inspections. In the case of installations in places of work the situation is somewhat different, as under the Electricity at Work Regulations 1989 the duty holder responsible for the maintenance of the installation is required to maintain all electrical systems so as to prevent danger. It should also be noted that in some cases there is a statutory requirement for periodic inspections (e.g. cinemas, petrol filling stations, public houses, church and village halls, bingo rooms, school assembly halls used for public concerts).

17.6.2   Intervals between periodic inspection and testing

All installations should be regularly inspected and tested in order that the person responsible for the safety of the installation can be reassured that it continues to give satisfactory and safe service. On completion of the initial verification of an installation it is the verifier who makes the recommendation as to when the first periodic inspection and testing should be made. The recommendation for a subsequent periodic inspection then rests with the test engineer carrying out the first periodic inspection, and so on. It is reasonable to expect the intervals to become progressively shorter as the installation ages, suffering as it surely must from normal wear and tear. Much will depend on the type of installation, its usage and its environment. The person responsible for making recommendations will need to take all these factors into consideration.

Regulation 622.1 states that the frequency of periodic inspection will depend on the type of installation and equipment, its use and operation, the frequency and quality of maintenance and the external influences to which it is subjected. Table 17.21 seeks to provide a listing of suggested intervals between inspections, but it should only be used as a guide, since all the relevant factors relating to the installation must be taken into account. Furthermore, where premises are required to be licensed, the licensing authority may

Table 17.21 Intervals between inspections

Ref Ref	Type of premises	Interval (years)	Code[a]	Other references[b]
General installations				
A	Private housing	10	R	–
B	Offices and other commercial premises	5	R	See also the Electricity at Work Regulations 1989
C	Schools and other educational establishments	5	R	See also the Electricity at Work Regulations 1989
D	Hospitals and other medical establishments	5	R	See also the Electricity at Work Regulations 1989
E	Factories and other industrial premises	3	M	See also the Electricity at Work Regulations 1989
Public building installations				
F	Churches and other places of worship with installations completed in the preceding 5 years	2	R	See also the Electricity at Work Regulations 1989. See Lighting and Wiring of Churches 1981
G	Churches and other places of worship with installations completed over 5 years ago	1	R	See also the Electricity at Work Regulations 1989. See Lighting and Wiring of Churches
H	Theatres	1	M	See also the Electricity at Work Regulations 1989. See Conditions of Licence imposed by the Licensing Authority
I	Cinemas	1	M	See also the Electricity at Work Regulations 1989. See clause 17 of The Cinematograph (Safety) Regulations 1955
J	Places of public entertainment	1	M	See also the Electricity at Work Regulations 1989. See Conditions of Licence imposed by the Licensing Authority
K	School halls, church halls, etc. used for public concerts	1	M	See also the Electricity at Work Regulations 1989. See Conditions of Licence imposed by the Licensing Authority.
L	Hotels	1	M	See also the Electricity at Work Regulations 1989. See Conditions of Licence imposed by the Licensing Authority
M	Restaurants	1	M	See also the Electricity at Work Regulations 1989. See Conditions of Licence imposed by the Licensing Authority
N	Sports and leisure centres	1	M	See also the Electricity at Work Regulations 1989. See Conditions of Licence imposed by the Licensing Authority
Special locations				
O	Farms and other agricultural premises	3	R	See also the Electricity at Work Regulations 1989

(continued overleaf)

Table 17.21   (Continued)

Ref Ref	Type of premises	Interval (years)	Code[a]	Other references[b]
P	Garden centres and other horticultural premises	3	R	See also the Electricity at Work Regulations 1989
Q	Caravan parks	1	M	See also the Electricity at Work Regulations 1989. See Conditions of Licence imposed by the Licensing Authority
R	Caravans	3	R	–
S	Construction and building sites and other temporary installations.	0.25	R	See also the Electricity at Work Regulations 1989.
T	Highway power supply and street lighting, etc.	6	M	See also the Electricity at Work Regulations 1989. See Conditions of Licence imposed by the Licensing Authority
Specialized installations				
U	Fire detection and alarm systems in buildings	1	M	See also the Electricity at Work Regulations 1989. See BS 5839. See Conditions of Licence imposed by the Licensing Authority
V	Emergency lighting systems	3	M	See also the Electricity at Work Regulations 1989. See BS 5266. See Conditions of Licence imposed by the Licensing Authority. Batteries and safety supply generators need further consideration
W	Laundries and launderettes	1	M	See also the Electricity at Work Regulations 1989
X	Petrol filling stations	1	M	See also the Electricity at Work Regulations 1989. See Conditions of Licence imposed by the Licensing Authority

[a]R: recommended; M: mandatory.
[b]Where electricity supplies are obtained from the public supply network, the ED will wish to ensure that the requirements of the Electricity Supply Regulations 1988 as amended are met.

seek to impose shorter time spans between inspections, particularly for those installations nearing the end of their useful service life. The table identifies the types of premises that are likely to be the subject of licensing conditions, or where there are implicit requirements in legislation, by the code 'M' and those that are not by the code 'R'.

Regulation 622.2 recognizes that, where an installation is under effective supervision in normal use, a regime of continuous monitoring may obviate the need for periodic inspection and testing, providing adequate records are maintained.

17.6.3   Approximate age of an installation

Most experienced installation engineers will have no difficulty in dating an installation within 5–10 years. Though perhaps unconsciously, features will be seen which will identify the period to within a decade or so, and a combination of features will reduce the

Table 17.22 Approximate age of an installation

Ref	Feature	Approximate dates[a]
A	Lead-sheathed cables	Pre 1948
B	Tough-rubber-sheathed (TRS) cables	1945–1962
C	'Capothene'- and 'Ashothene'-sheathed cables	1952–1960
D	PVC/PVC cables without cpc used on lighting circuits	1955–1966
E	PVC cables of imperial sizes	1955–1971
F	2.5 mm^2 PVC/PVC-sheathed cables with 1 mm^2 cpc	1971–1981
G	Black earthing conductors.-	Pre 1966
H	Absence of main equipotential bonding conductors	Pre 1966
I	Green protective conductor oversleeving	Pre 1966
J	2.5 mm^2 main equipotential bonding conductors – small installations	1971–1972
K	6 mm^2 main equipotential bonding conductors – small installations	1966–1983
L	10 mm^2 main equipotential bonding conductors – small installations	Post 1983
M	Twin-twisted flexible cords	Pre 1977
N	Fault-voltage-operated circuit-breakers	Pre 1981
O	Accessories mounted on wooden blocks	Pre 1966
P	Non-13 A socket-outlets	Pre 1955
Q	Double-pole fused switchgear on a.c. installations	Pre 1955
R	Red, yellow and blue line conductors and black neutral	Pre 2004

[a]Dates given should be treated with extreme caution.

period even further. In Table 17.22, an attempt has been made to collate some of the many installation features which will, hopefully, be an aid to making a more accurate assessment of the age of an installation. The dates given are necessarily very approximate.

17.6.4 Periodic inspection and testing

In an ideal world, the documentation (diagrams, circuit charts and design details) dealing with the initial verification of the electrical installation will be available to the person carrying out subsequent periodic inspections. The inspector should take particular note of any change in circumstances. This may, for example, include environmental conditions, alterations and additions to the building structure and electrical installation and any changes in loadings. As called for in Regulation 621.2, the inspection should comprise a detailed examination of the installation without wholesale dismantling. Partial dismantling will, of course, be necessary to establish, for example, whether wiring systems are generally in a good serviceable condition. It is important that an assessment is made to evaluate the general condition and whether it is safe to conduct tests without creating danger to persons and livestock. The inspector should take particular note of the suitability and condition of equipment in terms of safety, age, wear and tear, physical damage, corrosion, loading and overloading and any change of external influences (e.g. by a change of use of the premises).

Protective devices must be checked for their suitability, their ratings, in terms of short-circuit capacity and rated current, their accessibility and identification. Isolating and switching devices, including emergency switching arrangements, must also be checked for their suitability, accessibility, identification and performance.

Switchgear and enclosures generally should be checked for their suitability, condition and performance and that the intended degree of ingress protection is maintained. A check of the colour coding of conductors should also be made, together with a check of correct connection and an assessment of whether overloading is evident and that the insulation remains in a good serviceable condition. Where metallic-sheathed and/or armoured cables have been used, a check on the termination of their glands should be made to establish that a sound mechanical and electrical connection is evident. Conductors, including protective conductors, should be checked for correct sizing and protection against overcurrent and, where applicable, provide sufficient protection against electric shock by ADS in the event of a fault.

A check must also be made of the protective provisions employed for basic protection, and, for example, a note of any absent barriers should subsequently be made on the report. Particular attention should be paid to provision of safety features intended to prevent danger to persons and livestock (from shock, fire, burns and injury from mechanical movement) and damage to property; where deficiencies are evident, these matters should be noted on the report.

The marking and labelling of an installation is an important feature, and special attention should be paid to this aspect. Warning notices and labelling should be as set out in Chapter 51 of BS 7671 and addressed in Chapter 9 of this Guide. Again, any deficiencies should be reported.

As previously mentioned, testing must not commence before the inspector has satisfied themselves that tests can be made without danger, and this can only be done after a careful visual inspection of all equipment that can reasonably be assessed.

The sequence of testing for periodic testing is likely to have to be different to some extent from that of initial testing, which is shown in Figure 17.11. It is up to the person carrying out the testing to decide whether, and in what way, to change the sequence to suit the circumstances of the installation being periodically inspected and tested, by using his or her experience and knowledge and by consulting any available records of the installation. This person must also decide which of the tests referred to in Section 612 of BS 7671 (and listed in Figure 17.11) for initial verification are appropriate for the installation being periodically inspected and tested. Particular care is needed to identify equipment and circuits vulnerable to a particular test; these should be disconnected (after prior arrangement with the consumer) before tests are made and other means should be used to establish their safety.

Tests of the continuity of the protective conductors cannot be carried out without first isolating the installation from the supply source. It is important to recognize that a dangerous potential can exist between the MET and disconnected protective conductors under installation normal conditions (e.g. if the means of earthing provides functional earthing requirements) and under installation earth fault conditions. A potential can also exist between these parts under earth fault conditions of supply equipment. Main bonding conductors should, therefore, only be disconnected from the MET for testing after taking the precaution of isolating the installation *and* ensuring that simultaneous contact between the bonding conductors and the MET is prevented.

Polarity tests can now follow, again starting at the origin and working downstream through distribution boards and on to final circuits. Particular attention must be paid to ensure that single-pole devices (fuses, circuit-breakers, switches) are in the line conductors only. Socket-outlets, fused connection units, ES lampholders and lighting-track systems must also be checked for correct polarity. Any alterations and/or additions to the installation since the previous inspection should rate very close scrutiny.

Having established that the earthing and main protective bonding are satisfactory and also that polarity is correct, it is then necessary to carry out earth loop impedance tests in order to establish that the integrity of all the protective conductors is intact. Testing would normally start with a measurement of the external earth fault loop impedance Z_e and then progress to distribution circuits (sub-mains). From the results of these tests it can then be confirmed that all protective conductors, including cpcs, are satisfactory and that, where Z_s influences the operation of protective devices for fault protection, disconnection times are within permitted limits. More extensive earth fault loop impedance tests can now follow for final circuits, including all socket-outlets, lighting circuits and other radial circuits with exposed-conductive-parts.

Insulation testing at the main switch (with the installation isolated) may be all that is required provided that the required value of insulation resistance is obtained (≥ 1 MΩ for circuits up to 500 V). This must be done with all the fuse links in place and with circuit-breakers in the closed position and with lighting luminaires switched on and lamps removed. Before testing, care should be taken to disconnect any vulnerable circuits (e.g. electronic equipment) before applying the test voltage in order to avoid damage to such equipment and costly replacements. The value of insulation resistance should be compared with the reading taken on the previous occasion of testing. Where there is a significant difference, further investigation may be necessary. Any vulnerable equipment disconnected for test purposes should, if it contains exposed-conductive-parts, be tested separately by a suitable instrument in order to ensure that it complies with the safety requirements laid down in the British Standard for that equipment or, in the absence of a specified insulation value, to have an insulation resistance of not less than 1 MΩ for equipment rated at up to 1000 V and 0.5 MΩ for SELV and PELV equipment.

Where the necessary insulation resistance is not obtained at the main switch, further insulation tests will be necessary to identify the problem circuit(s) and to remedy the reasons for the poor insulation. It is worth remembering that overall insulation resistance is related to the individual circuit insulation resistances as given by

$$\frac{1}{R_o} = \frac{1}{R_1} + \frac{1}{R_2} + \frac{1}{R_3} + \cdots \frac{1}{R_n} \tag{17.5}$$

where R_o is the insulation resistance overall, R_1 = is the insulation resistance of circuit 1, R_2 = is the insulation resistance of circuit 2, R_3 = is the insulation resistance of circuit 3 and R_n = is the insulation resistance of n.

By way of example to illustrate the point, take three circuits with insulation resistances to earth of 6, 5 and 4 MΩ. By inserting these values into Equation (17.5), the reciprocal of resistance is obtained:

$$\frac{1}{R_o} = \frac{1}{6} + \frac{1}{5} + \frac{1}{4} = \frac{10 + 12 + 15}{60} = \frac{37}{60} \tag{17.6}$$

from which the overall insulation resistance R_o is $1.62\,\mathrm{M\Omega}$ ($60/37\,\mathrm{M\Omega}$).

According to Regulation 612.3.2 and Table 61 BS 7671, the minimum acceptable insulation resistance for the whole installation is $1\,\mathrm{M\Omega}$, and this relates to a leakage current on an installation, operating on 230 V to Earth, of 0.23 mA. On the larger installation this limiting value of insulation resistance may be applied to each distribution circuit.

Devices for switching and isolation must be checked for correct operation. Such devices include isolators, disconnectors, switch-disconnectors, devices for switching off for mechanical maintenance and emergency switches. The devices should be verified as being capable of providing their allocated function, and the indicators should also be checked where a visual indication is a requirement. A voltage indicator should be used to ensure that the device cuts off the equipment served from the supply, and devices used for safety purposes, such as emergency switches, should all be verified as being capable of satisfactory service by operating the device in the intended manner. Where the isolating or other device is remote from the equipment it serves and where interlocking arrangements have not been used, locking devices, removable handles, etc. should be checked and the interchangeability of the locking mechanisms should be verified as being satisfactory.

All RCDs should be tested by simulating earth fault conditions in accordance with the requirements laid down in the relevant product standards, which are summarized in Table 17.3 (see also Section 17.4.10 of this Guide), together with the requirements for time-delay RCDs. Where RCDs have been installed to provide additional protection (e.g. for socket-outlets or for cables concealed in walls), these must also be checked to ensure that their residual operating current $I_{\Delta n}$ does not exceed 30 mA.

It is not possible to test the effectiveness of overcurrent protective devices (e.g. MCBs and MCCBs) in situ because of the dangers that would be created by allowing high currents which would flow in order to check satisfactorily the overload protection and much higher currents for fault current operation. However, such devices should be checked by operating the devices by hand and by observing that they open and close as intended. The inspector would also note that the device had originally been correctly selected and installed. Where there is doubt regarding the suitability of the device, or it has suffered damage or deterioration, it should be replaced.

IEE Guidance Note No. 3 on inspection and testing recognizes that sample testing may be carried out during the periodic inspection and testing of an existing installation, with the percentage of testing being at the discretion of the person carrying out the testing, subject to a cautionary note that a percentage of less than 10% is inadvisable. The guidance note points out that considerable care and engineering judgement should be used when deciding on the extent of the installation that will be subjected to testing. In this respect, the tester should obtain any available electrical installation certificates, minor works certificates, previous periodic inspection reports, maintenance reports, site plans, drawings and data sheets relating to the installed equipment. Where the necessary information cannot be obtained relating to the design, maintenance and modification of the installation, Guidance Note No. 3 advises that it would be necessary to test a larger percentage, or in some circumstances 100%, of the installation. Also to be taken into consideration when deciding on the size of sample are the age, size, type, usage and general condition of the installation, as well as the environmental conditions affecting it, the effectiveness of ongoing maintenance (if any) and the length of time since any previous periodic inspection and test.

17.6.5 Reporting

Regulation 634.1 calls for the results of the periodic inspection to be recorded on a report by the person carrying out the inspection, or a person acting on their behalf, and given to the person ordering such an inspection. Explicitly, Regulation 634.2 calls for the report to contain information relating to any damage, deterioration, defects and dangerous conditions, and noncompliances with BS 7671 which may give rise to danger, and to identify the scope and limitations of the inspection. Any dangerous conditions should be reported to the duty-holder, or other person responsible for the installation, immediately by word of mouth and followed by written confirmation and, ideally, the defect remedied without delay. Figure 17.15 reproduces a periodic inspection report produced by the Electrical Safety Council based on the model form in Appendix 6 of BS 7671.

The compilation of the periodic inspection report form is fairly straightforward and the experienced inspector will have no difficulty in completing the form. However, for the benefit of the less experienced, the following guidance related to the form sections 'A' to 'L' and the necessary attachment schedules (see Figures 17.15a–f), may provide assistance.

It is important to note that the Periodic Inspection Report form is to be used only for reporting on the condition of an existing installation. It must not be used instead of an Electrical Installation Certificate for certifying a new electrical installation, or as a substitute for a Minor Electrical Installation Works Certificate for certifying an addition or an alteration to an existing circuit.

Section A: Details of client (Figure 17.15a). This section provides space for such client details in terms of name of person or organization and the associated address.

Section B: Purpose of the report (Figure 17.15a). This section is to enable the purpose of the report to be clearly identified and recorded. There are many reasons why a report may be required; for example, the normal periodic monitoring of an installation to confirm its continuing safety and serviceability, a report to support a building society mortgage application, or a report to satisfy the building duty-holder that the installation remains compliant with the relevant aspects of the *Electricity at Work Regulations, 1989* and other statutory requirements.

This report form is not intended for the specialized installations, such as fire detection and alarm systems, emergency lighting and installations in petrol filling stations, all of which are dealt with later in this Guide.

Section C: Details of the installation (Figure 17.15a). Again, this section is straightforward, with the occupier, occupier's address and the description of the premises requiring no further explanation. The estimation of the age of the installation and that of any alterations and additions will present no complication if proper records of the installation have been maintained and are available. However, should these records not be available, reference to Table 17.22 may assist the inspector to date the installation within a period of about 5 years. Of course, the older the installation, the less is the need to establish its age accurately. Where records are available, it will be rudimentary to determine the date

IPM4

PERIODIC INSPECTION REPORT
FOR AN ELECTRICAL INSTALLATION

Issued in accordance with *British Standard BS 7671- Requirements for Electrical Installations*

Original (To the person ordering the work)

A. DETAILS OF THE CLIENT

Client: Address:

B. PURPOSE OF THE REPORT   This Periodic Inspection Report must be used only for reporting on the condition of an existing installation.

Purpose for which
this report is required:

C. DETAILS OF THE INSTALLATION

Occupier: Domestic   Commercial   Industrial

Description of
premises:

Address: Other
                                                             (Please state)

                                                             Estimated age of the electrical installation: years

Postcode: Evidence of alterations
                                     or additions   If yes,
                                                                     estimated age   years

Date of previous inspection: Electrical Installation Certificate No or previous Periodic Inspection Report No:

Records of installation available: Records held by:

D. EXTENT OF THE INSTALLATION AND LIMITATIONS OF THE INSPECTION AND TESTING

Extent of the electrical installation covered by this report:

Agreed limitations (including the reasons), if any, on the inspection and testing:

This inspection has been carried out in accordance with BS 7671, as amended. Cables concealed within trunking and conduits, or cables and conduits concealed under floors, in inaccessible roof spaces and generally within the fabric of the building or underground, have not been visually inspected.

E. DECLARATION

I/We, being the person(s) responsible for the inspection and testing of the electrical installation (as indicated by my/our signatures below), particulars of which are described above (see C), having exercised reasonable skill and care when carrying out the inspection and testing, hereby declare that the information in this report, including the observations (see F) and the attached schedules (see H), provides an accurate assessment of the condition of the electrical installation taking into account the stated extent of the installation and the limitations of the inspection and testing (see D). I/We further declare that in my/our judgement, the said installation was overall in ✚   condition (see G) at the time the inspection was carried out and that it should be further inspected as recommended (see I).

✚ *(insert 'a satisfactory' or 'an unsatisfactory', as appropriate)*

INSPECTION, TESTING AND ASSESSMENT BY: **REPORT REVIEWED AND CONFIRMED BY:** † *See note below*

Signature: Signature:

Name: Name:
(CAPITALS)   (CAPITALS)

Position:

Date: Date:

† *The completed report should preferably be reviewed by another competent person to confirm that the declared overall condition of the electrical installation is consistent with the inspection and test results, and with the observations and recommendations for action (if any) made in the report.*   Page 1 of ▢

This form is based on the model shown in Appendix 6 of BS 7671.
© Copyright The Electrical Safety Council (Jan 2008)

Please see the 'Notes for Recipients'
on the reverse of this page.

IPM4/1

Figure 17.15a Periodic Inspection Report, page 1. Reproduced by kind permission of The Electrical Safety Council. (See plate 11)

IPM4

Original (To the person ordering the work)

F. OBSERVATIONS AND RECOMMENDATIONS FOR ACTIONS TO BE TAKEN

Referring to the attached schedules of inspection and test results, and subject to the limitations at D:

There are no items adversely affecting electrical safety. ☐

or

The following observations and recommendations are made. ☐

Item No		Code †
1		

Note: If necessary, continue on additional page(s), which must be identified by the Periodic Inspection Report date and page number(s).

† *Where observations are made, the inspector will have entered one of the following codes against each observation to indicate the action (if any) recommended:-*

1. *'requires urgent attention' or*
2. *'requires improvement' or*
3. *'requires further investigation' or*
4. *'does not comply with BS 7671'*

Please see the reverse of this page for guidance regarding the recommendations.

Urgent remedial work recommended for Items: ______ **Corrective action(s) recommended for Items:** ______

G. SUMMARY OF THE INSPECTION

General condition of the installation:

Note: If necessary, continue on additional page(s), which must be identified by the Periodic Inspection Report date and page number(s).

Date(s) of the inspection: ______ Overall assessment of the installation: ______

*(Entry should read either '**Satisfactory**' or '**Unsatisfactory**')*

Page 2 of ☐

This form is based on the model shown in Appendix 6 of BS 7671.

Please see the 'Guidance for Recipients on the Recommendation Codes' on the reverse of this page.

IPM4/3

SPECIMEN

Figure 17.15b Periodic Inspection Report, page 2. Reproduced by kind permission of The Electrical Safety Council. (See plate 12)

IPM4

H. SCHEDULES AND ADDITIONAL PAGES

Schedule of Items Inspected and Schedules of Items Tested: Page No 4

Additional pages, including additional source(s) data sheets: Page No(s)

Schedule of Circuit Details for the Installation: Page No(s)   5

Schedule of Test Results for the Installation: Page No(s)   6

The pages identified here form an essential part of this report. The report is valid only if accompanied by all the schedules and additional pages identified above.

I. NEXT INSPECTION

I/We recommend that this installation is further inspected and tested after an interval of not more than

(Enter interval in terms of years, months or weeks, as appropriate)

provided that any items at F which have been attributed a Recommendation Code 1 (*requires urgent attention*) and Code 2 (*requires improvement*) are remedied without delay and as soon as possible respectively. Items which have been attributed a Recommendation Code 3 should be actioned as soon as practicable (see F).

J. DETAILS OF ELECTRICAL CONTRACTOR

Trading Title:

Address: Telephone number:

                                                                        Number:

                            Postcode:

K. SUPPLY CHARACTERISTICS AND EARTHING ARRANGEMENTS   *Tick boxes and enter details, as appropriate*

✛ System Type(s)	✛ Number and Type of Live Conductors		Nature of Supply Parameters		✛ Characteristics of Primary Supply Overcurrent Protective Device(s)
TN-S	a.c.	d.c.	Nominal voltage(s): $U^{(1)}$ V U_0 $^{(1)}$ V		
TN-C-S	1-phase (2 wire)	1-phase (3 wire)	2 pole	Nominal frequency $f^{(1)}$ Hz *Notes:* (1) by enquiry	BS(EN)
TN-C	2-phase (3 wire)	3-pole	Prospective fault current I_{pf} $^{(2)}$ kA (2) by enquiry or by measurement		Type
TT	3-phase (3 wire)	3-phase (4 wire)	other	External earth fault loop impedance, Z_e $^{(3)(4)}$ Ω (3) where more than one supply, record the higher or highest values	Rated current A
IT	Other *Please state*		Number of supplies (4) by measurement		Short-circuit capacity kA

L. PARTICULARS OF INSTALLATION AT THE ORIGIN   *Tick boxes and enter details, as appropriate*

✛ Means of Earthing		Details of Installation Earth Electrode (where applicable)
Distributor's facility:	Type: (eg rod(s), tape etc)	Location:
Installation earth electrode:	Electrode resistance, R_A (Ω)	Method of measurement:

✛ Main Switch or Circuit-breaker		Maximum Demand (Load):	kVA / Amps *Delete as appropriate*	Protective measure(s) against electric shock:
Type BS(EN):	Voltage rating V			

Earthing and Protective Bonding Conductors

No of Poles		Earthing conductor		Main protective bonding conductors		Bonding of extraneous-conductive-parts (✓)	
Rated current, I_n A		Conductor material		Conductor material		Water service	Gas service
Supply conductors material		Conductor csa	mm²	Conductor csa	mm²	Oil service	Structural steel
RCD operating current, $I_{\Delta n}$ mA							
Supply conductors csa mm²	RCD operating time (at $I_{\Delta n}$)* ms	Continuity check (✓)		Continuity check (✓)		Lightning protection	Other incoming service(s)

✛ Where a number of sources are available to supply the installation, and where the data given for the primary source may differ from other sources, a separate sheet must be provided which identifies the relevant information relating to each additional source.

Page 3 of

Please see the 'Notes for Recipients' on the reverse of this page.

IPM4/5

Figure 17.15c   Periodic Inspection Report, page 3. Reproduced by kind permission of The Electrical Safety Council. (See plate 13)

IPM4

SCHEDULE OF ITEMS INSPECTED
† See note below

PROTECTIVE MEASURES AGAINST ELECTRIC SHOCK

Basic and fault protection

Extra low voltage

SELV | PELV

Double or reinforced insulation

Double or Reinforced Insulation

Basic protection

Insulation of live parts | Barriers or enclosures

Obstacles ** | Placing out of reach **

Fault protection

Automatic disconnection of supply

Presence of earthing conductor

Presence of circuit protective conductors

Presence of protective bonding conductors

Presence of earthing arrangements for combined protective and functional purposes

Presence of adequate arrangements for alternative source(s), where applicable

FELV

Choice and setting of protective and monitoring devices (for fault protection and/or overcurrent protection)

Non-conducting location **

Absence of protective conductors

Earth-free equipotential bonding **

Presence of earth-free equipotential bonding

Electrical separation

For one item of current-using equipment

For more than one item of current-using equipment

Additional protection

Presence of residual current device(s)

Presence of supplementary bonding conductors

** For use in controlled supervised conditions only

Prevention of mutual detrimental influence

Proximity of non-electrical services and other influences

Segregation of Band I and Band II circuits or Band II insulation used

Segregation of Safety Circuits

Identification

Presence of diagrams, instructions, circuit charts and similar information

Presence of danger notices and other warning notices

Labelling of protective devices, switches and terminals

Identification of conductors

Cables and Conductors

Selection of conductors for current carrying capacity and voltage drop

Erection methods

Routing of cables in prescribed zones

Cables incorporating earthed armour or sheath or run in an earthed wiring system, or otherwise protected against nails, screws and the like

Additional protection by 30mA RCD for cables concealed in walls (where required, in premises not under the supervision of skilled or instructed persons)

Connection of conductors

Presence of fire barriers, suitable seals and protection against thermal effects

General

Presence and correct location of appropriate devices for isolation and switching

Adequacy of access to switchgear and other equipment

Particular protective measures for special installations and locations

Connection of single-pole devices for protection or switching in line conductors only

Correct connection of accessories and equipment

Presence of undervoltage protective devices

Selection of equipment and protective measures appropriate to external influences

Selection of appropriate functional switching devices

SCHEDULE OF ITEMS TESTED
† See note below

External earth fault loop impedance, Z_e

Installation earth electrode resistance, R_A

Continuity of protective conductors

Continuity of ring final circuit conductors

Insulation resistance between live conductors

Insulation resistance between live conductors and Earth

Protection by SELV, PELV or by electrical separation

Basic protection by barrier or enclosure provided during erection

Insulation of non-conducting floors or walls

Polarity

Earth fault loop impedance, Z_s

Verification of phase sequence

Operation of residual current devices

Functional testing of assemblies

Verification of voltage drop

† **All boxes must be completed.**
✓ indicates that an inspection or a test was carried out and that the result was **satisfactory**
✗ indicates that an inspection or a test was carried out and that the result was **unsatisfactory**
'N/A' indicates that an inspection or a test was **not applicable** to the particular installation
'LIM' indicates that, that exceptionally, a **limitation** agreed with the person ordering the work (as recorded in Section D) **prevented** the inspection or test being carried out

Page 4 of [ ]

This form is based on the model shown in Appendix 6 of BS 7671
© Copyright The Electrical Safety Council (Jan 2008)

IPM4/7

Figure 17.15d Periodic Inspection Report, page 4. Reproduced by kind permission of The Electrical Safety Council. (See plate 14)

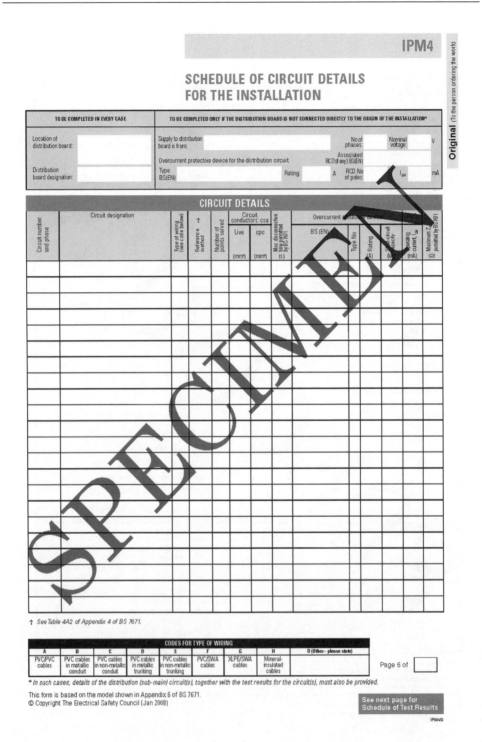

IPM4

SCHEDULE OF TEST RESULTS
FOR THE INSTALLATION

TO BE COMPLETED ONLY IF THE DISTRIBUTION BOARD IS NOT CONNECTED DIRECTLY TO THE ORIGIN OF THE INSTALLATION

Characteristics at this distribution board

Confirmation of supply polarity

* See note below

Z_s * Ω   Operating times   At $I_{\Delta n}$ ms
of associated
I_{pf} * kA   RCD (if any)   At 5$I_{\Delta n}$ ms
(if applicable)

Test instruments (serial numbers) used:

Earth fault loop impedance   RCD
Insulation resistance   Other
Continuity   Other

TEST RESULTS

Circuit number and phase	Circuit impedances (Ω)					Insulation resistance + Record lower or lowest value				Polarity	Maximum measured earth fault loop impedance, Z_s	RCD operating times	
	Ring final circuits only (measured end to end)			All circuits (At least one column to be completed)		Line/Line+	Line/Neutral+	Line/Earth+	Neutral/Earth			at $I_{\Delta n}$	at 5$I_{\Delta n}$ (if applicable)
	r_1 (Line)	r_n (Neutral)	r_2 (cpc)	$R_1 + R_2$	R_2	(MΩ)	(MΩ)	(MΩ)	(MΩ)	(✓)	(Ω)	(ms)	(ms)

* Note: Where the installation can be supplied by more than one source, such as a primary source (eg public supply) and a secondary source (eg standby generator), the higher or highest values must be recorded.

TESTED BY

Signature: Position: Page 6 of

Name: (CAPITALS)   Date of testing:

This form is based on the model shown in Appendix 6 of BS 7671.
© Copyright The Electrical Safety Council (Jan 2008)

See previous page for Schedule of Circuit Details

IPM4/11

Figure 17.15f Periodic Inspection Report, page 6. Reproduced by kind permission of The Electrical Safety Council. (See plate 16)

of the previous inspections and to identify any subsequent alterations and additions to the installation and, for the record, the person responsible for maintaining the documentation.

Section D: Extent of the installation and limitations of the inspection and testing (Figure 17.15a). This section must fully identify the extent of the installation covered by the report and any agreed limitations on the inspection and testing. This section is crucially important for both the author and recipient of the report. The scope, or extent, of the installation (possibly only part of the whole installation) which is the subject of the inspection and report needs to be clearly and unambiguously defined, as do the limitations of the periodic inspection and testing by virtue of the constraints laid down by the installation owner and by any physical restrictions imposed by the nature of the installation or its use. For example, for many commercial and domestic premises much of the wiring systems will be permanently concealed within the building fabric and it would, in most cases, be unreasonable to extract cables, etc. to determine their condition. Experience has shown that wiring systems are more likely to deteriorate at their terminations (especially at heat sources like tungsten-lamped luminaires), and these points will be the most productive to inspect. The owner or occupier of the premises may place restrictions on the extent to which they are agreeable in the dismantling of equipment and/or testing, which must be recorded on the report. Similarly, if procedures involve a sampling technique, then the basis of such a strategy must be recorded. There may, of course, be other limitations, but it is important that these are mutually agreed before carrying out the periodic inspection and testing and that an entry is made in this section. The absence of any stated limitations would, justifiably, lead the recipient to assume that all parts and aspects of the installation had been included. *The contractor should have agreed all such aspects with the client and other interested parties (licensing authority, insurance company, building society) before carrying out the inspection and testing.*

Section E: Declaration (Figure 17.15a). This section requires the signature of the person carrying out the periodic inspection and testing and assessment of the overall condition of the installation. This declaration of the overall condition of the installation must reiterate that given in Section G, which should summarize the observations and recommendations made in Section F. The inspection, testing and assessment by the inspector should preferably be reviewed by another competent person to confirm that the declared overall condition of the electrical installation is consistent with the inspection and test results, and with the observations and recommendations for action (if any) made in the report.

In the case of the NICEIC 'red' Periodic Inspection Report form or Domestic Periodic Inspection Report form, for use by NICEIC Approved Contractors, the inspection, testing and assessment by the inspector must be reviewed and confirmed by the NICEIC registered Qualified Supervisor. The same is true for the NICEIC 'purple' Domestic Periodic Inspection Report form, for use by NICEIC Domestic Installers meeting certain requirements. Where the Qualified Supervisor carries out the inspection personally, they should sign in both places.

A list of observations and recommendations for urgent remedial work and for corrective actions necessary to maintain the installation in a safe working order should be given in Section F, where appropriate.

Section F: Observations and recommendations for actions to be taken (Figure 17.15b). This section includes two data-entry boxes at the top in which the report compiler is required to confirm that 'There are no items adversely affecting electrical safety' *or* that 'The following observations and recommendations are made', as appropriate. In the latter case, the observations and recommendations are to be listed with a Code 1, 2, 3 or 4 (see below). At the bottom of the section, two data-entry boxes are provided, one for recording the deficiencies which, in the opinion of the report's compiler, need urgent remedial work and the other for those items requiring corrective actions.

Where the inspector classifies a recommendation as 'requires urgent attention' the client is to be advised immediately, in writing, to satisfy the duties imposed by the Electricity at Work Regulations 1989. It should be noted that, where a real and immediate danger is observed that puts the safety of persons or livestock at risk, Recommendation Code 1 (requires urgent attention) must be used. Where a potential danger is observed, Recommendation Code 2 (requires improvement) must be made. If the space available on the form for recording recommendations is insufficient, additional numbered pages are to be provided as necessary.

The codes for ranking the items are

- 1 – requires urgent attention;
- 2 – requires improvement;
- 3 – requires further attention;
- 4 – does not comply with BS 7671: 2008 (as amended).

Defects which represent an immediate safety hazard would necessarily warrant a Code 1 and would include, for example, bare live parts exposed to touch or absence of an effective means of earthing for the installation. A Code 2 would include such items that do not represent an instant danger but would benefit from an improvement embracing; for example, a 30/32 A ring final circuit discontinuous or cross-connected with another circuit. Where, because of limitations placed on the inspection or because some unforeseen difficulty arises, full diagnosis of some malfunction or apparent fault condition is not possible, this fact must be reported in this section in order that the person responsible for the installation is made aware that further investigation is needed by attributing a Code 3. Items that do not comply with BS 7671 also need to be recorded. Some deficiencies may seem to warrant more than one code; for example, live parts with missing barriers or insulation demand a Code 4 (contravention of Regulation 411.2) and a Code 1 signifying that urgent attention is required. However, only one code should be attributed to any one item, this being the code which is more or most appropriate in the circumstances.

Best Practice Guide Number 4, produced by The Electrical Safety Council in association with leading industry bodies, gives practical guidance for competent persons on the use of the Recommendation Codes when compiling periodic inspection reports on domestic and similar electrical installations. This publication may be downloaded free of charge from The Electrical Safety Council's website (www.esc.org.uk).

Section G: Summary of inspection (Figure 17.15b). This section must be completed with an accurate description of the general condition of the installation, together with the date(s) of the inspection and a succinct assessment of the condition of the installation, in terms of *a satisfactory condition* or *an unsatisfactory condition* entry. The entry for the date(s) of the inspection would, of course, be the actual date(s) on which the site inspection was carried out and may, in certain circumstances, be a number of dates and not necessarily consecutive days. The condition of the installation needs to be an overview of the overall state of the installation and should summarize the main findings of the inspection and testing. Evidence of wear and tear of the installation generally, and that of accessories, wiring systems and switchgear in particular, should be noted here, as would any change of use of the premises (which may render the installation unsuitable for the new environmental conditions). Any misuse and abuse of the installation, together with evidence of added loads (which may have resulted in overloading), should also be chronicled. The attention of the report reader should be drawn to a summary of any damage, deterioration, defects and dangerous conditions and entered in Section G.

Section H: Schedules and additional pages (Figure 17.15c). This section is intended to identify the page numbers of all the various schedules and additional pages which form part of the periodic inspection report. The *Schedule of items inspected* and the *Schedule of items tested* will always be page 4. The *Schedule of circuit details for the installation* and the *Schedule of test results for the installation* will always be pages 5 and 6 respectively, but space has been provided to record page numbers where there is a need to use more of these schedules. Space has also been provided to identify any additional pages which form part of the report, such as an additional page of observations and recommendations (see Section F). All pages forming part of the periodic inspection report must be numbered together with a reference to the total number of pages (e.g. 7 of 10).

Section I: Next inspection (Figure 17.15c). This section calls for the appropriate time interval before reinspection of the installation to be inserted. IEE Guidance Note 3 gives guidance on the maximum recommended intervals for various types of premises, but due account must be taken of the present condition of the installation. The recommendation for the interval to the next inspection is to be conditional on all deficiencies which have attracted a Recommendation Code 1 or 2 in Section F being remedied without delay. Additionally, the recommendation for the interval to the next inspection is also to be conditional on all deficiencies which have attracted Recommendation Code 3 being actioned as soon as practicable. The report compiler would need to exercise their professional judgement in deciding the date for the next inspection. In exercising this judgement, account should be taken of the condition of the installation, the level of maintenance received in the past and that expected in the future, and any factors which may materially affect the interval. In some cases (e.g. places of public entertainment) the maximum interval may be the subject of licensing conditions and the recommended reinspection interval would need to take this into account (see Table 17.21).

Section J: Details of electrical contractor (Figure 17.15c). This section is self-explanatory.

Section K: Supply characteristics and earthing arrangements (Figure 17.17c). The compilation of this section is straightforward and needs no explanation, except as regards the *Number of supplies*, which must be identified in the data-entry box even where there is only one. Where the installation can be supplied from more than one source, such as the public supply and a standby generator, the higher or highest value of prospective fault current I_{pf} and external earth fault loop impedance Z_e must be recorded in the data-entry boxes provided for this purpose. Where a number of sources are available to supply the installation, and where the data given for the primary source may differ from the other sources, an additional page must be provided which gives the relevant information relating to each source.

This section is almost identical to the similarly titled section in the Electrical Installation Certificate, where more detailed guidance is given.

Section L: Particulars of installation at the origin (Figure 17.15c). The compilation of this section is again straightforward and self-explanatory, except where a number of sources are available to supply the installation. Where the data given for the *Means of earthing* and the *Main switch or circuit-breaker* relating to the primary source may differ from that of other sources, the relevant information must be recorded for each additional source.

This section is identical to the similarly titled section in the Electrical Installation Certificate, where more detailed guidance is given.

Schedule of items inspected (Figure 17.15d). Except for the unique numbering prefix, this section is identical to the similarly titled section in the Electrical Installation Certificate, where more detailed guidance is given. All data-entry boxes on this schedule are to be completed, as appropriate for the particular installation, by inserting a '$\checkmark$' indicating that the installation inspection or test has been undertaken with satisfactory results. Where an inspection has revealed an unsatisfactory result, the entry of an '$\times$' should be made. It is unlikely that all items will apply, and the range of applicable inspections will depend on the particular installation covered by the report. If an inspection is not applicable, 'N/A' should be recorded in the box. Exceptionally, where a limitation on a particular inspection or test has been agreed, and has been recorded in Section D, the appropriate data-entry box(es) must be completed by inserting 'LIM', indicating that an agreed limitation has prevented the inspection of test being carried out.

Schedule of items tested (Figure 17.15d). Except for the unique numbering prefix, this section is identical to the similarly titled section in the Electrical Installation Certificate, where more detailed guidance is given. All data-entry boxes on this schedule are to be completed, as appropriate for the particular installation, by inserting a '$\checkmark$' indicating that the test has been undertaken with satisfactory results. Where the test result is unsatisfactory, this should be recorded by inserting an '$\times$'. It is unlikely that all items will apply, and the range of applicable tests will depend on the particular installation covered by the report. If a test is not applicable, 'N/A' should be recorded in the box. Exceptionally, where a limitation on a particular test has been agreed, and has been recorded in Section D, the appropriate data-entry box(es) must be completed by inserting 'LIM', indicating that an agreed limitation has prevented the inspection of test being carried out.

Schedule of circuit details for the installation (Figure 17.15e). Except for the unique numbering prefix, this schedule is identical to the similarly titled section in the Electrical Installation Certificate, where more detailed guidance is given.

Schedule of test results for the installation (Figure 17.15f). Except for the unique numbering prefix, this schedule is identical to the similarly titled section in the Electrical Installation Certificate, where more detailed guidance is given.

17.7   Alterations and additions

17.7.1   General

As indicated in Regulation 610.4, the inspection, testing and certification requirements apply equally to alterations and additions to an electrical installation. It is essential that any addition or alteration does not have an adverse effect on the safety of the existing installation and that any protective measures embodied in the existing installation and used to protect the addition or alteration are satisfactory and effective. For example, if the ADS measure of protection against electric shock is used and a new circuit made use of an existing MCB for this purpose, then it would be necessary to ensure that earthing and protective bonding were adequate and that the MCB was of the correct type and rated current to provide adequate protection for the new circuit. Normally, the designer/installer of an alteration and/or addition would need to ensure that the supply to the installation was adequate for the additional load. They would also need to ensure that the short-circuit capacity of protective devices, etc. was adequate, and this would be particularly important where there was a need to reinforce the supply. A check on the MET, the earthing conductor and protective bonding conductors would be a matter of course, as would a check of line–earth loop impedances Z_e and Z_s. In all cases, a systematic approach is required to ensure that any newly installed circuit(s) and alterations will operate safely under normal and fault conditions and that the newly installed work will not adversely affect the safety of the existing installation.

It is important to recognize that whilst installations constructed to an earlier edition of the Wiring Regulations may well have deviations from the current edition (BS 7671), these may or may not represent an unsafe condition and it will be a matter of judgement of how the inspector treats such deviations (e.g. reports as unsafe, undesirable or replacement needed in due course).

17.7.2   Inspection

The inspection required for an alteration or addition would be similar to that entailed in a completely new installation (restricted to the alteration or addition) and, as mentioned earlier, would also include inspection of any protective measures of the existing installation used to protect the new work (e.g. earthing and bonding and protective devices).

17.7.3   Testing

Testing of alterations and additions is essentially no different from that required for a whole installation, but would be restricted to the extent of the new work and/or alteration, except that any existing installation protective measures used to protect the new work would also need testing as if for initial verification.

17.7.4   Certification

The requirements for the certification of alterations and additions are the same as for any new installation and the procedure is as for initial inspection and testing of an installation (see Section 17.5 of this Guide). Defects observed in unrelated parts of the existing installation will need to be included in certification form.

17.8   Inspection, testing and certification of specialized installations

17.8.1   Fire detection and alarm systems in buildings

The installation of fire detection and alarm systems in buildings is required to comply with the BS 7671 and the British Standard CP BS 5839: Part 1 or, where applicable, Part 6. The installation of such a system will, of course, be the subject of an Electrical Installation Certificate, as detailed in Section 17.5.4, as well as the form of certification required by the CP.

17.8.2   Emergency lighting

The installation of the emergency lighting systems in buildings is required to comply with the BS 7671 *and* the British Standard CP BS 5266: Parts 1 and 7. The installation of such a system will, of course, be the subject of an Electrical Installation Certificate, as detailed in Section 17.5.4, as well as the form of certification required by the CP.

17.8.3   Petrol filling stations

In terms of initial inspection and testing of an installation in petrol filling stations, the requirements and practice for inspection and testing and certification are not too dissimilar to those relating to other installations, providing the testing is carried out before the stations are fuelled up. However, as soon as petrol or liquefied petroleum gas is delivered to the station, testing becomes more difficult and potentially dangerous because of the risks of igniting the fuel in its gaseous form. For this reason, any periodic testing can be extremely dangerous, and such work needs to be carried out by an engineer fully versed in the inspection and testing procedures relating to such installations.

Familiarization with *Guidance for the design, construction, modification and maintenance of petrol filling stations*, published jointly by the APEA and the Institute of Petroleum, is a necessary prerequisite to carrying out any installation work at a petrol filling station. Additionally, all persons undertaking such work would of necessity be experienced and thoroughly versed in the risks and hazards associated with these special environments. They should also be conversant with the Petrol (Regulations) Act 1928 and 1936 and the Health and Safety at Work etc. Act 1974 (including the Electricity at Work Regulations 1989). Furthermore, persons should be aware of, and have had practical experience of, the relevant parts of BS EN 60079 *Electrical Apparatus for Explosives Gas Atmospheres:* Parts 10, 14 and 17.

The model inventory and certificate given in *Guidance for the design, construction, modification and maintenance of petrol filling stations* is complex, but adequate guidance is contained therein to aid compilation by competent persons. In addition to the issue of the above forms, initial certification of a petrol filling station must be completed by the issue of a 'normal' Electrical Installation Certificate (see Figure 17.12).

Petrol filling stations are the subject of licensing (by local authorities) and are required to be periodically inspected and tested annually. The model forms may be used for both initial and periodic verification.

Appendix

Standards to which reference has been made

National standards (British Standards)

BS 31	Specification. Steel conduit and fittings for electrical wiring.
BS 67	Specification for ceiling roses.
BS 88	Cartridge fuses for voltages up to and including 1000 V a.c. and 1500 V d.c.
BS 196	Specification for protected-type non-reversible plugs, socket-outlets, cable couplers and appliance-couplers with earthing contacts for single phase a.c. circuits up to 240 volts.
BS 415	Specification for safety requirements for mains-operated electronic and related apparatus for household and similar general use.
BS 476	Fire tests on building materials and structures.
BS 546	Specification. Two-pole and earthing-pin plugs, socket-outlets and socket-outlet adaptors.
BS 559	Specification for electric signs and high voltage luminous-discharge-tube installations.
BS 646	Specification. Cartridge fuse-links (rated up to 5 amperes) for a.c. and d.c. service.
BS 731	Flexible steel conduit for cable protection and flexible steel tubing to enclose flexible drives.
BS 951	Specification for clamps for earthing and bonding purposes.
BS 1361	Specification for cartridge fuses for a.c. circuits in domestic and similar premises.
BS 1362	Specification for general purpose fuse links for domestic and similar purposes (primarily for use in plugs).
BS 1363	13 A plugs, socket-outlets, connection units and adaptors.
BS 1710	Specification for the identification of pipelines and services.
BS 2632	See BS EN 61 011.
BS 2754	Memorandum. Construction of electrical equipment for protection against electric shock.
BS 2848	Specification for flexible insulating sleeving for electrical purposes.

A Practical Guide to The Wiring Regulations: 17th Edition IEE Wiring Regulations (BS 7671:2008)
Fourth Edition Geoffrey Stokes and John Bradley
© Geoffrey Stokes and John Bradley. Published by John Wiley & Sons, Ltd

BS 3036	Specification. Semi-enclosed electric fuses (ratings up to 100 amperes and 240 volts to earth).
BS 3042	Test probes to verify protection by enclosures.
BS 3456	Specification for safety of household and similar electrical appliances.
BS 3535	Isolating transformers and safety isolating transformers.
BS 3676	Switches for household and similar fixed electrical installations.
BS 3858	Specification for binding and identification sleeves for use on electric cables and wires.
BS 3871	Specification for miniature and moulded case circuit-breakers.
BS 4066	Tests on electric cables under fire conditions.
BS 4099	Colours of indicator lights, push-buttons, annunciators and digital readouts.
BS 4293	Specification for residual current-operated circuit-breakers. (See also BS EN 61 008-1 and BS EN 61 009-1.)
BS 4343	See BS EN 60 309-2.
BS 4363	Specification for distribution assemblies for electricity supplies for construction and building sites.
BS 4444	Guide to electrical earth monitoring and protective conductor proving.
BS 4483	Specification for steel fabric for the reinforcement of concrete.
BS 4491	Appliance couplers for household and similar general purposes.
BS 4533	Luminaires.
BS 4568	Specification for steel conduit and fittings with metric threads of ISO form for electrical installations.
BS 4573	Specification for 2-pin reversible plugs and shaver socket-outlets.
BS 4579	Specification for performance of mechanical and compression joints in electric cable and wire connectors.
BS 4607	Non-metallic conduits and fittings for electrical installations.
BS 4662	Specification for boxes for the enclosure of electrical accessories.
BS 4678	Cable trunking.
BS 4737	Intruder alarm systems.
BS 4752	See BS EN 60 947.
BS 4884	Technical manuals.
BS 4940	Recommendation for the presentation of technical information about products and services in the construction industry.
BS 4941	See BS EN 60 947-4-1.
BS 5042	Specification for bayonet lampholders.
BS 5266	Emergency lighting.
BS 5306	Fire extinguishing installations and equipment on premises.
BS 5445	Components of automatic fire detection systems.
BS 5446	Components of automatic fire alarm systems for residential premises.
BS 5467	Specification for cables with thermosetting insulation for electricity supply for rated voltages of up to and including 600/1000 V and up to and including 1900/3300 V.
BS 5468	See BS 6469 and BS 6899.
BS 5486	Low-voltage switchgear and controlgear assemblies (see also BS EN 60 439).
BS 5490	See BS EN 60 529.
BS 5501	Electrical apparatus for potentially explosive atmospheres.
BS 5518	Specification for electronic variable control switches (dimmer switches) for tungsten filament lighting.
BS 5588	Fire precautions in the design, construction and use of buildings.

BS 5593	Specification for impregnated paper-insulated cables with aluminium sheath/neutral conductor and three shaped solid aluminium phase conductors (CONSAC), 600/1000 V, for electricity supply.
BS 5655	Lifts and service lifts.
BS 5733	Specification for general requirements for electrical accessories.
BS 5784	Safety of electrical commercial catering equipment.
BS 5839	Fire detection and alarm systems in buildings.
BS 6004	Specification for PVC-insulated cables (non-armoured) for electric power and lighting.
BS 6007	Specification for rubber-insulated cables for electric power and lighting.
BS 6053	Specification for outside diameters of conduits for electrical installations and threads for conduits and fittings. (See also BS EN 60 423.)
BS 6081	Specification for terminations for mineral-insulated cables.
BS 6099	Conduits for electrical installations. (See also BS EN 50 086-1.)
BS 6121	Mechanical cable glands.
BS 6141	Specification for insulated cables and flexible cords for use in high temperature zones.
BS 6207	Specification for mineral-insulated cables with a rated voltage not exceeding 750 V.
BS 6231	Specification for PVC-insulated cables for switchgear and controlgear.
BS 6266	Code of practice for fire protection for electronic data processing installations.
BS 6346	Specification for PVC-insulated cables for electricity supply.
BS 6351	Electric surface heating.
BS 6360	Specification for conductors in insulated cables and cords.
BS 6369	BS EN 61 011-1
BS 6458	Fire hazard testing for electrotechnical products.
BS 6467	Electrical apparatus with protection by enclosure for use in the presence of combustible dusts.
BS 6469	Insulating and sheathing materials of electric cables.
BS 6480	Specification for impregnated paper-insulated lead or lead alloy sheathed electric cables of rated voltages up to and including 33 000 V.
BS 6500	Specification for insulated flexible cords and cables.
BS 6651	Code of practice for protection of structures against lightning.
BS 6701	Code of practice for installation of apparatus intended for connection to certain telecommunication systems.
BS 6702	See BS EN 60 400.
BS 6708	Specification for flexible cables for use at mines and quarries.
BS 6713	Explosion protection systems.
BS 6724	Specification for armoured cables for electricity supply having thermosetting insulation with low emission of smoke and corrosive gases when affected by fire.
BS 6726	Specification for festoon and temporary lighting cables and cords.
BS 6739	Code of practice for instrumentation in process control systems: installation design and practice.
BS 6746	Specification for PVC insulation and sheath of electric cables.
BS 6765	Leisure accommodation vehicles: caravans.
BS 6776	See BS EN 60 238.
BS 6840	Sound system equipment.
BS 6883	Specification for elastomer insulated cables for fixed wiring in ships and in mobile and fixed offshore units.

BS 6899	Specification for rubber insulation and sheath of electric cables.
BS 6972	Specification for general requirements for luminaire supporting couplers for domestic, light industrial and commercial use.
BS 6977	Specification for insulated flexible cables for lifts and for other flexible connections.
BS 6991	Specification for 6/10 A, two-pole weather-resistant couplers for household, commercial and light industrial equipment.
BS 7001	Specification for interchangeability and safety of a luminaire supporting coupler.
BS 7002	See BS EN 60 950.
BS 7071	Specification for portable residual current devices.
BS 7211	Specification for thermosetting insulated cables (non-armoured) for electric power and lighting with low emission of smoke and corrosive gases when affected by fire.
BS 7288	Specification for socket-outlets incorporating residual current devices.
BS 7361	Cathodic protection.
BS 7375	Code of practice for distribution of electricity on construction and building sites.
BS 7430	Code of practice for earthing.
BS 7454	Method of calculation of thermally permissible short-circuit currents, taking into account non-adiabatic heating effects.
BS 7527	Classification of environmental conditions.
BS 7629	Thermosetting insulated cables with limited circuit integrity when affected by fire.
BS 7671	Requirements for electrical installations. IEE Wiring Regulations. Sixteenth edition.
BS 8450	Code of practice for installation of electrical and electronic equipment in ships.
BS 7909	Code of practice for design and installation of temporary distribution systems delivering a.c. electrical supplies for lighting, technical services and other entertainment related purposes.
PD 6531	Queries and interpretations on BS 5839: Parts 1 and 4 (as amended).
PD 6519	Effect of current passing through the human body.

European standards (CENELEC Standards)

BS EN 54	Fire detection and fire alarm systems.
BS EN 81-1	Safety rules for the construction and installation of lifts. Electric lifts.
BS EN 1648-1	12 V direct current extra-low voltage electrical installations. Caravans.
BS EN 1648-2	12 V direct current extra-low voltage electrical installations. Motor caravans.
BS EN 13636	Cathodic protection of buried metallic tanks and related piping.
BS EN 15112	External cathodic protection of well casing.
BS EN 50 014	Electrical apparatus for potentially explosive atmospheres.
BS EN 50016	Electrical apparatus for potentially explosive atmospheres.
BS EN 50 081	Electromagnetic compatibility. Generic emission standard.
BS EN 50 082	Electromagnetic compatibility. Generic immunity standard.
BS EN 50 085-1	Specification for cable trunking and ducting systems for electrical installations.
BS EN 50085-2-1	Cable trunking systems and cable ducting systems intended for mounting on walls and ceilings.
BS EN 50 086-1	Specifications for conduit systems for electrical installations (Part 1. General requirements).

BS EN 50107	Signs and luminous-discharge-tube installations operating from a no-load rated output voltage exceeding 1 kV but not exceeding 10 kV.
BS EN 50174	Installation technology. Cabling installation.
BS EN 50200	Method of test for resistance to fire of unprotected small cables for use in emergency circuits.
BS EN 50214	Flat polyvinyl chloride sheathed flexible cables.
BS EN 50262	Cable glands for electrical installations.
BS EN 50265-1	Common test methods for cables under fire conditions. Test for resistance to vertical flame propagation for a single insulated conductor or cable. Apparatus.
BS EN 50265-2.1	Procedures. 1 kW pre-mixed flame. 1 kW pre-mixed flame.
BS EN 50266	Common test methods for cables under fire conditions.
BS EN 50281	Electrical apparatus for use in the presence of combustible dust.
BS EN 50438	Requirements for the connection of micro-generators in parallel with public low-voltage distribution networks.
BS EN 60 079	Electrical apparatus for explosive gas atmospheres.
BS EN 60204	Safety of machinery. Electrical equipment of machines.
BS EN 60 238	Specification for Edison screw lampholders.
BS EN 60 269-1	Low voltage fuses (Part 1. General requirements – also numbered BS 88: Part 1: 1988).
BS EN 60269-2	Low-voltage fuses. Supplementary requirements for fuses for use by authorized persons (fuses mainly for industrial application).
BS EN 60 309	Plugs, socket-outlets and couplers for industrial purposes.
BS EN 60 309-1	Plugs, socket-outlets and couplers for industrial purposes (Part 1. General requirements – also numbered BS 4343:1968).
BS EN 60 309-2	Plugs, socket-outlets and couplers for industrial purposes (Part 2: Dimensional interchangeability requirements for pin and contact tube accessories of harmonised configurations).
BS EN 60320	Appliance couplers for household and similar general purposes.
BS EN 60332-1-2	Tests on electric and optical fibre cables under fire conditions. Test for vertfical flame propagation for a single insulated wire or cable. Procedure for 1 kW pre-mixed flame.
BS EN 60335-1	Household and similar electrical appliances.
BS EN 60335-2-53	Particular requirements. Sauna heating appliances.
BS EN 60335-2-71	Particular requirements. Electrical heating appliances for breeding and rearing animals.
BS EN 60335-2-96	Particular requirements for flexible sheet heating elements for room heating.
BS EN 60 400	Specification for lampholders for tubular fluorescent lamps and starter-holders.
BS EN 60 423	Conduits for electronic purposes. Outside diameters of conduits for electrical installations and threads for conduits and fittings (see BS 6053).
BS EN 60 439	Specification for low-voltage switchgear and controlgear assemblies.
BS EN 60 439-1	Type-tested and partially type-tested assemblies.
BS EN 60439-3	Distribution boards.
BS EN 60 439-4	Particular requirements for assemblies for construction sites (ACS).
BS EN 60 529	Specification for degrees of protection provided by enclosures (IP Code).
BS EN 60570	Electrical supply track systems for luminaires.

BS EN 60 598	Luminaires.
BS EN 60 598-1	General requirements and tests.
BS EN 60598-2-18	Particular requirements. Luminaires for swimming pools and similar applications.
BS EN 60 598-2-22	Particular requirements. Luminaires for emergency lighting
BS EN 60598-2-23	Particular requirements. Extra low voltage lighting systems for filament lamps.
BS EN 60598-2-24	Particular requirements. Luminaires with limited surface temperatures.
BS EN 60 617	Graphical symbols for diagrams. (Replaces BS 3939.)
BS EN 60664-1	Insulation coordination for equipment within low-voltage systems. Principles, requirements and tests.
BS EN 60669-1	Switches for household and similar fixed-electrical installations. General requirements.
BS EN 60669-2-1	Particular requirements. Electronic switches.
BS EN 60684	Flexible insulating sleeving.
BS EN 60702-1	Mineral insulated cables and their terminations with a rated voltage not exceeding 750 V. Cables.
BS EN 60704	Household and similar electrical appliances.
BS EN 60 742	See BS 3535: Part 1.
BS EN 60 898	Specification for circuit-breakers for overcurrent protection for household and similar installations.
BS EN 60 900	Specification for safety of information technology equipment, including electrical business equipment.
BS EN 60909	Short-circuit current calculation in three-phase a.c. systems.
BS EN 60 947	Specification for low-voltage switchgear and controlgear.
BS EN 60 947-1	General rules.
BS EN 60 947-2	Circuit-breakers.
BS EN 60 947-3	Switches, disconnectors, switch-disconnectors and fuse-combination units.
BS EN 60 947-4	Contactors and motor-starters.
BS EN 60 947-4-1	Electromechanical contactors and motor-starters.
BS EN 60 950	Specification for safety of information technology equipment, including electrical business equipment.
BS EN 61 008	Residual current operated circuit-breakers without integral overcurrent protection for household and similar uses (RCCBs) (Part 1. General rules).
BS EN 61 009	Residual current operated circuit-breakers with integral overcurrent protection for household and similar uses (RCBOs).
BS EN 61 011	Electric fence energisers safety requirements.
BS EN 61 011-1	Electric fence energisers safety requirements for battery-operated electric fence energisers suitable for connection to the supply mains.
BS EN 61010-2-031	Safety requirements for electrical equipment for measurement, control and laboratory use. Particular requirements for hand-held probe assemblies for electrical measurement and test.
BS EN 61034-2	Measurement of smoke density of cables burning under defined conditions. Test procedure and requirements.
BS EN 61140	Protection against electric shock. Common aspects for installation and equipment.
BS EN 61 184	Bayonet lampholders.

BS EN 61215	Crystalline silicon terrestrial photovoltaic (PV) modules. Design qualification and type approval.
BS EN 61238	Compression and mechanical connectors for power cables for rated voltages up to 36 kV (Um = 42 kV). Test methods and requirements.
BS EN 61241	Electrical apparatus for use in the presence of combustible dust.
BS EN 61347-1	Lamp controlgear. General and safety requirements.
BS EN 61347-2-2	Particular requirements for d.c or a.c. supplied electronic step-down convertors for filament lamps.
BS EN 61386-1	Conduit systems for cable management. General requirements.
BS EN 61386-21	Particular requirements. Rigid conduit systems.
BS EN 61535	Installation of couplers intended for permanent connections in fixed installations.
BS EN 61537	Cable management. Cable tray systems and cable ladder systems.
BS EN 61557	Electrical safety in low voltage distribution systems up to 1000 V a.c. and 1500 V d.c. Equipment for testing, measuring or monitoring of protective measures.
BS EN 61557-2	Insulation resistance.
BS EN 61557-3	Loop impedance.
BS EN 61557-4	Resistance of earth connection and equipotential bonding.
BS EN 61557-5	Resistance to earth.
BS EN 61557-7	Phase sequence.
BS EN 61558-1	Safety of power transformers, power supplies, reactors and similar products.
BS EN 61558-2-4	Particular requirements for isolating transformers for general use.
BS EN 61558-2-5	Particular requirements for shaver transformers and shaver supply units.
BS EN 61558-2-6	Particular requirements for safety isolating transformers for general use.
BS EN 61558-2-23	Particular requirements for transformers for construction sites.
BS EN 61034-2	Measurement of smoke density of cables burning under defined conditions. Test procedure and requirements.
BS EN 62020	Electrical accessories. Residual current monitors for household and similar uses (RCMs).
BS EN 62262	Degrees of protection provided by enclosures for electrical equipment against external mechanical impacts (IK code).
BS EN 62305	Protection against lightning.

International standards (IEC Standards)

IEC 60269-1	Low-voltage fuse. Part 1: General requirements.
IEC 60364	Electrical installations in buildings.
IEC 60439-1	Low-voltage switchgear and controlgear assemblies; Part 1: Type-tested and partially type-tested assemblies.
IEC 60621 series 2	Electrical installations for outdoor sites under heavy conditions (including open-cast mines and quarries).
IEC/TR 60755	General requirements for residual current operated protective devices.
IEC 60800	Heating cables with a rated voltage of 300/500 V for comfort heating and prevention of ice formation.
IEC 60947-2	Low-voltage switchgear and controlgear. Part 2: Circuit-breakers.
IEC 60949	Calculation of thermally permissible short-circuit currents, taking into account non-adiabatic heating effects.
IEC 61423-1	Heating cables for industrial applications.

Bibliography

The following is a list of useful publications.

Association of Petroleum Explosives Administration (APEA) and the Institute of Petroleum (IP) (2008) *Guidance for the Design, Construction, Modification and Maintenance of Petrol Filling Stations*. APEA Saffron Walden, Essex.

Church Housing Publishing (1988) *Lighting and Wiring of Churches*. Church Housing Publishing, London.

CIBSE (2005) *Guide K: Electricity in Buildings*. The Chartered Institution of Building Services Engineers, London.

Department of Health/Estates and Facilities Division (2007) *Health Technical Memorandum 06-01: Electrical Services Supply and Distribution*. TSO Online Bookshop www.tsoshop.co.uk.

Electricity Association (1996) *Engineering Recommendations P25/1*. Energy Networks Association, London.

The Electrical Safety Council *Best Practice Guide No 1: Replacing a consumer unit in domestic premises where lighting circuits have no protective conductor*.

The Electrical Safety Council *Best Practice Guide No 2: Guidance on safe isolation procedures for low voltage installations*.

The Electrical Safety Council *Best Practice Guide No 3: Connecting a microgeneration system to a domestic or similar installation (in parallel with the mains supply)*.

The Electrical Safety Council *Best Practice Guide No 4: Periodic inspection reporting – recommendation codes for domestic and similar electrical installations*.

The Electrical Safety Council *Best Practice Guide No 5: Electrical installation and their impact on the fire performance of buildings. Part 1 – Domestic single-family units*.

Health & Safety Agency for Northern Ireland (2001) *Memorandum of Guidance on the Electricity at Work Regulations (Northern Ireland) 1991*. HMSO, Belfast.

Health & Safety Executive (1995) *Guidance Note GS38: Electrical Test Equipment for Use by Electricians*. TSO Online Bookshop www.tsoshop.co.uk.

Health & Safety Executive (1987) *HS(G)34: Storage of Liquefied Petroleum Gas at Fixed Installations*. TSO Online Bookshop www.tsoshop.co.uk.

Health & Safety Executive (2007) *Memorandum of Guidance on the Electricity at Work Regulations 1989*. TSO Online Bookshop www.tsoshop.co.uk.

Health & Safety Executive (1997) *Electrical Safety for Entertainers*. TSO Online Bookshop www.tsoshop.co.uk.

Health & Safety Executive (1991) *Guidance Note GS50: Electrical Safety at Places of Entertainment*. TSO Online Bookshop www.tsoshop.co.uk.

IET *IEE Guidance Note No. 1: Selection and Erection*. The Institution of Engineering and Technology, London.

IET *IEE Guidance Note No. 2: Isolation and Switching*. The Institution of Engineering and Technology, London.

A Practical Guide to The Wiring Regulations: 17th Edition IEE Wiring Regulations (BS 7671:2008)
Fourth Edition Geoffrey Stokes and John Bradley
© Geoffrey Stokes and John Bradley. Published by John Wiley & Sons, Ltd

IET *IEE Guidance Note No. 3: Inspection and Testing*. The Institution of Engineering and Technology, London.

IET *IEE Guidance Note No. 4: Protection Against Fire*. The Institution of Engineering and Technology, London.

IET *IEE Guidance Note No. 5: Protection Against Electric Shock*. The Institution of Engineering and Technology, London.

IET *IEE Guidance Note No. 6: Protection Against Overcurrent*. The Institution of Engineering and Technology, London.

IET *IEE Guidance Note No. 7: Special Locations*. The Institution of Engineering and Technology, London.

IET *IEE Guidance Note No. 8: Earthing and Bonding*. The Institution of Engineering and Technology, London.

IET *IEE On-Site Guide*. The Institution of Engineering and Technology, London.

Institution of Lighting Engineers (2006) *Code of Practice for Electrical Safety in Public Lighting Operations*, 3rd edn. The Institution of Lighting Engineers, Rugby.

International Electrotechnical Commission *IEC 60364: Electrical Installations of Buildings*. International Electrotechnical Commission, Geneva.

LDSA (1990) *Fire Safety Guide No. 1 – Fire Safety in Section 20 Buildings*. London District Surveyor's Association, London.

NICEIC (2008) *Inspection, Testing and Certification Book*, 5th edn. NICEIC, Dunstable.

NICEIC (2008) *Domestic Electrical Installation Guide*. NICEIC, Dunstable.

NICEIC (2008) *Domestic Periodic Inspection, Testing and Reporting*. NICEIC, Dunstable.

NHBC (1993) *NHBC Standards*. National House-Building Council, Milton Keynes.

Oldham Smith, K. (2002) *Electrical Safety and the Law*, 4th edn. Blackwell Science, Oxford.

Index of Figures

A Practical Guide to The Wiring Regulations: 17th Edition IEE Wiring Regulations (BS 7671:2008)
Fourth Edition Geoffrey Stokes and John Bradley
© Geoffrey Stokes and John Bradley. Published by John Wiley & Sons, Ltd

Index of Tables

A Practical Guide to The Wiring Regulations: 17th Edition IEE Wiring Regulations (BS 7671:2008)
Fourth Edition Geoffrey Stokes and John Bradley
© Geoffrey Stokes and John Bradley. Published by John Wiley & Sons, Ltd

Index of regulation numbers

Note: reference is by section number; figure and table numbers are given in *italics*.

110.1	3.2.1, 13.2, 13.3, 16.2.10
110.2	3.2.1, 3.2.2
120.2	3.3.1
120.3	3.3.1
120.3	17.5.4
120.4	17.5.4
130.1.1	*Table 17.8*
131.1	2.2, 3.4.1
131.2	5.1
131.2.1	3.4.2
131.2.2	3.4.3
131.3.1	3.4.4
131.3.2	3.4.4
131.4	3.4.5, 3.4.6, 7.1
131.5	3.4.6, 7.1
131.6	3.4.7
131.6.1	3.4.7
131.6.2	3.4.7
131.6.3	3.4.7
131.6.4	3.4.7
131.7	3.4.8
131.8	3.4.9
132.1	*Table 3.4, Table 3.5*
132.2	*Table 3.5*
132.3	*Table 3.5*
132.4	*Table 3.5*
132.5	*Table 3.5*
132.5	*Table 17.6*
132.5.1	*Table 9.1*
132.6	*Table 3.5*
132.7	*Table 3.5*

A Practical Guide to The Wiring Regulations: 17th Edition IEE Wiring Regulations (BS 7671:2008)
Fourth Edition Geoffrey Stokes and John Bradley
© Geoffrey Stokes and John Bradley. Published by John Wiley & Sons, Ltd

Subject Index

A Practical Guide to The Wiring Regulations: 17th Edition IEE Wiring Regulations (BS 7671:2008)
Fourth Edition Geoffrey Stokes and John Bradley
© Geoffrey Stokes and John Bradley. Published by John Wiley & Sons, Ltd